Food Production and Processing: Fermentation, Nutritional Value and Quality Control

Food Production and Processing: Fermentation, Nutritional Value and Quality Control

Edited by Albert McCoy

SYRAWOOD
PUBLISHING HOUSE

New York

Published by Syrawood Publishing House,
750 Third Avenue, 9th Floor,
New York, NY 10017, USA
www.syrawoodpublishinghouse.com

Food Production and Processing: Fermentation, Nutritional Value and Quality Control
Edited by Albert McCoy

International Standard Book Number: 978-1-68286-762-4 (Hardback)

Cataloging-in-Publication Data

Food production and processing : fermentation, nutritional value and quality control / edited by Albert McCoy.
    p. cm.
Includes bibliographical references and index.
ISBN 978-1-68286-762-4
1. Food. 2. Food industry and trade. 3. Produce trade. 4. Farm produce. 5. Fermented foods.
6. Nutrition. 7. Food industry and trade--Quality control. I. McCoy, Albert.
TX353 .F66 2019
641.3--dc23

TABLE OF CONTENTS

PREFACE

The food industry includes the processes of agriculture, food manufacturing, processing, marketing, distribution, research and development. Food processing refers to the transformation of raw materials into finished products for consumption, using various methods such as mincing, liquefaction, cooking, pickling, pasteurization, preservation, etc. The process of conversion of carbohydrates into alcohol or organic acids with the help of microorganisms such as bacteria or yeasts, under anaerobic conditions is known as fermentation. It is a significant process in food processing. Some common fermented foods are cheese, vinegar, bread, etc. This book contains some path-breaking studies in the field of food production and processing. It discusses the fundamentals as well as modern approaches of these fields. This book will serve as a reference to a broad spectrum of readers.

This book is the end result of constructive efforts and intensive research done by experts in this field. The aim of this book is to enlighten the readers with recent information in this area of research. The information provided in this profound book would serve as a valuable reference to students and researchers in this field.

At the end, I would like to thank all the authors for devoting their precious time and providing their valuable contribution to this book. I would also like to express my gratitude to my fellow colleagues who encouraged me throughout the process.

Editor

An Assessment of Antioxidant and Antiproliferative Activities of Super Grain Quinoa

Sikha Bhaduri*

CUNY School of Public Health, NY 10035, USA

Abstract

Quinoa is known to be an excellent source of natural antioxidants and therefore the extract of quinoa seed was considered to have a significant anti-inflammatory activity. Two different varieties of quinoa seeds and six different solvents hexane, acetone, methanol, ethanol, ethyl acetate and water were used as solvent for extraction in the present study. Extracts from water, methanol and ethanol showed significant antioxidant and phytochemical activities. Water extract showed highest Phenol content (89.73 ± 1.74), antioxidant activity (1586 ± 41.42) and DPPH scavenging capacities (82.71 ± 0.03) compared to other solvents used for extraction. IC50 value for percentage DPPH scavenging capacities by water extract was 14.71 ± 0.02, compared to ascorbic acid (7.15 ± 0.13), which is a control. All extracts exhibit significantly high levels of flavonoid content. Ethyl acetate extract represented highest (88.41 ± 0.37) NO scavenging capacity. Lowest IC50 value (52.58 ± 0.14) for NO scavenging capacity was identified for ethanol extract compared to control (24.19 ± 3.53). Ascorbic acid used as control in both DPPH and Nitric oxide scavenging capacities measurement. Quinoa seed extracts from all six solvents found to have antimicrobial activities towards gram positive bacteria but not towards all gram negative bacteria. All extracts showed significant anti proliferative activities towards P 116 cells.

Keywords: Quinoa seed; Extract; Antioxidant activity; Extraction; Scavenging; IC_{50} value

Introduction

Normally, there is a balance between the quantities of free radicals generated in the body and the antioxidant mechanisms scavenge/quench these free radicals preventing them from causing deleterious effects in the body. But, many detrimental diseases such as periodontal disease, metabolic bone diseases, diabetes mellitus, atherosclerosis, neurodegenerative diseases are characterized by an enhanced state of oxidative stress and are associated with the overproduction of reactive oxygen species (ROS) and/or a decrease in antioxidant defenses, leading to impose oxidative stress in the body [1]. Oxidative stress can be defined as an excessive amount of Reactive species (RS), which is the net result of an imbalance between production and destruction of RS, and which is regulated by antioxidant defenses. RS is a collective term that includes both oxygen radicals and other reactive oxygen and nitrogen species (ROS/RNS). Reactive oxygen (ROS) and nitrogen (RNS) species are products of normal cellular metabolism, which, at high concentrations, thought to be important mediators of damage to cellular structures, such as nucleic acids, lipids and proteins. Oxidation in the living system is essential for generation of energy due to catabolism but prolonged oxidative stress result in the continuous production of free radicals and reactive oxygen species (ROS), which can lead to permanent damage of the body organs leading to chronic disorders such as heart diseases, diabetes, cirrhosis, malaria, neurodegenerative diseases, AIDS, cancer, and premature aging [2,3]. ROS includes free radicals such as superoxide anion radicals (O^{2-}), hydroxyl radicals (OH) and non-free-radical species such as H_2O_2 and singled oxygen are various forms of activated oxygen. Oxidative damage to biomolecules such as proteins, lipids, and DNA is also thought to accumulate with ageing, leading to the pathogenesis of many age-related diseases [4]. Oxidative damage to DNA is considered a critical step in cancer development [5].

It is well-known that diets rich in fruit and vegetables are protective against cardiovascular disease and certain forms of cancer and perhaps against other diseases also [6]. Celiac Disease (CD), an systematic autoimmune enteropathy, currently CD is thought to resemble multisystem immunological disorder rather than a disease restricted to the gastrointestinal tract [7,8], characterized by chronic inflammation of the intestinal mucosa, atrophy of intestinal villi and a permanent intolerance to gluten protein [9]. Plants have been used as traditional medicine system throughout the world for many years since they are rich in a wide variety of phytochemical compounds, which are good sources of antioxidants [10]. Following antioxidant hypothesis coined by Gey [11], it was thought that an increased intake of antioxidants found in biological system via dietary supplements might lower the risk of diseases that involve oxidants. A growing number of observational studies has examined the association between the intake of foods rich in polyphenols (onions, apples, tea, cocoa, red wine) and chronic diseases as well as the relation between the intake of individual dietary flavonoids and chronic diseases [12]. The antioxidant defense mechanism has led to the concept that increased antioxidant defenses might lower the risk of diseases that involve oxidants and free radicals generated from endogenous metabolic systems, such as, inflammation, aerobic respiration, exercise etc. and from exogenous systems such as, tobacco smoking, air/other pollution, radiation and UV-light and thus protect DNA, proteins and lipids from oxidative damage [4]. High antioxidant capacity is provided mainly by their phenolic contents and the beneficial effects of the polyphenols are known. Due to their antioxidant activities polyphenols might also enhance the antioxidant defense of the body. These properties have been demonstrated in various in vitro and ex vivo models [4].

Quinoa (*Chenopodium quinoa* Willd) a pseudo cereal belonging to the Chenopodiaceae family, has been considered to be a functional

---

Corresponding author: Sikha Bhaduri, CUNY School of Public Health, NY 10035, USA, E-mail: sbhaduri@hunter.cuny.edu food and it has gained increasing interest in recent years due to its high nutritional value. Quinoa also acts as a cell protector and presents an important source of antioxidants [13] due to its polyphenolic phytonutrient constituents, and therefore could be used for medicinal purposes in humans [14,15]. This gluten free pseudocereal recommended to be used as an ingredient in food such as bread, pasta or baby food and also in common diets [16], since; only effective treatment [17] for celiac disease (CD) is a life-long gluten-free diet. Despite its slightly bitter taste, quinoa flour produces acceptable gluten free baked products [18] that are high in fiber and rich in antioxidants. However, plants rich in phytonutrients may play an important role in inhibiting both free radicals and oxidative chain reactions within tissues and membranes [19].

Therefore due to its excellent nutritional, functional and technological properties, exceptional antioxidant activities and phytonutrient contents, *Chenopodium quinoa* might has a potential ability to be used as a naturally occurring food preservative also to inhibit microbial growth along with its anti-inflammatory activities, which probably needs a more in depth investigation. Development of cost effective isolation procedures that yield standardized extracts as well as safety and toxicology evaluation of quinoa requires a deeper investigation [20]. Due to the complex nature of phytochemicals, a single method cannot evaluate the antioxidant activities of the seed. Therefore several standard methods were used. Considering all these, the proposed study was aimed to evaluate the antioxidant and anti-inflammatory properties and activities of super grain quinoa phytochemicals.

Materials and Methods

Preparation of extracts

The quinoa seeds were powdered with a mechanical grinder to obtain a coarse powder, which were then subjected to extraction using a modified method [21] with hexane, acetone, methanol, ethanol, ethyl acetate and water. Briefly, 2 g of seed powder were extracted with 20 mL of solvent into a 50 mL polyethylene centrifuge tube. The mixture was kept on a rotary shaker at speed 5 for 24 hours at room temperature. After 24 hours, the filtrate was centrifuged at 5000 g for 10 minutes, the supernatant was collected. The extraction was done at least three times with the residues and supernatants collected each time. All the collected supernatants were pulled out together and were filtered through Whatman No. 1 paper filter and concentrated to a dry mass with the aid of a rotary evaporator. Two different varieties of quinoa seeds grain type I and grain type II from two different companies were selected to get extracts from hexane, acetone, methanol, ethanol, ethyl acetate and water as solvent. Therefore, two different sets of extracts from each of two different grains I and II were obtained. Each dried extracts were dissolved in 1 ml dimethyl sulfoxide (DMSO).

Determination of total phenolic content

The amount of total soluble phenolic content in different seed extracts was determined according to Folin-Ciocalteu method [22] with slight modifications. Briefly, 10 µL of the sample extract or a series of gallic acid standards (0, 20, 40, 60, 80, and 100 mg/l) from the stock solution was mixed with 100 µL of Folin-Ciocalteu reagent (Sigma Chemical Co., St. Louis, Mo., USA). After 10 min of incubation, 300 µL of 20% Na_2CO_3 solution was added and the volume was adjusted to 1 mL using deionized water. The mixture was incubated in dark for 2 hours at room temperature and the absorbance was measured at 750 nm using a microplate spectrophotometer (Bio-Rad Laboratories, Inc., 2000 Alfred Nobel Drive, CA 94547, USA) against blank sample. The

total phenolic content was measured as gallic acid equivalents per gram of dry mass (mg GAE/g dw) and the values were presented as means of triplicate analysis.

Determination of total flavonoid content

Total flavonoid content was estimated by a modified colorimetric method [23] by taking 20 µL of each extract and mixed with 500 µl de-ionized water and 30 µl of 5% sodium nitrite ($NaNO_2$) solution. After 5 min of incubation at room temperature, 60 µl of 10% Aluminium chloride ($AlCl_3$) solution was added. Subsequently, 350 µl of 1 M sodium hydroxide (NaOH) and 40 µl of de-ionized water were added to make the final volume 1 mL. Samples were further incubated for 15 min at room temperature and the absorbance of the samples was measured at 510 nm in a spectrophotometer (Spectrnics 20, Spectronics, CA). The total flavonoids were determined as qurecetin equivalents per gram of dry mass (mg QE/g dw) and the values were expressed as means of triplicate analysis.

Evaluation of antioxidant capacity

Total antioxidant activity was estimated by phosphomolybdenum assay method [24] using Molybdate reagent Solution (1 ml each of 0.6 M sulfuric acid, 28 mm sodium phosphate and 4 mm ammonium molybdate were added in 20 ml of distilled water and made up volume to 50 ml by adding distilled water). Extracts in different concentration ranging from 100 µl to 500 µl were added to each test tube individually containing 3 ml of distilled water and 1 ml of Molybdate reagent solution. These tubes were kept incubated at 95°C for 90 min. After incubation, these tubes were normalized to room temperature for 20-30 min and the absorbance of the reaction mixture was measured at 655 nm using a microplate spectrophotometer (Bio-Rad Laboratories, Inc., 2000 Alfred Nobel Drive, CA 94547, USA) against blank sample. Mean values from three independent samples were calculated for each extract. Ascorbic acid (AA) was used as positive reference standard and the total antioxidant capacity was expressed as milligrams of ascorbic acid equivalents per gram of dry mass (mg AAE/g dw).

Determination of ferric reducing antioxidant power (FRAP)

A modified method [25] was used to measure the ferric ions (Fe^{3+}) reducing antioxidant power (FRAP) of extracts. The FRAP method is based on a redox reaction in which an easily reduced oxidant (Fe^{3+}), the ferricyanide complex is used in stoichiometric excess and the extracts used as antioxidants, which acts as reductants to reduce the ferricyanide complex to the ferrous form. Briefly, an aliquot of 100 µl of different extracts (5-200 mg/ml) were mixed with 1 ml phosphate buffer (0.2 M, pH 6.6) and 1 ml of potassium ferricyanide (1% w/v), shaken well and incubated at 50°C for 20 min. After incubation, 1 mL of trichloroacetic acid (10% w/v) was added to stop the reaction, which was then centrifuged at 3000rpm for 10 min. After centrifugation, 1.5 ml supernatant was mixed with 1.5 ml deionized water and 0.1 ml ferric chloride (0.1%). The mixture was incubated for 10 min and absorbance was read at 700 nm in a spectrophotometer (Spectrnics 20, Spectronics, CA). The assays were carried out in triplicate and the results are expressed as mean values ± standard deviations. Ascorbic acid was used as standard. The reducing power of the extracts was represented as mg AAE/g of dry mass. Higher absorbance indicates higher reducing power.

DPPH radical scavenging activity

The antioxidant activities of quinoa seed extracts were assessed [26] on the basis of the radical scavenging effect using stable 1, 1-diphenyl-

2-picrylhydrazyl (DPPH). DPPH solution (0.004% w/v) was prepared in 95% methanol and serial dilutions were carried out with the stock solutions (20 mg/mL) of the extracts. Various concentrations of extracts were mixed with DPPH solution (900 μl), incubated in dark for 30 min and then absorbance was measured at 517 nm (Spectrnics 20, Spectronics, CA). Methanol (95%), DPPH solution and ascorbic acid (AA) were used as blank, control and reference standard respectively.

DPPH scavenging ability (%) = $[A_{517 \text{ nm of control}} - A_{517 \text{ nm of sample}} / A_{517 \text{ nm of control}}]$ X 100

Nitric oxide scavenging activity

Nitric oxide scavenging activity was determined according to Griess Illosvoy reaction [27] by sodium nitroprusside method [28] in a 96-well microplate. The reaction solution (50 μl) containing 10 mm sodium nitroprusside in PBS (pH 7.0) was mixed with 50 μl of different concentration (50–200 μg/ml) of sample extracts, followed by incubation at 37°C for 20 min under light. After incubation, 100 μL of Griess (Promega) reagent (1% sulfanilamide, 2% H_3PO_4) added to each well. Incubate at room temperature for 5-10 minutes, protected from light. A purple/magenta color will begin to form immediately. Absorbance was noted within 30 minutes in a microplate spectrophotometer (Bio-Rad Laboratories, Inc., 2000 Alfred Nobel Drive, CA 94547, USA) at 540 nm and the results were expressed as per cent of scavenged nitric oxide with respect to the negative control without addition of any antioxidant. Ascorbic acid was used as a positive control.

Nitric oxide scavenging ability (%) = $[A_{540} \text{ nm of control} \times A_{540} \text{ nm of sample} / A_{540} \text{ nm of control}] \times 100$.

Antimicrobial activity

Briefly, the extracts were tested against the reference strains for antimicrobial activity containing different concentrations of extracts using micro dilution method [29-31] in 96 well micro plates with minor modifications. The antimicrobial activity of the extracts was evaluated against two gram positive (*Enterococcus faecalis* ATCC 43062, *Staphylococcus epidermidis* ATCC 33501) and two gram negative (*Escherichia coli* ATCC 25922, *Pseudomonas aeruginosa* ATCC 27853) bacterial strains. The culture suspension (100 μl) were seeded into 96-well plate at 1×10^4 cells/well density and treated with all extracts of quinoa seeds for two different concentrations, 50 μg/ml and 200 μg/ml and incubated for 48 h at $37 \pm 1°C$ and then 10 μl of MTT (Promega) (5 mg/ml) was added to each well. In control experiments, sterile distilled water was added in place of quinoa extracts and only sterile broth was used in place of suspension cultures (without inoculums) as blank. Then absorbance was measured at 570 nm in a microplate spectrophotometer (Bio-Rad Laboratories, Inc., 2000 Alfred Nobel Drive, CA 94547, USA). All experiments were performed in triplicate.

The percentage inhibition of the extracts was calculated using the formula:

% Inhibition = $(A_{control} - A_{sample}) / A_{control} \times 100$

Anti-proliferative activity

Cytotoxic activity of the extract was determined by MTT (Cell Titer 96TM, Promega) assay against a lymphocyte as a preliminary step of experiment by colorimetric assay. Briefly, P 116 T lymphocyte cells were plated at 10^4 cells/well in a 96-well plate. The culture cells were incubated in a humidified incubator at 37°C, atmosphere of 5% CO_2 and 95% air for 24 h. After incubation for 24h at 37°C, the medium was discharged and cells were treated by all extracts of quinoa seeds with two different concentrations, 50 μg/ml and 200 μg/ml and incubated for 24 h. MTT reagent (0.5 mg/ml) was added to each well and further incubated for 4 hours at 37°C. Viable cells react with MTT to produce purple formazan crystals. After 4 h, the stop solution was added. The cells were then incubated for 2 h in room temperature and protected from light. After incubation, the cells were shaken. Optical density was read with a microplate spectrophotometer (Bio-Rad Laboratories, Inc., 2000 Alfred Nobel Drive, CA 94547, USA) at a wavelength of 570 nm. The experimental data was absorbance of each well represents viability of cells in each well and then converted to percentage inhibition of growth. All experiments were performed in triplicate.

The percentage inhibition of the extracts was calculated using the formula:

% Inhibition = $(A_{control} - A_{sample}) / A_{control} \times 100$

Statistical analysis

Microsoft® Excel 2013 programs were used for statistical analyses and calculations, Data were expressed as mean (M) ± standard deviation (SD) for 3 samples in each group. Student's t-test and one way ANOVA were used to compare means, presenting their respective p-value for determination of statistical significance. The difference between compared groups was considered to be significant when $p<0.05$. Lower case letters a, b, c, d, e, f and g indicate statistical significance.

Results and Discussion

All extracts and their abbreviations

Table 1 explained all abbreviations used for different types of extracts. They were designated as QHX-I and QHX-II (for hexane extract), QAC-I and QAC-II (for acetone extract), QMT-I and QMT-II (for methanol extract), QET-I and QET-II (for ethanol extract), QEA-I and QEA-II (for ethyl acetate extract), and QWA-I and QWA-II (for water extract). The dried extracts were dissolved in 1 ml dimethyl sulfoxide (DMSO).

Antioxidant and phytochemical activities

The results of total phenolic content of two different quinoa seed extracts (I and II) have been shown in Table 2. The total phenol content in the quinoa seed extracts (QHX, QAC, QMT, QET, QEA and QWA) expressed as gallic acid equivalent (GAE). The total phenol content in different extracts varied with different solvents and the range was found from 23.74 ± 1.63 mg GAE/g dw for QHX-II to 89.73 ±1.74 mg GAE/g dw for QWA-II, which is the highest total phenolic content. The results were also indicated that the methanol (QMT I and II) and water extracts (QWA I and II) possess a higher level of total phenol content compared to all other extracts. These data indicate major phenolics of quinoa are located in the water derived extracts.

All extracts from different solvents showed a moderately high level of total flavonoid content , the highest level of flavonoid content (106.49

Extracts	Quinoa grain type I	Quinoa grain type II
Hexane	QHX-I	QHX-II
Acetone	QAC-I	QAC-II
Methanol	QMT-I	QMT-II
Ethanol	QET-I	QET-II
Ethyl acetate	QEA-I	QEA-II
Water	QWA-I	QWA-II

Table 1: All abbreviations used.

Type of extract	Total Phenols[X]	Total flavonoid[Y]	Total antioxidant activities[Z]	Ferric reducing antioxidant power[Z]
QHX-I	24.38 ± 1.28[b]	41.72 ± 4.91[f]	39.13 ± 0.89[e]	60.06 ± 2.32[d]
QHX-II	23.74 ± 1.63[c]	67.37 ± 16.68[c]	48.33 ± 0.58[f]	61.34 ± 0.84[d]
QAC-I	27.16 ± 1.82[f]	80.85 ± 5.89[c]	69.78 ± 2.37[f]	133.29 ± 2.11[b]
QAC-II	37.21 ± 1.98[d]	106.49 ± 10.14[c]	88.64 ± 2.37[g]	134.58 ± 2.24[b]
QMT-I	49.09 ± 2.11[c]	51.83 ± 6.68[a]	318 ± 13.98[d]	176.65 ± 12.84[a]
QMT-II	66.63 ± 1.23[d]	72.38 ± 3.03[a]	438 ± 6.98[h]	183.08 ± 6.42[b]
QET-I	38.39 ± 1.15[e]	46.81 ± 3.45[a]	139 ± 2.25[d]	155.78 ± 1.69[d]
QET-II	39.05 ± 2.25[b]	65.85 ± 8.32[a]	176 ± 5.32[e]	156.63 ± 0.49[e]
QEA-I	25.02 ± 0.44[c]	45.74 ± 3.48[b]	53.28 ± 11.85[g]	44.01 ± 0.64[b]
QEA-II	29.09 ± 3.45[e]	50.63 ± 3.98[b]	78.03 ± 4.97[g]	44.64 ± 0.32[b]
QWA-I	57.54 ± 9.01[c]	66.94 ± 13.03[b]	879 ± 15.31[g]	37.25 ± 0.64[c]
QWA-II	89.73 ± 1.74[f]	89.43 ± 10.75[c]	1586 ± 41.42[g]	37.47 ± 0.81[c]

X gallic acid, Y quercetin, Z ascorbic acid
Results are expressed as Mean ± Standard deviation, where n = 3. Values with different superscripts (a, b, c, d, e, f, g, h) in a column differ significantly (p < 0.05).

Table 2: Quantitative representation of antioxidant and phytochemical activities.

± 10.14 mg QE/mg dw) was found for acetone extract (QAC-II) and the lowest level (41.72 ± 4.91 mg QE/mg dw) was for hexane extract (QHX-I).

Both methanol extracts (I and II) water extracts (I and II) showed significantly high level of anti-oxidant activities, however the water extracts showed higher level of total antioxidant activities compared to other solvent extracts and the highest level (1586 ± 41.42 mg ascorbic acid/mg dw) of total antioxidant property was observed for QWA-II, as represented in Table 2. These data indicate major phenolics of quinoa are located in the water and also in the methanol derived extracts. Water extracts of quinoa might be considered to have a high level of antioxidant defensive action against reactive oxygen species.

Ferric reducing antioxidant power were observed high for acetone, methanol and ethanol extracts and the highest level (183.08 ± 6.42 mg ascorbic acid/mg dw) was for QMT-II, showed in Table 2. Both the extracts of water showed lowest level of ferric reducing antioxidant activities.

DPPH radical scavenging activity

DPPH scavenging activities are represented in Table 3. The DPPH assay measured hydrogen atom (or one electron) donating activity and hence provided an evaluation of antioxidant activity due to free radical scavenging, 2,2-Diphenyl-lpicrylhydrazyl radical (DPPH) a purple-colored stable free radical is reduced into the yellow colored diphenyl picryl hydrazine [26]. Very significant antioxidant activities were found in all the six extracts for two different seeds at higher concentration 200 μg/ml (Table 3). Highest DPPH scavenging activities 82.39 ± 0.03 and 82.71 ± 0.03 were for water extracts. All other extracts (QHX, QAC, QMT, QET and QEA) showed similar DPPH scavenging capacity and the range was 66.56 ± 0.04 to 77.47 ± 0.03. IC_{50} values (concentration of the extract in μg/ml that was able to scavenge half of the DPPH radical) was 14.71 ± 0.31 and 14.93 ± 0.06, the lowest for water extracts (QWA I and II).

Since, water extracts from both type of seeds showed lowest IC_{50} value compared to other solvents, this indicates water extracts of quinoa has a strong proton donating ability, which could serve as a free radical scavenger and can neutralize the reactive oxygen species originate due to prolonged oxidative stress in living organisms.

Nitric oxide scavenging activity

Percentage free radical nitric oxide scavenging activities against concentrations of the extracts have been tabulated in Table 4. The scavenging activities increased with an increase in concentration of all extracts. Nitric oxide generated from aqueous sodium nitroprusside (SNP) solution interacts with oxygen to produce nitrite ions at physiological pH, which may be quantified and determined according to Griess Illosvoy reaction [28]. All the extracts of quinoa exhibited significant NO scavenging activity in a concentration dependent manner (Table 4). The results clearly identify QEA-II as better NO scavenger where percentage inhibition reached to 88.41 ± 0.37 at a concentration 200 μg/ml with an IC50 value of 74.31 ± 0.27. Lowest nitric oxide scavenging activity (26.91 ± 0.37) was observed with water extracts (QWA-II) at a concentration 200 μg/ml.

The NO radicals play an important role in inducing inflammatory responses Nitric oxide (NO) is a diffusible free radical that plays many roles as an effectors molecule in diverse biological systems including neuronal messenger, vasodilatation, and antimicrobial and antitumor activities, but Chronic exposure to nitric oxide radical is associated with various carcinomas and inflammatory conditions including juvenile diabetes, multiple sclerosis, arthritis, and ulcerative colitis [32].

Antimicrobial activity

All extracts of quinoa had shown moderate to significant antimicrobial activities against to both gram positive bacteria *E. faecalis* ATCC 43062, *S. epidermidis* ATCC 33501 shown in Figures 1(a) and 1(b) and against one of the two gram negative bacteria used, which is *P. aeruginosa* ATCC 27853 at a concentration 200 μg/ml as shown in Figure 1(c). The extracts from methanol, ethyl acetate and water showed significant antimicrobial activities against *E. coli* ATCC 25922 presented in Figure 1(d) at a concentration 200 μg/ml compared to control. The extracts from hexane, acetone and ethanol did not show any significant activities at this concentration. However, a higher concentration over 200 μg/ml was not tried in this study.

Table 5 representing IC_{50} values of all cells. All extracts besides extracts from hexane showed moderately low IC_{50} values against *E. faecalis* ATCC 43062. Methanol extract showed lowest IC_{50} value (18.75 ± 0.03) against *E. faecalis* ATCC 43062, whereas, IC_{50} values were very low and were within the range of 14.55 ± 0.02 to 15.59 ± 0.05 for all the extracts against *S. epidermidis* ATCC 33501.

Both ethanol and methanol extracts showed lowest IC_{50} values 19.19 ± 0.04 and 16.67 ± 0.03 respectively against *P. aeruginosa* ATCC 27853, and only methanol extract showed lowIC_{50} values 20.24 ± 0.02 against *E. coli* ATCC 25922. This observation might conclude quinoa seed extracts have antimicrobial activities towards gram positive

Concentration µg/ml	% Cavenging of DPPH												
	Ascorbic acid	QHX-I	QHX-II	QAC-I	QAC-II	QMT-I	QMT-II	QET-I	QET-II	QEA-I	QEA-II	QWA-I	QWA-II
5	45.04 ± 0.04^g	12.32 ± 0.02^d	11.06 ± 0.01^f	12..54 ± 0.03^g	14.57 ± 0.11^f	10.05 ± 0.01^f	12.33 ± 0.02^e	10.12 ± 0.02^f	10.11 ± 0.01^f	10.01 ± 0.03^f	10.05 ± 0.03^f	10.09 ± 0.08^f	10.06 ± 0.02^g
10	57.06 ± 0.05^f	28.43 ± 0.04^d	26.55 ± 0.03^f	20.53 ± 0.02^d	23.34 ± 0.03^c	22.07 ± 0.04^e	31.02 ± 0.02^d	33.95 ± 0.04^c	32.27 ± 0.07^d	22.02 ± 0.02^e	24.04 ± 0.02^e	36.17 ± 0.04^c	35.15 ± 0.03^f
25	70.69 ± 0.11^d	55.43 ± 0.06^b	48.91 ± 0.36^c	42.78 ± 0.07^c	45.06 ± 0.05^c	47.03 ± 0.04^d	62.15 ± 0.30^b	69.94 ± 0.91^b	65.49 ± 0.08^a	29.42 ± 0.07^b	30.18 ± 0.07^b	75.18 ± 0.19^d	72.23 ± 0.22^b
50	84.28 ± 0.32^d	59.01 ± 0.51^b	55.33 ± 0.28^c	55.76 ± 0.26^a	57.07 ± 0.39^b	66.67 ± 0.76^a	63.89 ± 0.10^c	71.26 ± 0.26^a	69.18 ± 0.27^a	39.42 ± 0.43^b	40.36 ± 0.34^d	74.21 ± 0.25^a	74.91 ± 0.48^b
100	87.86 ± 0.05^e	64.63 ± 0.04^c	62.91 ± 0.06^b	70.99 ± 0.07^b	71.98 ± 0.08^b	73.31 ± 0.27^a	74.07 ± 0.39^a	73.09 ± 0.11^a	72.45 ± 0.42^c	58.44 ± 0.05^a	59.86 ± 0.80^c	77.11 ± 0.09^a	79.82 ± 0.74^b
200	90.69 ± 0.07^e	67.91 ± 0.09^b	66.56 ± 0.04^d	73.91 ± 0.03^c	74.71 ± 0.03^c	76.73 ± 0.04^c	77.47 ± 0.03^c	75.25 ± 0.03^c	73.12 ± 0.02^c	71.34 ± 0.02^c	72.31 ± 0.03^c	82.39 ± 0.03^b	82.71 ± 0.03^d
IC$_{50}$	7.15 ± 0.13^d	22.56 ± 0.05^b	26.58 ± 0.38^c	38.75 ± 0.21^a	39.03 ± 0.25^a	28.88 ± 0.34^c	20.13 ± 0.15^c	17.87 ± 0.24^a	19.08 ± 0.38^d	73.91 ± 0.42^d	74.83 ± 0.84^d	14.71 ± 0.02^a	14.93 ± 0.06^d

Table 3: Scavenging of DPPH by the extracts of Quinoa.

Results are expressed as Mean ± Standard deviation, where n = 3. Values with different superscripts (a, b, c, d, e, f, g, h) in a row differ significantly ($p < 0.05$).

Concentration µg/ml	% Scavenging of Nitric oxide												
	Ascorbic acid	QHX-I	QHX-II	QAC-I	QAC-II	QMT-I	QMT-II	QET-I	QET-II	QEA-I	QEA-II	QWA-I	QWA-II
5	21.92 ± 1.12^b	10.22 ± 0.25^a	7.91 ± 0.86^a	10.86 ± 0.11^c	11.08 ± 0.11^a	10.74 ± 0.02^a	13.84 ± 0.95^b	15.27 ± 1.11^a	18.54 ± 0.44^b	11.48 ± 0.48^a	13.06 ± 0.05^b	8.33 ± 0.55^b	7.69 ± 0.80^b
10	43.09 ± 0.37^c	13.81 ± 0.73^a	12.11 ± 0.12^b	14.01 ± 0.48^a	15.27 ± 0.25^a	13.8 ± 0.66^b	18.12 ± 0.11^b	23.7 ± 0.64^b	28.75 ± 0.90^c	13.59 ± 0.37^d	18.65 ± 0.30^c	8.85 ± 0.35^a	8.11 ± 0.09^d
25	54.16 ± 0.19^e	18.12 ± 0.33^a	16.54 ± 0.16^b	20.12 ± 0.39^a	22.31 ± 0.42^a	21.35 ± 0.57^a	22.02 ± 0.47^a	38.25 ± 0.21^b	44.41 ± 0.52^c	21.49 ± 0.31^b	29.39 ± 0.53^d	10.22 ± 0.25^a	9.79 ± 0.38^e
50	65.51 ± 0.51^d	25.52 ± 0.81^a	22.12 ± 0.21^d	32.13 ± 0.32^b	37.41 ± 0.39^c	33.61 ± 0.55^c	35.93 ± 0.39^a	57.11 ± 0.84^b	65.11 ± 0.67^c	30.66 ± 1.11^d	44.02 ± 0.02^e	11.08 ± 0.35^d	10.64 ± 0.33^e
100	77.63 ± 0.71^c	37.61 ± 0.34^a	34.66 ± 0.29^d	48.78 ± 1.33^b	57.01 ± 0.48^b	54.05 ± 0.06^a	61.01 ± 0.48^c	65.44 ± 0.71^c	79.66 ± 0.29^d	54.57 ± 0.39^d	65.85 ± 0.78^d	22.24 ± 0.27^a	21.61 ± 0.53^d
200	88.12 ± 0.82^d	58.37 ± 0.33^a	54.83 ± 0.15^d	70.81 ± 0.73^b	80.92 ± 0.51^a	78.92 ± 0.82^a	83.14 ± 0.53^a	84.93 ± 0.89^a	86.82 ± 0.36^d	70.07 ± 0.06^d	88.41 ± 0.37^d	27.49 ± 0.21^a	26.91 ± 0.37^d
IC$_{50}$	24.19 ± 3.53^c	136.31 ± 0.60^c	152.09 ± 1.01^d	98.51 ± 0.98^a	84.23 ± 0.22^c	91.17 ± 0.28^c	80.81 ± 0.73^c	62.13 ± 0.81^c	52.58 ± 0.14^d	87.87 ± 0.87^b	74.31 ± 0.27^b	>200	>200

Table 4: Scavenging of Nitric oxide by the extracts of Quinoa.

Results are expressed as Mean ± Standard deviation, where n = 3. Values with different superscripts (a, b, c, d, e, f, g, h) in a row differ significantly ($p < 0.05$).

Quinoa extracts	Type of Cell lines	Bacteria			Lymphocyte
Type of solvents	Enterococcus fecalis	Staphylococcus epidermidis	Pseudomonas aeruginosa	Escherichia coli	P 116
QHX-I	119.78 ± 0.27^d	15.26 ± 0.03^b	129.74 ± 0.07^e	257.06 ± 0.03^f	50.02 ± 0.03^c
QHX-II	112.75 ± 0.28^e	14.55 ± 0.02^a	110.25 ± 0.03^f	196.76 ± 0.05^e	47.57 ± 0.03^d
QAC-I	26.89 ± 0.18^b	14.97 ± 0.15^c	22.55 ± 0.11^c	178.83 ± 0.04^e	42.69 ± 0.07^d
QAC-II	24.73 ± 0.03^a	14.86 ± 0.15^a	20.08 ± 0.04^a	156.65 ± 0.04^g	35.04 ± 0.01^e
QMT-I	22.98 ± 0.79^a	14.94 ± 0.19^a	20.45 ± 0.04^c	21.08 ± 0.02^d	21.04 ± 0.04^a
QMT-II	18.75 ± 0.03^a	14.71 ± 0.16^a	19.19 ± 0.04^c	20.24 ± 0.02^h	20.98 ± 0.02^e
QET-I	47.53 ± 0.08^c	15.28 ± 0.05^a	17.06 ± 0.01^b	200.08 ± 0.03^g	31.26 ± 0.03^d
QET-II	40.19 ± 0.23^b	15.26 ± 0.05^a	16.67 ± 0.03^f	118.67 ± 0.01^f	20.84 ± 0.04^c
QEA-I	34.45 ± 0.11^d	15.16 ± 0.06^a	40.05 ± 0.04^c	104.45 ± 0.02^d	22.74 ± 0.04^d
QEA-II	23.75 ± 0.02^b	14.89 ± 0.05^b	35.11 ± 0.08^d	100.05 ± 0.05^d	18.19 ± 0.01^e
QWA-I	37.51 ± 0.05^c	15.59 ± 0.05^c	23.75 ± 0.02^d	113.03 ± 0.02^c	20.03 ± 0.04^d
QWA-II	38.08 ± 0.07^a	15.35 ± 0.04^a	22.51 ± 0.03^f	110.04 ± 0.04^h	16.67 ± 0.02^f

Table 5: IC$_{50}$ values of different cells against different extracts

Results are expressed as Mean ± Standard deviation, where n = 3. Values with different superscripts (a, b, c, d, e, f, g, h) in a column differ significantly ($p < 0.05$).

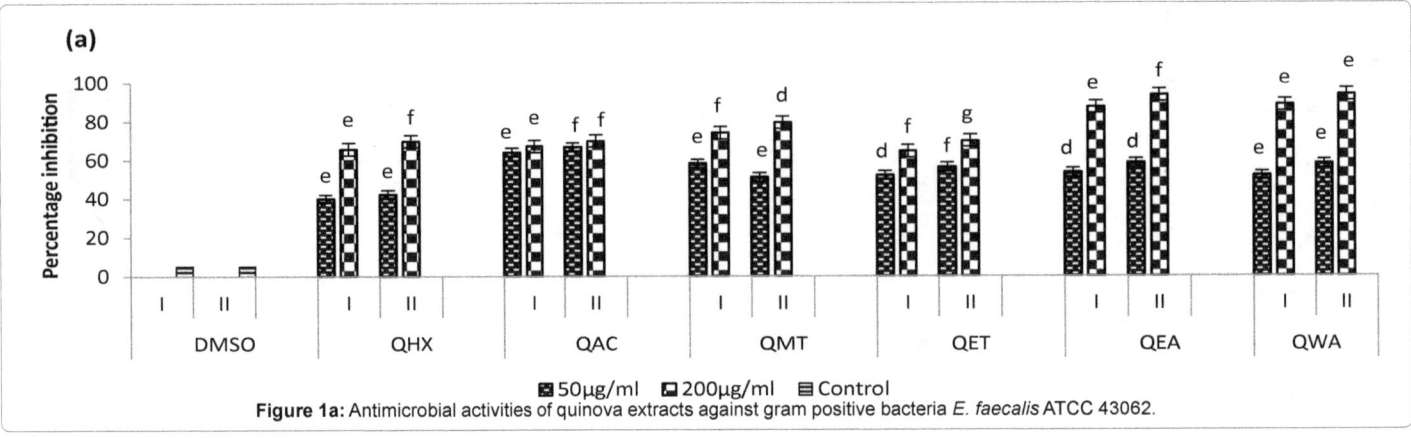

(a)

Figure 1a: Antimicrobial activities of quinova extracts against gram positive bacteria *E. faecalis* ATCC 43062.

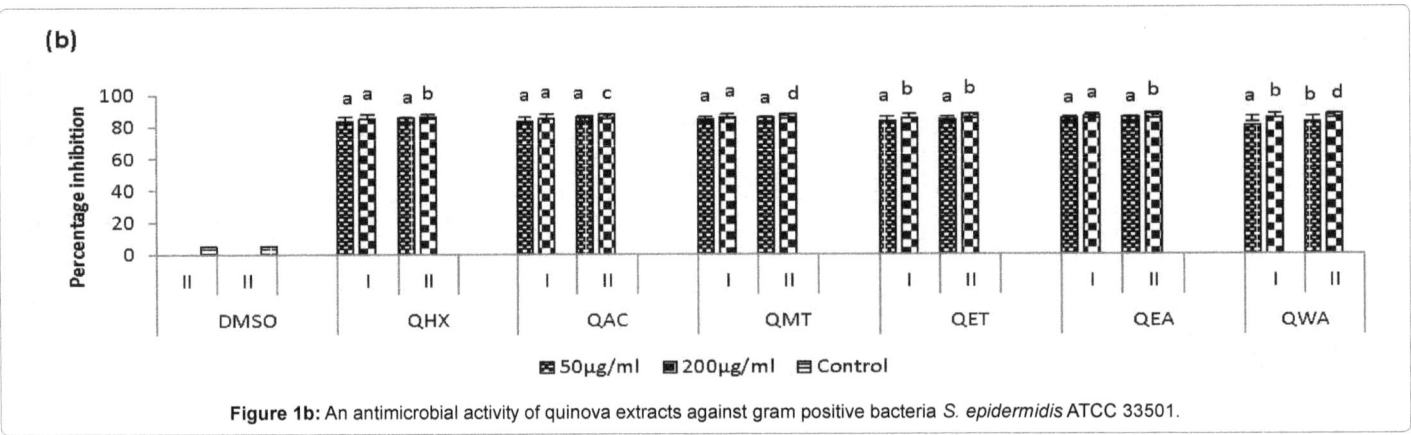

(b)

Figure 1b: An antimicrobial activity of quinova extracts against gram positive bacteria *S. epidermidis* ATCC 33501.

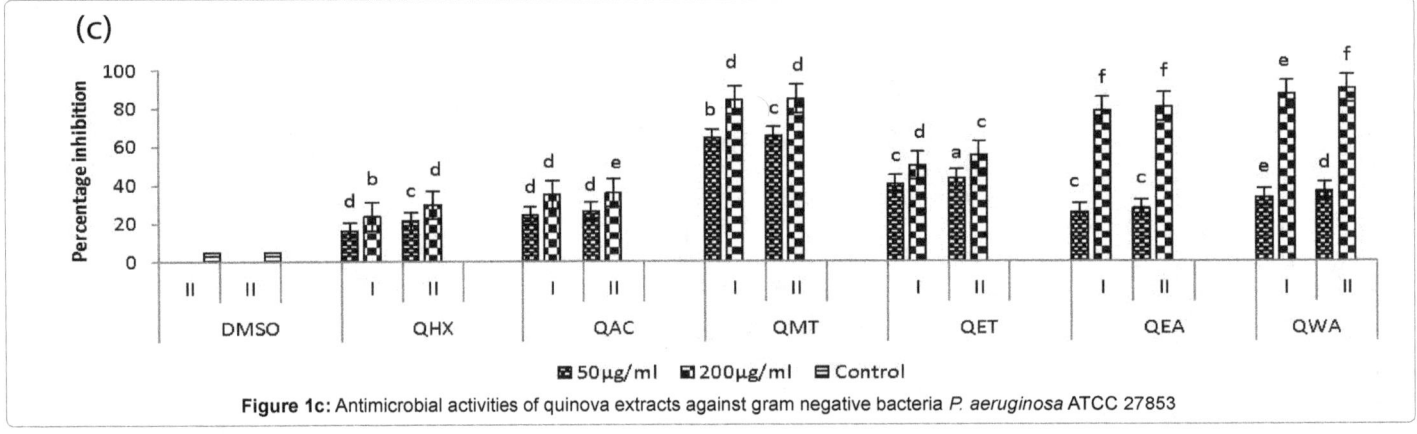

(c)

Figure 1c: Antimicrobial activities of quinova extracts against gram negative bacteria *P. aeruginosa* ATCC 27853

bacteria but not to all gram negative bacteria. However, IC_{50} values are higher for extracts from hexane against *E. faecalis* ATCC 43062 (119.78 ± 0.27) and *P. aeruginosa* ATCC 27853 (129.74 ± 0.03).These results also indicate hexane extracts might have limited antimicrobial activities towards both the gram positive and gram negative bacteria. The mechanism could be a further subject of studies.

Anti-proliferative activity

All extracts of quinoa extracts from all six solvents hexane, methanol, ethanol, ethyl acetate and water showed significant anti proliferative activities against P 116 cells at a concentration of 200 μg/ml, represented in Figure 2. However, methanol (QMT-I and QMT-II), ethanol (QET-I and QET-II), ethyl acetate (QEA-I and QEA-II)

and water (QWA-I and QWA-II) showed significantly low IC_{50} values ranging from 16.67 ± 0.02 (for water extract) to 31.26 ± 0.03 (for ethanol extract), as in Table 5. For the crude extracts an IC_{50} value lower than 30 μg/mL established as the criterion for cytotoxicity [33] by the National Cancer Institute (NCI). Since, water extracts QWA I and QWA II showed lowest IC_{50} values 20.03 ± 0.04 and 16.67 ± 0.02, water extracts for quinoa should be further studied for its cytotoxic activities.

Conclusion

The present study clearly demonstrated that the extract of quinoa seeds contains considerable amount of total phenols and flavonoids and exhibits high antioxidant and free radical scavenging activities. In addition, it has been demonstrated that the quinoa seed extracts also is

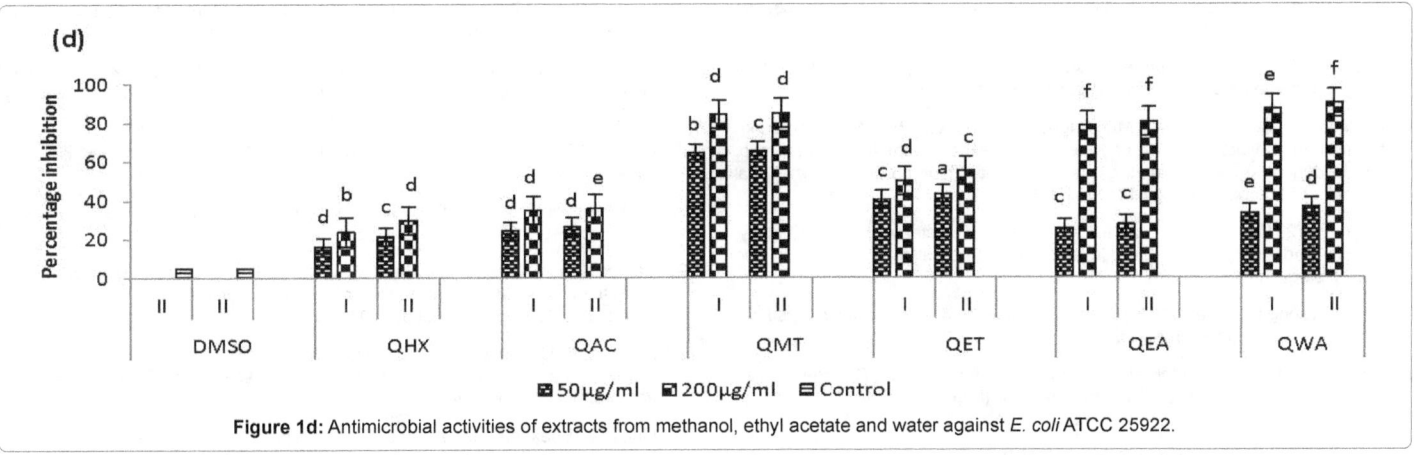

Figure 1d: Antimicrobial activities of extracts from methanol, ethyl acetate and water against *E. coli* ATCC 25922.

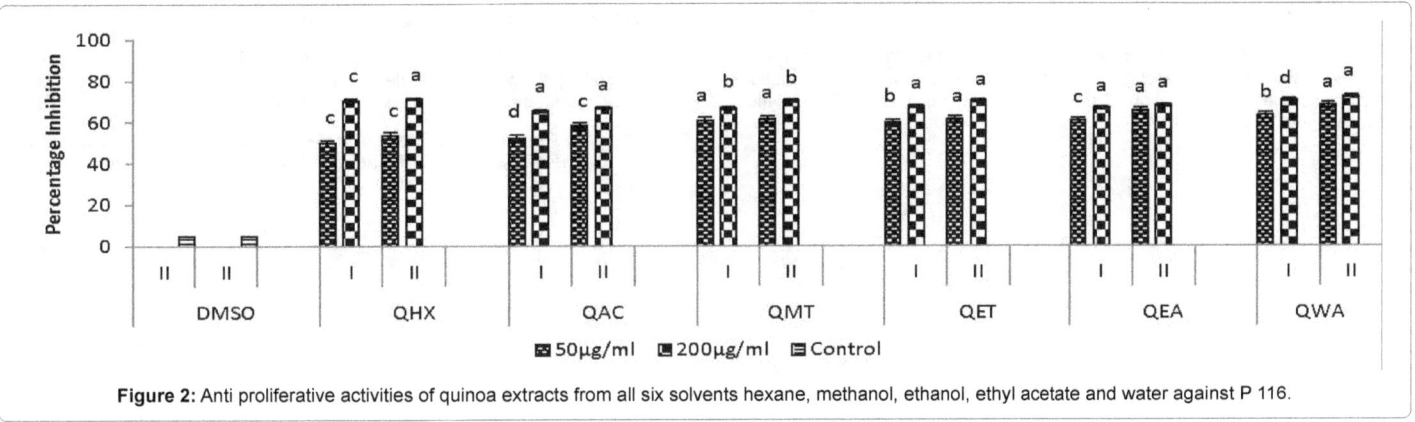

Figure 2: Anti proliferative activities of quinoa extracts from all six solvents hexane, methanol, ethanol, ethyl acetate and water against P 116.

a potential antiproliferative and antimicrobial agent. The antioxidant and biological activities might be due to the synergistic actions of bioactive compounds present in them. However, it is still unclear which components and mechanisms are playing vital roles for these activities. Therefore further investigation is needed to identify the exact moiety responsible for antioxidant and antiproliferative activities of quinoa seeds and also for other gluten free pseudocereals.

Acknowledgments

The work was supported by PSC-CUNY-TRADA-43-29 grant.

References

1. Lee SH, Kim SJ, Kim SJ (2014) Anti-oxidant activity with inhibition of osteoclastogenesis by Atractylodes Rhizoma extract. European Review for Medical and Pharmacological Sciences 18: 1806-1812.

2. Nagmoti DM, Khatri DK, Juvekar PR, Juvekar AR (2011) Antioxidant activity and free radical-scavenging potential of *Pithecellobium dulce* Benth seed extracts. Free Radical and Antioxidants 2: 37-43.

3. Agarwal A, Aponte-Mellado A, Premkumar BJ, Shaman A, Gupta S, et al. (2012) The effects of oxidative stress on female reproduction, a review. Reproductive Biology and Endocrinology.

4. Hollman PCH, Cassidy A, Comte B. Heinonen M, Richelle M, et al. (2011) The Biological Relevance of Direct Antioxidant Effects of Polyphenols for Cardiovascular Health in Humans Is Not Established. J Nutr 141.

5. Guyton KZ, Kensler TW (1993) Oxidative mechanisms in carcinogenesis. Brit Med Bull 49: 523-544.

6. Catherine AR, Nicholas JM, George P (1996) Structure-antioxidant activity relationships of Flavonoids and phenolic acids. Free Radical Biology &

Medicine 20: 933-956.

7. Green PH, Rostami K, Marsh MN (2005) Diagnosis of coeliac disease. Best Pract. Res. Clin Gastroenterol.

8. Murray JA, Van Dyke C, Plevak MF, Dierkhising RA, Zinsmeister AR, et al. (2003) Trends in the identification and clinical features of celiac disease in a North American community. Clin. Gastroenterol. Hepatol 1: 19-27.

9. Saturni L, Ferretti G, Bacchetti T (2010) The Gluten-Free Diet: Safety and Nutritional Quality. Nutrients 2: 16-34.

10. Bellik Y, Boukraâ L, Alzahrani HA, Bakhotmah BA, Abdellah F, et al. (2013) Molecular Mechanism Underlying Anti-Inflammatory and Anti-Allergic Activities of Phytochemicals: An Update. Molecules 322-353.

11. Gey KF (1986) On the antioxidant hypothesis with regard to arteriosclerosis. Bibl Nutr Dieta 37: 53-91.

12. Arts ICW, Hollman PCH (2005) Polyphenols and disease risk in epidemiological studies. Am J Clin Nutr 81: S317-S325.

13. Laus MN, Gagliardi A, Soccio M, Flagella Z, Pastore D, et al. (2012) Antioxidant Activity of Free and Bound Compounds in Quinoa (*Chenopodium quinoa* Willd.) Seeds in Comparison with Durum Wheat and Emmer. J Food Science 77: 1150-1155.

14. Bhargava A, Shukla S, Ohri D (2006) Chenopodium quinoa-an Indian perspective. Indus Crops Prodn 23: 73-87.

15. Vega-Galvez A, Miranda M, Vergara J, Uribe E, Puente L, et al. (2010) Nutrition facts and functional potential of quinoa (*Chenopodium quinoa* willd.), an ancient Andean grain: a review. J Sci Food Agric 90: 2541-2547.

16. Hirose Y, Fujita T, Ishii T, Ueno N (2010) Antioxidative properties and flavonoid composition of *Chenopodium quinoa* seeds cultivated in Japan. Food Chem 119: 1300-1306.

17. Hischenhuber C, Crevel R, Jarry B, Maki M, Moneret-Vautrin DA, et al. (2006)

Review article: safe amounts of gluten for patients with wheat allergy or coeliac disease. Alimentary Pharmacology and Therapeutics.

18. Bhaduri S (2013) A Comprenensive study on physical properties of two Gluten-free flour fortified muffins. J Food Processing and technology.

19. Rizzello CG, Coda R, Angelis MD, Cagno RD, Carnevali P, et al. (2009) Long-term fungal inhibitory activity of water-soluble extract from Amaranthus spp. seeds during storage of gluten-free and wheat flour breads. International Journal of Food Microbiology 131: 189-196

20. Negi PS (2012) Plant extracts for the control of bacterial growth: Efficacy, stability and safety issues for food application. International Journal of Food Microbiology 156: 7-17.

21. Choi Y, Jeong HS, Lee J (2007) Antioxidant activity of methanolic extracts from some grains consumed in Korea. Food Chemistry 103: 130-138.

22. Yang J, Paulino R, Janke-Stedronsky S, Abawi F (2007) Free-radical-scavenging activity and total phenols of noni (Morindacitrifolia L.) juice and powder in processing and storage. Food Chemistry 102: 302-308.

23. Barreira JCM, Ferreira ICFR, Oliveira MBPP, Pereira JA (2008) Antioxidant activities of the extracts from chestnut flower, leaf, skins and fruit. Food Chemistry 107: 1106-1113.

24. Prieto P, Pineda M, Aguilar M (1999) Spectrophotometric Quantitation of Antioxidant Capacitythrough the Formation of a Phosphomolybdenum Complex: Specific Application to the Determination of Vitamin E. Analytical Biochemistry 269: 337-341.

25. Oyaizu M (1986) Studies on products of browning reaction prepared from glucoseamine. Jpn J Nutri 44: 307-315.

26. Braca A, Sortino C, Politi M, Morelli I, Mendez J, et al. (2002) Antioxidant activity of flavonoids from Licanialicaniaeflora. Journal of Ethnopharmacology 79: 379-381.

27. Garratt DC (1964) The quantitative analysis of Drugs. Chapman and Hall Ltd, Japan 3: 456-458.

28. Sreejayan NR (1997) Nitric oxide scavenging by curcuminoids. J Pharm Pharmacol 49: 105-107.

29. Camporese A, Balick MJ, Arvigo R, Esposito RG, Morsellino N, et al. (2003) Screening of anti-bacterial activity of medicinal plants from Belize (Central America). J Ethnopharmacol 87: 103-107.

30. NCCLS (2001) National Committee for Clinical Laboratory Standards: Performance Standards for Anti-Microbial Susceptibility Testing. Eleventh Informational Supplement.

31. Gul MZ, Bhakshu LM, Ahmad F, Kondapi AK, Qureshi IA, et al. (2011) Evaluation of Abelmoschusmoschatusextracts for antioxidant, free radical scavenging, antimicrobial and antiproliferative activities using in vitro assays. BMC Complementary and Alternative Medicine 17: 11-64.

32. Boora F, Chirisa E, Mukanganyama S (2014) Evaluation of Nitrite Radical Scavenging Properties of Selected Zimbabwean Plant Extracts and Their Phytoconstituents. Journal of Food Processing.

33. Suffness M, Pezzuto J (1990) Assays related to cancer drug discovery. Methods in Plant Biochemistry: Assay for Bioactivity, Academic Press, London.

2

Development, Quality Evaluation and Shelf Life Studies of Probiotic Beverages using Whey and *Aloe vera* Juice

Sasi Kumar R*

Department of Agri-Business Management and Food Technology, North-Eastern Hill University, NEHU Tura Campus, Chandmari, Meghalaya, India

Abstract

The investigation was aimed to develop a probiotic beverage using whey and *Aloe vera* juice. *Bifidobacterium bifidus* (BB) was used as the probiotic organism, the level of whey and *Aloe vera* juice, addition of probiotic inoculums and fermentation time were optimized on the basis of sensory quality evaluation. The blend ratio 70:30 of whey and *Aloe vera* juice and whey fermented for 9 hrs using inoculums 1% BB resulted with highest sensory scores for overall acceptability. The developed probiotic beverage could be recommended for the large scale production at industrial level.

Keywords: Whey; Probiotic beverage; *Aloe vera* whey beverage; Sensory quality; Shelf life studies

Introduction

Milk whey drink can replace much of the lost organic and inorganic to the extracellular fluid. Whey, which is so rapidly assimilable, forms an ideal metabolic substrate. Whey is one of the highly nutritious by-products obtained from the dairy industry producing cheese, *chhanna* and *paneer*. It constitutes almost 45-50 percent of total milk solids, 70 percent of milk sugar mainly lactose, 20 percent of milk proteins, 70-90 percent of milk minerals and almost all the water soluble vitamins originally present in milk [1,2]. It resulted into unravelling the secrets of whey proteins and other components and established a sound basis for their nutritional and functional value [3]. Several authors have investigated the possibility of utilizing the milk whey in the fruit beverage preparation [4]. The manufacture of whey based beverage requires the mixing of appropriate fruit juices and minimally processed whey with selection of suitable stabilizers and acidulants to develop acceptable whey based fruit beverages [5].

The functional beverages covers probiotic beverages, dairy drinks, drinking yoghurts, functional and functionally fortified soft drinks, juices, energy drinks, sports drinks, functional waters and ready-to-drink tea. The global functional beverages market drives due to the growth in health and wellness concerns increase in disposable incomes, consumer awareness, introduction of new flavoured products meeting various nutritional and health requirements, obesity concerns, aging population and changing lifestyles [6]. Energy drinks is the fastest growing segment in functional drinks market followed by functional dairy products and nutraceutical drinks.

Functional Drinks Reassessing future potential, determining consumer targets, and delivering credible products provides a comprehensive overview of the functional drink landscape, analyzing the regulatory and consumer drivers to identify the best opportunities and strategies [7]. The global regulatory environment is examined and key hubs for innovation are highlighted. The report investigates the opportunities that the functional drinks market provides for beverage manufacturers as well as the challenges that diversification into functional beverages presents.

Fresh *Aloe vera* leaves used to obtain two components, firstly bitter yellow latex from peripheral bundle sheath of aloe, called *Aloe vera* sap and a mucilaginous gel from parenchymatous tissue. The interest and use of gel has increased dramatically in the field of health care and cosmetics [8]. It can be utilized as a valuable ingredient for food application due to its biological activities and functional properties [9]. *Aloe vera* has a bitter taste which can be unpleasant in raw state and its palatability could be enhanced with addition of some other fruit juices.

The conception of probiotic came into focus in early 1900's but the term "probiotic" was suggested by Lilly and Stillwell. Probiotic cultures are live microbial food ingredients that are beneficial for human health [10], which includes improvement of intestinal microbial balance which results in the inhibition of bacterial pathogens, reducing the risk of colon cancer, improving the immune system, lowering serum cholesterol levels [11], alleviation of lactose intolerance and nutritional enhancement [12].

The probiotic culture like *Bifidobacterium bifidum* specific organism used as a probiotic and its health promoting effects [13,14] and in recent years, which has been considered important bacteria for human health; the main probiotic effects attributed to these bacteria include: improvement in lactose utilisation, prevention of diarrhoea, colon cancer, hypercholesterolemia, improvement of vitamin synthesis and calcium absorption [15], development of longer villi and significantly production of substances of low molecular mass with antimicrobial activity [16].

The manufacture of whey-based beverages requires the mixing of appropriate fruit juices and minimally processed whey with selection of suitable stabilizers and acidulants to develop acceptable whey based fruit beverages [17]. Recently, the key growth sector in utilization of whey has been probiotic drinks and their role in fermentation processes, some probiotic lactic acid bacteria have been studied as dietary sources of live microorganisms destined to promote a positive impact in the host by improving the properties of the indigenous beneficial microflora [18]. Based on these facts, the present investigation was conducted to develop a probiotic beverage using whey and *Aloe vera* juice and to study its storage stability.

*Corresonding author:** Sasi Kumar R, Department of Agri-Business Management and Food Technology, North-Eastern Hill University, NEHU Tura Campus, Chandmari-794002, Meghalaya, India, E-mail: sashibiofood@yahoo.co.in

Materials and Methods

Sample preparation

Skim milk was heated in a stainless steel vessel to 95°C followed by cooling to 70°C. The hot milk was acidified by adding 2% of citric acid solution followed by continuous stirring which resulted in the complete coagulation of the milk protein (casein). The liquid (whey) was filtered using muslin cloth. The prepared whey was heated to 85°C before blending with *Aloe vera* juice.

The freeze dried probiotic culture *Bifidobacterium bifidus* NCDC 255 was obtained from National Dairy Research Institute, Karnal, India. Culture was cultivated in 10% non-fat dry milk, autoclaved at 115°C for 15 min, inoculated under sterile conditions, incubated at 42 ± 1°C and kept in refrigerator at 4 ± 1°C until used.

Aloe vera juice: *Aloe vera* juice, chemical composition of juices depends majorly upon the method of juice extraction [19]. *Aloe vera* juice was extracted using cold extraction method and processed into juice as per the method reported by Williams [20]. Freshly harvested *Aloe vera* leaves were dipped into 500 ppm of potassium meta bisulphite (KMS) solution and washed thoroughly with tap water and kept for flash cooling to 5°C for juice stabilization. Further leaves were cut vertically into two half and juice was separated using stainless steel knife, it was allowed to settle for 12 hrs and then homogenized using mixer grinder and enzymatically treated with 1% pectolytic enzyme at 50°C for 20 minutes. Then it was filtered and pH was adjusted to 3.0 by adding citric acid and ascorbic acid to control browning while high heat treatment. Further it was dearated, pasteurized, flash cooled and stored. During the pasteurization pectolytic enzyme was inactivated. The obtained juice was stored refrigerated temperature until further use.

Probiotic beverage process: The combination of whey and *Aloe vera* juice were optimized with different blends ratio, such that sample A (65wy:35Av), sample B (70wy:30Av), sample C (75wy:25Av) and sample D (80wy:20Av). The beverage incubation period was optimized by inoculating 1% of probiotic culture (*Bifidobacterium bifidus*) to whey and sample with addition of sugar level 15% and fermented for 5, 9, 13, 17 and 21 and 24 hrs then the beverages were evaluated for sensory characteristics, total viable count, pH, titratable acidity, storage stability and shelf life studies carried out for sample with different blend and fermented periods.

Analytical methods

Bio-chemical and microbiological methods: Total soluble solid was measured using a hand refractometer of 0-32°B (ERMA make). The pH of the beverages was determined using the digital pH meter (Model No. 5633, Electronics Corporation of India Ltd., Hyderabad). Titratable acidity was determined according to the AOAC [21] method. Viable counts in the samples were determined according to A.P.H.A. [22] procedure using lactic agar [23].

Statistical analysis

Statistical procedures as described by Snedecor and Cochran [24] were used to analyze the data for the interpretation of results. Mean, standard deviation and analysis of variance (ANOVA) were used to describe the results.

Sensory quality evaluation

The beverage samples were evaluated as described by Larmond [25] for their sensory characteristics namely colour and appearance, taste and flavour, body or consistency and overall acceptability by a trained panels. The panellists were asked to record their observations on the sensory sheet based on a 9 points hedonic scale (9 and 1 points showing like extremely and dislike extremely).

Results and Discussion

Optimization of whey and *Aloe vera* juice blend ratio levels for beverage preparation

Whey and *Aloe vera* juice were blended in five different proportions i.e. A (65wy:35Av), B (70wy:30Av), C (75wy:25Av) and D (80wy:20Av) and evaluated for sensory attributes namely colour and appearance, consistency, flavour and overall acceptability. The sample B was highest score for flavour, consistency, colour and appearance to beverage preparation. The mean scores of this blend for colour and appearance, consistency, flavour and overall acceptability were 7.93, 7.81, 8.37 and 8.93 respectively. Therefore, sample B was chosen for the further investigation and differed significantly (P<0.05) and rated best among others samples.

Optimization of growth conditions for *Bifidobacterium bifidus* (BB) in beverage

Whey and sample B were supplemented with 15% sucrose and fermented for different time intervals. Variations in fermentation time were studied in terms of overall acceptability, pH, titratable acidity and total viable counts of the beverage. The effect of *Bifidobacterium bifidus* (BB) on sensory attributes of whey with or without addition of *Aloe vera* juice during fermentation has been shown in Table 1. Colour and appearance of the beverage was significantly (P≤0.05) affected by the incubation period. The mean scores for colour and appearance ranged from 7.56 to 5.87 for whey. The mean score was highest (7.56) for whey fermented for 9 hrs. The sensory scores for consistency reduced significantly (P ≤ 0.05) with increase in fermentation time for both whey and sample B.

The sensory score for consistency of fermented whey ranged from 7.55 to 5.97 but the sensory scores for consistency of sample B was ranged from 8.12 to 6.15. The sensory scores for flavour ranged from 7.21 to 5.17 for whey fermented alone and from 7.91 to 6.97 for sample B. The mean score for flavour decreased significantly with increasing fermentation time irrespective of the medium. The mean score for overall acceptability ranged from 7.31 to 5.80 for whey and 8.10 to 6.34 for sample B. Highest score for overall acceptability was seen in case of sample B fermented for 9 hrs.

Enumeration of total viable counts

The viable counts were assessed after 5, 9, 13, 17, 21 and 24 hrs of incubation at 37°C. It was observed that whey gave higher viable counts (9.54×10^8 cfu/ml) up to 24 hrs of fermentation. But after 24 hrs, the sample B fermented alone, gave higher viable counts ranging from 3.35×10^7 to 8.10×10^8 cfu/ml. Both whey and sample B attained a total viable count of more than 5.0×10^8 cfu/ml within 13 hrs of fermentation (Table 2). The total viable count increased significantly with increase in fermentation time from 5 to 24 hrs. The total viable count for whey ranged from 2.23×10^7 to 9.54×10^8 cfu/ml. Sample B showed the stable increase total viable count with increasing fermentation time as controlled way but the whey alone showed uncontrolled fermentation with increasing fermentation time.

Effect of probiotic on pH and titratable acidity

Table 3 shows the effect of incubation period on the pH of whey and sample B, pH of the samples (whey and sample B) reduced significantly

Sr.No	Name of the sample	Sensory Scores *			
		Colour and Appearance	Consistency	Flavour	Overall Acceptability
1	Sample A	7.15^B	7.21^{AB}	7.38^B	7.23^B
2	Sample B	7.91^A	7.86^A	8.21^A	8.54^A
3	Sample C	6.95^B	6.81^B	7.11^C	7.12^C
4	Sample D	6.72^C	6.68^C	6.92^D	6.96^D
5	S.E.M	0.189	0.186	0.153	0.123
6	CD at 5%	0.533	0.525	0.452	0.361

Sample A (65wy:35Av), Sample B (70wy:30Av), Sample C (75wy:25Av) and Sample D (80wy:20Av)
*Average of four trials
* Means by different letters (A, B, C, D) as superscripts in a column differ significantly at 5% level
Table 1: Effect of different blends of whey and *Aloe vera* juice on the sensory characteristics of the beverage.

Fermentation Time (hrs)	Sensory Scores							
	Colour and Appearance		Consistency		Flavour		Overall Acceptability	
	Wy	B	Wy	B	Wy	B	Wy	B
5	7.15	7.66	7.18	7.55	7.10	7.55	7.11	7.60
9	7.56	8.12	7.55	8.11	7.21	7.91	7.31	8.10
13	6.95	7.25	7.10	7.25	6.75	7.45	6.87	7.50
17	6.34	7.10	6.30	7.14	6.14	7.24	6.31	7.20
21	6.21	6.95	6.11	6.21	5.43	7.13	5.91	6.98
24	5.87	6.17	5.97	6.15	5.17	6.97	5.80	6.34
	S.E.M	CD at 5%	S.E.M	CD at 5%	S.E.M	CD at 5%	S.E.M	CD at 5%
Medium (A)	0.043	0.123	0.045	0.093	0.053	0.135	0.53	0.135
Fermentation Time (B)	0.069	0.187	0.053	0.156	0.087	0.213	0.073	0.198
AXB	0.081	0.267	0.076	0.213	0.123	0.293	0.121	0.287

*Wy-Whey, Sample B (70wy:30Av)
Table 2: Effect of *Bifidobacteriumbifidus*(BB) on sensory characteristics of whey and sample B at 37 ± 1°C.

(P<0.05) with increasing fermentation time. The mean value of pH obtained in case of whey fermented alone ranged from 4.85 to 3.43. The mean value of pH obtained in case of sample B ranged from 4.65 to 3.44. Higher viable counts during the last stage of fermentation (24 hrs) resulted in comparative lowering of pH for whey and sample B. The sample B showed gradual decrease pH value with increase of fermentation time because of controlled fermentation during entire 24 hrs, but whey alone showed uncontrolled fermentation process for 24 hrs.

Table 3 shows the effect of incubation period on the titratable acidity of whey and sample B. Titratable acidity increased significantly (P ≤ 0.05) with increasing fermentation time irrespective of the medium. The mean values obtained for whey ranged from 0.390 to 1.293. The mean values obtained for sample B ranged from 0.453 to 0.945. Sample B showed balanced titratable acidity from 5 to 24 hrs of fermentation process, because of *Aloe vera* juice blend with whey, whereas there was a significant increase in the titratable acidity with respect of increasing fermentation time (5 to 24 hrs) for whey and sample B. The commercial probiotic beverage should possess a minimum viable count of 4.11×10^7 cfu/ml with optimum fermentation time (9 hrs) cfu/ml and should also have an acceptable flavour. The commercial probiotic beverage should possess a minimum viable count of 106 cfu/ml [26] and should also have an acceptable flavour.

Keeping in view these aspects, experiments were carried out to appraise the suitability of whey with or without the addition of *Aloe vera* juice as a growth medium for *Bifidobacterium bifidus* (BB) for the preparation of probiotic beverage. It was observed that sample B for 9 hrs gave the highest scores for overall acceptability (8.10) as compared to whey fermented alone (7.31). The scores for overall acceptability gradually declined with increasing fermentation time irrespective of the medium. Table 4 shows that the pH and titratable acidity affects colour and appearance, consistency, flavour profile and overall accepbility of a beverage. In the present study, it was observed that sample B having a pH of 4.50 and titratable acidity of 0.525% gave the best colour and appearance, consistency, flavour profile and overall acceptability to the probiotic beverage. These observations exhibited an acceptable beverage by fermenting a blend of whey and *Aloe vera* juice (70:30) that is sample B for a period of 9 hrs. The prepared product had desired health benefits due to the probiotic organisms.

Storage studies of probiotic beverage

The changes in sensory attributes on storage studies of the probiotic beverage have been shown in Table 4 for refrigerated and ambient storage temperatures. The whey and *Aloe vera* based probiotic beverage did not show sensory differences for the first 15 days at refrigerated storage. But after the 15 days, difference was perceived in colour and appearance and flavour. According to a consensus made with the panelists during sensory evaluation, it was determined that the main descriptors that characterized the product were acidity and sweetness, with acidity being the attribute responsible for the sensory difference perceived by the panelists. Even though a slight acidification was detected by the sensory panels and agreed that the beverage was acceptable for a period of 30 days at 4°C ± 1°C and 6 days at 35°C ± 1°C.

The pH of the fresh beverage prepared from blend of whey and *Aloe vera* juice fermented for 5 hr was 4.40. The pH of the samples gradually declined during 30 days at refrigerated storage. The pH ranged from 4.40 to 4.00 after 30 days of storage. During storage of the samples at ambient temperature, pH lowered significantly (P ≤ 0.05) after 6 days. The initial acidity of the fermented beverage prepared

Sr.No	Fermentation Time (hrs) @ 37°C	Medium	
		Whey (cfu/ml)	Sample B (cfu/ml)
1	5	2.23×10^7	3.35×10^7
2	9	3.23×10^7	4.11×10^7
3	13	5.90×10^8	5.18×10^8
4	17	8.23×10^8	6.88×10^8
5	21	9.10×10^8	7.64×10^8
6	24	9.54×10^8	8.10×10^8
		Medium (A)	Fermentation Time (B)	AXB
	SEM	3.11×10^4	4.31×10^4	6.55×10^4
	CD at 5%	8.91×10^4	1.22×10^5	2.14×10^5

Table 3: Effect of fermentation period on the total viable count of whey with and without *Aloe vera* juice.

Sr.No	Fermentation Time	Medium			
		Whey		Sample B	
		pH	T.A %	pH	T.A %
1	5	4.85	0.390	4.65	0.453
2	9	4.55	0.496	4.50	0.525
3	13	4.13	0.915	4.25	0.680
4	17	3.98	1.023	4.00	0.796
5	21	3.60	1.165	3.87	0.815
6	24	3.43	1.293	3.44	0.945
		Medium (A)		Fermentation Time (B)	AXB
	SEM	0.0250		0.00051	0.0716
	CD at 5%	0.0564		0.0219	0.0303

* Incubation temperature: 37 ± 1°C.

Table 4: Effect of fermentation period on the pH and titratable acidity of whey with and without *Aloe vera* juice.

Sr.No	S.P (d) at 4°C	pH	T.A. (%) La	TPC (cfu)	S.P (d) at 35°C	pH	T.A (%) La	TPC (cfu)
1	0	4.40	0.553	3.5×10^7	0	4.40	0.553	3.5×10^7
2	5	4.39	0.551	3.1×10^7	1	4.35	0.610	5.2×10^8
3	10	4.38	0.563	2.7×10^7	2	4.30	0.723	8.7×10^8
4	15	4.37	0.580	2.7×10^7	3	4.26	0.793	6.1×10^7
5	20	4.35	0.594	2.1×10^7	4	4.10	0.843	4.4×10^7
6	25	4.35	0.610	1.7×10^7	5	4.00	0.910	3.1×10^7
7	30	4.00	0.870	1.3×10^7	6	3.83	1.120	2.3×10^7

Table 5: Changes in pH, total viable count and titratable acidity of the sample B probiotic beverage during storage.

SP(d) at 4°C	C and A	Cons	Flav	OAA	SP(d) at 35°C	C and A	Cons	Flav	OAA
0	8.53[a]	8.65[a]	8.74[a]	8.63[a]	0	8.58[a]	8.60[a]	8.73[a]	8.67[a]
5	8.51[a]	8.64[a]	8.73[a]	8.63[a]	1	8.45[a]	8.50[a]	8.68[a]	8.58[a]
10	8.36[a]	8.40[b,c]	8.51[a]	8.43[a]	2	8.00[a]	8.15[a,b]	8.14[b]	8.15[a]
15	8.11[a]	8.00[b,c]	7.91[b]	8.13[b,c]	3	7.83[b]	7.95[b]	7.75[c]	7.88[b]
20	7.93[b]	7.95[c]	7.74[c]	7.73[c]	4	7.13[c]	7.15[c]	7.00[d]	7.14[c]
25	7.20[c]	7.15[d,e]	7.00[d]	7.17[d]	5	6.51[d]	6.65[d]	6.14[e]	6.60[d]
30	6.55[d]	6.34[e]	6.14[e]	6.51[e]	6	6.00[e]	5.95[e]	5.74	5.93[e]
S.E.M	0.059	0.071	0.062	0.051	S.E.M	0.128	0.103	0.148	0.102
CD at 5%	0.171	0.027	0.184	0.157	CD at 5%	0.364	0.326	0.396	0.215

*Means by different superscript as letters (a,b,c,d,e) in a column differ significantly at 5% level SP(d) - Storage Period (day); C&A - Colour and Appearance; Cons. -Consistency; Flav-Flavour: OAA - Overall acceptability.

Table 6: Changes in sensory characteristics of the sample B probiotic beverage during storage

from whey and *Aloe vera* juice was 0.553% (La). The acidity increased during the refrigerated storage from 0.546 to 0.870% after 30 days. The increase in acidity was more prominent in case of storage at ambient temperature wherein the acidity reached 1.120% after 6 days of storage (Table 5). It showed that the titratable acidity values were significantly influenced by the days of storage of whey beverage both at refrigeration and room temperature. Rodas et al. [27] showed that that the titratable

acidity values were significantly influenced by the day of storage whey beverage both at refrigerated and room temperature.

Total viable counts of probiotic beverage during storage

The initial total viable count of the beverage was 3.5×10^7 cfu/ ml which decreased to 1.3×10^7 at refrigerated storage (4°C ± 1°C). Although the viability of *Biofidobacterium bifidum (B.p)* population decreased, the viable count of the probiotic beverage did not fall below 10^7 cfu/ml (Table 6). During storage at 35°C ± 1°C the total viable count first increased to 8.7×10^7 cfu/ml (in 2nd day) and then gradually declined to 2.3×10^7 cfu/ml after 6 days. Our results are in confirmation with other researchers [28] who also reported a decline in total viable count of *Lactobacillus acidophilus, Lactobacillus reuteri* and *Biofidobacterium bifidum* of whey based probiotic beverage stored at 4°C ± 1°C. Wang et al. [28], studied the survival *Biofidobacterium bifidum (B.p)* of in commercial yoghurt during refrigerated storage.

Conclusion

Above study has revealed satisfactorily good quality probiotic beverage with therapeutic value prepared by using a 70:30 blend of whey and *Aloe vera* juice inoculated with 1 percent inoculum of *Biofidobacterium bifidum (B.p)* with a shelf life of 30 days at 4°C ± 1°C and 6 days at 35°C ± 1°C.

References

1. Yalcin S, Wade VN, Hassan Md N (1994) Utilization of chhana whey for the manufacture of soft drinks. J Food 19: 351-355.

2. Horton BS (1995) Whey processing and utilization. Information Bulletin -International Dairy Federation 308: 2-6.

3. Smithers GW (2008) Whey and whey proteins - From gutter-to-gold. Int Dairy J 18: 695-704.

4. Shukla FC, Sharma A, Singh B (2004) Studies on the preparation of fruit beverages using whey and buttermilk. J Food Sci Technol 41: 102-104.

5. Singh S, Singh AK, Patil GR (2005) Whey utilization for health beverage. Indian Food Industry 21: 38-41.

6. Yadav RB, Yadav BS, Kalia N (2010) Development and storage studies on whey-based banana herbal (*Menthaarvensis*) beverage. Am J Food Technol 5: 121-129.

7. Katz F (2001) Active cultures add function to yogurt and other foods. Food Technol 55: 46-49.

8. Devi, Radha Y, Rao YM (2005) Cosmeceutical applications of aloe gel. Natural Product Radiance 4: 322-327.

9. Eshun K, He Q (2004) Aloe vera: a valuable ingredient for the food, pharmaceutical and cosmetic industries–A review. Crit Rev Food Sci Nutr 44: 91-96.

10. Salminen S, Ouwehand A, Benno Y, Lee YK (1999) Probiotics: how should they be defined? Trends Food Sci Tech 10: 107-110.

11. Saarela M, Lähteenmäki L, Crittenden R, Salminen S, Mattila-Sandholm T (2002) Gut bacteria and health foods-The European perspective. Int J Food Microbiol 78: 99-117.

12. Alizadeh A, Ehsani MR (2008) Probiotic survival in yogurt made from ultrafiltered skim milk during refrigeration storage. Research Journal of Biological Sciences 3: 1163-1165.

13. Vinderola CG, Bailo N, Reinheimer JA (2000) Survival of probiotic microflora in argentinian yoghurts during refrigerated storage. Food Res Int 33:97-102.

14. Heller JK (2001) Probiotic bacteria in fermented foods: Product characteristics and starter organisms.Am J Clin Nutr 73: 374-379.

15. Sakhale BK, Pawar VN, Ranveer RC (2012) Studies on the development and storage of whey based RTS beverage from Mango v. Kesar. J Food Process Technol 3: 1000148.

16. Shah NP (2000) Probiotic bacteria: Selective enumeration and survival in dairy foods. J Dairy Sci 83: 894-907.

17. Sasikumar R, Ray RC, Paul PK, Suresh CP (2013) Development and storage studies of therapeutic ready to serve (RTS) made from blend of *Aloe vera, Aonla* and ginger juice. J Food Process Technol 4: 232-237.

18. Ramachandra CT, Srinivasa PR (2008) Processing of *Aloe Vera* leaf gel: A review. American Journal of Agricultural and Biological Sciences 3: 502-510.

19. Williams S (1984) Official methods of analysis. Association of Official Analytical Chemists, (14thedn), Washington DC, USA.

20. APHA, Vanderzant C (1992) Splits Toesser, ed. compendium of methods for microbiological examination of foods. Washington. Amer Public Health Assoc 14: 919-927.

21. Elliker PR, Anderson AW, Hennenson G (1956) An agar culture medium for lactic acid *Streptococci* and *Lactobacilli*. J Dairy Sci 39: 1611-1612.

22. Snedecor GW, Cochron WG (1989) Statistical methods. (6thEdn), Oxford and IBH Publishing Co, India.

23. Larmond E (1977) Laboratory methods for sensory evaluation of foods. Department of Agriculture. Ottawa, Canada.

24. Divya, Kumari A (2009) Effect of different temperatures, timings and storage periods on physico-chemical and nutritional characteristics of whey guava beverage. World J Dairy Food Sci 4: 118-122.

25. Shah NP, Lankaputhra WEV, Britz ML, Kyle WAS (1995) Survival of Lactobacillus acidophilus and *Bifido bacteriumbifidum* in commercial yoghurt during refrigeration.

26. Rodas BA, Angulo JO, Jde la Cruz, Garcia HS (2002) Preparation of probiotic buttermilk with *Lactobacillus reuteri*. Milchwissenschaft 57: 26-28.

27. Heller JK (2001) Probiotic bacteria in fermented foods: Product characteristics and starter organisms. Am J Clin Nutr 73: 374-379.

28. Wang YC, Yu RC, Chou CC (2002) Growth and survival of bifidobacteria and lactic acid bacteria during the fermentation and storage of cultured soymilk drinks. Food Microbiol 19: 501-508.

3

Characterization and Kinetics of Growth of Bacteriocin like Substance Produced by Lactic Acid Bacteria Isolated from Ewe Milk and Traditional Sour Buttermilk in Iran

Mahdieh Iranmanesh[1]*, Hamid Ezzatpanah[1], Naheed Mojgani[2] and Torshizi MAK[3]

[1]Department of Food Science and Technology, Science and Research Islamic Azad University, Tehran, Iran

[2]Biotechnology Department, Razi Vaccine and Serum Research Institute, Karaj, Iran

[3]Department of Poultry Science, Faculty of Agriculture, Tarbiat Modares University, Tehran, Iran

Abstract

Ethnic people of Iran consume variety of traditional fermented milk products including buttermilk made from ewe's milk. The purpose of this study was to isolate and characterize bacteriocin producing by Lactic acid bacteria from these products, and to exploit their potential as bio preservative. Ten strains of Lactic acid bacteria isolated from ewe's milk, traditional yoghurt and sour buttermilk from different areas in Azarbayjan-e-sharqi, Iran were screened for their ability to produce bacteriocin like inhibitory substances (BLIS). According to results, *Lactobacillus pentosus*, *Lactobacillus paracasei*, *Lactobacillus brevis*, *Pediococcus acidilactici* were shown to produce proteinaceous substances inhibitory against a number of Gram positive and negative bacteria including *Staphylococcus aureus*, *Listeria monocytogenes* and *Salmonella enteritidis*. The inhibitory activities of two Lactic acid bacteria (*Lactobacillus paracasei* and *Pediococcus acidilactici* isolated from ewe milk and buttermilk respectively) were unaffected by the action of pH neutralization and hydrogen peroxide while completely inhibited in the presence of proteolytic enzymes. The kinetic of bacteriocin like inhibitory substances against *Staphylococcus aureus* indicated a direct relationship between the growth rate and the amount of bacteriocin produced. The inhibitory activity of these lactic acid bacteria started in the early logarithmic phase and continued to the end of exponential phase. During ultrafiltration studies, bacteriocins produced by *Pediococcus acidilactici*, *Lactobacillus paracasei* were able to pass through the cellulose membranes with10 and 30 KDa. Titre of bacteriocins produced by *Lactobacillus pentosus*, *Lactobacillus paracasei*, *Lactobacillus brevis* were estimated 1600 AU/mL while the titre for bacteriocin which produced by *Pediococcus acidilactici* was calculated as 3200 AU/mL.

Keywords: Antagonistic activity; Bacteriocin like inhibitory substances; Ewe milk; Lactic acid bacteria

Introduction

A great number of Gram (+) bacteria and Gram (-) bacteria produce substances of protein structure during their growth with antimicrobial activities, called bacteriocins. Among the Gram positive (+) bacteria, lactic acid bacteria (LAB) have gained particular attention nowadays, due to the production of bacteriocins [1]. Lactic acid bacteria (LAB) play an essential role in the majority of food fermentations, and a wide variety of strains are routinely employed as starter cultures in the manufacture of dairy, meat, vegetable, and bakery products. In addition, the growth of spoilage and pathogenic bacteria in food containing LAB is inhibited. This can be due to pH reduction by the organic acids produced or their ability to produce a variety of antimicrobial substances like ethanol, formic acid, acetoin, hydrogen peroxide, diacetyl and bacteriocins. The latest synthesized ribosomally, peptides or proteins which inhibit microorganisms that are usually closely related to the producer strain [2]. On the other hand, artificial chemical preservatives are employed to limit the number of microorganisms capable of growing within foods, but increasing consumer awareness of potential health risks associated with some of these substances has led researchers to examine the possibility of using bacteriocins produced by LAB as bio-preservatives [3,4]. In this study our aim was to investigate the properties of Lactic acid bacteria present in ewe's milk and traditional sour buttermilk. We analyzed the antimicrobial spectrum and physico-chemical properties of bacteriocin like substances (BLIS) produced by these locally isolated LAB strains.

Material and Methods

Sample preparation

Ewe milk (n:20), yoghurt (n:20) and buttermilk (n:20) were collected from 30 sheep herds in Myaneh (15 herds) and Hashrood (15 herds) two cities in Azarbayjan-e-sharqi (north-west of Iran). All samples were collected according to EN ISO 707:2001 in sterile bottles of 250 mL and transported to the laboratory under refrigeration (4°C) within 36 h [5]. The sour buttermilk preparation operation is shown in Figure 1.

Isolation and identification of LAB

All collected samples were screened for the presence of LAB by morphological and biochemical tests. All isolated pure colonies were subjected to Gram staining and catalase test. To identify the species, the carbohydrate fermentation profiles using API 50 CHL medium (Bio-Meriux, France) according to the manufacture's instruction was used.

Antimicrobial spectrum

The antimicrobial effects of selected LAB against Gram positive and negative pathogens were examined by agar well diffusion method described earlier [6]. Three pathogens namely *Staphylococcus aureus*

*Corresonding author: Mahdieh Iranmanesh, Research Associate, Department of Food Science and Technology, Science and Research Islamic Azad University, Tehran, Iran, E-mail: mi_manesh@yahoo.com

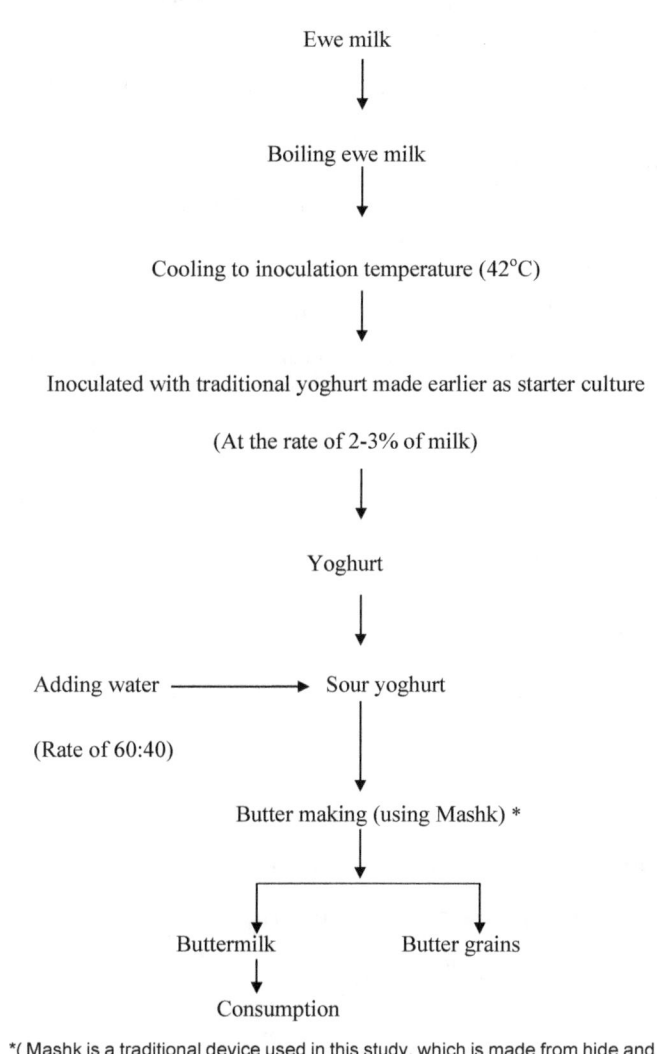

Ewe milk

↓

Boiling ewe milk

↓

Cooling to inoculation temperature (42°C)

↓

Inoculated with traditional yoghurt made earlier as starter culture

(At the rate of 2-3% of milk)

↓

Yoghurt

↓

Adding water ⟶ Sour yoghurt

(Rate of 60:40)

↓

Butter making (using Mashk) *

↓

Buttermilk   Butter grains

↓

Consumption

*(Mashk is a traditional device used in this study, which is made from hide and is used for making butter and buttermilk from yoghurt or keeping fermented dairy products).

Figure 1: Preparation of traditional sour buttermilk.

(RTCC 1112), *Listeria monocytogenes* (RTCC 1298), and *Salmonella enteritidis* (local isolate) were used as indicator culture in the study. The antimicrobial activity was recorded as appearance of clear zone around the wells.

Sensitivity to enzymes

The cell free supernatants of the selected LAB strains was obtained by centrifuging, and adjusted to pH 6.5 with 1 N NaOH followed by filtration through a 0.22 μm pore size filters. The remaining activity was determined by agar well diffusion assay using *Staphylococcus aureus* as indicator strain. The supernatant fluid of the isolates which resisted pH neutralization were treated with catalase, lysozyme, lipase, pronase E, trypsin and proteinase K (Sigma) at a final concentrations of 1 mg/ml in phosphate buffer (pH 7.0). The remaining activity was determined as described earlier, after 2 h of incubation of the treated samples at 37°C [7].

Molecular size estimation of bacteriocin-like substances

Approximate molecular weight of the antagonistic peptides was determined by ultrafilteration. A 2 ml culture free supernatant fluid was ultrafiltered through cellulose membranes with 5, 10 and 30 KDa exclusion units (Centricon, Micro concentrations, USA). Bacteriocin activity in retained and eluted fractions was determined by well-diffusion agar [8].

Bacteriocin titration

Bacteriocin-like activities of selected LAB strains were examined by the dilution assay. Serial two-fold dilutions of bacteriocin were tested by the well-diffusion method. The antimicrobial activity of the bacteriocin was defined as the reciprocal of the highest dilution showing inhibition zone and was expressed in activey units per milliliter (AU/ml) [9].

Kinetics of bacteriocin-like production

20 μl of overnight culture supernatants of LAB strains were inoculated into 50 ml of MRS broth, then the incubated at 37°C. At time intervals of one hour, the growth of cells was measured by the absorbance (OD 660 nm) and the antimicrobial activity [10].

Antibiotic resistance

The antimicrobial susceptibilities of the isolates and reference strains were preliminarily assessed by the disk diffusion method on MRS [11]. Ten antibiotics used in the treatment of the most common hospital infections were chosen, belonging to different categories: ampicillin, penicillin, tetracycline, vancomycin, erythromycin, gentamicin, clindamycin, Chloramphenicol, Kanamycin and Sterptomycin. The assay was performed as follows: swabs imbibed with bacterial cultures, adjusted to 0.5 McFarland, were plated, and paper disks with antibiotics were placed aseptically on the inoculated plates and incubated for 24 h at 37 1C in 5% CO_2. The diameters of the inhibition zones were measured (mm).

Result

Isolation and Identification of LAB

Seventy-seven bacterial isolates from the traditionally made dairy samples including yogurt and buttermilk were identified as LAB. All isolates were identified as catalase negative and Gram-positive bacilli or cocci in pairs or long chains, and or cocobacilli. All isolates showed different level of activity against the tested pathogens. Ten isolates with higher activey against *Staphylococcus aureus, Listeria monocytogenes*, and *Salmonella enteritidis* were identified to species level as *Lactobacillus pentosus, Lactobacillus paracasei* (two strains), *Lactobacillus brevis, Pediococcus acidilactici* (four strains) and *Lactococcus lactis* (two strains) (Table 1).

Antimicrobial spectrum

Ten selected strains showing antibacterial activity against all pathogens. In addition, all isolates showed maximum antimicrobial zone against *Salmonella enteritidis* (Table 2).

Strains	Samples
Lactobacillus pentosus	Milk
Lactobacillus brevis	Milk
Lactobacillus paracasei (LP1,2)	Milk
Lactococcus lactis (LL1,2)	Milk
Pediococcus acidilactic (PA1,2)	Milk
Pediococcus acidilactic (PA 3,4)	Buttermilk

Table 1: Identification of the selected LAB isolates to species level using standard API 50CHL identification kit.

Sensitivity to enzymes

The antagonistic activity shown by *Lactococcus lactis* (LL1), *Lactobacillus paracasei* (LP1) and *Pediococcus acidilactici* (PA 1, 2, 3) appeared to be due to acid production as their activity was completely lost on pH neutralization of their supernatant fluids. The antibacterial activity of *Lactobacillus pentosus*, *Lactobacillus paracasei* (LP2), *Pediococcus acidilactici* (PA4) and *Lactobacillus brevis* are BLIS, as their antimicrobial activity were completely eliminated after treatment with lysozyme, pronase and proteinase (Table 3).

Molecular size estimation of bacteriocin-like substances

The fractions of the bacteriocins produced by *Lactobacillus pentosus* and *Lactobacillus brevis* showing antibacterial activity corresponded to peptide molecules in the range of 5-10 KDa while *Lactobacillus paracasei* (LP2) and *Pediococcus acidilactici* (PA4) range of 10-30 KDa which was subsequently confirmed by subjecting the fractions to ultra-filtration using filtron membranes with 5, 10 and 30 KDa molecular weight cut off.

Bacteriocin titration

The highest titration of bacteriocin of *Lactobacillus pentosus*, *Lactobacillus paracasei* (LP2) and *Lactobacillus brevis* was 1600 AU/ml though in *Pediococcus acidilactici* (PA4) was 3200 AU/ml.

Kinetics of bacteriocin-like production

Figures 2 and 3 show the growth curves and the profiles of bacteriocin production of the *Lactobacillus paracasei* (LP2) and

Pathogens / strains	*Listeria monocytogenes*	*Staphylococcus aureus*	*Salmonella enteritidis*
L.b pentosus	+	+	+++
L.b brevis	++	+	+++
L.c lactis 1	+++	+++	+++
L.c lactis 2	+	+++	+++
L.b paracasei 1	+	+++	+++
L.b paracasei 2	+++	+++	+++
Ped. acidilactici 1	+++	+	+++
Ped. acidilactici 2	+	++	+++
Ped. acidilactici 3	+	+	++
Ped. acidilactici 4	+++	+	++

+: 3mm< zone
++: 3mm< zone <5mm
+++: 5mm< zone <7mm

Table 2: Antagonistic activity cell free supernatant of selected strains against pathogens by agar well diffusion assay.

LAB isolates	pH	Catalase	Lysozyme	Pronase E	Proteinase K	Trypsin
L.b pentosus	+	+	-	-	-	+
L.b brevis	+	+	-	-	-	+
L.c lactis 1	-	Acid	Acid	Acid	Acid	Acid
L.c lactis 2	+	+	+	-	-	+
L.b paracasei 1	-	Acid	Acid	Acid	Acid	Acid
L.b paracasei 2	+	+	-	-	-	+
Ped. acidilactici 1	-	Acid	Acid	Acid	Acid	Acid
Ped. acidilactici 2	-	Acid	Acid	Acid	Acid	Acid
Ped. acidilactici 3	-	Acid	Acid	Acid	Acid	Acid
Ped. acidilactici 4	+	+	-	-	-	+

+: antimicrobial activity remaining
-: no antibacterial activity

Table 3: Inhibitory activity of selected LAB isolates against *Staphylococcus aureus* after pH neutralization and enzyme treatments.

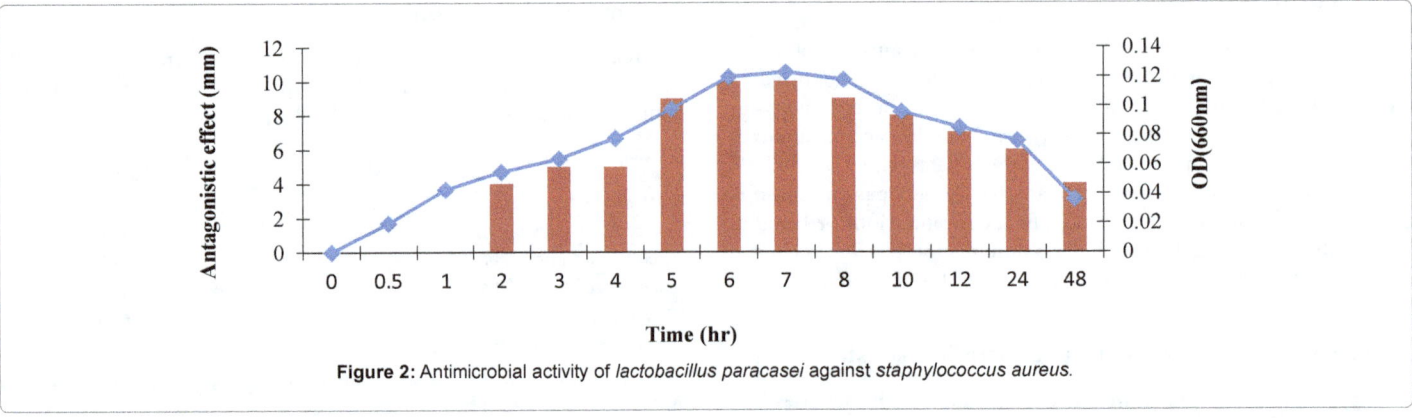

Figure 2: Antimicrobial activity of *lactobacillus paracasei* against *staphylococcus aureus*.

Pediococcus acidilactici (PA4) strains, respectively. In all these cases, bacteriocin production was initially detected in the logarithmic phase of growth, and the maximum levels of antimicrobial activity were found at different stages of the stationary phase of growth, depending on both the producer strain and the indicator microorganism. Both bacteriocins were decreased after 24 and 48 h.

Antibiotic resistance

Ten selected isolates were resistance to tetracycline and chloramphenicol. Furthermore all isolates were sensitive to kanamycin, sterptomycin, vancomycin, clindamycin (Table 4).

Discussion

The traditional fermented dairy products can potentially be a good source of potential probiotic organisms. The microbial ecology and beneficial health effects of fermented dairy products such as buttermilk have not been reported earlier. These products are popular especially in the rural areas because of their good natural tastes and flavors. During our results, we isolated *Lactobacillus pentosus, Lactococcus lactis, Lactobacillus paracasei, Lactobacillus brevis* from ewe milk and *Pediococcus acidilactici* from buttermilk. Similar to our result, Tajabadi et al. [12] isolated LAB from different kinds of yogurt, cheese, fermented milk, dough and kashk. As a functional probiotic, anti-pathogen activity is one of important properties to be considered. The spectrum of activity of the antagonistic compound produced by these isolates indicated their action against certain important food borne pathogens. *Staph.aureus* is considered a potential public health risk due to its production of enterotoxins that cause food poisoning [7]. In addition, food borne transmission of *L.monocytogenes* has been implicated of listeriosis in human involving the consumption of various foods [13]. Our study indicated that the selected LAB strains were more effective in inhibiting the growth of *Salmonella enteritidis* compared to other pathogens. A number of metabolites such as acid, Hydrogen peroxide and bacteriocin are believed to contribute to antimicrobial activity. The sensitivity of the inhibitory agent to lysozyme, pronase, trypsin and proteinase K indicates of their proteinaceous nature and thus might be considered as bacteriocin. Similar to this result, inhibition of pathogens like *Listeria monocytogenes* and *Staphylococccus aureus* by LAB was reported [14,15]. In this research, Bacteriocins were detected early in the logaritnic growth phase. Studies produced continuously during this phase supporting the fact that bacteriocins display primary metabolite kinetics. Bacteriocin activity showed a maximum level at the stationary growth phase. A decline in inhibitory activity during the late stationary growth phase was observed. This may be due to proteolytic degradation, adsorption to cells or bacteriocin aggregation [16]. In contrast to our result, production of bacteriocin by *Lactobacillus paracasei* occurred throughout logarithmic growth [17]. Also in contrast to our result, Ogunbanwo showed that bacteriocin of *Lactobacillus brevis* was unable to pass through 1,000 and 10,000 KDa molecular weight cut-off membranes. A tendency to aggregate with other proteins has been

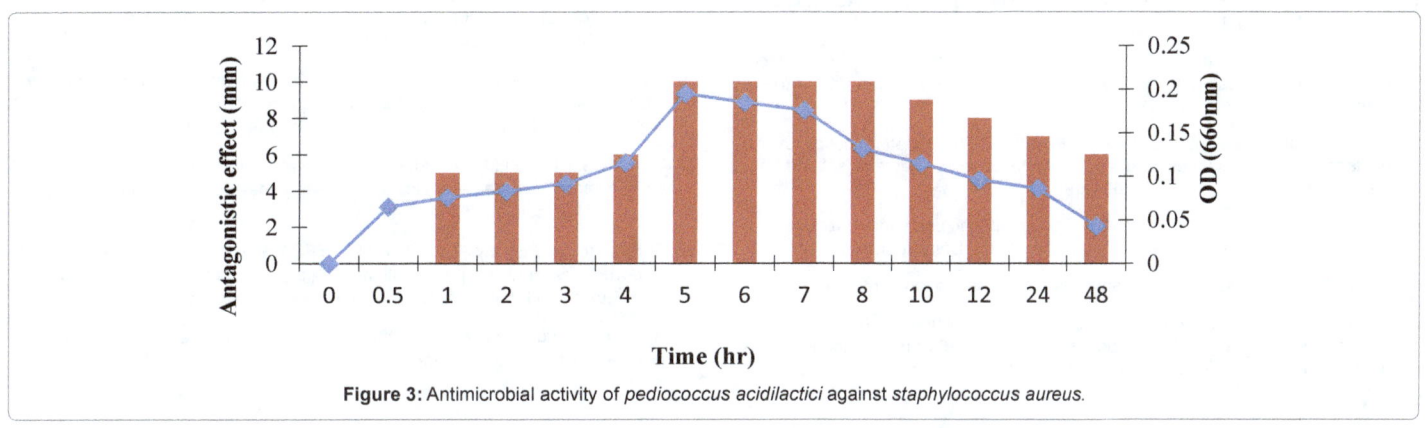

Figure 3: Antimicrobial activity of *pediococcus acidilactici* against *staphylococcus aureus*.

LAB isolates	Kanamycin	Gentamicin	Sterptomycin	Ampicillin	Vancomycin	Tetracycline	Chloramphenicol	Penicillin	Clindamycin	Erythromycin
L.b pentosus	-	-	-	5	-	8	10	5	-	9
L.b brevis	-	5	-	6	-	8	10	10	-	10
L.c lactis 1	-	-	-	-	-	7	5	-	-	-
L.c lactis 2	-	-	-	5	-	5	6	5	-	5
L.b paracasei 1	-	-	-	5	-	6	8	8	-	9
L.b paracasei 2	-	-	-	-	-	7	10	10	-	10
Ped. acidilactici 1	-	-	-	5	-	6	6	5	-	6
Ped. acidilactici 2	-	-	-	6	-	8	10	6	-	10
Ped. acidilactici 3	-	-	-	5	-	6	10	10	-	10
Ped. acidilactici 4	-	-	-	5	-	5	7	7	-	10

mm : zone of inhibition

− : no antibiotic resistances

Table 4: Antibiotic resistances of selected LAB isolates by agar well diffusion assay.

reported in bacteriocins produced by other lactic acid bacteria, and may have been a contributing factor why the bacteriocins could not pass through the membrane with low molecular weight cut-off.

The EFSA considers antibiotic resistances, especially transferable resistances, a safety concern and a decision criterion for determining a strain's QPS status [18]. Antibiotic resistance genes carried by LAB can be transferred to human pathogenic bacteria either during food manufacture or during passage through the GIT [19]. In this research showed that *Lactococcus lactis* (LL2) was the most sensitive and *Lactobacillus brevis* and *Pediococcusa cidilactici* (PA2) were the most resistance to different antibiotics. In compare to our result, Hummel et al., showed that lactic acid bacteria like *lactobacillus* and *pedicoccus* were unaffected by gentamicin, streptomycin.

Conclusion

There is a wide variety of traditional dairy products especially in rural areas of Iran. These products mostly made from unpasteurized milk are appreciated by local people regarding its proven health benefits. This study showed the presence of viable probiotic LAB micro flora in these products. The antagonistic activity possessed by these isolates might be used for the control of unwanted pathogens mainly in dairy products, and could be exploited further for use in fermented dairy products.

References

1. Zacharof MP, Lovitt RW (2012) Bacteriocins Produced by Lactic Acid Bacteria. APCBEE Procedia 2 :50-56.

2. Ogunbanwo ST, Sanni AI, Onilude A (2003) Characterization of bacteriocin produced by *Lactobacillus plantarum* F1 and *Lactobacillus brevis* OG1. African Journal of Biotechnology 2: 219-227.

3. Aslim B, Yukesekdag Z, Sarikaya E, Beyatli Y (2005) Determination of the bacteriocin-like substances produced by some lactic acid bacteria isolated from Turkish dairy products. LWT 38: 691-694

4. Avonts L, Uytven EV, Vuyst LD (2004) Cell growth and bacteriocin production of probiotic lactobacillus strains in different media. Int. Dairy Journal 14: 947-955.

5. AOAC (2012) Official method of analysis of the association of official.

6. Iranmanesh M, Ezzatpanah H, Mojgani N (2013) Antibacterial activity and cholesterol assimilation of lactic acid bacteria isolated from traditional Iranian dairy products. LWT-Food Science and Technology.

7. Bromberg R, Moreno I, Lopez ZC, Regina DR, Oliveria DJ, et al. (2004) Isolation of bacteriocin-producing Lactic Acid Bacteria from meat and meat products and spectrum of inhibitory activity. Brazilian Journal of Microbiology 35: 137-144.

8. Mojgani N, Sabiri G, Ashtiani M, Torshizi M (2008) Characterization of Bacteriocins Produced by Lactobacillus brevis NM 24 and L.fermentum NM 332 Isolated from Green Olives in Iran. The Internet Journal of Microbiology.

9. De Vuyst L, Vandamme EJ (1994). Antimicrobial potential of lactic acid bacteria in bacteriocins of lactic acid bacteria: Microbiology, Genetics and Applications. London Blackie Academic and professional.

10. Gulahamadov S.G, Batdorj B, Dalgalarrondo M, Chobert J.M, Kuliev A, et al. (2006) Characterization of bacteriocin-like inhibitory substances (BLIS) from lactic acid bacteria isolated from traditional Azarbaijani cheeses. Eur Food Res Technology 224: 229-235

11. Klare I, Konstabel C, Muller-Bertling S, Reissbrodt R, Huys G, et al. (2005) Evaluation of new media formicrodiluition antibiotic susceptibility testing of lactobacilli, lactococci, pediococci, and bifidobacteria. Applied and Environmental Microbiology 71: 8982- 8986.

12. Tajabadi EM, Ouwehand AC, Hejazi AM, Jafari P (2011) Traditional Iranian dairy products: A source of potential probiotic lactobacilli. African Journal of Microbiology Research 5: 20-27

13. Fleming DW, Cochi SL, MacDonald KL, Brondum J, Hayes PS, et al. (1985) Pasteurized milk as a vehicle of infection in an outbreak of listeriosis. New England Journal of Medicine 312: 404-407.

14. Harris J, Daeschel MA, Stiles ME, Klaenhammer TR (1989) Antimicrobial activity of lactic acid bacteria against Listeria monocytogenes. Journal of food protection 52: 533-537.

15. Saranya S, Hemashenpagam N (2011) Antagonistic activity and antibiotic sensitivity of Lactic acid bacteria from fermented dairy products. Advances in Applied Science Research 2: 528-534.

16. Aasen IM, Møretrø T, Axelsson L, Storo I (2000) Influence ofcomplex nutrients, temperature and pH on bacteriocin production byLactobacillus sakei CCUG 42687. Applied Microbiology and Biotechnology 53: 159-166

17. Todorov SD, Dicks LMT (2004) Partial characterization of bacteriocins produced by four lactic acid bacteria isolated from regional South African barley beer. Annals of Microbiology 54: 403-413

18. European Food Safety Authority (2005) EFSA Scientific Colloquium Summary Report. QPS: qualified presumption of safety of microorganisms in food and feed. European Food Safety Authority, Brussels, Belgium.

19. Ammor MS, Florez AB, Mayo B (2007) Antibiotic resistance in non-*enterococcal* lactic acid bacteria and bifidobateria. *Food Control* 24: 559-570.

4

Chemical Composition, Physico-Chemical Properties, and Acceptability of Instant 'Ogi' from Blends of Fermented Maize, Conophor Nut and Melon Seeds

Ojo DO* and Enujiugha VN

Department of Food Science and Technology, Federal University of Technology, Akure, Nigeria

Abstract

This study was carried out to evaluate the nutrient, anti-nutrient composition, physico-chemical properties and acceptability of 'ogi' from blends of maize, conophor nut and melon seed flours at different proportions. Results of analysis of fermented maize: melon: conophor nut -90:5:5, 80:10:10, 70:15:15, 100:0:0 showed an increase in protein, ash, fat and crude fiber contents with increased supplementation with conophor nut and melon seed flour. The physico-chemical properties also varied: pH ranged from 5.70-6.20, viscosity ranged from 0.61-0.71 dPa, bulk density ranged from 0.66-0.91 g/ml, water and oil absorption capacities ranged from 660% to 680% and 820% to 870% respectively. Emulsification capacity, reconstitution index, foaming capacity, foaming stability and least gelation concentration ranged from 50.20% to 78.15%, 3.61-5.05 ml/g, and 1.38% to 10.00%, 1.38% to 5.63% and 6% to 20% respectively. The solubility index of the flour increased as supplementation levels increased. There were variations in the pasting properties of the supplemented 'ogi'. Peak viscosity ranged from 161.17-213.83 RVU, breakdown ranged from 28.17-106.76 RVU, final viscosity ranged from 145.25-247.34 RVU. Peak time was averagely at 5 min and pasting temperature from 83.65°C to 94.75°C. Result of mineral contents showed a significant increase in iron, magnesium, copper and phosphorus contents while calcium and sodium contents decreased significantly. Increased supplementation with conophor and melon seed flour increased the anti-nutrient contents. Tannin, oxalate, and phytate contents ranged from 4.65-5.85 mg/g, 2.48-2.65 mg/g, and 5.25-5.96 mg/g respectively. Consumer acceptability of the instant 'ogi' was rated best at 5% supplementation level with conophor and melon seed flours (90:5:5) when compared with the control (100% fermented maize).

Keywords: Fermented maize; Instant 'ogi'; Supplementation; Acceptability; Nutritional value

Introduction

Maize (*Zea mays*) also referred to as corn, is the most important cereal in the world after wheat and rice with regard to cultivation areas and total production [1]. Apart from being consumed by humans, it is also used to prepare animal feeds, and useful in the chemical industry. Maize can be cooked, roasted, fried, ground, pounded or crushed [2].

The conophor nut plant (*Tetracarpidum conophorum*) commonly called the African Walnut, is a perennial climbing shrub found in the moist forest zones of Sub-Saharan Africa. It is cultivated principally for the nuts, which are cooked and consumed as snacks, along with boiled corn [3]. Conophor nut commonly called 'Ukpa', 'asala', and 'awusa' in some parts of southern Nigeria is one of the several high nutrients dense foods with the presence of protein, fiber, carbohydrate and vitamins [4]. Conophor nut is a rich source of minerals such as calcium, magnesium, sodium, potassium, and phosphorus [5]. A bitter after taste is usually observed upon drinking water immediately after eating conophor nut and this could be attributed to the presence of alkaloids and other anti-nutritional and toxic factors. Ripe conophor nuts are mostly consumed in the fresh or toasted form or used in cakes, desserts and confectionaries.

Melons are food crops with several varieties which serve as a major food source. Melon seeds are generally rich in oil and are a good source of protein. The seed contains about 44% oil and 32% protein [6]. It has both nutritional and cosmetic importance and is rich in vitamin C, riboflavin and carbohydrates. Melon seed is a good source of amino-acids such as isoleucine and leucine [7]. It also contains palmitic, stearic, linoleic and oleic acids important in protecting the heart. It can serve as an important supplementary baby food, helping to prevent malnutrition. The present study examines the effect of supplementing fermented maize flour at different levels with conophor nut and melon seed flours in the production of instant 'ogi'.

Materials and Methods

Raw material source and collection

White maize (*Zea mays*), melon seed (*Citrullus lanatus*) and conophor nut (*Tetracarpidum conophorum*) were obtained from the local Oba market in Akure, Ondo State, Nigeria.

Sample preparation

Four formulations were made in the following proportions (maize: melon seeds: conophor nut); 90:5:5, 80:10:10, 70:15:15, 100:0:0. The sample consisting of 100% 'ogi' flour was used as the control. The samples were then analyzed and subjected to sensory evaluation.

Processing of fermented maize flour: The maize grains were cleaned and sorted by removing the pest-infested grains and discolored ones. It was then steeped for 72 h at room temperature and the steep water was decanted while the fermented grain was washed with portable water and wet-milled. It was then wet-sieved and the slurry could ferment for 24 h. It was afterwards decanted, dried at 70°C for 4 h and milled using hammer mill. The fermented maize flour was then sieved to obtain a finer particle (630 μm mesh size) and packaged in air-tight containers prior to analysis. The production chart is presented on Figure 1.

*Corresponding author: Ojo DO, Department of Food Science and Technology, Federal University of Technology, Akure, Nigeria, E-mail: doojo@futa.edu.ng

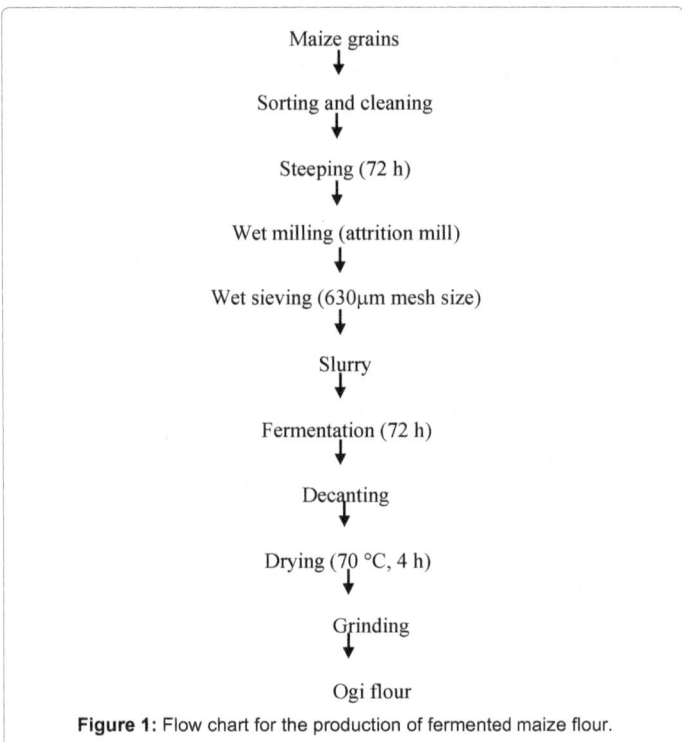

Figure 1: Flow chart for the production of fermented maize flour.

Processing of melon seed flour: The melon seeds were sorted to remove the discolored ones and then dried at 65°C for 6 h, milled with hammer mill and was defatted using n-hexane as the solvent for 6 h. The defatted melon was air dried and milled using hammer mill. The melon flour was then sieved to obtain a finer particle and packaged in air-tight containers prior to analysis. The production chart is shown on Figure 2.

Processing of conophor nut flour: The conophor nuts were cleaned to remove debris and dirt and cooked at 100°C for 1 h. It was then shelled to obtain the kernels. The kernels were dried at 50°C for 8 h, milled with hammer mill and defatted using n-hexane as solvent for 6 h. The defatted conophor nut cake was afterwards air dried at 70°C for 4 h and milled using hammer mill. The conophor nut flour was then sieved to obtain a finer particle and packaged in air- tight containers prior to analysis. The production chart is shown in Figure 3.

Analysis

Proximate chemical composition analysis: Proximate chemical composition of the samples was determined using the methods of AOAC [8]. Carbohydrate content was determined by subtracting the sum of the percentage weight of crude protein, crude fiber, ash, fat from 100%.

Functional properties analysis: For the determination of functional properties, the method of Onwuka [9] was used for the determination of Water/Oil absorption capacity. Bulk density and Swelling index were determined by the method described by Ukpabi and Ndimele [10]. The rotating spindle method described in the Encyclopedia of Industrial Chemical Analysis (E.I.C.A, 1971) was employed in viscosity determination. pH was determined using a Fischer Science Education pH meter (Model S90526, Singapore). The method of Jitngarmkusol et al. [11] with some slight modifications was used for the determination

of the foaming capacity and stability of the instant 'ogi' flour blends. Emulsion capacity was determined by the method of Yasumatsu et al. [12]. Least gelation concentration (LGC) of the flour blends was determined using the modified method of Coffman and Garcia [13]. Solubility index were determined as described by Takashi and Sieb [14] using SPECTRA, UK (Merlin 503) centrifuge. Reconstitution index were also determined as described by Banigo and Akpapunam [15].

Pasting properties analysis: The pasting properties of the samples were determined using a Rapid Visco-analyser (Newport Scientific Australia) as described by Adeyemi et al. [16]. The peak, viscosity, trough, breakdown, final viscosity, set back, peak time and pasting

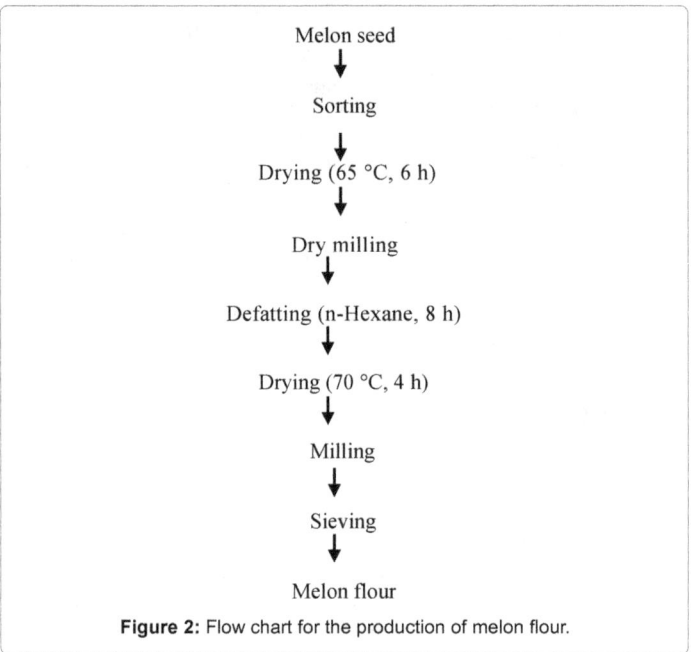

Figure 2: Flow chart for the production of melon flour.

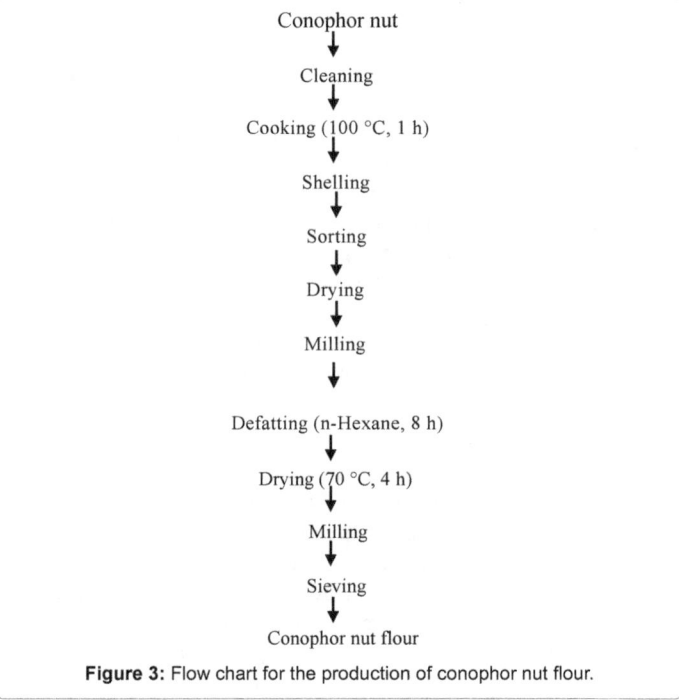

Figure 3: Flow chart for the production of conophor nut flour.

temperature were read with the aid of Thermocline for Windows Software connected to a computer.

Mineral elements analysis: The mineral elements Ca, Mg, Fe, Zn, Cu, were determined using atomic absorption spectrophotometer (AAS Model: PYE UNICAMSP9). Flame photometer was also used to measure the values of Na and K in all the samples and Phosphorus (P) was determined using a spectrophotometer (Model: Lemfield Spectrulab 23A) as described by AOAC [8].

Anti- nutrients content analysis: For anti-nutrients content analysis, tannin content was determined by the method of Makkar and Goodchild [17]. Oxalate content was determined by the method of Nwika et al. [18] and phytate content was determined by the method of Latta and Eskin [19].

Sensory evaluation: The instant 'ogi' was made into slurry by adding water till it formed a paste and boiled water was added to it and stirred continuously till it became viscous and formed a gruel. The products were evaluated for taste, appearance, aroma, and overall acceptability by a panel of ten members using a 9-point Hedonic scale. The rating of the samples ranged from 1 (dislike extremely) to 9 (Like extremely).

Statistical analysis: The data obtained were analyzed using a one-way Analysis of Variance and the means separated by Duncan New Multiple Range Tests (DMNRT) at 5% significance level (SPSS version 19 computer software) [20].

Results and Discussion

Proximate chemical composition of the instant 'Ogi' flour blends

The proximate composition of instant 'ogi' from blends of fermented maize, conophor nut and melon seed flours is presented in Table 1. The increase in the protein value of the flour was due to the supplementation of the maize flour with melon seeds and conophor nuts. Low fat content of the flour coupled with the low moisture content of the flour blends is an indication that the samples will be stable during storage. According to Adeyeye and Adejuyo [21], the low moisture content of the samples would hinder the growth of micro-organism and increase the shelf life of the samples. Sample SPB had the highest crude fiber content. According to Norman and Joseph [22], fiber has an important function in providing roughage or bulk that aids in digestion, softens stool and lowers plasma cholesterol level in the body. Increased melon/conophor nut flour substitution gave progressively higher protein, crude fiber and ash contents of the samples while fat and carbohydrate contents were reduced. Crude protein, ash and crude

Samples	BPO	SPB	BPC	POS
Moisture	4.73 ± 1.09ᵇ	4.73 ± 1.09ᵇ	4.73 ± 1.09ᵇ	4.73 ± 1.09ᵇ
Ash	9.01 ± 0.72ᵃ	1.80 ± 0.00ᵇ	2.44 ± 0.00ᵃ	2.98 ± 0.00ᶜ
Crude fiber	5.94 ± 0.87ᵇ	1.14 ± 0.12ᶜ	3.96 ± 0.01ᶜ	3.96 ± 0.01ᵇ
Fat	5.55 ± 1.11ᵇ	0.44 ± 0.12ᵈ	5.20 ± 0.29ᵈ	5.20 ± 0.29ᵃ
Crude fiber	9.01 ± 0.72ᵃ	1.80 ± 0.00ᵇ	2.44 ± 0.00ᵃ	2.98 ± 0.00ᶜ
Carbohydrate	5.94 ± 0.87ᵇ	1.14 ± 0.12ᶜ	3.96 ± 0.01ᶜ	3.96 ± 0.01ᵇ

Values with different superscript in a row are significantly different ($p<0.05$).
Values are means ± standard deviation of triplicate determinations.
Key: BPO = 70% fermented maize flour; 15% Melon flour; 15% Conophor nut flour; SPB = 80% fermented maize flour; 10% Melon flour; 10% Conophor nut flour; BPC = 90% fermented maize flour; 5% Melon flour; 5% Conophor nut Flour; POS = 100% fermented maize flour.

Table 1: Percentage proximate composition of instant 'ogi' from blends of fermented maize, conophor nut and melon seed flours.

fiber were significantly different in the four samples; however, there were no significant differences in the fat content of samples BPO and SPB (Table 1).

Functional properties of the instant 'Ogi' flour blends

According to Oyerekua and Adeyeye [23], high water absorption capacity (WAC) is desirable for the improvement of mouthfeel and viscosity reduction in food products. According to Afoadek and Sefa-Dedeh [24], WAC and OAC in the blended flour might be due to the thickness of interfacial bi-layer model of protein to protein interaction. Sample BPC had the lowest oil absorption capacity. The reduced value of OAC in Sample BPC might be due to collapse of the flour blend proteins thereby increasing the contact between protein molecules leading to coalescence and thus reduce stability of the samples. Bulk density is an important factor in food products handling, packaging, storage, processing and distribution. It is particularly useful in the specification of products derived from size reduction or drying processes. Bulk densities of the samples were similar to that reported by Adeyemi and Becky [25]. The bulk densities ranged from 0.66-0.90 g/ml with sample POS having the highest value which indicate that its packaging would be economical. Plaami [26] reported that higher bulk density is desirable, since it helps to reduce the paste thickness which is an important factor in convalescent and child feeding. Viscosity ranged from 0.61-0.70 dPa with samples SPB and BPC having the highest value. pH is important in determining the acid factor which is an indicator of the rate of conversion of starch to dextrin. The pH value ranged from 5.70-6.60. The foaming capacity ranged from 1.38% to 10.00% with BPO having the highest value. The increase in foaming capacity with melon and conophor nut supplementation might be due to soluble proteins and higher emulsion capacity; this might make it a better flavor retainer and enhance mouthfeel [23]. It has also been reported that foam capacity is related to the rate of decrease of the surface tension of the air/water interface caused by absorption of protein molecules [27]. The foaming stability of the flour increased with increment in the supplementation level of the flour though sample SPB had the highest value. Sample BPC had the highest value for least gelation capacity. The emulsion capacity ranged from 50.20% to 78.15%, with sample BPO having the highest value. High level of least gelation capacity means less thickening capacity of food; the contents ranged from 6.0% to 18.0%. Reconstitution index ranged from 3.61-5.05 ml/g with Sample POS (control) having the highest value. The functional properties of the instant 'ogi' flour blends are shown in Table 2.

Pasting characteristics of the instant 'Ogi' flour blends

Table 3 shows the results of the pasting characteristics of the instant 'ogi' flour blends. The pasting properties of the samples BPO, SPB, BPC and control (POS) were significantly different ($p<0.05$). Peak viscosity of the instant 'ogi' samples ranged from 133.51-213.83 RVU, the values were observed to reduce with increase in supplementation levels. Final viscosity is a measure of stability of the cooked sample [28]. The final viscosity ranged from 145.67-243.59 RVU with BPO having the highest value; this implies that highly viscous paste can be formed during cooking. The setback value is a measure of retrogradation the cooked sample and it ranged from 60.17-108.58 RVU with SPB having the highest value. Pasting temperature is also a measure of the temperature at which flour viscosity begins to rise during cooking, it provides information on the cost of energy required to cook the instant 'ogi'. The pasting temperature of the instant 'ogi' ranged from 83.65°C to 94.75°C with BPC having the highest value. The pasting time ranged from 5.36-5.85 sec, with POS having the highest value (Table 3).

Samples	BPO	SPB	BPC	POS
Ph	6.60 ± 0.00[a]	6.20 ± 0.00[b]	5.70 ± 0.00[c]	6.20 ± 0.00[b]
WAC (%)	660.00 ± 0.70[a]	680.00 ± 0.28[a]	660.00 ± 0.56[a]	665.00 ± 0.63[a]
OAC (%)	830.00 ± 0.70[a]	870.00 ± 0.21[a]	800.00 ± 0.28[a]	820.00 ± 0.00[a]
EC (%)	78.15 ± 0.21[a]	50.20 ± 0.28[b]	75.25 ± 0.35[a]	53.25 ± 2.47[b]
Viscosity (dPa)	0.61 ± 0.01[b]	0.70 ± 0.01[a]	0.70 ± 0.01[a]	0.61 ± 0.01[b]
RI (ml/g)	3.61 ± 0.02[c]	3.62 ± 0.03[c]	4.35 ± 0.00[b]	5.05 ± 0.07[a]
BD (g/ml)	0.66 ± 0.00[d]	0.71 ± 0.00[c]	0.76 ± 0.00[b]	0.90 ± 0.00[a]
FC (%)	10.00 ± 0.01[a]	9.85 ± 0.03[b]	4.28 ± 0.02[c]	1.38 ± 0.01[d]
FS (%)	4.28 ± 0.03[a]	5.63 ± 0.02[a]	1.42 ± 0.01[c]	1.38 ± 0.01[d]
SI (v/v)	3.61 ± 0.02[c]	3.62 ± 0.03[c]	4.35 ± 0.07[b]	5.05 ± 0.07[a]
LGC (%)	18.00 ± 0.00	14.00 ± 0.00	20.00 ± 0.00	6.00 ± 0.00

Values with different superscript in a row are significantly different (p<0.05).
Values are means ± standard deviation of triplicate determinations.
WAC: Water Absorption Capacity; OAC: Oil Absorption Capacity; EC: Emulsion Capacity; RI: Reconstitution Index; BD: Bulk Density; FC: Foaming Capacity; FS: Foaming Stability; SI: Swelling Index; LGC: Least Gelation Capacity.
Key: BPO = 70% fermented maize flour, 15% Melon flour, 15% Conophor nut flour; SPB = 80% fermented maize flour, 10% Melon flour, 10% Conophor nut flour; BPC = 90% fermented maize flour, 5% Melon flour, 5% Conophor nut Flour; POS = 100% fermented maize flour.

Table 2: Functional properties of instant 'ogi' from blends of fermented maize, conophor nut and melon seed flours.

Samples	BPO	SPB	BPC	POS
Peak viscosity (RVU)	163.17 ± 0.007[b]	161.17 ± 0.007[c]	133.51 ± 0.14[d]	213.83 ± 0.28[a]
Trough (RVU)	135.00 ± 0.014[a]	123.25 ± 0.010[b]	85.08 ± 0.007[d]	107.08 ± 0.014[c]
Breakdown (RVU)	28.17 ± 0.007[d]	37.93 ± 0.014[c]	48.41 ± 0.007[b]	106.76 ± 0.014[a]
Final viscosity (RVU)	243.59 ± 0.014[b]	247.34 ± 0.007[a]	145.25 ± 0.007[d]	196.67 ± 0.010[c]
Setback (RVU)	108.58 ± 0.014[b]	124.09 ± 0.010[a]	60.17 ± 0.010[d]	89.57 ± 0.007[c]
Pasting time (sec)	5.36 ± 0.010[d]	5.68 ± 0.007[b]	5.58 ± 0.007[c]	5.85 ± 0.007[a]
Pasting temperature (°C)	86.05 ± 0.014[b]	85.95 ± 0.028[c]	94.75 ± 0.007[a]	83.65 ± 0.010[d]

Values with different superscript in a row are significantly different (p<0.05).
Values are means ± standard deviation of triplicate determinations.
Key: BPO = 70% fermented maize flour; 15% Melon flour; 15% Conophor nut flour; SPB = 80% fermented maize flour; 10% Melon flour; 10% Conophor nut flour; BPC = 90% fermented maize flour; 5% Melon flour; 5% Conophor nut flour; POS = 100% fermented maize flour.

Table 3: Pasting characteristics of instant 'ogi' from blends of fermented maize, conophor nut and melon seed flour blends.

Mineral content of the instant 'Ogi' flour blends

Table 4 shows the mineral contents of the instant 'ogi' flour blends. Calcium value decreased with increase in supplementation level but magnesium content increased. Magnesium is well known to be important in cellular energy production and enzyme activity; its value ranged from 106.06-126.03 mg/100 g. Iron (Fe) ranged from 9.87-11.70 mg/100 g with the sample BPC having the lowest value. The instant 'ogi' flour blends provides a good amount of iron that is needed in the production of haemoglobin which carries oxygen in the blood. Zn ranged between 2.25 mg/100 g and 2.91 mg/100 g. The potassium content ranged from 195.68-198.37 mg/100 g and there were no significant differences between samples BPO, BPC and POS (p<0.05). A major function of potassium is to maintain the excitability of nerve and muscle tissue (Table 4).

Anti-nutrient content of the instant 'Ogi' flour blends

Table 5 shows the level of anti-nutrients in the instant 'ogi' flour blends. The phytate content of the 'ogi' flour blends ranged from 5.25-5.96 mg/g. Phytates are known to form complexes with iron, zinc, calcium and magnesium making them less available and thus inadequate in food samples especially for children [29]; however, the phytate content of the 'ogi' flour blends are far lower than the minimum amounts of phytic acid reported by Siddhuraju and Becker [30] to hinder the absorption of iron and zinc. Oxalates are also known to make complexes with calcium to form an insoluble calcium-oxalate salt. Siddhuraju and Becker [30] reported a safe normal range of 4-9 mg/ 100 g for oxalates. The oxalate content of the samples which range from 2.48-2.67 mg/100 g is quite lower than the reported value. Tannin content range from 4.65-5.85 mg/100 g. Tannins have been implicated in the interference of iron absorption; it usually forms insoluble complexes with proteins, thereby interfering with their bioavailability [31-33] (Table 5).

Sensory evaluation of the instant 'Ogi' flour blends

Table 6 shows the sensory evaluation results of the instant 'ogi' flour blends. Sensory evaluation was carried out by ten (10) untrained panelists and the parameters evaluated were taste, flavor, appearance and overall acceptability. Consumer evaluation of taste showed that there were no significant differences between sample BPC and SPB. Sample BPC were also rated higher than SPB in terms of appearance and aroma. In terms of overall acceptability, Samples BPC and SPB compared favorably with the control (POS) and there were no significant differences between them (Table 6).

Samples	BPO	SPB	BPC	POS
Iron	11.70 ± 0.63[a]	11.53 ± 0.58[a]	9.87 ± 0.17[b]	10.87 ± 0.40[ab]
Zinc	2.91 ± 0.10[a]	2.64 ± 0.21[b]	2.25 ± 0.14[ab]	2.30 ± 0.35[ab]
Calcium	140.68 ± 0.45[d]	144.05 ± 0.63[c]	145.77 ± 0.26[b]	150.46 ± 0.37[a]
Magnesium	126.03 ± 0.09[a]	123.66 ± 0.43[b]	120.06 ± 0.12[c]	106.06 ± 0.38[d]
Potassium	196.59 ± 0.50[b]	198.37 ± 0.53[a]	195.83 ± 0.25[b]	195.68 ± 0.37[b]
Sodium	111.88 ± 0.24[d]	115.90 ± 0.14[c]	122.65 ± 0.35[b]	145.84 ± 0.44[a]
Copper	1.87 ± 0.11[a]	2.27 ± 0.35[a]	2.07 ± 0.10[a]	0.97 ± 0.35[b]
Phosphorus	64.97 ± 0.25[a]	63.67 ± 0.60[b]	58.89 ± 0.21[c]	56.57 ± 0.81[d]

Values with different superscript in a row are significantly different (p<0.05).
Values are means ± standard deviation of triplicate determinations.
Key: BPO = 70% fermented maize flour; 15% Melon flour; 15% Conophor nut flour; SPB = 80% fermented maize flour; 10% Melon flour; 10% Conophor nut flour; BPC = 90% fermented maize flour; 5% Melon flour; 5% Conophor nut Flour; POS = 100% fermented maize flour.

Table 4: Mineral composition of instant 'ogi' from blends of fermented maize, conophor nut and melon seed flours (mg/100 g).

Samples	Phytate (mg/100 g)	Oxalates (mg/100 g)	Tannin (mg/100 g)
BPO	5.64 ± 0.01[c]	2.66 ± 0.01[a]	4.66 ± 0.02[d]
SPB	5.76 ± 0.01[b]	2.59 ± 0.02[b]	5.37 ± 0.03[a]
BPC	5.96 ± 0.03[a]	2.48 ± 0.01[c]	4.86 ± 0.01[b]
POS	5.26 ± 0.02[b]	2.67 ± 0.02[a]	4.78 ± 0.01[c]

Values with different superscript in a row are significantly different (p<0.05).
Values are means ± standard deviation from triplicate determinations.
Key: BPO = 70% fermented maize flour; 15% Melon flour; 15% Conophor nut flour; SPB = 80% fermented maize flour; 10% Melon flour; 10% Conophor nut flour; BPC = 90% fermented maize flour; 5% Melon flour; 5% Conophor nut Flour; POS = 100% fermented maize flour.

Table 5: Anti-nutritional properties of instant 'ogi' from blends of fermented maize, conophor nut and melon seed flours (mg/100 g).

Samples	Taste	Appearance	Aroma	Overall Acceptability
BPO	4.50 ± 1.26[c]	6.10 ± 0.99[b]	6.30 ± 0.94[b]	5.30 ± 0.67[c]
SPB	6.50 ± 0.70[b]	6.10 ± 1.19[b]	7.30 ± 0.94[a]	6.30 ± 0.82[b]
BPC	5.80 ± 0.42[b]	7.00 ± 0.94[a]	7.10 ± 0.31[a]	6.80 ± 0.42[b]
POS	7.50 ± 0.52[a]	7.50 ± 1.26[a]	7.70 ± 0.94[a]	7.70 ± 1.15[a]

Values with different superscript in a row are significantly different ($p < 0.05$).
Values are means ± standard deviation from triplicate determinations.
Key: BPO = 70% Fermented maize flour; 15% Melon flour; 15% Conophor nut flour; SPB = 80% fermented maize flour; 10% Melon flour; 10% Conophor nut flour; BPC = 90% fermented maize flour; 5% Melon flour; 5% Conophor nut Flour; POS = 100% fermented maize flour.

Table 6: Sensory evaluation of instant 'ogi' from blends of fermented maize, conophor nut and melon seed flours.

Conclusion

The study has shown that conophor nut/melon/ogi flour with improved nutrient composition, sensory quality and pasting properties that is comparable to the traditional fermented maize 'ogi' flour can be obtained up to 80:10:10 ratio. Conophor and melon seeds which are under-utilized are suitable for use in instant 'ogi' flour production for improved nutritional and pasting characteristics.

References

1. Osagie AU, Eka OU (1998) Nutritional quality of plant foods. University of Benin, Benin.

2. Abdulraharan AA, Kolawole OM (2006) Traditional preparations and uses of maize in Nigeria. Ethnobot leaf 10: 219-227.

3. Enujiugha VN (2008) *Tetracarpidium conophorum* conophor nut. In. the encyclopedia of fruit and nuts. CABI Publishing, Oxfordshine, UK.

4. Savage GP, Mc Niel DL, Dutta PC (2001) Some nutritional advantage of walnuts. J Acta Horticultur 544: 557-563.

5. James NR (2009) Volatile components of green walnuts husks. J Agri Food Chem 48: 2858-2861.

6. Enujiugha VN, Ayodele-Oni O (2003) Evaluation of nutrients and some anti-nutrients in lesser known, underutilized oil seeds. Int J Food Sci Technol 38: 525-528.

7. Olaofe O, Adeyemi FO, Adediran GO (1994) Amino acid, mineral composition, functional properties of some oil seeds. J Agri Food Chem 42: 878-884.

8. AOAC (2012) Association of Official Analytical Chemists. Official Methods of Analysis.

9. Onwuka GI (2005) Food analysis and Instrumentation theory and practice. Naphtali prints, Lagos, Nigeria.

10. Ukpabi UJ, Ndimele C (1990) Evaluation of the quality of garri produced in Imo state. Nigeria Food J 8: 105-108.

11. Jitngarmkusol S, Hongsuwankul J, Tananuwong K (2008) Chemical compositions, functional properties and microstructure of defatted macadamia flours. J Food Chem 110: 23-30.

12. Yasumatsu K, Sawada K, Maritaka S, Mikasi M, Toda J, et al. (1972) Whipping and emulsifying properties of soybean products. J Agricul Biol Chem 36: 719-727.

13. Coffman CW, Garcia VV (1977) Functional properties of flours prepared from Chinese indigenous legume seed. J Food Chem 61: 429-433.

14. Takashi S, Sieb PA (1988) Paste and gel properties of prime corn and wheat starches with and without native lipids. J Cereal Chem 65: 474-480.

15. Banigo P, Akpapunam S (1987) Physico-chemical and nutritional evaluation of protein-enriched fermented maize flour. Nigeria Food J 5: 30-36.

16. Adeyemi IA, Adabiri BO, Afolabi OA, Oke OL (1992) Evaluation of some quality characteristics and baking potentials of Amaranth flour. Nigeria Food J 10: 8-15.

17. Makkar AOS, Goodchild S (1996) Quantification of tannins. A laboratory manual. Internal Centre for Agricultural Research in Dry Areas (ICARDA). Aleppo 25.

18. Nwika N, Ibe G, Ekeke G (2005) Proximate composition and level of toxicants in four commonly consumed spices. J Appl Sci Environ Manag 9: 150-155.

19. Latta M, Eskin M (1980) A simple and rapid colorimetric method for phytate determination. J Agri Food Chem 28: 1213-1315.

20. Steel R, Torrie J, Dickey D (1997) Principles and procedures of statistics: A biometrical approach. (3rdedn), McGraw Hill Book Co., New York, USA.

21. Adeyeye EI, Ayejuyo OO (1994) Chemical composition of *Cola accuminata* and *Grarcina kola* seed grown in Nigeria. Int J Food Sci Nutri 45: 223-230.

22. Norman NP, Joseph HH (1995) Food Science (5thedn.) Chapman and Hall Publishers, New York, USA.

23. Oyerekua MA, Adeyeye EI (2009) Comparative evaluation of the nutritional quality, functional properties and amino acid profile of co-fermented maize/cowpea and sorghum/cowpea ogi as infant complementary food. Asia J Clini Nutri 1: 31-39.

24. Afoakwa EO, Sefa-Dedeh S (2001) Viscoelastic properties and changes in pasting characteristics of trifoliate yam (*Dioscorea dumentorum* pax) starch after harvest. J Food Chem 77: 85-91.

25. Adeyemi IA, Beckley O (1986) Effect of period of maize fermentation and souring on chemical properties and amylograph viscousity of Ogi. J Cereal Sci 4: 353-360.

26. Plaami SP (1997) Content of dietary fibre in foods and its physiological effects. Food Rev Int 13: 27-76.

27. Kiin Kabari OB, Eke Ejiofor J, Giami SY (2015) Wheat/plantain flour enriched with bambara groundnut protein concentrate. Int J Food Sci Nutri Eng 5: 75-81.

28. Chinma CE, Adewuyi AO, Abu JO (2009) Effect of germination on the chemical, functional; and pasting properties of flour from brown and yellow varieties of tiger nut. Food Res Int 42: 1004-1009.

29. Aletor VA (1993) Nutritional, biochemical and physiological aspects in animal production. J Vert Human Toxicol 35: 57-67.

30. Siddhuraju P, Becker K (2001) Effect of various domestic processing methods on anti-nutrients and *in-vitro* protein and starch digestibility of two indigenous varieties of Indian tribal pulse (*Mucuna pruriens var. utilis*). J Agri Food Chem 49: 3058-3067.

31. Enujiugha VN, Agbede P (2000) Nutritional and anti-nutritional characteristics of African oil bean (*Pentaclethra macrophylla*) seeds. Appl Tropic Agri 5: 11-14.

32. EICA (1971) Encyclopedia of industrial chemical analysis. Interscience publishers, New York, London, Sydney, Toronto.

33. Enujiugha VN (2003) Chemical and functional characteristics of conophor nut. Pak J Nutri 2: 335-338.

5

Determination of Essential Amino Acids in *Pangasius bocourti*

Danuwat P, Rimruthai P, Phattanawan C and Peerarat D*

Institute of Product Quality and Standardization, Maejo University, Nonghan, Sansai, Chiang Mai, Thailand

Abstract

The purpose of this research was to analyze for the determination of essential amino acid in *Pangasius bocourti* which was aged between1-12 months. Essential amino acid were extracting by 6 M hydrochloric acid (HCl) using the EZ-Fast technology technique and afterwards the quality and quantity was analyzed by using the GC-MS. The highest amounts of essential amino acid were found in *Pangasius bocourti* aged 1 month. The highest quantities of essential amino acids were Lysine 8.41% and Leucine 8.30%. Other essential amino acid such as Phenylalanine, Methionine, Isoleucine and Tryptophan were found at 4.54%, 4.35%, 4.25% and 2.36% respectively. In month 9, Threonine and Histidine were found at the highest amounts of essential amino acid, this was 5.85% and 2.96%, respectively. Valine had the highest amounts at 7 months and 15 days for 6.79%.

Keywords: Catfish; Essential amino acids; Extracting; EZ-Fasst; GC-MS; *Pangasius bocourti*

Introduction

Essential amino acids are amino acids that cannot be synthesized by organism. There is need to obtain these amino acids by consuming food which contain them. The essential amino acids that cannot be synthesized are Isoleucine, Leucine, Lysine, Methionine, Phenylalanine, Threonine, Tryptophan and Valine. In infants, there are addition 2 essential amino acids which are Arginine and Histidine. Amino acids are important in the chemical composition of fish. They are the critical component of proteins in the fish body, which help increase the growth rate. *Pangasius bocourti* or Mekong catfish is one of the important species for cage culture in the Mekong river basin, especially Vietnam and Thailand [1,2]. Recently, the market of Pangasius frozen fillets has been increasing due to its white tender meat, low-fat content and easily digestible protein [3]. The Pangasius fish production was estimated at approximately 1.1 million tons in 2010 [4]. The purpose of this study was to analyze the quantities of essential amino acids in *Pangasius bocourti*, comparison between fish aged 1-12 months. The data will be used to provide basic information about fish nutrition and to promote as the distinctive point for pangasius fish market. The data obtained will also be important for the consumers, as they can obtain sufficient essential amino acids through consumption of *Pangasius bocourti* fish containing these amino acids.

Materials and Methods

Materials

Pangasius fish was raised in two net cages. The first net cages were used to culture the Pangasius fish aged 1-6 month and also other selected fish that had similar weight and size which had been measured before released into the water. In month 7, the fish were moved to other net cages. Samples of the fish were collected every 15 days until they reached 12 months. Samples were stored in freezer before being extracted.

Extraction of the amino acids from *Pangasius bocourti*

Fish samples were chopped into small pieces and homogenized by using Homogenizer. After that, samples were dried in a freeze dryer. The dry fish meats were grounded into powder and kept in desiccators [5]. Samples were extracted by using Hydrochloric acid 6M and Thioglycolic acid 4% at a ratio of 1:1:1 (fish powder : HCl : Thioglycolic acid). Stirring continued in a shaking water bath at 50°C for 20 minutes.

The samples were kept in a hot air oven at 110°C for 22 hours. The solids and liquids phase were separated with centrifuges at 1,300 rpm for 20 minutes [6-9]. The remaining liquids were collected, this aqueous was used in extracted amino acid by extracting kit: EZ-Faast [6,10].

Derivatization

The amino acid analysis kit, EZ-Faast, was used to extract the amino acid. The liquid sample (100 µL) was mixed with 200 µL of sodium carbonate solution in a test tube. The sorbent tip was used to collect upper phase; eluting was added as a medium (sodium hydroxide and N-propanol, 3:2) 200 µL and rinse with 0.6 mL syringe. Reagent 4 (Iso-octane) in 50 µL and reagent 5 in (Iso-octane) 100 µL were added and shaken to allow the phases to separate. The upper phase was collected then evaporated in Nitrogen-evaporator. Reagent 6 (Iso-octane 80% and Chloroform 20%) was added to final sample in 100 µL before being injected into the GC-MS.

GC-MS

The column of EZ: Faast GC with 10 m length and 0.25 m diameter was used in this study. The oven temperature was initially set at 110°C and programmed to 320°C at 30°C/min for 1 min. The gas carrier (He) flow was constant at 1.1 mL/min. The injection volume was 2.5 µL in split less mode. Injector and transferred temperature were 250°C. All samples were injected into the GC-MS and then analyzed for 8 min.

Results and Discussion

Figure 1 showed the chromatogram of the essential amino acids from pangasius fish, compared with standard essential amino acids (Figure 2). The Pangasius fish contained 9 essential amino acids; Valine, Leucine, Isoleucine, Threonine, Methionine, Phenylalanine, Lysine,

---

***Corresponding author:** Peerarat D, Institute of Product Quality and Standardization, Maejo University, Nonghan, Sansai, Chiang Mai, Thailand
E-mail: peeraratdoungtip@gmail.com

Histidine and Tryptophan. Quantitative results of essential amino acid are shown in Table 1. Percentage values of essential amino acid among fish groups were significantly different (p<0.05). The data were plotted in linear graphs as divided in type of amino acid (Figures 3-11).

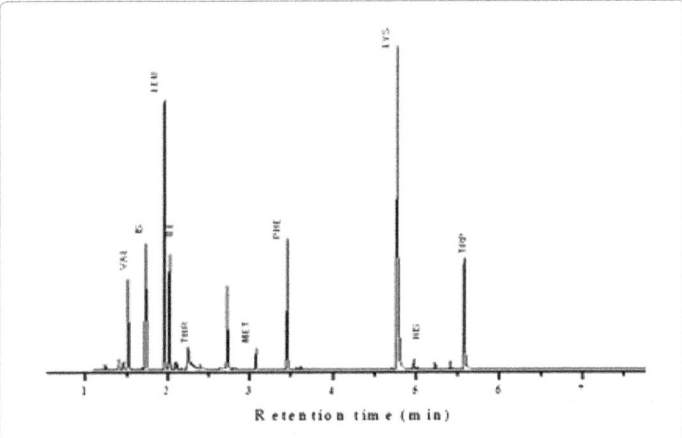

Figure 1: Chromatogram of the essential amino acids in Pangasius boucourti.

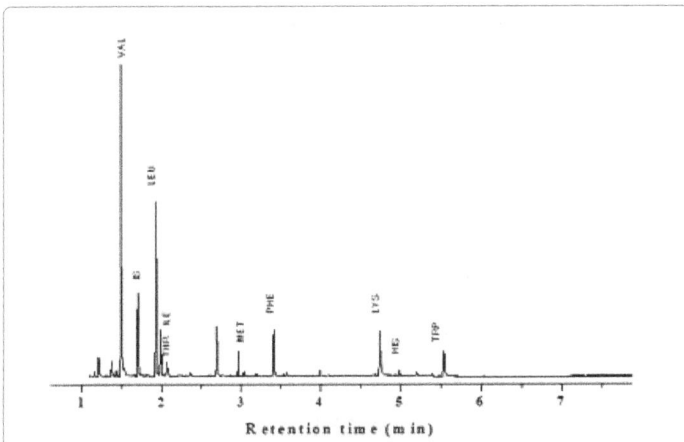

Figure 2: Chromatogram of standard essential amino acids.

Figure 3 percentage of Valine was significantly different (p<0.05). It was found highest in fish aged 7.5 month with 6.79%. Fish aged 7 month and 8 month also had high quantities of Valine at 5.64% and 5.14%. However, after 7.5 month, percentage of Valine was slightly decreased and remained stable when fish aged 9 month and 15 days. As showed in figures 4-8, Pangasius fish in aged 1 month had the most highest of Leucine, Isoleucine, Methionine, Phenylalanine and Lysine with respective 8.30% 4.25% 4.35% 4.53% and 8.41%. Threonine was found at the highest level when aged 9 months, 5.85%. Fish aged 7.5 month also had high Threonine with 5.73%. But they were significantly different (p<0.05) in mean value of amino acid percentage (Figure 9). The quantity of Histidine was the same as Threonine, it was found at its highest level in fish aged 9 month, approximately 2.96%. In addition, Histidine values were high in fish 2.5, 10 and 1 month with 2.81% 2.81% and 2.78% respectively (Figure 10). Tryptophan was different from the other amino acid. It appeared that between 1-12 months, they were not significantly different in mean values of amino acid (p≥0.05), except for the first month of the fish. It has high percentage of Tryptophan (Figure 11) but after the second month, the amino acid remained at a steady level. Amino acid values were changed but none of them were lost during the experimental period. The highest percentage of Valine was found in fish aged 7.5 month with 6.79%. The highest amounts of Leucine, Isoleucine, Methionine, Phenylalanine and Lysine were found in fish aged 1 month with 8.30%, 4.25%, 4.35%, 4.53% and 8.41% respectively. It appeared that most of the essential amino acid was found in young fish. But 1 month old fish were not appropriated for extracting the amino acids due to the fish size and economical reason. They were too small and inconvenient when preparing samples. The recommended aged of pangasius fish should be around 7-8 months because they had high amino acid after the first month. However, Threonine and Histidine were found at the highest level in fish aged 9 month with 5.73% and 2.96%, respectively. The percentage of Tryptophan was not significantly different (p≥0.05) during 1-12 month. It stayed at a steady level during the experimental period.

Quantities of the essential amino acid in pangasius fish depend on

Month	Valine	Leucine	Isoleucine	Threonine	Methionine	Phenylalanine	Lysine	Histidine	Tryptophan
1	4.63 ± 0.03[d]	8.30 ± 0.03[a]	4.25 ± 0.04[a]	2.63 ± 0.14[mn]	4.35 ± 0.03[a]	4.53 ± 0.37[a]	8.41 ± 0.06[a]	2.78 ± 0.07[abc]	2.36 ± 0.26[a]
1.5	2.86 ± 0.07[gh]	4.29 ± 0.04[jk]	3.04 ± 0.04[e]	3.21 ± 0.05[j]	2.69 ± 0.03[cd]	2.46 ± 0.15[ef]	4.46 ± 0.05[fg]	2.58 ± 0.03[defg]	2.02 ± 0.01[bc]
2	2.88 ± 0.10[gh]	4.32 ± 0.07[jk]	2.57 ± 0.02[jkl]	2.71 ± 0.03[m]	2.55 ± 0.09[e]	2.26 ± 0.01[gh]	3.90 ± 0.04[h]	2.47 ± 0.06[fghi]	2.00 ± 0.02[c]
2.5	3.16 ± 0.04[fg]	5.12 ± 0.08[e]	2.75 ± 0.10[gh]	3.58 ± 0.14[gh]	2.71 ± 0.16[cd]	2.66 ± 0.03[c]	4.51 ± 0.11[f]	2.82 ± 0.12[ab]	2.03 ± 0.00[bc]
3	2.76 ± 0.40[gh]	5.44 ± 0.17[d]	2.54 ± 0.01[klm]	3.34 ± 0.03[i]	2.77 ± 0.12[c]	2.33 ± 0.02[g]	4.50 ± 0.04[f]	2.50 ± 0.10[fghi]	1.99 ± 0.00[c]
3.5	3.38 ± 0.50[ef]	4.27 ± 0.09[jk]	2.57 ± 0.01[jkl]	3.53 ± 0.02[h]	2.64 ± 0.16[d]	2.30 ± 0.01[g]	4.38 ± 0.05[g]	2.65 ± 0.11[bcdef]	2.00 ± 0.01[c]
4	3.62 ± 0.34[e]	5.02 ± 0.07[f]	2.81 ± 0.06[fg]	3.65 ± 0.2[g]	2.56 ± 0.08[e]	2.46 ± 0.03[ef]	3.74 ± 0.12[i]	2.54 ± 0.04[efgh]	2.05 ± 0.02[bc]
4.5	2.75 ± 0.01[gh]	4.07 ± 0.01[lm]	2.49 ± 0.01[mn]	2.63 ± 0.08[mn]	2.43 ± 0.02[gh]	2.07 ± 0.04[li]	3.18 ± 0.06[n]	2.36 ± 0.02[hi]	1.99 ± 0.01[c]
5	4.56 ± 1.39[d]	4.75 ± 0.10[h]	2.76 ± 0.02[g]	3.87 ± 0.08[e]	2.51 ± 0.07[efg]	2.26 ± 0.05[gh]	3.94 ± 0.01[h]	2.72 ± 0.04[bcde]	2.01 ± 0.02[bc]
5.5	3.19 ± 0.20[fg]	4.91 ± 0.07[g]	2.76 ± 0.10[g]	3.77 ± 0.07[f]	2.43 ± 0.01[gh]	2.17 ± 0.07[hi]	5.02 ± 0.16[c]	2.59 ± 0.08[cdefg]	2.01 ± 0.01[bc]
6	2.69 ± 0.08[h]	4.01 ± 0.04[m]	2.45 ± 0.02[n]	3.36 ± 0.03[i]	2.43 ± 0.01[gh]	2.09 ± 0.01[li]	3.31 ± 0.06[m]	2.45 ± 0.03[fghi]	1.99 ± 0.01[c]
6.5	3.12 ± 0.26[fgh]	4.34 ± 0.02[j]	2.85 ± 0.03[f]	4.68 ± 0.05[c]	2.70 ± 0.09[cd]	2.64 ± 0.03[cd]	5.95 ± 0.06[b]	2.48 ± 0.05[fghi]	2.04 ± 0.01[bc]
7	5.64 ± 0.07[b]	4.12 ± 0.06[l]	3.50 ± 0.10[c]	4.67 ± 0.18[c]	2.94 ± 0.07[b]	2.80 ± 0.05[b]	3.41 ± 0.14[l]	2.57 ± 0.18[efgh]	2.02 ± 0.02[bc]
7.5	6.79 ± 0.06[a]	6.18 ± 0.04[b]	3.66 ± 0.02[b]	5.73 ± 0.06[b]	2.66 ± 0.012[d]	2.74 ± 0.13[bc]	4.92 ± 0.08[d]	2.55 ± 0.04[efgh]	2.04 ± 0.02[bc]

*Mean ± SD in the same column with different superscript letters are significant differences (p<0.05).

Table 1: Content of essential amino acid in *Pangarius boucourti* (1-12 months)

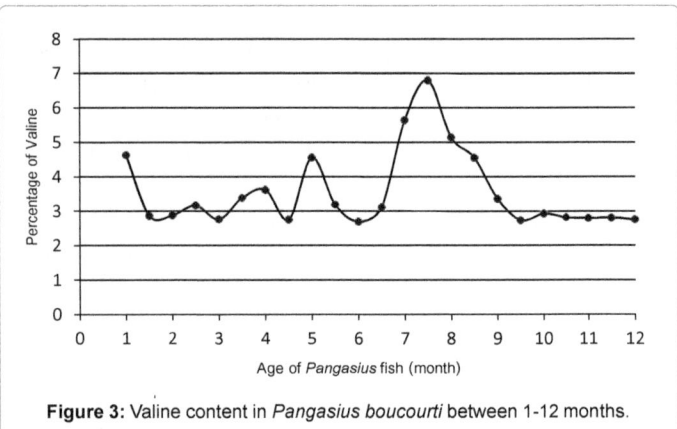

Figure 3: Valine content in *Pangasius boucourti* between 1-12 months.

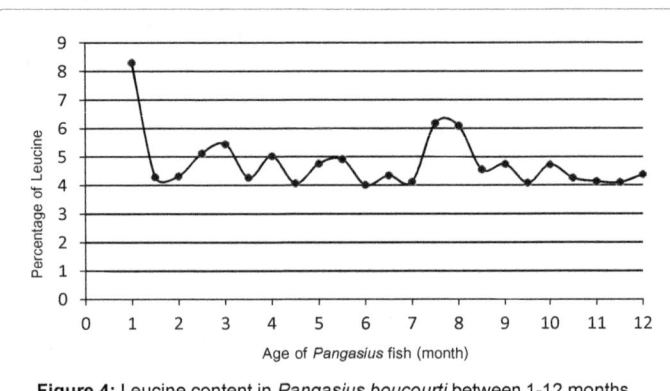

Figure 4: Leucine content in *Pangasius boucourti* between 1-12 months.

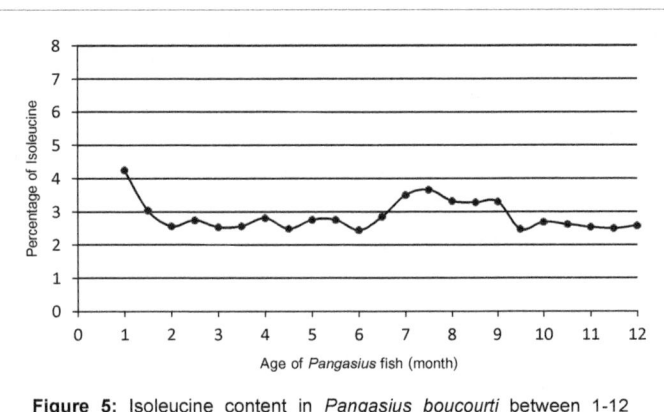

Figure 5: Isoleucine content in *Pangasius boucourti* between 1-12 months.

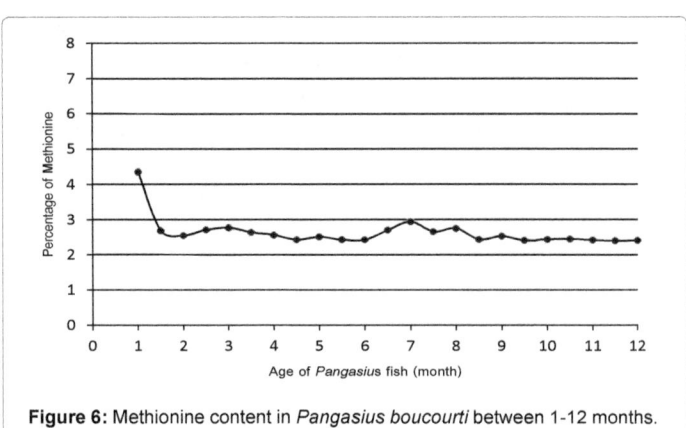

Figure 6: Methionine content in *Pangasius boucourti* between 1-12 months.

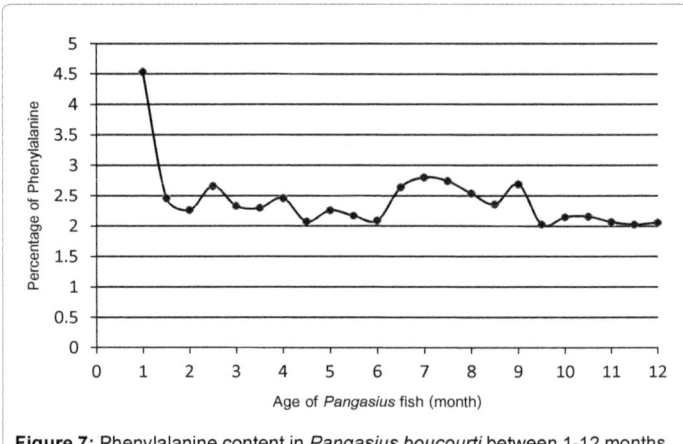

Figure 7: Phenylalanine content in *Pangasius boucourti* between 1-12 months.

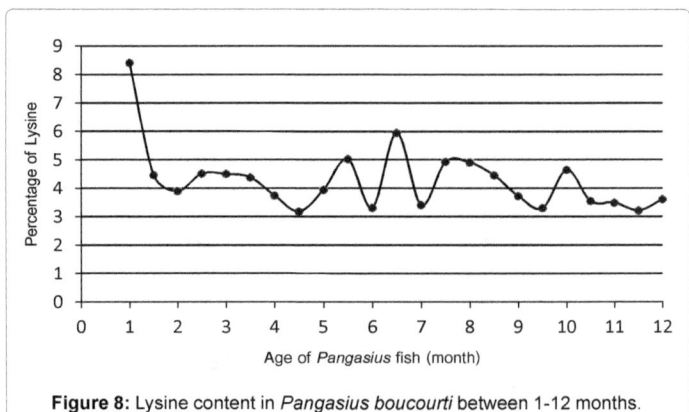

Figure 8: Lysine content in *Pangasius boucourti* between 1-12 months.

many factors such as fish diet. This is because they cannot synthesize the essential amino acid by themselves, fish also need amino acid through foods too. Now-a-days, supplement fish diet products are popular in fisheries industry. Amino acid in fish diet should be adequate and appropriate for fish. It has an influence towards the fish's growth rate. Fish that are feeding by nutritive diet will grow faster in a shorter period and it will be beneficial to the export industry. For recommendation, this study should observe requirement of amino acid in fish by feeding them with purified diet or semi purified diet to control the nutrition balance [11] and in addition the quantities of amino acid in fish diet should be analyzed to control growth rate.

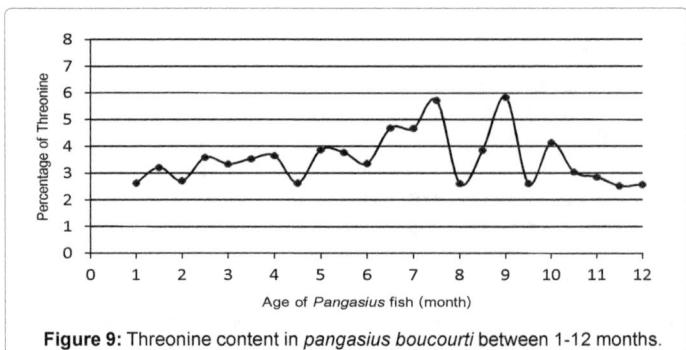

Figure 9: Threonine content in *pangasius boucourti* between 1-12 months.

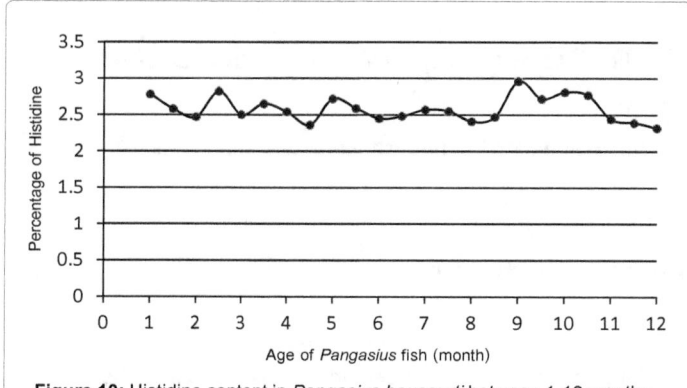

Figure 10: Histidine content in *Pangasius boucourti* between 1-12 months.

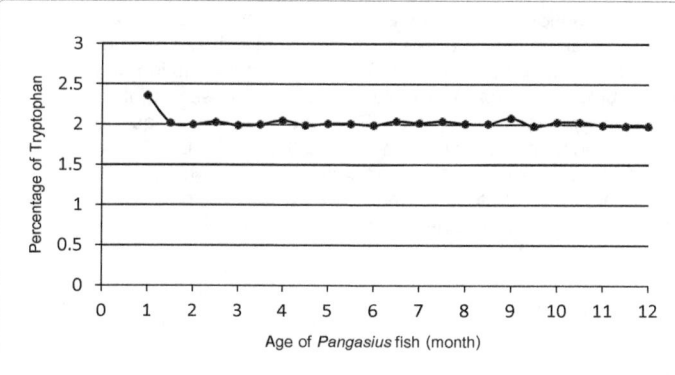

Figure 11: Tryptophan content in *Pangasius boucourti* between 1-12 months.

Conclusion

Pangasius bocourti, aged between 1-12 months, were containing 9 types of amino acids: Valine, Leucine, Isoleucine, Threonine, Methionine, Phenylalanine, Lysine, Histidine and Tryptophan. Pangasius fish aged 1 month have the highest amount of Lysine, Leucine, Phenylalanine, Methionine, Isoleucine and Tryptophan. Threonine and Histidine are highest in fish 9 month and Valine is highest in fish 7 month. However, the appropriate age of pangasius fish for extraction of essential amino acids should be around 7-8 months.

Acknowledgements

The authors wish to thank the National Food Institute (NFI) of Thailand for the financial support.

References

1. Jiwyam W (2010) Growth and Compensatory Growth of Juvenile *Pangasius bocourti* Sauvage, 1880 relative to ration. Aquaculture 306: 393-397.

2. Cacot P, Legendre M, Dan TQ, Tung LT, Liem PT, et al. (2002) Induced Ovulation of Pangasius. Aquaculture 213: 199-206.

3. Thammapat P, Raviyan P, Siriamornpun S (2010) Proximate and Fatty Acids Composition of the Muscles and Viscera of Asian Catfish (*Pangasius bocourti*). Food Chemistry 122: 223-227.

4. Van DH, Doolgindachbaporn S, Suksri A (2014) Effect of Low Molecular Weight agar an *Lactobacillus plantarum* on Growth Performance, Immunity and Disease Resistance of Basa fish (*Pangasius bocouti*, Sauvage 1880). Fish&Shellfish Immunology 41: 340-345.

5. AOAC (2005) Official Methods of Analysis. Association of Official Analytical Chemists. Washington DC.

6. Mustafa A, Aman P, Anderson R, Kamal-Eldin A (2007) Analysis of Free Amino Acids in Cereal Products. Food Chemistry 105: 317-324.

7. Fountoulakis M, Lahm HW (1998) Hydrolysis and amino acid composition of proteins. J Chromatogr A 826: 109-134.

8. Xian Z, Chao Z, Liang Z, Hongbin C (2008) Amino Acids Production from Fish Proteins Hydrolysis in Subcritical Water. Chinese Journal of Chemical Engineering 16: 456-460.

9. Usydus Z, Szlinder-Richert J, Adamczyk M (2009) Protein Quality and Amino Acid Profiles of Fish products available in Poland. Food Chemistry 112: 139-145.

10. Phenomenex (2001) EZ-Faast Easy-fast Amino Acid Sample Testing Kit Manual. Torrance, Phenomenex, CA, USA.

11. Tantikitti C, Chimsung N (2001) Dietary Lysine Requirement of Fresh Water Catfish (*Mystus nemurus* Cuv. & Val.). Aquaculture Research 32: 135-141.

6

A Predictive Model Based on Surface Electromyography to Assess the Easiness of Deglutition of Dysphagia Diets

Yuki Kayanuma[1], Reiko Ueda[1], Michiko Minami[2], Arata Abe[3], Kazumi Kimura[3], Junko Funaki[4,5], Yoshiro Ishimaru[1] and Tomiko Asakura[1*]

[1]*Graduate School of Agricultural and Life Sciences, The University of Tokyo, Tokyo, Japan*
[2]*Faculty of Education, Tokyo Gakugei University, Koganei, Tokyo, Japan*
[3]*Department of Neurological Sciences, Graduate School of Medicine, Nippon Medical School, Sendagi, Bunkyo-ku, Tokyo, Japan*
[4]*International College of Arts and Sciences, Fukuoka Women's University, Higashi-ku, Fukuoka, Japan*
[5]*Graduate School of Health and Environmental of Sciences, Fukuoka Women's University, Higashi-ku, Fukuoka, Japan*

Abstract

Dysphagia diet is used for the people who have disorder of swallowing caused by aging or cerebral arterial diseases. The current standards for dysphagic diets are based on their physical characteristics. However, parameters that reflect the easiness of swallowing are also critical. Here, we developed a method to objectively evaluate the easiness of deglutition. First, we collected 68 terms that describe food textures related to easiness of deglutition, and selected 54 commercial dysphagia diets as samples. Using these terms and samples, we conducted a texture-perception questionnaire survey, and the results were subjected to a correspondence analysis. Referring to the results of this analysis, 10 textures that represent the easiness of deglutition were selected and dysphagia diets corresponding to each texture were selected as well. Then, sensory evaluation and surface electromyography (sEMG) of the anterior triangle of the neck (submental triangle) were recorded using these samples. We developed a predictive model for the easiness of deglutition by applying a partial least squares (PLS) regression technique to the sensory evaluation and sEMG data. Parameter fitting of the cross-validation model was significant (R^2, 0.87; RMSE, 0.34). The model accuracy was further investigated by fitting the model to test data, and results were again significant (R^2, 0.89; RMSE, 0.10). This indicates that our predictive model using sEMG measurements was highly accurate. Evaluating the easiness of deglutition with this predictive model will help identify and develop new foods that make swallowing easier for patients with dysphagia.

Keywords: Dysphagia; Sensory evaluation; Surface electromyography; Partial least square regression analysis; Aging; Submental triangle

Introduction

Recently, instances of aspiration pneumonia have led to an increased mortality rate in people of advanced age. As this is a disease caused by weakened swallowing of bolus, studying dysphagia–difficulty swallowing–has become especially important. In Japan, the standard diets for dysphagia are determined by the Japanese Society of Dysphagia Rehabilitation, which has two specifications: physical characteristics such as viscosity (assessed by the Line Spread Test) and thickness attributes. The Japan Consumer Affair Agency and Japan Care Food Conference have determined specifications for dysphagia diets that include desirable physical characteristics such as thickness attributes and ability to be swallowed. In the United States, the national standard for dietary treatment of dysphagia is based on food-texture and liquid levels [1]. However, for developing dysphagia diets, evaluating whether or not the food can be easily swallowed is crucial. Conventionally, a sensory test is carried out to evaluate the ease of swallowing, although reproducibility is difficult because of biased testing conditions and variation in individual evaluators. Here, we developed an objective method for more precisely assessing the ease with which diets designed for dysphagia can be swallowed.

By combing sensory data with the results of instrumental analysis, sensory evaluations can be correlated with objective factors. Partial least square (PLS) regression has been used to predict several lingual sensations. The bitterness of dairy protein hydrolysates was corresponded to the degree of hydrolysis which is analyzed by size-exclusive and reverse-phase chromatography elution patterns [2]. Additionally, PLS regression has shown that attributes of flavored mineral water are correlated with an electronic tongue device [3]. Thus, PLS analyses have been successful in providing datasets of instrumental measurement for variables that explain the sensations in the mouth during eating. Although the relationships between physical properties of food components and sensations have been investigated, reports of these relationships in physiological responses in human are limited.

Surface electromyography (sEMG) has been reported to provide a non-invasive method of assessing certain aspects of complex muscle activity for deglutition [4]. Miura et al. [5] reported that the carbonated or cooled beverages have effect on the submental muscles and these responses can be examined by sEMG. The aim of our present study was to develop a sEMG-based predictive model of easiness of deglutition by applying a PLS analysis to submental muscle contraction and sensory evaluations.

Materials and Methods

The schematic representation of study protocol is shown in Figure 1. All experiments were allowed by the ethics committee of the University of Tokyo. The application number is 16-26.

Participants

Forty-seven healthy young adults who had no current or past swallowing abnormalities (male, 9; female, 38; age, 19-30 years) participated in developing the predictive model and the cross validation dataset for sensory evaluation, and 10 different individuals (male, 5;

*Corresponding author: Tomiko Asakura, Graduate School of Agricultural and Life Sciences, The University of Tokyo, 1-1-1 Yayoi, Bunkyo-ku, Tokyo 113-8657, Japan
E-mail: asakura@mail.ecc.u-tokyo.ac.jp

female, 5; aged 22-24 years) participated in the sEMG measurement. A secondary test, validating the predictive model was carried out with test data set by 50 new participants (male, 24; female, 26; aged 22-25 years) for sensory evaluation and another 10 participants (male, 5; female, 5; aged 22-25 years) for sEMG measurement. Ten dysphagia patients (male, 5; female, 5; aged 57-86 years) participated to a sensory evaluation with the 4 experimental samples used for the test dataset. The objective of the experiment was explained to each participant, and an informed consent for use in publication was obtained for sensory evaluation and sEMG measurements.

Texture terms and dysphagia diets

Sixty-eight food-texture terms (Supplementary Table 1) that were related to the easiness of deglutition were selected from the list of 271 terms reported by Hayakawa et al. [6]. Fifty-four commercially available dysphagia diets (Supplementary Table 2) were used as candidate experimental samples. These included four categories of products (Class 1: easily masticated foods; Class 2: foods that can be mashed by gums; Class 3: foods that can be mashed by the tongue; Class 4: foods that do not need to be masticated and unidentified items). All foods were specified by the Japan Care Food Conference. Class-2 foods were not included as candidates because healthy evaluators found them difficult to eat.

Representative textures and experimental samples

Using the 54 commercially available dysphagia diets (Supplemantary Table 2), a texture perception questionnaire for the 68 textures was answered by five women who are well trained with sensory teste engaged in food-science research. The perception of each texture, such as sticky, was scored on 0 (imperceptible) or 1 (perceptible). The texture with total score more than 3 was selected for further analysis. 51 textures and 54 dysphagia diets were input into the correspondence analysis. To calculate the statistical distance between variables with each texture and diet sample, we transformed the data into a three dimensional scatter plot graph and examined their interrelations. Referring to this scatter plot, 10 representative textures which explain the deglutition comprehensively, were selected. And 10 dysphagia diet samples that were closely plotted to these textures were selected as experimental samples. All statistical analyses were conducted using JMP Pro 11.0.0 (SAS Institute Inc., SAS Campus Drive, Cary, NC, USA).

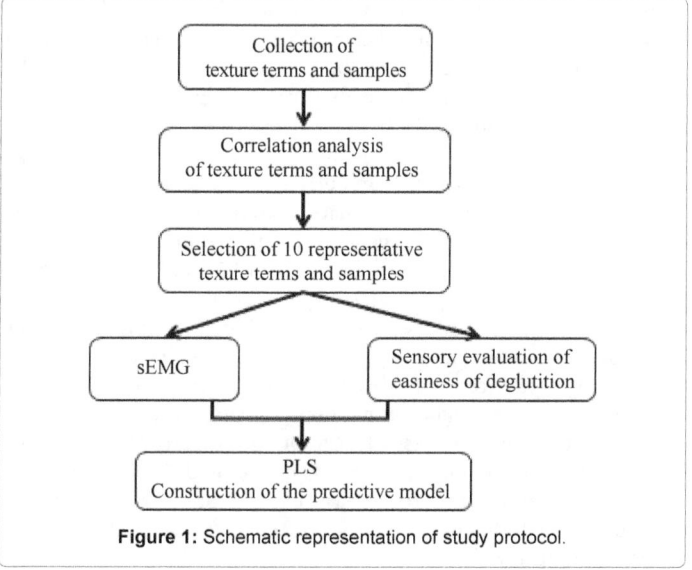

Figure 1: Schematic representation of study protocol.

Sensory evaluation

Using the 10 experimental samples, we conducted a sensory evaluation for the easiness of deglutition. In detail, experimental samples were given random code numbers and independently presented in a random order. The code numbers were attached to white paper cups (3 ounces, 90 ml) and 10 grams of each sample were put into each cup. The samples were tasted using a clear plastic spoon. All samples were provided at room temperature.

The easiness of deglutition was assessed by scoring in a 9-point scale. (-4, imperceptible; -3, very weak; -2, weak; -1, rather weak; 0, flat; 1, rather strong; 2, strong; 3, very strong; 4, extremely strong). Breaks were provided between evaluations, and participants were asked to drink rinse their mouths.

Surface electromyography

sEMGs were recording with a Personal EMG plus 8-channel computer-based EMG unit with Oisaka software (Oisaka Electronic Equipment, Hiroshima, Japan). Wet sensors (BLUE SENSOR, Ambu, Denmark) were taped to the right submental triangle where movements could be clearly detected. To ensure that the sensor was accurately attached, the skin was wiped with cotton dampened with 70% ethanol to remove sebum. The attachment points along the same muscle fiber were searched for during swallowing. Electrodes were taped in tandem at 1 cm distances. As a reference, a sensor was attached to the elbow bone on the opposite side of the dominant hand. Since the data were affected by the position of electrode, one series of test was performed at the same sensor position. The sampling speed was 3 kHz and the signals from the pair of electrodes were amplified 5,000 times. The original EMG was processed with full-wave rectification. Muscle activity was obtained from the integrated EMG wave. The duration of swallowing was defined as the onset of the rapidly increasing peak amplitude until the return of the trace to baseline surrounding the peak signal. The highest peak voltage (V_{p-p}) was obtained from the integrated EMG wave. Power frequency analysis was applied to the original EMG and power spectrum was carried out using the MemCalc/Win program (Suwa Trust, Sapporo, Japan) and the spectral density was obtained. Power spectral density (PSD) was obtained from the whole spectral area of spectral density. The low-frequency components were obtained from the spectrum area between 0.2-10 Hz spectrum area divided by the whole spectrum area, and high-frequency components are the spectrum area greater than 100 Hz divided by whole area (Figure 2).

Development of the predictive model using the cross-validation dataset

To develop a predictive model for the easiness of deglutition, a PLS analysis was conducted using the sEMG measurements as the explanatory variables, and the sensory evaluations as the response variable. NIPALS (Non-linear Interactive Partial Least Squares) and K-fold cross validation (K = 7) was applied to find the combination of the parameters that could precisely explain the objective variable. The precision of the predictive model was evaluated using the coefficient of determination (R^2) and root mean square error (RMSE). PLS-VIP was applied for the selection of the explanatory variables using the variable importance in the projection (VIP) value and weight of the model (W) as the threshold values.

Validation using the test dataset

To verify the predictive model, we selected test samples from the 54 dysphagia diets, excluding the experimental samples used for

Sample ID	Ease of deglutition score	Muscle activity (mV·sec)	V_{p-p} (mV)	Low-frequency component	High-frequency component	PSD (μV^2/Hz)
1	1.02 ± 2.18	1625.53 ± 174.31	0.62 ± 0.09	0.10 ± 0.02	0.42 ± 0.02	0.04 ± 0.01
2	1.93 ± 1.94	1637.50 ± 155.80	0.64 ± 0.06	0.13 ± 0.02	0.36 ± 0.02	0.04 ± 0.01
3	1.34 ± 1.90	1826.35 ± 164.57	0.72 ± 0.08	0.12 ± 0.02	0.39 ± 0.02	0.05 ± 0.01
4	2.20 ± 1.87	1512.43 ± 164.89	0.68 ± 0.09	0.10 ± 0.01	0.39 ± 0.02	0.04 ± 0.01
5	1.70 ± 2.19	1942.27 ± 189.01	0.76 ± 0.08	0.13 ± 0.02	0.37 ± 0.02	0.05 ± 0.01
6	2.54 ± 1.68	1559.80 ± 195.65	0.70 ± 0.10	0.12 ± 0.01	0.38 ± 0.01	0.04 ± 0.01
7	0.00 ± 1.88	2057.19 ± 175.97	0.80 ± 0.09	0.09 ± 0.02	0.39 ± 0.02	0.05 ± 0.01
8	0.72 ± 2.29	1590.52 ± 160.45	0.69 ± 0.09	0.09 ± 0.01	0.42 ± 0.02	0.04 ± 0.01
9	2.39 ± 1.77	1421.29 ± 156.56	0.67 ± 0.08	0.12 ± 0.01	0.41 ± 0.02	0.04 ± 0.01
10	3.23 ± 0.97	1367.46 ± 158.25	0.66 ± 0.10	0.10 ± 0.01	0.40 ± 0.02	0.04 ± 0.01

Table 1: Mean scores for the ease of deglutition (mean ± SD) was assessed by scoring in a 9-point scale (-4 to +4), muscle activity, V_{p-p}; the highest peak voltage, low-frequency component, high-frequency component, and PSD; power spectral density (mean ± SE) for each sample. Sample IDs refer to Figure 3.

Sample ID	Samples and textures	C1	C2	C3
a'	Smooth	0.64	-0.34	0.07
1'	Carrot paste	0.43	-0.6	0.12
b'	Residual feeling on the tongue; the material is viscous, sticky or piquant	-0.17	-0.16	0.05
2'	Shiso (Japanese basil) and seaweed mousse	0.07	-0.16	0.04
c'	Sticky, elastic, and chewy	-0.71	-0.02	0.52
3'	Rice gruel with vegetables and seaweed	-0.57	-0.04	0.48
d'	Jelly-like	1.03	0.52	0.17
4'	Pineapple-flavored jelly	0.86	0.32	0.15

Table 2: The four texture terms and products that made up the test dataset: texture terms (a'-d') correspond to products (1'-4'). (C1, C2, C3) are the 3D correspondence-analysis coordinates.

the primary examination. The samples which were closely plotted to the representative textures determined in section 2.3 were selected. Furthermore, four samples were collected and used for the predictive model development. Using the experimental samples, we evaluated sensations and sEMG measurements as described above. These resulting values were applied to the predictive model and the precision was evaluated by looking at the R^2 and RMSE values.

Results

Selected representative textures and experimental samples

The following food samples were selected from the correlation analysis: powdery, smooth, residual feeling on the tongue, melting, juicy, creamy, elastic, sticky, smooth, and jelly-like. The following dysphagia diets that best represented these textures were: sweetened green peas paste, macaroni and scallop gratin, rice porridge, pineapple and apple sauce, simmered cod dumplings, corn soup with kernels, meatloaf with brown sauce, bread mousse, strained pork-curry stew, and apple-flavored jelly (Figure 3).

Sensory evaluation

The jack-knife technique was used to detect outliers, which were then removed (upper control limit: UCL, 4.51 at α = 0.05). The mean values ± SD after removing the outliers are shown in Table 1. A Tukey-Kramer HSD test was applied to the sensory evaluation data, and the degrees to which experimental samples easy to swallow were compared. The highest gap was observed between apple-flavored jelly and meatloaf ($p < 0.00001$). This result suggests that jellies, which have high water content and cohesiveness, are perceived to be easy to swallow. In contrast, meatloaf, which feels like it absorbs saliva and clogs the throat, is thought to be difficult to swallow. Next, a sensory evaluation was conducted with dysphagia patients (male, 5; female, 5; aged 57-86 years) with the 4 experimental samples used for the test dataset. A correlation analysis showed that the senses of deglutition for healthy participants correlated significantly with those of dysphagia patients ($r = 0.94$; $p = 0.02$).

sEMG measurement

The jack-knife technique was used to detect outliers, which were then removed (UCL, 3.38 at α = 0.05). The mean values ± SD after removing the outliers are shown in Table 1. Because the relationship between physical stimuli and perceptions is logarithmic according to the Weber-Fechner law, the sEMG data were transformed to logarithmic values, which were then used in the following PLS-VIP analysis.

Development of the predictive model by PLS-VIP analysis

A predictive model for the easiness of deglutition was generated using a PLS-VIP analysis that incorporated the sEMG data as the explanatory variable. The number of latent variables was three when the cumulative contribution of the response was 87.47% and the cumulative contribution of the explanatory variable was 91.48%. All sEMG factors; muscle activity, low-frequency component, V_{p-p}, PSD and high-frequency component, satisfied the VIP threshold of greater than 0.8 except for high-frequency components. The VIP value obtained via PLS was used to estimate the response variable from the explanatory variables [7]. In general, predictive models using PLS-VIP are conducted by applying factors with VIP values greater than 0.8. However, because the R^2 value of the predictive model was higher when we included all sEMG factors, we included the high-frequency component in the explanatory variables. The weight of the model corresponds to the centralized data (Figure 4). As shown in Figure 4, the VIP value was highest for muscle activity, suggesting that it is most relevant to the easiness of deglutition. The model was applied to the dataset to validate the accuracy of the model. Figure 5 shows the relationship between the measured and predicted easiness of deglutition. The fitting parameters for the cross validation dataset were $R^2 = 0.87$ and RMSE = 0.34.

Validation using the test dataset

Four dysphagia diets, paste-formed carrot, shiso and seaweed mousse, rice gruel with vegetables and seaweed, and pineapple-flavored jelly were selected as experimental samples for obtaining the test dataset. These samples were closely plotted with the following food-textures: smooth, residual feeling on the tongue, sticky, and jelly-like (Table 2). The jack-knife technique was used for outlier detection and outliers were removed (sensory evaluation: UCL, 4.59; α = 0.05; sEMG: UCL, 3.43; α = 0.05). The easiness of deglutition and sEMG data is shown in Table

3. The model was applied to the test dataset to validate the accuracy of the model. Figure 6 shows the relationship between the measured and predicted values of the easiness of deglutition. The fitting parameters for the cross validation dataset were $R^2 = 0.89$ and RMSE = 0.10.

Discussion

Here, we combined sEMG data recorded from the submental triangle during deglutition with sensory evaluations to determine factors that make swallowing easier. The VIP value obtained from the PLS regression analysis was highest for muscle activity and the correlation coefficient (r) between the easiness of deglutition and muscle activity was -0.86 ($p = 0.07$). Motor unit recruitment and muscle activity have been reported to be dependent on muscle contraction [8]. When participants feel that swallowing is easy, muscle activity gets likely decreased. Low-frequency components (VIP = 0.97), PSD (VIP = 0.92), and V_{p-p} (VIP = 0.96) were also sEMG factors that satisfied the criterion of VIP > 0.8 (Figure 4). When muscles are exposed to continuous contraction, the wave of sEMG signals gradually becomes flat. This is thought to result from muscle fatigue, and frequency-spectrum analysis can be used to estimate the degree of fatigue. Although the physiological significance of frequency-spectrum analysis is not yet clear, an increase in low-frequency components (5-30 Hz) has been reported to be the most reliable index of fatigue [9]. PSD has also been reported to decrease with muscle fatigue [10]. sEMG amplitude is known to increase with muscle fatigue, and this is thought to result from an increase of motor units and impulse firing frequency, as well as synchronization of firing activity [11]. Although low-frequency components, PSD, and V_{p-p} were found to contribute to the easiness of deglutition in our present study, the easiness of deglutition was only found to be significantly correlated with PSD

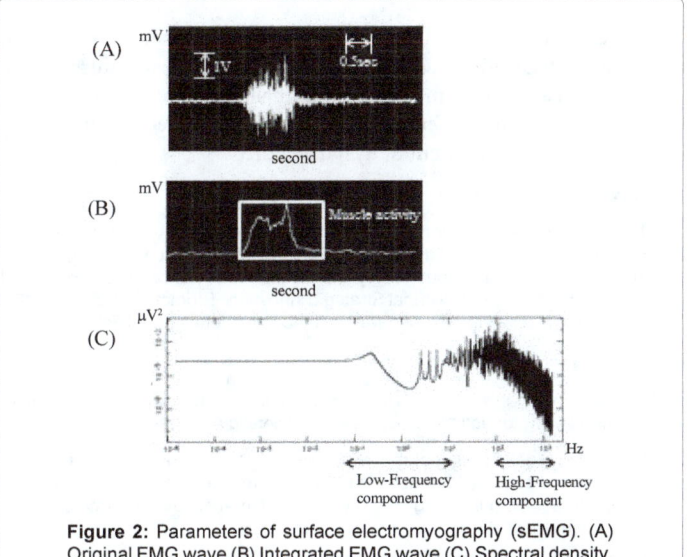

Figure 2: Parameters of surface electromyography (sEMG). (A) Original EMG wave (B) Integrated EMG wave (C) Spectral density.

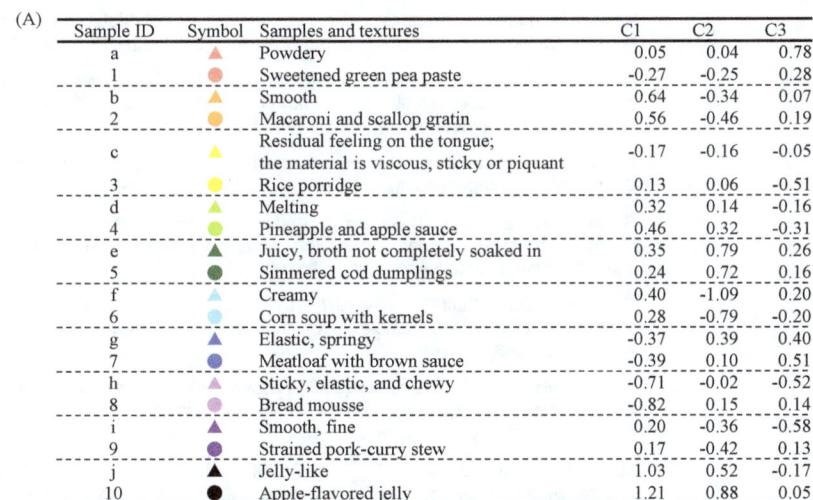

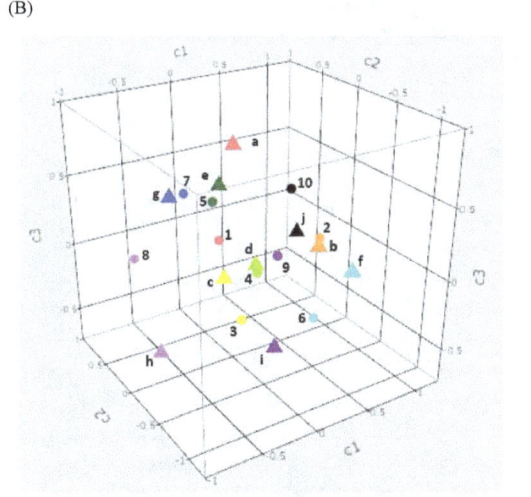

(A)

Sample ID	Symbol	Samples and textures	C1	C2	C3
a	▲	Powdery	0.05	0.04	0.78
1	●	Sweetened green pea paste	-0.27	-0.25	0.28
b	▲	Smooth	0.64	-0.34	0.07
2	●	Macaroni and scallop gratin	0.56	-0.46	0.19
c	▲	Residual feeling on the tongue; the material is viscous, sticky or piquant	-0.17	-0.16	-0.05
3	●	Rice porridge	0.13	0.06	-0.51
d	▲	Melting	0.32	0.14	-0.16
4	●	Pineapple and apple sauce	0.46	0.32	-0.31
e	▲	Juicy, broth not completely soaked in	0.35	0.79	0.26
5	●	Simmered cod dumplings	0.24	0.72	0.16
f	▲	Creamy	0.40	-1.09	0.20
6	●	Corn soup with kernels	0.28	-0.79	-0.20
g	▲	Elastic, springy	-0.37	0.39	0.40
7	●	Meatloaf with brown sauce	-0.39	0.10	0.51
h	▲	Sticky, elastic, and chewy	-0.71	-0.02	-0.52
8	●	Bread mousse	-0.82	0.15	0.14
i	▲	Smooth, fine	0.20	-0.36	-0.58
9	●	Strained pork-curry stew	0.17	-0.42	0.13
j	▲	Jelly-like	1.03	0.52	-0.17
10	●	Apple-flavored jelly	1.21	0.88	0.05

Figure 3: 10 representative food-quality terms and products. (A) List of samples (1-10) and their texture terms (a-j). (B) Three dimensional correspondence analysis scatter plot made from the coordinates (C1, C2 and C3).

Sample ID	Ease of deglutition score	Muscle activity (mV·sec)	V_{p-p} (mV)	Low-frequency component	High-frequency component	PSD (μV²/Hz)
1'	2.43 ± 1.36	2910.00 ± 183.71	1.40 ± 0.11	0.09 ± 0.03	0.46 ± 0.02	0.10 ± 0.01
2'	0.73 ± 1.63	2890.50 ± 204.99	1.52 ± 0.09	0.07 ± 0.02	0.46 ± 0.01	0.12 ± 0.01
3'	-1.57 ± 1.39	3345.93 ± 280.26	1.69 ± 0.16	0.08 ± 0.03	0.46 ± 0.02	0.13 ± 0.01
4'	2.36 ± 2.04	2551.65 ± 174.79	1.38 ± 0.09	0.06 ± 0.02	0.45 ± 0.01	0.09 ± 0.01

Table 3: Mean scores for the easiness of deglutition (mean ± SD) was assessed by scoring in a 9-point scale (-4 to 4), muscle activity, Vp-p; highest peak voltage, low-frequency component, high-frequency component, PSD: power spectral density (mean ± SE) for each test data sample. Sample IDs refer to Table 2.

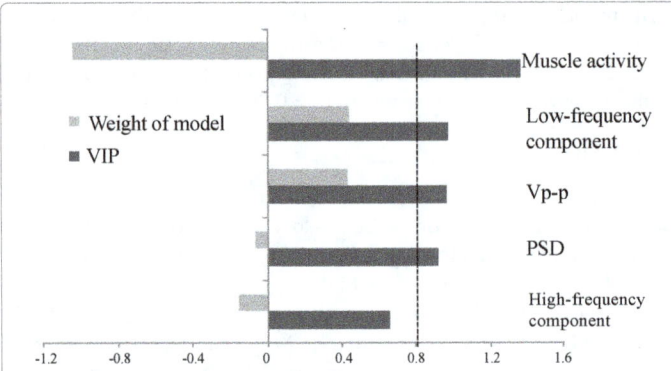

Figure 4: Degree of five sEMG factors contributed to the easiness of deglutition (threshold value; VIP = 0.80). (V_{p-p}: The highest peak voltage; PSD: Power Spectral Density and VIP: Variable Importance in the Projection).

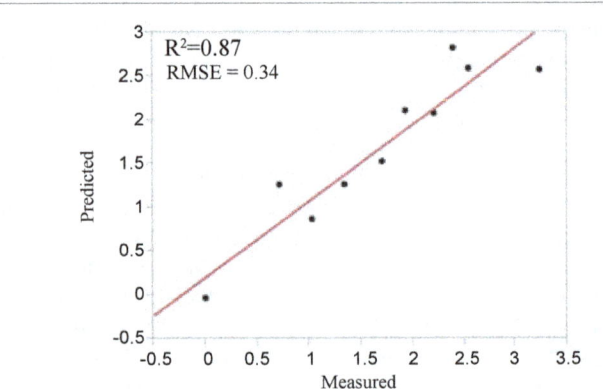

Figure 5: Relationship between the measured and predicted values for the easiness of deglutition using the cross-validation dataset. Red line: linear regression line, each plots were derived from 10 representative samples.

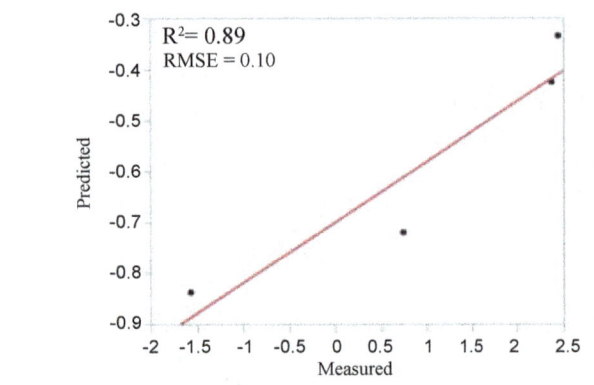

Figure 6: Relationship between the measured and predicted values for the ease of deglutition using the test dataset. Red line: linear regression line, each plots were derived from test samples.

(r = -0.91, p = 0.03) and V_{p-p} (r = -0.99, p = 0.0008). The finding that PSD and V_{p-p} are negatively correlated with the easiness of deglutition suggests that when swallowing becomes difficult, a response similar to muscle fatigue is taking place. The fitting parameters for the cross validation dataset (R^2, 0.87; RMSE, 0.34) and the test dataset (R^2, 0.89; RMSE, 0.10) signify the high accuracy of the predictive model. Thus, the predictive model was able to objectively evaluate the easiness of deglutition in healthy participants. However, to utilize this predictive

model for developing dysphagia diets, we first need to determine whether it can be applied as accurately to dysphagia patients who are the primary beneficiary of these products. A sensory evaluation was thus conducted with dysphagia patients with the 4 samples used for the test dataset as a preliminary experiment. A correlation analysis showed that the senses of deglutition for healthy participants correlated significantly with those of some dysphagia patients. This suggests that the easiness of deglutition felt by dysphagia patients might be accurately predicted from sEMG measurements in healthy participants. Further experimentation with sufficient numbers of patients is needed in the future.

We conclude that sEMG measurements can be used for predictive modeling of deglutition ease in dysphagia diets. The procedure can be performed in a shorter time and with a smaller number of participants than are needed for the sensory evaluation, and can contribute to the evaluation and development of dysphagia diets.

Acknowledgements

This work was supported by a Grant-in-Aid for Scientific Research 20259013 to T.A. from the Ministry of Education, Culture, Sports, Science and Technology of Japan. This work was supported by the Council for Science, Technology and Innovation (CSTI), Cross-Ministerial Strategic Innovation Promotion Program (SIP), "Technologies for creating next-generation agriculture, forestry and fisheries".

References

1. National Dysphagia Diet Task Force (2002) National dysphagia diet: Standardization for optimal care. American Dietetic Association.

2. Newman J, Egan T, Harbourne N, O'Riordan D, Jacquier JC, et al. (2014) Correlation of sensory bitterness in dairy protein hydrolysates: Comparison of prediction models built using sensory, chromatographic and electronic tongue data. Talanta 126: 46-53.

3. Sipos L, Gere A, Szöllősi D, Kovács Z, Kókai Z, et al. (2013) Sensory evaluation and electronic tongue for sensing flavored mineral water taste attributes. J Food Sci 78: S1602-S1608.

4. Vaiman M, Eviatar E, Segal S (2004) Evaluation of normal deglutition with the help of rectified surface electromyography records. Dysphagia 19: 125-132.

5. Miura Y, Morita Y, Koizumi H, Shingai T (2009) Effects of taste solutions, carbonation, and cold stimulus on the power frequency content of swallowing submental surface electromyography. Chem Sens 34: 325-331.

6. Hayakawa F, Kazami Y, Nishinari K, Ioku K, Akuzawa S, et al. (2013) Classification of Japanese texture terms. J Tex Studi 44: 140-159.

7. Shiga K, Yamamoto S, Nakajima A, Kodama Y, Imamura M, et al. (2014) Metabolic profiling approach to explore compounds related to the umami intensity of soy sauce. J Agric Food Chem 62: 7317-7322.

8. Cashaback JGA, Cluff T, Potvin JR (2013) Muscle fatigue and contraction intensity modulates the complexity of surface electromyography. J Electro Kinesiol 23: 78-83.

9. Dolan P, Mannion AF, Adams MA (1995) Fatigue of the erector spinae muscles: A quantitative assessment using "frequency banding" of the surface electromyography signal. Spine 20: 149-159.

10. Maton B, Rendell J, Gentil M, Gay T (1992) Masticatory muscle fatigue: endurance times and spectral changes in the electromyogram during the production of sustained bite forces. Archiv Oral Biol 37: 521-529.

11. Sadoyama T, Miyano H (1981) Frequency analysis of surface EMG to evaluation of muscle fatigue. Eur J Appl Physiol Occup Physiol 47: 239-246.

Texture terms	Japanese term	Texture terms	Japanese term
Rough, coarse	*Arai*	High water content	*Mizuke ga ōi*
Loose, primarily for foods consisting of many particles	*Barabara*	Juicy	*Mizumizushii*
Dry and rough, stale, implies the material is not compact	*Basabasa*	Watery	*Mizuppoi*
Sticky, viscous and watery; implies unpleasantness; the area where the material adheres is bigger than *Bechobecho/Bechot*	*Bechabecha*	Sticky, elastic and chewy	*Mocchiri*
Sticky, viscous and watery; implies unpleasantness; the amount of water is slightly larger than *Bechabecha/ Bechari/Bechot*	*Bechobecho*	Dry and crumbly, implying a lack of moisture	*Mosomoso*
Sticky, the area where the material adheres is bigger than *Betobeto/Betot/Betotsuku/Bettori*	*Betabeta*	Thick and viscous, resistant to flow	*Mottari*
Sticky, implies a little bit of unpleasantness	*Betobeto*	Smooth	*Nameraka*
Crumbly, not compact and easily crumbled	*Boroboro*	Sticky, viscous and spinnable	*Nebaneba*
Dry, crumbly and not compact; implies staleness	*Bosoboso*	Sticky and viscous, implies wateriness; the area where the material adheres is bigger than *Nechonecho*	*Nechanecha*
Slippery, smooth and wet surface. Sounds and feels like when long thin noodles are slurped	*Churuchuru*	Sticky and viscous, implies wateriness	*Nechonecho*
Cream like	*Cream-jō no*	Pasty, glue-like	*Norijō no*
Creamy	*Creamy*	Slimy, surface covered with slime or mucus	*Numeri ga aru*
Elastic, springy	*Danryoku ga aru*	Slimy and slippery	*Nurunuru*
Thick, heavier than *Taratara*	*Daradara*	Loose, primarily for granular foods or many tiny particles; the particles are smaller and lighter than *Barabara*	*Parapara*
Thick and viscous	*Dorodoro*	Dry, stale	*Pasapasa*
Liquid	*Ekijyō no*	Thick, resistant to flow	*Potteri*
Soft and limp, flexible	*Funyafunya*	Soft elastic and resilient	*Purin*
Mushy; having lost its original shape through cooking, mixing, or mashing	*Guchagucha*	Soft elastic and slightly wobbly	*Purupuru*
Mushy; soft and watery; having lost its original shape through cooking, mixing, or mashing	*Guchogucho*	Light, thin, a state or behavior like flowing powders or thin liquids	*Sarasara*
Crumbly and soft	*Horohoro*	Fibrous	*Sen'ijō no*
Jelly-like	*Jelly-jō no*	Firm, solid	*Shikkari*
Hard, firm, stiff, tough, rigid	*Katai*	Juicy, broth not completely soaked in	*Shiruke ga ōi*
Lumpy, chunky	*Katamarijō no*	Residual feeling on the tongue; the material is viscous, sticky or piquant	*Shita ni nokoru*
Porridge-like	*Kayujō no*	Smooth and slippery	*Suberu*
Smooth, fine	*Kimekomakai*	Fluid, dripping	*Taratara*
Homogeneous	*Kin'itsuna*	Melting	*Torokeru*
Powdery	*Konappoi*	Slightly viscous	*Toromi ga aru*
Feels smooth in the mouth	*Kuchiatari ga yoi*	Smooth and viscous	*Torotoro*
Sticky	*Kuttsuku*	Beady, grainy	*Tsubujō no*
Mellow and soft	*Maroyaka*	Easy to crush or mash	*Tsubureyasui*
Sticky, difficult to remove material that adheres to eating utensils and teeth	*Matowaritsuku*	Beady, grainy, smaller and harder than *Tsubujō no*	*Tsubutsubu*
Thick, viscous and creamy; primarily for cream-like foods	*Mattari*	Smooth surface, slippery	*Tsurutsuru*
Honey-like	*Mitsujō no*	Soft, tender	*Yawarakai*
Dense	*Mitsuna*	Thin, loose, easy to deform	*Yurui*

S1 Table: Textures and their Japanese terms: 68 textures related to the easiness of deglutition extracted from 271 textures [6].

Sample	Classification (Japan Care Food Conference)	Sample	Classification (Japan Care Food Conference)
Strawberry-flavored jelly with pulp	3	Stewed soybean paste	4
Stewed sardines and plum mousse	3	Simmered cod dumplings	1
Sweetened green pea paste	4	Pineapple-flavored jelly	Unknown
Broiled eel in egg soup	3	Fried rice with chicken and ketchup	2
Pineapple-flavored pudding	3	Tuna-and-vegetable paste	4
Creamed shrimp and scallop	2	Corn soup with kernels	Unknown
Rice gruel with vegetables and seaweed	1	Tofu and eggs in starchy sause	2
Rice porridge with salmon	3	Chicken-and-vegetables paste	4
Macaroni and scallop grati	1	Simmered meat and potatoes	3
Rice porridge with crab	3	Meatballs in onion soup	1
Stewed pumpkin mousse	3	Stewed meatloaf	1
Stewed flatfish and Japanese radish	3	Carrot soup	Unknown
Simmered beef and burdock	2	Meatloaf in brown sauce	3
Pumpkin gratin	2	Bread mousse	3
Japanese wheat noodles with vegetables and tofu	2	Vegetable and seaweed mousse	3
Stewed freeze-dried tofu mousse	3	Pumpkin boiled in sugary syrup	3
Rice porridge, wheat, millet, beans and barnyard millet	Unknown	Broccoli mousse	3
Salmon and vegetables in egg sou[5]	2	Strained pork-curry stew	4
Salmon mousse	3	Tofu and minced meat in spicy sauce	2
Shiso (Japanese basil) and seaweed mousse	3	Purple sweet potato mousse	3
Rice porridge	Unknown	Melon-flavored jelly	Unknown
Thin beef strips cooked with vegetables, tofu, and eggs	2	Corn paste	4
Pineapple and apple sause	4	Carrot paste	4
Steamed rice with red beans	3	Steamed rice	3
Grape-flavored jelly	4	Yuzu (Japanese citron) flavored jelly	4
Pear-flavored jelly	Unknown	Apple-flavored jelly	4
Japanese radish with minced chicken and starchy sauce	3	Chestnut-flavored mousse	3

S2 Table: Samples and classifications: 54 commercially available food samples. Classifications are based on the Japanese Care Food Conference. All samples were purchased from food companies in Japan.

7

Drying Kinetics of Banana Peel

Sravan Kumar K*

Department of Food Engineering and Technology, Sant Longowal Institute of Engineering and Technology, Longowal, India

Abstract

Drying of blanched banana peel was carried out in tray dryer. The drying experiments were performed at two different temperatures and a constant air velocity of 0.5 m/s. To select a best model, five different thin layer drying models were fitted to experimental data. Nonlinear regression procedure was used to fit five different models. The models were compared with experimental data of banana peel drying at air temperature of 60°C and 70°C. The best thin layer drying model was selected using the coefficient of determination (R^2), chi-square (X^2) and Root Mean Square Error (RMSE). The highest value of R^2 (0.99640, 0.99652), the lowest of chi-square(X^2) (0.000218, 0.000231) and RMSE (0.014778, 0.015177) at temperature of 60°C and 70°C indicated that the Logarithmic and Henderson and Pabis model is the best mathematical model to describe the drying behaviour of banana peel.

Keywords: Drying kinetics; Banana peel; Mathematical modelling; Tray dryer

Introduction

Bananas (Musa sp.) are one of the most important tropical fruits consumed worldwide by people of all age groups. The nutritional and functional properties of bananas are known to provide good health. Nutritionally, bananas contain available carbohydrates which provide energy, vitamins B and C, and significant amounts of potassium and magnesium and amino acids. The peels of banana also contain carbohydrates, proteins, and fiber in significant amounts, making it an ideal substrate for production of value added products [1,2]. Assuming that banana peels account for only 30% of the total fruit weight and have 20% dry matter, it contains minerals and various amino acids like aspartic acid, threonine, serine, glutamicacid, proline, glycine, alanine, cystine, valine, methionine, isoleucine, leucine, phenylalanine, lysine and arginine. To the best of our knowledge no past research was conducted to investigate the drying kinetics of banana peel in hot air drying [3-5]. In the view of the health promoting properties and high nutritional benefits of banana peel, the present study was carried out to observe the effect of different temperatures on drying characteristics of banana peel and to select the best mathematical model to observe the drying behavior of the banana peel.

Materials and Methods

Drying experiment

The peels of ripened bananas were collected from a vendor of local market. The peels was cut into the slices and blanched. The drying experiments carried out in a tray dryer at selected two air drying temperatures (60°C and 70°C) to obtain the drying characteristics of banana peel. Air velocity was set at 0.5 m/s. When the dryer reached steady state conditions (i.e. when desired temperature was reached) the sliced banana peels were distributed onto the trays. Moisture loss was recorded at every 15 min intervals during drying [1,6]. Drying was continued until the moisture content of the sample reached to the equilibrium moisture content about 13-15% (wb). The drying was continued till weight became constant and experiments were conducted in duplicates.

Mathematical modeling of drying curves

The moisture ratio (MR) of banana peel during drying experiments was calculated using the following Equation

$$MR = (M - M_e) / (M_0 - M_e)$$

Where M, Mo, and M_e are moisture content at any drying time, initial and equilibrium moisture content (kg water/kg dry matter), respectively.

Data obtained from the measurements of weight in a test was used for the analysis of drying kinetics of materials need to be changed first in the form of moisture content data. The moisture content was expressed as a percentage wet basis and then converted to moisture ratio [7,8]. The experimental drying data for banana peel were fitted to the exponential model thin layer drying models as shown in Table 1 by using non-linear regression analysis.

The regression analysis was performed using the STATISTICA computer program. In Non-linear regression, the parameters used to evaluate goodness of fit of the mathematical models to the experimental data are coefficient of determination (R^2) and the reduced chi-square (X^2) was used for data analysis. The higher value for R^2 and the lower values for X^2 and root mean square error analysis (RMSE) indicate the better fitness of model [9-12]. These parameters were calculated as follows:

$$R^2 = 1 - \left[\frac{\sum_{i-1}^{N} (MR_{pre,i} - MR_{exp,i})^2}{\sum_{i-1}^{N} (\overline{MR}_{pre} - MR_{exp,i})^2} \right]$$

Where, n = no. of unknown and N= Data point measured

$$Chi\ Square = \chi^2 = \sum_{i=1}^{N} \left[\frac{\left(MR_{Experimental\ Value} - MR_{predicted\ value} \right)^2}{(N - n)} \right]$$

The mean relative deviation E (%) is an absolute value that was used

*Corresonding author: Sravan Kumar K, Department of Food Engineering and Technology, Sant Longowal Institute of Engineering and Technology, Longowal, India, E-mail: sravankumark39@yahoo.com

S.NO	Model Name	Type	Reference
1	Page	MR=exp(-kt^n)	Diamante and Munro [3]
2	Newton	MR=exp(-kt)	Mujumdar [7]
3	Wang and Singh	MR=1+at+bt^2	Wang and singh [10]
4	Logarithamic	MR=a exp(-kt) +c	Yagcioglu et al. [11]
5	Henderson and Pabis	MR=a exp(-kt)	Zhang and Litchfleld [12]

Table 1: List of models with references.

Model	Temperature °C	Coefficient	Coefficient Of Determination (R^2)	Chi-Square ($\mathcal{X}^2$)	RSME	E%
Newton	60	K=0.002258	0.98510	0.000905	0.238748	7.727946
	70	K=0.002427	0.98289	0.001133	0.033654	5.213993
Page	60	K=0.002494 n=0.983696	0.98518	0.009021	0.03002	7.262423
	70	K=0.000929 n=1.161446	0.99268	0.000484	0.022007	3.371934
Logarithmic	**60**	K=0.003234 a=0.981314 c=0.113281	**0.99640**	**0.000218**	**0.014778**	2.795789
	70	K=0.002713 a=1.091012 c=0.002017	0.99652	0.000231	0.015179	2.497142
Wang and Singh	60	a=0.002006 b=0.001001	0.98817	0.000718	0.026801	7.080818
	70	a=0.002047 b=0.000012	0.99004	0.000659	0.025673	3.136054
Henderson and pabis	60	K=0.002347 a=1.033488	0.98680	0.000802	0.028318	8.48627
	70	K=0.002703 a=1.092349	**0.99652**	**0.000231**	**0.015177**	2.519419

Table 2: The fitness of different models at different temperatures

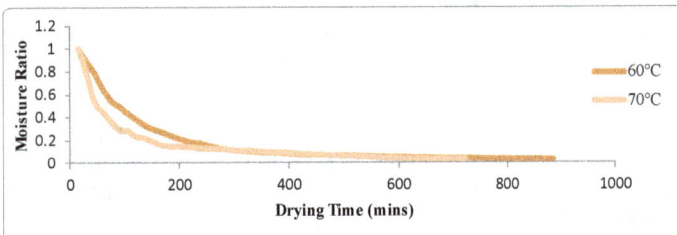

Figure 1: Moisture ratio versus drying time of banana peel at different temperatures.

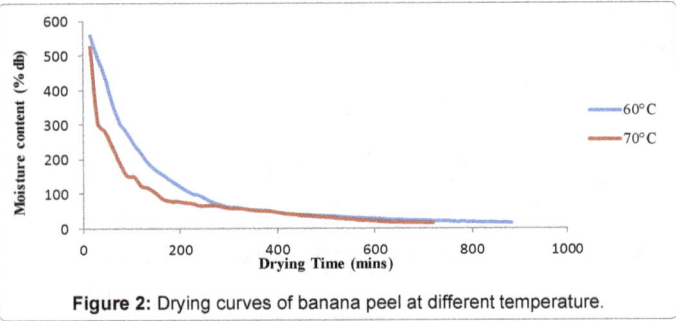

Figure 2: Drying curves of banana peel at different temperature.

because it gives a clear idea of the mean divergence of the estimated data from the measured data.

$$RMSE = \text{Root mean square error} = \sum_{i=1}^{N} \sqrt{\frac{\left(MR_{Experimental\ value} - MR_{predicted\ value} \right)^2}{N}}$$

The values of E less than 5.0 indicate an excellent fit, while values greater than 10 are indicative of a poor fit.

Results and Discussion

The drying process was stopped no change in the weight of sample of banana peel was observed. At this point moisture content was decreased. Moisture content data was converted to moisture ratio then fitted to the five thin layer drying models. Table 2 showing the results of fitting the experimental data to the thin layer drying models with best fitting model in bold type for temperature of 60°C and 70°C. The drying rate decreased continuously throughout the drying period. The constant rate period was absent and the drying process of banana peel took place in falling rate period. The moisture ratio reduced exponentially as the drying time increased [4]. Continuous decrease in moisture ratio indicates that diffusion has governed the internal mass transfer. A higher drying air temperature decreased the moisture ratio faster due to the increase in air heat supply rate to banana peel and the acceleration of moisture migration [2].

From Table 2 it can be concluded that the best models for the temperature 60 and 70°C are Logarithmic and Henderson and Pabis with 0.99640, 0.000218, 0.014778 and0.99652, 0.000231, 0.015177 values for R2, and RMSE respectively.

It can be seen that different initial moisture contents resulted in longer drying times for samples dried at 60 and 70°C. Greater initial drying rate for the banana with the higher moisture content. Thus, higher initial moisture content in food resulted in a higher drying rate as expected. This causes rapid decrease in moisture content of the banana peel. High temperature accelerated the evaporation of moisture near the surface better than low temperature thus drying time could be reduced. The results of drying at 60°C agreed with the reports of Sabarez, Price and co-workers (1997) for drying plum of different initial moisture contents at 70°C. Thus, a higher drying air temperature produced a higher drying rate and consequently the moisture ratio decreased (Figures 1 and 2).

Conclusion

Drying using a hot-air chamber was tested on samples of banana peel. Drying kinetics curves of drying banana peel demonstrated that drying at 60°C and 70°C were the optimum values for drying banana peel with the appropriate equations using the logarithmic model and Henderson and pabis model with R^2 of 0.9964 and R^2 of 0.9965. According to the results which showed the highest average values of R^2 and the lowest average values of chi-square and RMSE, it can be stated that the logarithmic model and Henderson and pabis model was best fitted model to describe the drying characteristics of the samples.

References

1. Aghbashlo M, Kianmehr MH, Samimi Akhljahani H (2008) Influence of drying conditions on the effective moisture diffusivity energy of activation and energy consumption during the thin-layer drying of barberries fruit (Berberidaceae) Energy Conversion and Management 49: 2865-2871.

2. Demir V, Gunhan T, Yagcioglu AK, Degirmencioglu A (2004) Mathematical modeling and the determination of some quality parameters of air-dried bay leaves. Bio systems Engineering 88: 325-335.

3. Diamante LM, Munro PA (1993) Mathematical modeling of the thin layer solar drying of sweet potato slices. Solar Energy 51: 271-276.

4. Doymaz I (2007) Air-drying characteristics of tomatoes. Journal of Food Engineering 78: 1291-1297.

5. Doymaz I (2004) Convective air drying characteristics of thin layer carrots. Journal of Food Engineering 61: 359-364.

6. Henderson SM, Pabis S (1961) Grain drying theory. II: temperature effects on drying coefficients. Journal of Agricultural Engineering Research 6: 169-174.

7. Mujumdar AS (1987) Handbook of industrial drying. New York: Marcel Dekker.

8. Sabarez H, Price WE, Back PJ, Woolf LA (1997) Modelling the kinetics of drying of DAgen Plums (Prunus Domestica) Food Chemistry 60: 371-382.

9. Sarsavadia PN, Sawhney RL, Pangavhane DR, Singh SP (1999) Drying behaviour of brined onion slices. Journal of Food Engineering 40: 219-226.

10. Wang C, Singh RP (1978) A single layer drying equation for rough rice. ASAE, St. Joseph MI.

11. Yagcioglu A (1999) Drying technique of agricultural products. Ege University Faculty of Agriculture Publications, Bornova Izmir.

12. Zhang Q, Litchfleld JB (1991) An optimization of intermittent corn drying in a laboratory scale thin layer dryer. Drying Technology 9: 383-395.

8

Cooking Treatment Effects on Sugar Profile and Sweetness of Eleven-Released Sweet Potato Varieties

Owusu-Mensah E[1,2]*, Oduro I[1], Ellis WO[1] and Carey EE[2]

[1]Food Science and Technology Department, College of Science, Kwame Nkrumah University of Science and Technology, Kumasi, Ghana
[2]International Potato Centre (CIP), Kumasi, Ghana

Abstract

Cooking can significantly alter sugar content of sweet potato roots. Sweet potato roots were processed using three different cooking treatments, with the aim of investigating the effects of these methods on sugar profile and sweetness levels. Significant contribution of the cooking treatment and genotype, and their interaction on levels of the sugars were also determined. Moreover, sugar values were converted to relative sweetness per sucrose equivalent. The results revealed that cooking treatment produced the highest effect on sugar except fructose. Variability due to the interactions was significant and ranged from 2.60% to 11.74%. Whilst sucrose was the predominant in the raw form, maltose increased dramatically during cooking. Sweetness level increased substantially upon cooking and was highly dependent on initial sugar content, amylase activity and cooking treatment. Thus, evaluation of sweetness levels in sweet potato clones should not only be on the uncooked samples but should take into account the cooking methods employed.

Keywords: Cooking treatments; Sugar profile; Sweetness level; Amylase activity; Maltose

Introduction

Sweetness, derived from sugars in the raw sweet potato root and maltose formed during cooking, is the predominant attribute controlling the taste of cooked sweet potato products [1,2]. The level of sweetness in the root determines the type of product or formulation that can be developed. A number of factors including maturity period, storage, amylase potential, curing and baking treatment significantly influence sweetness/sugar content of sweet potato roots [3-5]. Baking treatment and the amylolytic potential nonetheless have the greatest effect on sugar content of the final product [6-8] Baking generally increases sugar content of sweet potato roots [9,10]. Increase in sugar content during baking can be dramatic, leading to a very sweet product [9]. Though effect of baking treatment on sugars of sweet potato roots has been extensively investigated, limited data is available on other cooking treatment such as steaming and microwaving. Nevertheless, sweet potato roots are cooked by different treatments including microwaving; baking, steaming and boiling prior to consumption with the aim to increase the culinary properties and enhance digestibility [11] Temperature, time and mode of heat transfer differentiate these cooking methods. Conventional baking usually lasts for 60-90 min at 180-220°C, depending on the genotype and tuber size [9]. Baking temperature as reported by Simkovic [12] and Chan [6] can however cause sucrose caramelisation, a phenomenon, which results in conversion of sucrose to oligomers and polymers. Microwave cooking employs a high temperature, short time heating mechanism to cook food products [10]. Heat is transferred by convection and conduction during baking whilst electromagnetic waves penetrate food materials causing agitation and friction to produce heat for cooking during microwaving [5]. The effect of steaming on quality characteristics of sweet potato root has not been widely reported.

Although effects of some cooking methods, especially baking, on quality attributes of sweet potatoes have been evaluated comparative studies with the view of understanding the effects of different cooking treatments on sugar profiles, sweetness and utilisation of sweet potatoes are limited. Moreover the influence of cooking treatments on sugars of eleven officially released sweet potato varieties in Ghana has not been investigated. To better understand the contribution of different cooking methods on sugar formation and sweetness of sweet potato roots, individual sugar and sweetness levels of eleven released varieties were determined following baking, microwaving and steaming.

Methodology

Experimental design

Triplicates of eleven sweet potato varieties released by Council for Scientific and Industrial Research (CSIR) – Crops Research Institute (CRI) were planted in a randomized complete block design on May 2014 at the CSIR-CRI experimental station, Fumesua, Ghana [13-15]. Harvesting was done four months after planting (September, 2014) and each plot was treated as a separate sample during laboratory evaluations. Harvested roots were stored for a week at room condition (25 to 30°C) prior to processing.

Sample preparation

Four medium-size intact roots of each variety were washed with clean water, rinsed and air-dried. The clean roots were then quartered, rinsed with de-ionised water and dried using paper towels. Each quarter was sliced across its longitudinal axis to approximately 1.0 cm thickness and composite samples from each plot, divided into four groups of 50 g. One group was designated as raw and the rest were subjected to three different processing methods; baking, steaming and microwaving. For baking, one group of the sliced samples was wrapped in aluminium foil and placed in a forced air oven (Genlab MINI/50/DKG), which has been preheated to 205°C, for 30 mins. For steaming, another group of

*Corresponding author: Owusu-Mensah E, Food Science and Technology Department, College of Science, Kwame Nkrumah University of Science and Technology, Kumasi, Ghana, E-mail: e.owusu@cgiar.org

root samples was placed in a Kitchen steamer with boiling water and cooked for 10 min. The third group of the root samples was wrapped in paper towel and moistened with about 5 mL of portable water and microwaved (sharp microwave model R-228H) for 5 min inside a plastic microwaveable food container. Cooked samples were allowed to cool to room temperature for about 20 min, transferred to whirl-Pak polyethylene bags and frozen at –20°C before drying using the freeze dryer (True Ten, Ind, YK18-50, Taiwan). Dried samples were milled and sieved as described in chapter four (under methodology) prior to sugars determination.

Sugar determination

Freeze-dried and milled sweet potato samples were sent to the Quality Plant Product Laboratory (Department of Crop Science, University of Gottingen, Germany) for sugar analysis. Water extract of the freeze-dried sweet potato samples (0.1 g in 100 mL) was used. The samples were incubated in a water bath at 60°C for 1 h and treated with 0.2 mL Carrez I and Carrez II solution to remove proteins. Samples were purified by centrifugation (Sorvall RC-5B Refrigerated Superspeed, GMI, Ramsay, USA) at 10,000 rpm for 10 min at 20°C. Sugars were determined from the membrane-filtered supernatant (pores size 0.45 μm). Glucose, fructose, sucrose, and maltose were separated using a LiChrospher 100 NH$_2$ (5 μm) 4 x 4 mm pre-column in combination with a LiChrospher 100 NH$_2$ (5 μm) 4 x 250 mm separation column (Merck KGaA, Darmstadt, Germany) and an acetonitrile: pure water solution (80:20 v/v) as mobile phase at a flow rate of 1.0 mL min^{-1} at 20°C and an injection volume of 20 μL. Sugars were detected with a Knauer differential refractometer 198.00 (Knauer, Berlin, Germany).

Determination of amylase activity

The 3,5-dinitrosalicyclic acid (DNSA) method for reducing sugars was employed to determine the total amylase activity of the freeze-dry sweet potato roots [16,17].

A unit (U) of amylase activity was defined as the amount of enzymes required to release reducing sugars equivalent to one μmole of maltose/min under the above stated conditions [16].

Calculation of sweetness level

In order to ascertain and compare sweetness levels among the varieties, sweetness (sucrose equivalent) was calculated from the equation: Sucrose Equivalent (SE) = 1.2 fructose + 1 sucrose + 0.64 glucose + 0.43 maltose [1,18]. Based on the SE values obtained, the varieties were classified into four categories: non sweet (SE ≤ 12 g/100g dry weight); low sweet (SE 13 – 20 g/100 g); moderate sweet (SE 21 – 28 g/100 g); and high sweet (SE29 – 37 g/100g) [1].

Statistical analysis

Experimental means were calculated from triplicate values of each variety per treatment. Data obtained were subjected to analysis of variance using Statistical Analysis System (SAS) [19]. Significant differences among means were assessed using Least Significant Difference (LSD) at probability level of 5%.

Results and Discussion

Effect of cooking treatment, genotype and interaction on sugars of cooked sweet potato roots

The effect of cooking, genotype and their interaction were significant on all sugars (maltose, sucrose, glucose and fructose), though the percentage contributions varied considerably (Tables 1 and 2).

Variety	Skin Colour	Skin Shape	Flesh colour	Yield (t/ha)
Apomuden	Reddish brown	Obovate	Reddish orange	48.9
Bohye	Purple	Obovate	Pale orange	16.8
Dadanyuie	Dark purple	Round elliptic	White	10.5
Faara	Deep purple	Long elliptic	Cream	16.9
Hi-Starch	Creamy	Elliptic	Cream	14.7
Ligri	Cream	Round elliptic	Pale yellow	16.3
Okumkom	Cream	Long elliptic	Cream yellow	19.91
Ogyefo	Purple	Long elliptic	White	25.9
Otoo	Cream	Long elliptic	Light orange	30.7
Patron	Dark yellow	Long elliptic	Dark yellow	15.9
Sauti	Cream	Long elliptic	Yellow	15.4

Table 1: Phenotypic attributes and yield of the sweet potato varieties used for assessment of changes in sugar content [3-5].

Source of Variation	*Variance (%)			
	Maltose	Sucrose	Glucose	Fructose
Genotype (G)	7.26"	16.93**	38.82**	45.68**
Cooking treatment (CT)	90.12**	79.04**	52.60**	43.12**
GxCT	2.60**	4.03**	8.65**	11.47**
**Significant at p < 0.05. *Calculated from sum of squares.				

Table 2: Percentage variability of cooking treatment, genotype, and interactions on sugars of cooked sweetpotato roots.

Cooking treatment showed the highest effect of the total variability on the sugars except fructose. The effect was more profound on maltose content with percentage variability of 90.12%. Nearly 80% and 53% of the total variation in sucrose and glucose contents of the cooked roots were due to the cooking treatment. Effect of genotype was highest on fructose relative to the other sugars. While 45.68% of the variation in fructose resulted from the genotypic composition of the roots, only 7.26% of the difference in maltose content was due to genotypic effect. Percentage variability resulting from genotypic effect on sucrose and glucose was 16.93% and 38.82% respectively. Overall variation from interactions between cooking treatment and genotype ranged from 2.60% to 11.47% of the entire differences noticed. Although it was significant, it contributed the least of the total variation.

The results from the analysis of variation depict that changes in sugar concentrations during cooking are significantly dependent on cooking treatment, genotype and interaction. Among these factors cooking treatment exerted the highest effect. Its effect was more profound on maltose content, which increased from 7.26% prior to cooking to 90.12% afterward. Cooking increases temperature intensity and penetration, and also facilitates breakdown of hydrolytic bonds holding starch granules and other compounds. Such conditions enhance the activity of native amylase resulting in starch degradation and the production of sugars mainly maltose as observed in the study [8,20]. Apart from fructose, changes in individual sugars were remarkable. Response from fructose was higher for genotype effect rather than cooking treatment.

Effect of cooking treatment on sugars of sweet potato roots

Table 3 shows the means and ranges in sugars as a result of the different cooking treatments. Wide variation existed among the sugars of the cooked sweet potato roots, with maltose and sucrose showing the highest variability. Maltose was hardly present in the raw form whilst sucrose (10.58%) predominated. This finding agrees with Morrison [8] and Sun [10] who reported that sucrose is the major sugar in raw forms and the most important sugar for predicting sweetness in sweet potatoes [6]. Sucrose concentration, generally, increased slightly

Individual Sugars (% DM)	Cooking Treatment			
	Raw	Baking	Microwaving	Steaming
Sucrose	10.58 (9-23)ᵃ	11.01 (6-20)ᵃ	10.72 (7-16)ᵃ	4.30 (0-8)ᵇ
Glucose	2.69 (1-4)ᵃ	1.10 (0-3)ᵇ	1.63 (0.4-5)ᵇ	1.55 (0-5)ᵇ
Fructose	1.58 (0-3)ᵃ	0.84 (0-2)ᵃ	0.92 (0-2)ᵃ	0.95 (0-4)ᵃ
Maltose	0.63 (0-1)ᵃ	20.13 (5-36)ᵇ	5.07 (2-15)ᶜ	14.35 (2-27)ᵈ

Ranges of maeans are presented in brackets. a,b,c Figures in rows with the same superscripts are not significantly different (p < 0.05).

Table 3: Means and ranges of individual sugars in raw and cooked sweet potato roots.

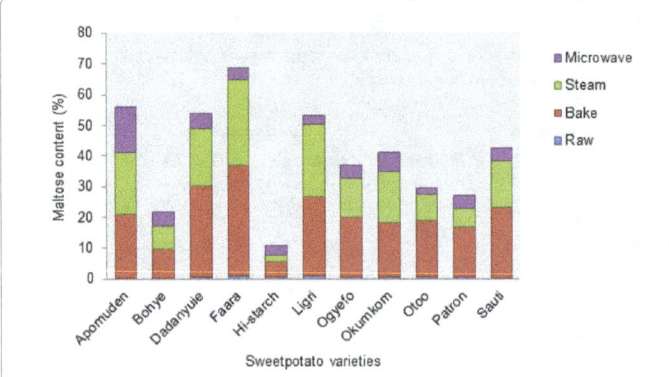

Figure 1: Changes in maltose content of sweetpotato roots as affected by different cooking treatments; microwaving, steaming and baking. LSD=0.30.

when baked, though it was not significant compared to the raw, but remained constant at microwaving and decreased significantly during steaming. Glucose and fructose contents were not significantly affected by the different cooking treatments, although the levels were generally lower compared to raw roots. Maltose content rose from 0.63% before cooking to 20.13%, 14.35% and 5.07% after baking, steaming and microwaving respectively. It became the principal sugar following baking and steaming. Increase in maltose content following cooking has been observed in several sweet potato varieties [7,8,10]. Changes in maltose and sucrose (the major sugars) concentrations per variety during cooking were also assessed and results presented in Figures 1 and 2 respectively. Maltose, which was not detected in most of the varieties prior to cooking increased dramatically after baking and steaming (Figure 1). Faara, Dadanyuie, Ligri Sauti and Apomuden had the highest increase and Hi-Starch the lowest in maltose content following baking and steaming. Though the effect of microwave cooking was also positive and significant on maltose content for all the varieties, it was comparatively much lower to both baking and steaming. In contrast, sucrose content decreased in some of the varieties while increasing slightly or remaining the same in others during cooking (Figure 2). Apomuden, Dadanyuie, and Hi-starch recorded a decrease whilst Bohye, Faara, Otoo, Sauti and Ligri showed an increase after baking. Sucrose contents in Ogyefo, Okumkom and Patron were not significantly affected by baking treatment. Steaming reduced sucrose content in all the varieties. The magnitude of reduction was extremely high in Faara, which lost almost 96% of its sucrose content. Effect of microwave treatment on sucrose was similar to that of baking. While negatively affecting sucrose content in Apomuden, Bohye, Hi-Starch, Ligri, Patron, and Sauti, microwaving enhanced sucrose levels in Dadanyuie, Faara, and Otoo. Sucrose content in Ogyefo, and Okumkom were not significantly affected.

Concentration of sugars in sweet potato roots varies significantly

during cooking, with the extent of variability being highly dependent on; 1) initial sugar concentration, 2) amylase activity and 3) cooking method employed. The impact of cooking treatment on sugar content is related to temperature, time, and mode of heat transfer. Baking treatment resulted in the highest sugar (maltose) formation mainly due to the long cooking period (30 min) coupled with the high temperature (205°C) employed. Moreover there was no direct contact between the sample and the heating medium, a system that prevented possible leaching of soluble sugars, during baking. Heat is transferred from the periphery to the centre of the root by conduction in baking as compared to microwaving for instance where electromagnetic radiation penetrates the entire root causing agitation and friction to produce heat for cooking instantaneously [5]. Hence baking utilises more time, a system that allows adequate starch gelatinisation and subsequent conversion to maltose by amylases [21,22]. It has been demonstrated that increasing heating temperature over a time frame increases starch degradation and maltose production [8,10]. Baking treatment at higher temperatures can however cause sucrose caramelisation, a phenomenon, which results in conversion of sucrose to oligomers and polymers as reported by Simkovic [12] and Chan [6]. Hence the reduction in sucrose content of some of the varieties (Figure 2) may be attributed to this effect. This finding corresponds with Chan [6] and Morrison [8] who reported a decrease in sucrose content of several sweet potato cultivars during baking. The rapid heating mechanism of microwaving deactivated the native amylases responsible for maltose formation, and consequently the reduction in

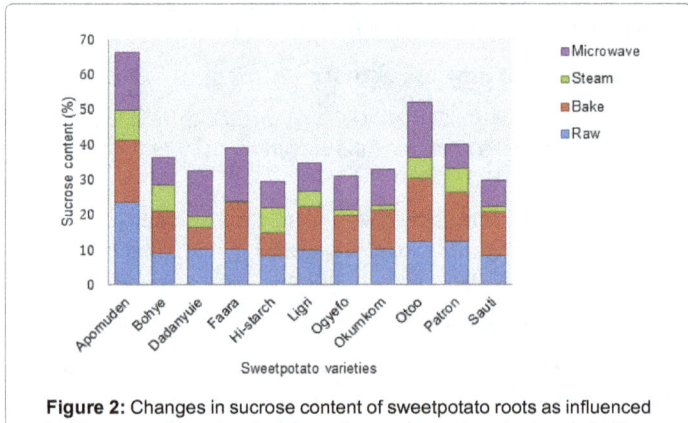

Figure 2: Changes in sucrose content of sweetpotato roots as influenced by three cooking treatments; microwaving, steaming and baking. LSD=0.86.

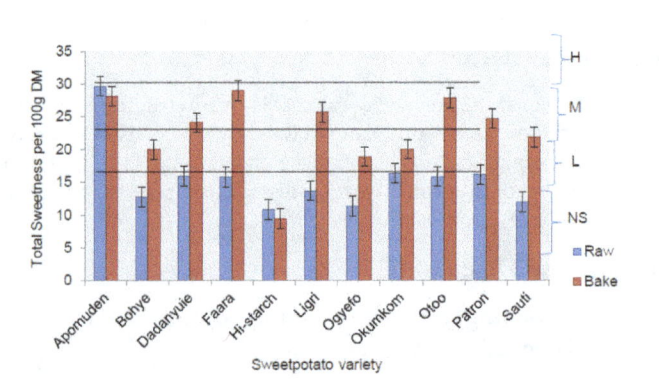

Figure 3: Changes in sweetness levels of sweet potato roots after baking. Standard error bars represent LSD at p<0.05.

its levels [6,10]. Moreover, the short heating period of microwaving does not enhance starch gelatinisation, a rate-determining step in initial stages of hydrolysis [7,21]. Whereas baking resulted in a dramatic increase in maltose content of Jewel, microwaving inhibited its formation, reducing the total sugar content of the cooked product [10]. Microwave cooking can therefore be an ideal method for food preparations where high sugar content is not a desirable attribute. In regions like Sub-Sahara Africa where less sweet potato varieties are perceived to be the preferred choice Tumwegamire [23], microwave cooking could be the recommended choice.

Steaming treatment resulted in an increase in maltose content in all the varieties. On the contrary, it caused a reduction in sucrose content in all the varieties compared to the raw roots. The heat transfer mechanism of steaming treatment allowed direct contact between the roots and the heat source. Such heat exchange technique allows movement of soluble substances; where solutes move from high concentration to low concentration. Sucrose, which was initially high in the raw roots, may have consequently moved from the roots to the steam. Hence the reduction in sucrose content observed in the roots after steaming.

Increase in sugars, particularly maltose, levels of sweet potato root can also be attributed to the hydrolytic ability of native amylases present in the uncooked roots. Sweet potato roots contain high levels of amylases, mainly α- and β-amylase, which significantly influence levels of sugar in processed sweet potatoes [24]. Amylases hydrolyse gelatinised starch into maltose and short-chain branched oligosaccharides (limit dextrins) during cooking resulting in a sweet taste [8,22]. The amylase activity of the varieties was therefore determined to ascertain the general hypothesis that amylases are also responsible for the increase in sugar content.

Table 4 presents amylase activity of the sweet potato varieties investigated. It ranged from 927.14 U/g in Ligri to 387.06 U/g in Hi-starch. Based upon levels of activity found, Ligri, Dadanyuie, Sauti, Ogyefo and Okumkom were grouped as very high amylase varieties. Faara, and Otoo are high-class varieties whilst Patron, Apomuden, Bohye and Hi-Starch are considered moderate types. The level of amylase activity correlated positively with the formation of maltose after cooking (Figure 1). Most of the high amylase varieties including Dadanyuie, Ligri, and Faara of low initial total sugar content (Figure 1) showed very high increase in maltose content after baking and steaming. Similarly, Hi-starch with a lower amylase activity but similar initial sugar content as that of Ligri for instance produced little extra maltose, and was not significantly different from the uncooked roots. Apomuden with moderate amylase potential produced moderate maltose content though it had the highest content prior to cooking. This result supports previous findings that maltose content in cooked sweet potato is a function of amylase activity of the roots [7,8]. However, it should be noted that different cooking treatments produced significantly different effects on sugar content of the cooked roots (Figure 1). Baking treatment however results in the highest final sugar contents.

Baking treatment and sweetness of sweet potato roots

To study the effect of cooking treatment on sweetness levels of the varieties, baking treatment, which resulted in the highest increase in sugars, was selected. Individual sugars in raw and baked roots were first converted to sucrose equivalent (SE) based on sweetness factors [25]. Such conversion allows easy comparison of sweetness among sweet potato varieties. Kays [1] employed this method to evaluate the

sweetness levels of 272 baked sweet potato clones and categorised the clones into five main groupings based on SE: Very high ≥38; high 29-37; moderate 21-28; low 13-20 and non-sweet ≤ 12 g per 100 g dry mass.

Sweetness among the sweet potato varieties prior to and after baking is presented in Figure 3. The levels increased significantly after baking in majority of the varieties, and the effect was more pronounced in the high amylase types (Table 4); Faara, Ligri, Otoo and Sauti. The increase also corresponded well with the maltose content after baking (Figure 1). Apomuden had the highest sweetness value of 29.79 SE, and Hi-Starch the lowest of 10.79 SE prior to baking (Figure 3). The other varieties had values in the range of 12 to 16 SE. Using the grouping by Kays [1], the varieties fell under the following classes prior to baking: Apomuden–High sweet; Bohye, Dadanyuie, Faara, Ligri, Okumkom, Otoo, Patron and Sauti–Low sweet; and Hi-starch, Ogyefo and Sauti–non sweet. However the levels of sweetness and subsequently the sweetness categories of the varieties changed significantly following baking. Whereas Apomuden dropped slightly, but not significant, from high sweet category (29.79 SE) to moderate sweet (28 SE), majority of the varieties including Dadanyuie, Faara, Ligri, Otoo and Patron moved from low sweet to moderately sweet category. The increase in SE of Bohye and Okumkom were not significant enough to place them in the moderate class. Whilst Ogyefo and Sauti increased in SE values and were categorised as low and moderate sweet respectively, Hi-starch, remained in the same non-sweet category following baking [26,27].

Sweetness in sweet potatoes is a function of cultivar, amylase activity, storage condition, and cooking treatment [1,5,6]. Nonetheless, amylase activity, initial sugar concentration and maltose formed during cooking are the most critical in determining the final sweet sensation of cooked root [1,8]. These factors can completely change the sweetness status of a variety as observed in Dadanyuie, Faara, Ligri, Sauti, Otoo, Patron and Ogyefo (Figure 3) which were low or non-sweet prior to cooking, but changed to moderate sweet when baked.

The sweet potato varieties in this study were also classified into four general groups based on initial sucrose equivalent (SE) and starch hydrolytic potential [8]. These are low initial SE/low starch hydrolysis; Low initial SE/high starch hydrolysis; High initial SE/low starch hydrolysis and High initial SE/high starch hydrolysis. Figure 4

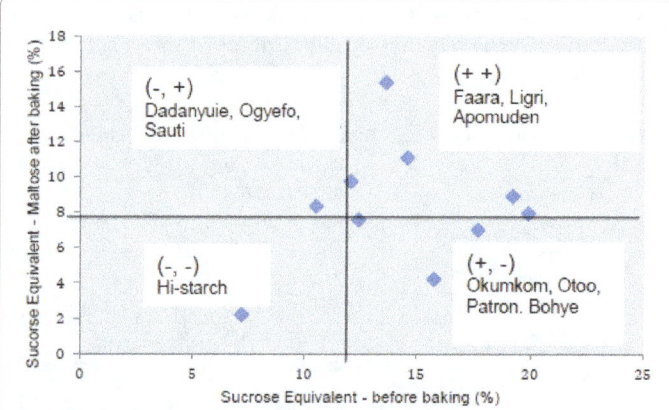

Figure 4: Classification of eleven sweetpotato varieties based on sucrose equivalent (SE) derived from starch hydrolysis (using maltose as indicator) during bakingand endogenous sugars (sucrose, glucose and fructose).

(-, -) – Low initial SE/low starch hydrolysis; (-, +) - Low initial SE/high starch hydrolysis; (+, -) – High initial SE/low starch hydrolysis; (+, +) – High initial sugar/High starch hydrolysis [11,14].

shows the classification of the sweet potato varieties assessed under this grouping.

Hi-starch was the only variety belonging to the class of low initial SE content coupled with low starch hydrolysis (-, -). It produced small amount of maltose upon cooking (Figure 1) as a result of its low amylase activity (Table 4). Natural inhibitors and starch-based structural resistance to hydrolysis are also probably inhibitory mechanisms for the low starch hydrolysis [8]. This lack of activity has been attributed to a recessive allele called β-amy for which the variety Satsumahikari was homozygous [8]. It is probable that Hi-Starch is the same variety since it was introduced to Ghana from Japan. Amylase activity in this variety was detected in vitro, but apparently was below the threshold required for effective hydrolysis during baking. Dadanyuie, Ogyefo and Sauti had low initial SE but produced significant amounts of maltose when baked (-, +) whilst Okumkom, Otoo, Patron and Bohye have moderate to high initial sugar content and produced low levels of maltose upon baking (+, -). The last group, Faara, Ligri and Apomuden, had relatively high initial SE and moderate to high starch hydrolytic (+, +) potential following baking. The outcome of this investigation establishes that final sweetness of cooked sweet potato roots is a function of initial sugar content and amylase potential of the raw root. Hence it would be unreliable to classify sweet potato clones in terms of sweetness prior to cooking.

Conclusion and Recommendations

The findings of this study indicate that cooking method, genotype and their interactions significantly influences sugars and sweetness of sweet potato root. Among these factors cooking treatment showed the highest variability. Baking which lasted for longer time resulted in the highest maltose formation. Maltose was barely absent in raw roots but increased considerably after cooking. The amount of maltose synthesized was however dependent on the level of amylase present in the raw root. Activity of amylases was facilitated by temperature, time, and mode of heat penetration by the cooking method. Whilst baking conditions enhances hydrolysis, electromagnetic radiation generated by microwave cooking deactivates amylases, suppressing maltose formation and rendering the product less sweet. Sweetness was found to be dependent on initial sugar content, amylase activity and cooking method. Cooking treatment should therefore be considered as a key criterion when evaluating quality attributes of sweet potatoes for appropriate utilization.

Sweet potato varieties	Total amylase activity	Groupings
Ligri	927.14 (40.56)	Very high
Dadanyuie	882.05 (26.82)	"
Sauti	809.24 (30.45)	"
Ogyefo	804.10 (30.67)	"
Okumkom	779.25 (37.76)	"
Faara	687.32 (50.34)	High
Otoo	650.67 (20.45)	"
Patron	489.81 (15.56)	Moderate
Apomuden	454.10 (21.56)	"
Bohye	414.26 (13.24)	"
Hi-starch	387.06 (25.67)	"

Grouping was based on ranges of amylase activity found: Very High (≥ 750), High (749-550), moderate (549- 350), low (≤ 349). Standard deviations are presented in brackets. LSD = 14.45

Table 4: Means and levels of amylases in sweet potato varieties.

References

1. Kays SJ, Wang Y, McLaurin JW (2005) Chemical and geographical assessment of the sweetness of the cultivated sweet potato clones of the world. J Amer Soc Hort Sci 130: 591-591.

2. Winklund T (2012) Amylolytic activity in selected sweet potato (Ipomoea batatas Lam) varieties during development and in storage. Food and Nutrition Sciences 3: 660-668.

3. Adu-Kwarteng E, Sakyi-Dawson EO, Ayernor GS, Truong VD, Shih FF, et al. (2014) Variability of sugars in staple-type sweet potato (Ipomoea batatas) cultivars. The effect of harvest time and storage. Intern J of Food Properties. 17: 410-420.

4. Dziedoave NT, Graffham AJ, Westby A, Otoo J, Komlaga G (2010) Influence of variety and growth environment on ß-amylase activity of flour from sweet potato (Ipomoea batatas). Food control 21: 162-165.

5. Wang Y, Kays SJ (2000) Effect of cooking method on the aroma constituents of sweet potato [Ipomoea batatas (L.) Lam]. J Food quality 24: 67-78.

6. Chan CF, Chiang CM, Lai CY, Huang CF, Kao SC, et al. (2012) Changes in sugar composition during baking and their effects on sensory attributes of baked sweet potatoes. J Food Sci Technol 51: 4072-4077.

7. Takahata Y, Noda T, Nagata T (1994) Effect of ß-amylase stability and starch gelatinization during heating on varietal differences in maltose content in sweet potatoes. J Agric Food Chem. 42: 2564-2569.

8. Morrison TA, Pressey R, Kays SJ (1993) Changes in α- and ß- amylase during storage of sweet potato lines with varying starch hydrolytic potential. J Amer Soc Hort Sci 118: 236 - 242.

9. Hagenimana V, Simard RE, Vezina LP (1996) Method for the hydrolysis of starchy materials by sweet potato endogenous amylases.

10. Sun JB, Severson RF, Kays SJ (1993) Effect of heating temperature and microwave pretreatment on the formation of sugars and volatile in jewel sweet potato. J Food quality 17: 447-456.

11. Woolfe JA (1992) Sweet potato: An untapped food resource. Cambridge University Press, Cambridge.

12. Simkovic I, Surina I, Vrican, M (2003) Primary reactions of sucrose thermal degradation. J Anal Appl Pyrol 70:493-504.

13. CSIR-CRI (2012) Technical report on sweet potato genotypes proposed for release. Council for Scientific and Industrial Research-Crops Research Institute.

14. CSIR-CRI (2005) Technical report on sweet potato genotypes proposed for release. Council for Scientific and Industrial Research-Crops Research Institute.

15. CSIR-CRI (1998) Technical report on sweet potato genotypes proposed for release. Council for Scientific and Industrial Research-Crops Research Institute.

16. Owusu-Mensah E, Oduro I, Sarfo KJ (2010) Steeping: A way of improving the malting of rice grain. Journal of food biochemistry 35: 80-91.

17. Osman AM (2002) The advantages of using natural substrate-based methods in assessing the roles and synergistic and competitive interactions on barley malt starch degrading enzymes. J Inst Brew 108: 204-214.

18. Kumagai T, Umemura Y, Baba T, Iwanaga M (1990) The inheritance of Î²-amylase null in storage roots of sweet potato, Ipomoea batatas (L.) Lam. Theor Appl Genet 79: 369-376.

19. SAS (2007) Statistical Analysis Software. SAS Institute Inc. Cary, North Carolina, USA.

20. Hagenimana V, Simard E (1994) Amylolytic activity in germinating sweet potato (Ipomea batatas L.)roots. J Amer Soc Hort Sci 119: 313-320.

21. Sawai J, Nakai T, Shimizu M (2009) Reducing sugar production in sweet potatoes heated by electromagnetic radiation. Food Sci Tech Int 15: 89-95.

22. Lewthwaite SL, Sutton KH, Triggs CM (1997) Free sugar composition of sweet potato cultivars after storage. New Zealand J Crop and Hort Sci 25: 33-41.

23. Tumwegamire S, Kapinga R, Patrick RR, LaBonte DR, Grüneberg WJ, et al. (2011) Evaluation of dry matter, protein, starch, sucrose, ß-carotene, Iron, zinc, calcium, and magnesium in east African sweet potato (Ipomea batatas) germplasm. Hort science 46: 348-357.

24. Minervini WP (2010) Characteristics of highway storm water runoff in Los Angeles: Metals and polycyclic aromatic hydrocarbons. In: S. Lau, Y. Han, J. Kang, M. Kayhanian, M. K. Stenstrom (eds.) 81: 308-318 (2009). Water Environ Res 82: 861-862.

25. Shallenberger RS (1993) Taste chemistry. Blackie Academic, London.

26. Grüneberg WJ, Manrique K, Zhang D, Hermann M (2005) Genotype-environment interactions for a diverse set of sweet potato clones evaluated across varying eco geographic conditions. Peru Crop Sci 45: 2160-2171.

27. Hashimoto A, Yamazaki Y, Shimizu M, Oshita S (1994) Drying characteristics of gelatinous materials irradiated by infrared radiation. Drying Technology 12: 1029-1052.

9

Durum Wheat Whole-meal Spaghetti with Tomato Peels: How By-product Particles Size Can Affect Final Quality of Pasta

Padalino L[1], Conte A[1], Lecce L[1], Likyova D[1], Sicari V[2], Pellicanò TM[2], Poiana M[2] and Del Nobile MA[1]*

[1]University of Foggia, Services Center of Applied Research - Via Napoli 25 Foggia, Italy
[2]Mediterranean University of Reggio Calabria, Agricultural Department, Reggio Calabria, Italy

Abstract

The goal of the study is to investigate the impact of the incorporation of by-product (tomato peels) on durum wheat whole-meal spaghetti. To the aim, different amounts of tomato peels flour were added to pasta dough until the overall sensory quality reached its threshold (peels flour at 15% TP). Moreover, the effect of different particle sizes of tomato peels addition on sensory quality of pasta was also evaluated. The increase of particle sizes determined a decline of pasta sensory quality. So, samples enriched with fine particles showed high sensory quality, a more acceptable cooking quality and the lowest value of starch digestibility. The utilization of fine particles of tomato peels seems to be useful to enhance the spaghetti quality. Therefore, fine particles allowed obtained fortified pasta with acceptable sensory properties.

Keywords: Whole-meal flour; Spaghetti; By-product; Particle sizes; Sensory properties

Introduction

Over the last decades consumer food demands changed considerably. For this reason, foods today are not intended only to satisfy hunger and to provide necessary nutrients, but also to prevent nutrition-related diseases and enhance physical and mental well-being of consumers [1,2]. In this regard, functional foods offer an outstanding opportunity to improve the quality of products. Pasta, in particular, is an important basic food widely consumed across the world and was among the first food to be authorized by the FDA (Food and Drug Administration) as a good vehicle for the addition of bioactive compounds, such as antioxidant compounds and dietary fibre [3,4]. However, pasta enriched with bioactive compounds of vegetable origin is still very limited [5,6]. Padalino [7] carried out studies to improve nutritional properties of pasta by adding artichoke, asparagus, pumpkin, zucchini, tomato, carrot, broccoli, spinach, eggplant and fennel, all very rich in phenolics and carotenoids that can impart health benefits being able to scavenge reactive oxygen species and protect against degenerative diseases like cancer and cardiovascular diseases.

Tomatoes (*Lycopersicon esculentum* L.) are known as an excellent source of many nutrients and secondary metabolities, as minerals, vitamins C and E, β-carotene, lycopene, flavonoids, organic acids, phenolics and chlorophyll [8] especially in the peels. Al-Wandawi [9] reported that tomato peels contain high levels of lycopene and β-carotene compared to pulp and seeds. When tomatoes are processed into products like Catsup, salsa and sauces, 10-30% of their weight becomes waste or pomace [10]. In fruit and vegetable industry, generally processing leads to one third of the product to be discarded. This can be costly for the manufacturer and also may have a negative impact on the environment. Many researches have shown that by-products generally exert high nutritional value, could be used as food ingredients, gelling and water binding agents and could provide a valid solution for pollution problems connected with food processing [11]. To the best of our knowledge, no reports are available on the use of tomato peels-based flour in pasta processing. Hence, the aim of this work was to study the impact of tomato peels addition on chemical composition, cooking and sensory quality of whole-meal durum wheat spaghetti. Specifically, the study was organized in the following steps. In the first one, the tomato peels flour amount added to the dough was continuously increased until reaching the sensory threshold (15% of

flour addition). The next experimental step was aimed to investigate the influence of peels particles size on texture properties, cooking quality, sensory and nutritional characteristics of final enriched pasta.

Material and Methods

Raw materials

Durum wheat seeds Pr22 were provided from the C.R.A. (Foggia, Italy). The whole-meal flour was produced from grinding of the seeds with a stone mill (Mod MB250 Partisani). Tomato skins of different cultivars (Ulisse, Docet, Ercole, Player, Herdon, Fuzzer and Komolix), obtained in the crop year 2012-2013 in Campania and Apulia (Southern Italy) industries, were used. Tomato skins were dehydrated by exposure to sunlight and then in the oven (40-50°C) and the flour was produced by hammer mill (16/BV-Beccaria s.r.l. Cuneo).

After the flour was sieved by Sieve Shakers (Mod AS 300 Retsch) in different particles sizes: fine particles size (FPS - ≥ 63 μm), medium particles size (MPS - ≥ 125 μm) and coarse particles size (CPZ - ≥ 250 μm).

Spaghetti preparation

Whole-meal flour of durum wheat was mixed with water (30% w/w) in the rotary shaft mixer (Namad, Rome, Italy) at 25°C for 20 minutes to uniformly distribute water. In the first experimental phase, the tomato peels flour (particles size<500μm) was added to the wheat flour at various concentrations: 10%, 15%, 20% and 25% (w/w). In a subsequent experimental phase the sample with 15% addition (15-TP) was prepared by tomato peels flour to different particle sizes: 63 μm (15-TP/FPS), 125 μm (15-TP/MPS) and 250 μm (15-TP/CPS). Spaghetti

*Corresponding author: Matteo Alessandro Del Nobile, University of Foggia, Services Center of Applied Research - Via Napoli 25 Foggia, Italy
E-mail: matteo.delnobile@unifg.it

based only on whole-meal flour were also manufactured and used as the reference sample (CTRL). In all the steps, dough was extruded with a 60VR extruder (Namad). Subsequently, the pasta was dried in a dryer (SG600; Namad). The process conditions were in according Padalino [12].

Sensory analysis

Dry spaghetti samples were submitted to a panel of fifteen trained tasters (six men and nine women, aged between 28 and 45) in order to evaluate the sensory attributes. The panelists were also trained in sensory vocabulary and identification of particular attributes by evaluating durum wheat commercial spaghetti [13]. They were asked to indicate color and resistance to break of uncooked spaghetti. Elasticity, firmness, bulkiness, adhesiveness, fibrous nature, color, odor and taste were evaluated for cooked spaghetti. To this aim, a nine-point scale, where one corresponded to extremely unpleasant, nine to extremely pleasant and five to the threshold acceptability, was used to quantify each attribute [14]. On the basis of the above-mentioned attributes, panelists were also asked to score the overall quality of the product using the same scale.

Chemical determination

Dry spaghetti samples were ground to fine flour on a Tecator Cyclotec 1093 (International PBI, Hoganas, Sweden) laboratory mill (1-mm screen – 60 mesh). Moisture and ash content (%) were measured according to AACC method [15]. Protein content (% N × 5.7) was analyzed using the micro-Kjeldahl method according to AACC method [15]. Total dietary fiber (TDF), soluble water dietary fiber (SDF) and insoluble water dietary fiber (IDF) contents were determined by means of the total dietary fiber kit (Megazyme International Ireland Ltd., Wicklow, Ireland) based on the method of Lee [16]. The available carbohydrates (ACH) were determined according to the method of McCleary [17] as described in the ACH assay kit (Megazyme). All nutritional analyses of the flour and spaghetti samples were made in triplicate.

For the carotenoids determination spaghetti were homogenized in a blender and an aliquot of 10 g was added of 100 ml of solvent mix (esano:acetone:methanol; 2:1:1; v/v/v) and sonication continuously for 10 min (Misonix Ultrasonic Liquid Processor, NY, U.S.A). The extraction was repeated until sample became colorless. The combined extract was transferred to a separating funnel and 5 ml of distilled water was added to separate polar and nonpolar phases. The nonpolar hexane layer containing carotenoids was collected and concentrated in a rotary evaporator (Heidolph, Germany) till dryness. Residue was dissolved in 10 ml of hexane. Lycopene and β-carotene were determined according to Fish [18] by a spectrophotometric method using an Agilent 8453 UV-Vis spectrophotometer. The concentration of lycopene was calculated at λ=503 nm using the molar extinction coefficient β=17.2 × 10⁴/M/cm. For β-carotene, the absorbance was measured at λ=450 nm and the quantification were carried out using a standard curve. All the nutritional analyses were made in triplicate and the results were expressed as mean ± standard deviation (SD).

Cooking quality

The optimal cooking time (OCT) was evaluated in according to the AACC approved method [15]. The cooking loss, the amount of solid substance lost to cooking water, was determined according to the AACC approved method 66-50. The swelling index and the water absorption of the cooked pasta (grams of water per gram of dry pasta) were determined according to the procedure described by Padalino [12].

Moreover, the cooked spaghetti samples to OCT were submitted to hardness and adhesiveness analysis, by means of a Zwick/Roell model Z010 Texture Analyzer (Zwick Roell Italia S.r.l., Genova, Italia) equipped with a stainless steel cylinder probe (2 cm diameter). The hardness (mean maximum force, N) and adhesiveness (mean negative area, Nmm) were measured in according to the procedure described by Padalino [12]. Six measurements for each spaghetti sample were performed.

In vitro digestion

The digestion was carried out as described by Chillo [19] with slight modifications. Briefly, dry spaghetti samples (5 g) were broken into 5.0 × 1.0 cm lengths and weighed accurately. Fifty milliliters of boiling water was immediately placed in a covered boiling water bath to cook the spaghetti to the OCT. The spaghetti were tipped into a digestion vessel with 50 ml of distilled water and 5 ml maleate buffer (0.2 M pH 6.0, containing 0.15 g CaCl₂ and 0.1 g sodium azide per liter) in an block at 37°C (GFL 1092; GFL Gesellschaft für Labortechnik, Burgwedel, Germany) and allowed to equilibrate for 15 min. Digestion was started by adding 0.1 ml amyloglucosidase (A 7095; Sigma Aldrich, Milan, Italy) and 1 ml of 2 g per 100 g pancreatin (P7545; Sigma Aldrich) in quick succession and the vessels were stirred at 130 rpm. An amount of 0.5 ml of the digested samples was taken at 0, 20, 60 and 120 min for the released glucose analysis. The sample digested to 120 min was homogenized through an Ultra Turrax (Ika Werke, Staufen, Germany).

Analysis of digested starch

The samples removed during digestion were added to 2.0 ml of ethanol ethanol and mixed. After 1 h, the ethanolic sub-samples were centrifuged (2000 g, 2 min) (Biofuge fresco; Heraeus, Hanau, Germany). Finally, the reducing sugar concentration was measured colorimetrically (k=530 nm) using a Shimadzu UV-Vis spectrophotometer (model 1700; Shimadzu corporation, Kyoto, Japan). Glucose standards of 10 mg/ml were used. Amyloglucosidase (0.25 ml) (EAMGDF, 1 ml per 100 ml in sodium acetate buffer 0.1M, pH 5.2; Megazyme International 205 Ireland Ltd., Wicklow, Ireland) was added to 0.05 ml of the supernatant and incubated at 20°C for 10 min. Afterwards, 0.75 ml DNS solution (10% 3,5-dinitrosalicylic acid, 16% NaOH and 30% Na-K tartrate – Sigma Aldrich) was added to the above solution, heated to 100°C for 15 min and allowed to cool at 15°C for 1 h. Then, 4 ml of distilled water (15°C) were added to the solution. The results were plotted as glucose release (mg) per g of sample vs. time. The starch digestibility was calculated as the area under the curve (0–120 min) for the tested products, and expressed as the percentage of the corresponding area for white bread [19].

Statistical analysis

Experimental data were compared by a one-way variance analysis (ANOVA). A Duncan's multiple range test, with the option of homogeneous groups ($P<0.05$), was carried out to determine significant differences between spaghetti samples. STATISTICA 7.1 for Windows (StatSoft, Inc, Tulsa, OK, USA) was used for this aim.

Results and Discussion

As reported above, the experimental plan has been organized in two subsequent steps, the first aimed to find the better concentration of tomato peels amount to be added to the dough and the second one, to study the effects of peel-particles size on quality of spaghetti. Results of each step were detailed in different paragraphs.

Step 1 - Optimization of tomato peels flour addition

The sensory properties of dry spaghetti samples are listed in Table 1. Results highlighted that in general the overall quality of spaghetti made with whole-meal flour (CTRL) without any peels addition was higher in comparison to the samples supplemented with tomato peels flour, above all at concentrations higher than 15%. In particular, in the uncooked spaghetti poor colour and break to resistance were found. Regarding the cooked spaghetti, the addition of tomato peels flour influenced pasta elasticity and firmness, due to the high fibres content. Incorporation of vegetable matter rendered a firmer texture to pasta sample due to the non-starchy nature of vegetables. In addition, the TDF content of tomato peels (mainly insoluble fibers) was found higher than that reported in other vegetables [20]. Similar results were observed by Yavad et al. [6], who found an increase of pasta firmness (100% durum wheat) enriched with vegetable flour. The low elasticity value was also due to the inclusion of tomato peels fibres that promoted the formation of discontinuities or cracks inside the pasta strand, which weakened its structure. Spaghetti fortified with tomato peels flour resulted less adhesive than the CTRL sample, even if the differences between samples were not significant. Stickiness did not increase in the different pasta samples, most probably because fibres addition is generally recognized to have a positive effect on stickiness [21]. Spaghetti exclusively made with whole-meal wheat or containing amounts of tomato peels flour up to 15% (w/w) appeared with a pleasant brown colour, whereas, the spaghetti samples made by using more than 15% tomato peels flour present an intense orange colour, which is different from the common pasta and considered unacceptable. Rekna [5] also observed reduction in colour intensity of cooked pasta enriched with vegetable flour, probably due to the pasta swelling and to the conversion of pigments resulting in a yellowness increase. Colour is the key factor for assessing the visual quality and market value of food products [22]. Svec [23] also studied the colour impact of non-traditional cereals and reported that until 10% addition the colour remained acceptable. In addition, spaghetti samples enriched with tomato peels flour had very intense taste and odour as compared to the CTRL sample. Therefore, on the basis of the sensory acceptability, the spaghetti samples enriched with tomato peels flour at 15% were selected for the subsequent work of pasta optimization.

Step 2 Effects of tomato peels particles size on spaghetti quality

Sensory analysis: The results of sensory properties of dry spaghetti are listed in Table 2. As can be inferred, the overall quality of uncooked and cooked spaghetti declined as the particles size increased, thus demonstrating that spaghetti enriched with the fine particles (15-TP/FPS) exerted the greatest overall quality. In terms of firmness no differences among samples were recorded, even though a little increase was found in pasta with fine particles because these particles contain more protein than coarse particles of tomato peels. Padalino [12] also found that the high protein content of pea flour increased pasta hardness due to low hydration of starch granules. Most probably, during cooking, the protein can link to most of the water molecules, leaving less water to swell the starch phase [24]. From data reported in Table 2 it is also clear that particles size has a marked effect on the fibrous nature of spaghetti. As compared to the other samples, the 15-TP/FPS showed a low fibrous sensation, due to the less dietary fibres content (Table 3), and present low adhesiveness and bulkiness. One possible explanation of the observed results is that with fine particles a more stable network can be realized, able to bind starch granules and vegetable flour and avoid solids loss during cooking [5]. No effects of particles size were underlined on pasta odour. Samples 15-TP/FPS and 15-TP/MPS in particular showed a pleasant orange colour. As concern the taste, the most prized samples were again 15-TP/FPS and 15-TP/MPS, due to the low fibrous sensation during mastication.

Chemical composition: Table 3 summarizes the protein and the dietary fibres content of spaghetti. It is clear that particles size of peels had no effects on protein and available carbohydrates content. A certain drop in dietary fibres for samples supplemented with fine particles (12.69%), in comparison to medium and coarse sizes (14.78%) was observed. It is plausible to suggest that the use of flour with coarse particles better corresponds to more functional compounds, such as dietary fibres [25]. Table 3 also reports the results of the *in vitro* starch digestibility. Results suggested that the particles size influenced the starch digestibility (SD). It can be seen a significant decline of SD in spaghetti enriched with coarse particles size. These differences could be due to the fact that the sample 15-TP/CSP had the greatest dietary fibres content that are known to reduce the glycaemic response of pasta [26]. These results also are also in agreement with Padalino [7], who

	Uncooked Spaghetti			Cooked Spaghetti								
	Color	Break to Resistance	Overall Quality	Elasticity	Firmness	Fibrous	Bulkiness	Adhesiveness	Color	Odor	Taste	Overall Quality
CTRL	7.26 ± 0.35ᵃ	6.20 ± 0.34ᵃ	7.09 ± 0.30ᵃ	7.09 ± 0.33ᵃ	6.83 ± 0.30ᵃ	6.83 ± 0.25ᵃ	6.37 ± 0.28ᵃ	6.09 ± 0.28ᵃ	7.00 ± 0.23ᵃ	7.31 ± 0.41ᵃ	7.07 ± 0.44ᵃ	7.09 ± 0.33ᵃ
10% TP	6.48 ± 0.42ᵇ	5.91 ± 0.38ᵃ·ᵇ	6.20 ± 0.27ᵇ	5.57 ± 0.33ᵃ	6.50 ± 0.31ᵃ·ᵇ	5.78 ± 0.26ᵇ	6.51 ± 0.40ᵃ	6.02 ± 0.40ᵃ	6.80 ± 0.33ᵃ·ᵇ	6.21 ± 0.40ᵇ	6.26 ± 0.34ᵇ	6.30 ± 0.34ᵇ
15% TP	6.08 ± 0.20ᵃ·ᵇ	5.80 ± 0.24ᵃ	6.05 ± 0.25ᵃ·ᵇ	5.37 ± 0.31ᵃ	6.31 ± 0.23ᵃ·ᵇ	5.68 ± 0.28ᵇ	6.60 ± 0.39ᵃ	6.11 ± 0.40ᵃ	6.30 ± 0.36ᵇ	6.01 ± 0.21ᵇ	6.11 ± 0.28ᵇ	6.04 ± 0.38ᵇ·ᶜ
20% TP	5.65 ± 0.36ᶜ·ᵈ	5.65 ± 0.27ᵃ·ᵇ	5.62 ± 0.31ᶜ·ᵈ	5.17 ± 0.36ᵃ	6.01 ± 0.34ᵇ	5.28 ± 0.28ᵇ·ᶜ	6.65 ± 0.34ᵃ	6.17 ± 0.24ᵃ	6.01 ± 0.23ᶜ	5.81 ± 0.33ᵇ	5.91 ± 0.21ᵇ	5.64 ± 0.37ᵇ·ᶜ
25 % TP	5.22 ± 0.27ᵈ	5.36 ± 0.28ᵃ	5.33 ± 0.30ᵈ	5.07 ± 0.36ᵃ	5.25 ± 0.37ᶜ	5.01 ± 0.34ᶜ	6.67 ± 0.33a	6.24 ± 0.27a	5.95 ± 0.24ᶜ	6.20 ± 0.33ᵇ	5.55 ± 0.40ᵇ	5.46 ± 0.38ᶜ

ᵃ⁻ᶜMean in the same column followed by different superscript letters differ significantly (P<0.05).

Table 1: Sensory characteristics of uncooked and cooked dry spaghetti samples obtained in step 1.

	Uncooked Spaghetti			Cooked Spaghetti								
	Color	Break to Resistance	Overall Quality	Elasticity	Firmness	Fibrous	Bulkiness	Adhesiveness	Color	Odor	Taste	Overall Quality
15% TP	6.08 ± 0.20ᵇ	5.80 ± 0.24ᵇ	6.05 ± 0.25ᵇ	5.37 ± 0.31ᵇ	6.31 ± 0.23ᵇ	5.68 ± 0.28ᵃ·ᵇ	6.60 ± 0.39ᵃ·ᵇ	6.11 ± 0.40ᵃ·ᵇ	6.30 ± 0.36ᵇ	6.01 ± 0.21ᵇ	6.11 ± 0.28ᵇ	6.04 ± 0.38ᵇ
15% TP/FPS	7.21 ± 0.27ᵃ	6.32 ± 0.34ᵃ	6.84 ± 0.32ᵃ	6.04 ± 0.34ᵃ	6.35 ± 0.36ᵃ	6.26 ± 0.34ᵃ	6.85 ± 0.34ᵃ	6.63 ± 0.35ᶜ	7.16 ± 0.32ᵃ	6.04 ± 0.23ᵃ	6.81 ± 0.24ᵃ	6.98 ± 0.27ᵃ
15% TP/MPS	7.18 ± 0.23ᵃ	6.10 ± 0.31ᵃ	6.59 ± 0.31ᵃ·ᵇ	5.69 ± 0.40ᵃ	6.20 ± 0.33ᵃ	6.10 ± 0.28ᵃ·ᵇ	6.56 ± 0.40ᵃ·ᵇ	6.02 ± 0.40ᵃ·ᵇ	7.12 ± 0.24ᵃ	6.06 ± 0.23ᵃ	6.73 ± 0.22ᵃ	6.53 ± 0.43ᵃ·ᵇ
15% TP/CMS	6.09 ± 0.39ᵇ	5.94 ± 0.35ᵇ	6.11 ± 0.24ᵇ	5.64 ± 0.35ᵃ	6.07 ± 0.26ᵃ	5.54 ± 0.28ᵇ	6.00 ± 0.28ᵇ	5.75 ± 0.31ᵇ	6.39 ± 0.48ᵇ	6.08 ± 0.26ᵃ	6.09 ± 0.28ᵇ	6.15 ± 0.30ᵇ

ᵃ⁻ᶜMean in the same column followed by different superscript letters differ significantly (P<0.05).

Table 2: Sensory characteristics of uncooked and cooked dry spaghetti samples obtained in step 2.

also found that the addition of pepper flour (containing a high level of dietary fibres) reduced the glycemic index of maize-based pasta.

Table 4 shows the lycopene and β-carotene content in spaghetti samples enriched with tomato peels flour. Data were expressed as mg of β -carotene and lycopene/100 g spaghetti. There were differences (P < 0.05) among studied samples, thus highlighting that particles of tomato peels decreased concentration of carotenoids mainly for β -carotene. As one could expect, the spaghetti 15-TP have the highest value of lycopene (2.32 mg/100g) and β-carotene (13.64 mg/100g) compared to the other ones. It is also worth noting that even with fine particles, a significant increment of carotenoids can be obtained, if compared to the control samples.

Cooking quality: The optimum cooking time, the cooking loss, the swelling index and the water absorption of spaghetti samples are presented in Table 5. From this table emerges that particles size of peels flour also affects cooking quality. In fact, the OCT decreased as the particles size increased. Specifically, the OCT for samples supplemented with fine and medium particles was higher (9.00 and 8.30 min respectively) than 15-TP and 15-TP/CPS samples, likely due to the protein matrix-starch granule network, which was affected by the vegetable flour fibres. The physical disruption of gluten matrix caused by fibres addition and the reduction in gluten content due to tomato peels flour addition may facilitate the water penetration into pasta core. The less cooking time of pasta supplemented with vegetable flour could be explained by the faster reconstitution of fine vegetable matter distributed in pasta matrix [5]. Table 5 also shows a decline in cooking loss for samples 15-TP/FPS and 15-TP/MPS, as compared to the other pasta samples. As reported for sensory analysis, these results could be mainly due to the better binding of starch granules and vegetable flour with fine particles size in gluten network [5]. Concerning the swelling index, significant differences were observed with the increase of particles size. Specifically, the 15-TP/CPS sample showed the highest

swelling index. The 15-TP/CPS presented agreater water absorption than the other samples, but without any significant differences. These results could be explained by the reduction of protein, as resulting from the increment of the average particles size. In fact, the sample with the coarse particles showed a slight drop in protein content that is known to counteract starch granule swelling during cooking, due to competition between protein and starch for water availability [24]. Regarding the adhesiveness, the 15-TP/FPS sample recorded the smallest value (0.51 N), and concerning the hardness, the same sample recorded the highest value, in accordance to sensory quality and cooking loss. This result also suggests that the high protein content of sample with fine particles of peels increased the hardness of pasta because of the low hydration of starch granules.

Conclusions

In this work, the impact of tomato peels-based flour addition on chemical composition, cooking and sensory quality of whole-meal durum wheat spaghetti was studied. In the first experimental step, tomato peels flour amount added to pasta dough was continuously increased until the sensory quality reached the threshold (tomato peels flour concentration 15%). In a second step, the influence of particles size on sensory quality of pasta with 15% tomato peels flour was investigated. The results indicated that the increase of particles size determined a decline of overall quality of samples; even a slight better nutritional composition was recorded. Specifically, the spaghetti enriched with the fine particles showed the greatest sensory score, due to the low fibrous, low adhesiveness, low bulkiness and high hardness values and showed a significant increase of starch digestibility. Therefore, our findings suggest that whole-meal spaghetti with fine particles represent fortified pasta with good sensory properties, very comparable to the control samples, and good cooking quality. This example of pasta fortification can offer a broad spectrum of new products with desired properties and encourage the use of agronomic by-products for further studies and new food applications.

Sample	Protein (%)	ACH (g/100g)	IDF (%)	SDF (%)	TDF (%)	SD
15% TP	10.29 ± 0.18[a,b]	59.10 ± 0.10[a,b]	14.95 ± 0.00[a]	8.00 ± 0.28[a]	22.95 ± 0.14[a]	58 ± 0.30[b]
15% TP/FPS	10.35 ± 0.30[a]	58.03 ± 0.90[b]	12.69 ± 0.01[d]	7.06 ± 0.00[c]	19.75 ± 0.00[d]	62 ± 0.90[a]
15% TP/MSP	10.20 ± 0.06[a,b]	60.20 ± 0.00[a]	13.01 ± 0.61[c]	7.42 ± 0.00[b]	20.48 ± 0.00[c]	61 ± 1.00[a]
15% TP/CSP	9.95 ± 0.05[b]	59.02 ± 1.20[a,b]	14.78 ± 0.09[b]	7.33 ± 0.20[b,c]	22.11 ± 0.10[b]	53 ± 0.20[c]

TDF: Total Dietary Fibre; SDF: Soluble Water Dietary Fibre; IDF: Insoluble Water Dietary Fibre; ACH: Available Carbohydrates; SD: Starch Digestibility
[a-c] Mean in the same column followed by different superscript letters differ significantly (P<0.05).

Table 3: Chemical composition of semolina and pea flour and dry spaghetti samples obtained in step 2.

	lycopene	β-carotene
	(mg/ 100g spaghetti)	(mg/100g spaghetti)
cnt	0.03 ± 0.01[d]	4.23 ± 0.23[e]
15% TP	2.32 ± 0.4[6]a	13.64 ± 1.15[a]
15% TP/FPS	1.31 ± 0.10[b]	10.46 ± 0.10[d]
15% TP/MSP	1.00 ± 0.12[c]	11.60 ± 0.91[b]
15% TP/CSP	1.36 ± 0.13[b]	10.87 ± 1.31[c]

[a-c] Mean in the same column followed by different superscript letters differ significantly (P<0.05).

Table 4: Carotenoids content of dry spaghetti obtained in step 2.

Sample	OCT (min)	Cooking Loss (%)	Swelling Index	Water Absorption (%)	Adhesiveness (Nmm)	Hardness (N)
15% TP	7.30	8.47 ± 0.07[a]	1.72 ± 0.10[b]	138 ± 3.65[a]	0.58 ± 0.08[a]	14.60 ± 1.13[a]
15% TP/FPS	9.00	6.91 ± 0.31[c]	1.71 ± 0.09[b]	139 ± 8.48[a]	0.41 ± 0.06[b]	13.93 ± 0.67[a,b]
15% TP/MSP	8.30	7.94 ± 0.40[b]	1.70 ± 0.05[b]	140 ± 4.20[a]	0.61 ± 0.09[a]	13.03 ± 0.92[a,b]
15% TP/CSP	8.00	8.53 ± 0.14[a]	1.86 ± 0.01[a]	144 ± 4.00[a]	0.57 ± 0.10[a]	12.61 ± 0.81[b]

[a-c]Mean in the same column followed by different superscript letters differ significantly (P<0.05).

Table 5: Cooking quality of spaghetti enriched with tomato peels flours at different particle size studied in step 2.

References

1. Betoret E, Betoret N, Vidal D, Fito P (2011) Functional foods development: trends and technologies. Trends Food Sci Tech 22: 498-508.

2. Brouns F, Vermeer C (2000) Functional ingredients for reducing the risks of osteoporosis. Trends Food Sci Tech 11: 22-33.

3. Chillo S, Laverse J, Falcone PM, Protopapa A, Del Nobile MA (2008) Influence of the addition of buckwheat flour and durum wheat bran on spaghetti quality. J Cereal Sci 47: 144-152.

4. Gallegos-Infante JA, Rocha-Guzman NE, Gonzalez-Lared RF, Ochoa-Martínez LA, Corzo N, et al. (2010) Quality of spaghetti pasta containing Mexican common bean flour (Phaseolus vulgaris L.). Food Chem 119: 1544-1549.

5. Rekha MN, Chauhan AS, Prabhasankar P, Ramteke RS, Venkateswara Rao G (2012) Influence of vegetable pastes on quality attributes of pastas made from bread wheat (*T. aestivum*). CyTA J Food 11: 142-149.

6. Deep NY, Yadav M, Sharma N, Chikara T, Anand SB, et al. (2014) Quality Characteristics of Vegetable-Blended Wheat-Pearl Millet Composite Pasta. Agr Res 3: 263-270.

7. Padalino L, Mastromatteo M, Lecce L, Cozzolino F, Del Nobile MA, et al. (2013a) Manufacture and characterization of gluten free spaghetti enriched with vegetable flour. J Cereal Sci 57: 333-342.

8. Giovanelli G, Paradise A (2002) Stability of dried and intermediate moisture tomato pulp during storage. J Agr Food Chem 50: 7277-7281.

9. Al-Wandawi H, Abdul Rehman MH, Al Shaikhly KA (1985) Tomato processing wastes as essential raw materials source. J Agr Food Chem 33: 804-807.

10. King AJ, Zeidler G (2004) Tomato pomace may be a good source of vitamin E in broiler diets. Cal Agr 58: 59-62.

11. Lario Y, Sendra E, Garciá-Pérez J, Fuentes C, Sayas-Barberá E, et al. (2004) Preparation of high dietary fiber powder from lemon juice by-products. Innov Food Sci Emerg Tech 5: 113-117.

12. Padalino L, Mastromatteo M, Lecce L, Spinelli S, Contò F, et al. (2013) Chemical composition, sensory and cooking quality evaluation of durum wheat spaghetti enriched with pea flour. Int J Food Sci Tech 49: 1-13.

13. ISO 11036/7304-2 (2008) Alimentary pasta produced from durum wheat semolina-Estimation of cooking quality by sensory analysis - Part 2: Routine Method.

14. Petitot M, Boyer L, Minier C, Micard V (2010) Fortification of pasta with split pea and faba bean flours: pasta processing and quality evaluation. Food Res Int 43: 634-641.

15. AACC (2000) Approved Methods of the American Association of Cereal Chemists. Technology and Engineering 1-2: 1-1200.

16. Lee SC, Prosky L, DeVries JW (1992) Determination of total, soluble, and insoluble, dietary fiber in foods-enzymatic gravimetric method, MES-TRIS buffer: collaborative study. J AOAC Int 75: 395-416.

17. McCleary BV, Rossiter PC (2006) Dietary fibre and glycemic carbohydrates. In: Gordon DT, Goda T (eds.) Dietary Fiber and its Energy Value.

18. Fish WW, Perkins-Veazie P, Collins JKA (2002) Quantitative assay for lycopene that utilizes reduced volumes of organic solvents. J Food Comp Anal 15: 309-317.

19. Chillo S, Ranawana DV, Henry CJK (2011) Effect of two barley beta-glucan concentrates on in vitro glycaemic impact and cooking quality of spaghetti. LWT-Food Sci Tech 44: 940-948.

20. Navarro-González I, García-Valverde V, García-Alonso J, Periago MJ (2011) Chemical profile, functional and antioxidant properties of tomato peel fiber. Food Res Int 4: 1528-1535.

21. Cleary L, Brennan C (2006) The influence of a (1-3)(1-4)-β-D-glucan rich fraction from barley on the physico-chemical properties and in vitro reducing sugars release of durum wheat pasta. Int J Food Sci Tech 41: 910-918.

22. Navarro MJ, Retamales J, Defilippi B (2001) Efecto del arreglo de racimo y aplicación de CPPU en la calidad de uva de mesa Sultanina tratada con dos fuentes de giberelinas. Agricultura Técnica 61: 15-21.

23. Švec I, Hrušková M, Vítová M, Sekerová H (2008) Colour evaluation of different pasta samples. Czech J. Food Sci, 26: 421-427.

24. Cocci E, Sacchetti M, Vallicelli M, Angiolini A, Dalla RM (2008) Spaghetti cooking by microwave oven: Cooking kinetics and product quality. J Food Eng 85: 537-546.

25. Laudadio V, Bastoni E, Introna M, Tufarellia V (2013) Production of low-fiber sunflower (*Helianthus annuus L.*) meal by micronization and air classification processes. CyTA-J Food 4: 398-403.

26. Yokoyama WH, Hudson CA, Knuckles BE, Chiu MCM, Sayre RN, et al. (1997) Effect of barley β-glucan in durum wheat pasta on human glycemic response. Cereal Chem 74: 293-296.

10

Chemical and Nutritional Evaluation of Biscuit Processed from Cassava and Pigeon Pea Flour

Ashaye OA[1]*, Olanipekun OT[1] and Ojo SO[2]

[1]Institute of Agricultural Research and Training P.M.B 5029, Moor-Plantation Ibadan, Nigeria
[2]Federal College of Agriculture, Ibadan, Nigeria

Abstract

Cassava is an important crop in the tropics. The use of cassava flour and pigeon pea flour in the manufacture of biscuit is uncommon. The chemical and nutritional evaluation of biscuit processed from cassava and pigeon pea flour was investigated. Commercial and cassava based biscuit were evaluated for chemical and nutritive properties. Sensory evaluation was done by a ten member panel randomly selected from male and female adults. One hundred percent cassava biscuit was significantly higher than other biscuit samples in Hydro-cyanide (HCN) at $P<0.05$. The crude protein and ash content of 30% cassava pigeon pea biscuit was significantly higher than other biscuit samples. Commercial biscuit was higher in crude fat (13.54%), crude fibre (0.85%) and moisture content (4.8%). Sensory evaluation showed that commercial biscuit, 30% cassava-pigeon pea biscuit and 100% cassava biscuit were not significantly different from each other in colour. Higher scores were given to 30% cassava-pigeon biscuit. The taste, texture, flavour and general acceptability of 100% cassava biscuit and commercial biscuit were not significantly different from each other. Generally, acceptable biscuit was processed from 100% cassava flour and 30% cassava pigeon flour. Thirty percent cassava pigeon biscuit gave better nutrient attributes and sensory scores than commercial biscuit.

Keywords: Biscuit; Cassava pigeon pea; Chemical; Nutritional; Sensory

Introduction

Cassava is sometimes classified as a crop for developing countries and for consumption only by rural people, whereas the large crop of cassava grown annually in the tropics is actually consumed in all its forms at nearly all income levels [1]. It is a very important crop in the tropics where an estimate of a billion people depends on it as a major staple crop. It is an important crop in the farming system because it can be intercropped with many other crops, and yield well in poor soils. Cassava is the fourth most important staple crop in the world after rice, wheat and maize [2]. It is a chief source of edible carbohydrate that could be processed into different forms of human foods, e.g., fufu, eba, gari, lafun, pupuru, abacha, etc. The cultivation of cassava requires minimal input but the processing is laborious and time consuming. Cassava contains 1% protein, 97% starch, 1% fibre with traces of fat and other minerals [2].

Pigeon pea (*Cajanus cajan* (L) *Mill* sp.) belongs to the family of Cajanicae. It is the only domesticated species well distinguished by the presence of vascular glands on the leaves, calyx and pods [3]. Pigeon pea seed is one of the tropical legume seeds which have been highly favourable because of its rich protein composition, energy and mineral content and widespread distribution in the tropics. However, its nutritive value is masked by the occurrence of anti-nutritional factors. Some of the anti-nutritional factors are known to have negative effect on the haematological parameters, but can easily be removed through processing by heat [3]. Biscuits are an ideal source of food as they possess the following attributes: they are highly palatable and acceptable, they have long storage and good sources of energy, they provide a good source of energy and can easily be modified to suit specific nutritional needs of any target population, and they can be produced in convenient sizes and different forms. Biscuit is therefore one of the popular food sources in any emergency feeding programme [4]. There is limited information about the production of biscuit from cassava and pigeon pea flour. Hence this work aims at evaluating the nutritional properties and consumer acceptability of biscuit made from cassava and pigeon pea flour.

Materials and Methods

Raw materials

The cassava tubers were procured from the Federal College of Agriculture Farm in Ibadan, while the pigeon pea was purchased from Apata Market in Ibadan, Oyo state Nigeria. The variety of cassava tubers was the sweet variety (TMS 20001 Variety) and other ingredients such as baking ingredients were also purchased from Apata Market.

Preparation of cassava and pigeon pea flour

The cassava tubers were harvested fresh, peeled manually, washed and grated. The grated mash was packed into perforated sack and pressed using the hydraulic press to drain out water after which it was sun-dried and dry-milled. Pigeon pea was par-boiled in hot water for 5 minutes, peeled, sun-dried and ground using electric blender.

Preparation of cassava based biscuits

Cassava based biscuits were processed from 100% cassava flour, 10%, 20%, 30% cassava/pigeon pea flour using commercial biscuit as control. Biscuit from cassava and cassava pigeon flour were processed by mixing about 100 gm of the flour with the following ingredients 0.15 g of salt, 15 g of sugar, 10 g of fat, 0.25 g of baking powder, 0.25 g of vanilla flavor. About 100 ml of water was gradually added while mixing together for 6 minutes to obtain slightly textured dough. The dough was kneaded on a flat rolling board and was manually rolled into sheet and

*Corresonding author: Ashaye OA, Institute of Agricultural Research and Training P.M.B 5029, Moor-Plantation Ibadan, Nigeria
Email: kayodeashaye@yahoo.com

cut into shapes using a suitably shaped tin. The cut pieces of dough were placed into a greased oven tray, baked at 180°C for 20 minutes using the modified method of Eneche [5] (Figure 1).

Determination of Proximate Composition of biscuit samples

The biscuit samples were analysed for moisture, dry matter, crude protein, crude fat, crude fibre and ash using the standard methods of AOAC [6].

Dry matter and moisture determination

The sample (2g) was weighed into a previously weighed crucible. The crucible plus sample taken was then transferred into the oven set at 100°C to dry to a constant weight for 24 hours overnight. At the end of the 24 hours, the crucible plus sample was removed from the oven and transferred to desiccator, cooled for ten minutes and weighed [6].

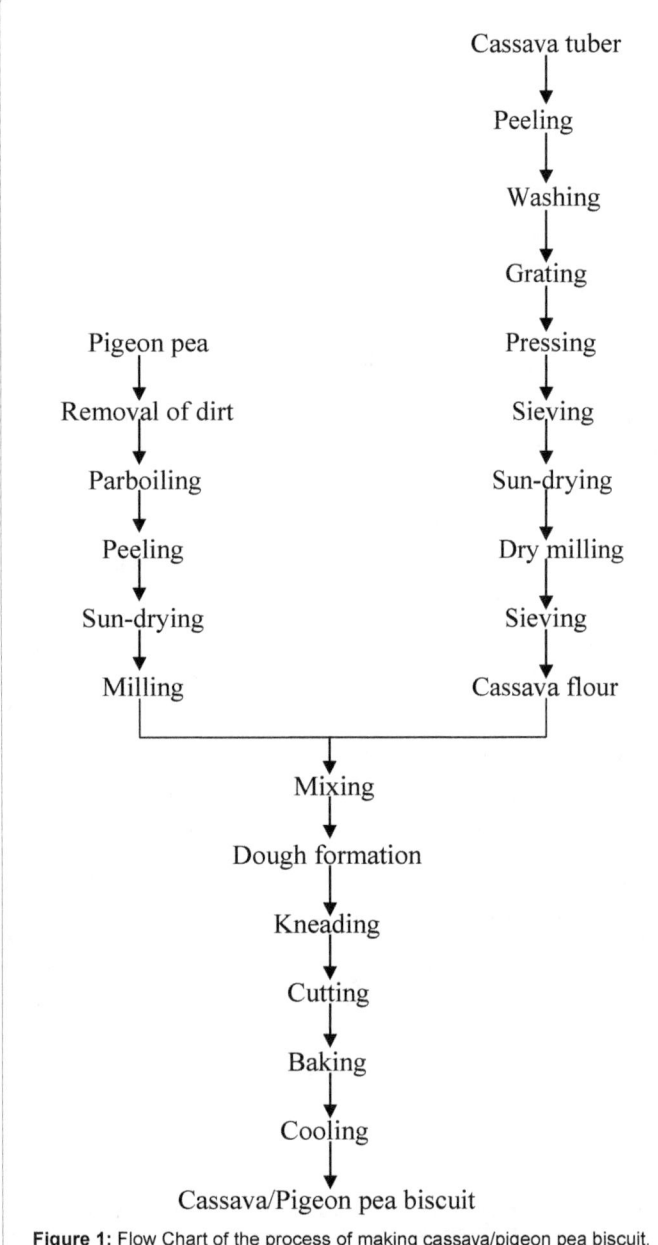

Figure 1: Flow Chart of the process of making cassava/pigeon pea biscuit.

If the weight of empty crucible is W_0

Weight of crucible plus sample is W_1

Weight of crucible plus oven dried sample W_3

$$(\% \, DM) \, \% \, Dry \, Matter = \frac{W_3 - W_0}{W_1 - W_0} \times 100$$

$$\% \, Moisture = \frac{W_1 - W_3}{W_1 - W_0} \times 100$$

Determination of Ash

The sample (2g) was weighed into a porcelain crucible. This was transferred into the muffle furnace set at 550°C and left for about 4 hours. About this time it had turned to white ash. The crucible and its content were cooled to about 100°C in air, then room temperature in a desiccator and weighed [6].

The percentage ash was calculated from the formula below:

$$\% \, Ash \, content = \frac{Weight \, of \, ash}{Original \, weight \, of \, sample} \times 100$$

Determination of crude protein

The micro-Kjeldahl method for protein determination is employed for protein determination. This is based on three principles:

Digestion: $RNH_2 + 2H_2SO_4 \rightarrow (NH_4)_2 + SO_4 + CO_2 + H_2O$

Distillation: $(NH_4)_2 \, SO_4 + 2NaOH \rightarrow NH_3 + H_2O + Na_2SO_4$

Absorption: $3NH_3 + H_3 \, BO_3 \rightarrow (NH_4)_3 \, BO_3$

Titration: $(NH_4)_3 \, BO_3 + HCl \rightarrow H_3BO_3 + 3NH_4Cl$

Procedure

The bread sample (0.5g) was weighed into the micro-Kjeldahl flask. To this were added 1 Kjeldahl catalyst tablet and 10 ml of conc. H_2SO_4. These were set in the appropriate hole of the digestion block heaters in a fume cupboard. The digestion was left on for 4 hours after which a clear colourless solution was left in the tube. The digest was carefully transferred into 100 ml volumetric flask, thoroughly rinsing the digestion tube with distilled water and the volume of the flask made up to the mark with distilled water. 5 ml portion of the digest was then pipetted to Kjeldahl apparatus and 5 ml of 40% ($^w/_v$) NaOH added.

The mixture was then steam distilled and the liberated ammonia collected into a 50 ml conical flask containing 10 ml of 2% boric acid plus mixed indicator solution. The green colour lsolution was then titrated against 0.01 NHCl solution. At the end point, the green colour turns to wine colour, which indicates that, all the nitrogen trapped as ammonium borate have been removed as ammonium chloride. The percentage nitrogen was calculated by using the formula:

% N = Titre value x atomic mass of nitrogen x normality of HCl used x 4

The crude protein is determined by multiplying percentage nitrogen by a constant factor of 6.25 [6].

Crude fat determination

The dried sample (1g) was weighed into fat free extraction thimble and plug lightly with cotton wool. The thimble was placed in the extractor and fitted up with reflux condenser and a 250 ml soxhlet flask which has been previously dried in the oven, cooled in the dessicator and weighed. The soxhlet flask is then filled to ¾ of it volume with petroleum either (B.Pt. 40-60°C) and the soxhlet flask extractor plus condenser set was placed on the heater. The heater was put on for six hours with constant running water from the tap for condensation of ether vapour. The set is constantly watched for ether leaks and the heat sources is adjusted appropriately for the either to bril gently. The ether is left to siphon over several times at least 10-12 times until it is short of siphoning. It is after this is noticed that any ether content of the extractor is carefully drained into the ether stock bottle. The thimble-containing sample is then removed and dried on a clock glass on the bench top. The extractor flask with condenser is replaced and the distillation continues until the flask is practically dried. The flask which now contains the fat or oil is detached, its exterior cleaned and dried to a constant weight in the oven [6]. If the initial weight of dry soxhlet flask is W_o and the final weight of oven dried flask + oil/fat is W_1, percentage fat/oil is obtained by the formula:

$$\text{percentage of fat / oil} = \frac{W_1 - W_o}{\text{Weight of sample taken}} \times 100$$

Crude fibre determination

The sample (2g) was accurately weighed into the fibre flask and 100 ml of 0.25N H_2SO_4 added. The mixture was heated under reflux for 1 hour with the heating mantle. The hot mixture was filtered through a fibre sieve cloth. The filtrate obtained was thrown off and the residue was returned to the fibre flask to which 100 ml of (0.31N NaOH) was added and heated under reflex for another 1 hour.

The mixture was filtered through a fibre sieve cloth and 10 ml of acetone added to dissolve any organic constituent. The residue was washed with about 50 ml hot water twice on the sieve cloth before it was finally transferred into the crucible. The crucible and the residue were oven-dried at 105°C overnight to drive off moisture. The oven-dried crucible containing the residue was cooled in a desiccator and later weighed to obtain the weight W_1. The crucible with weight W_1 was transferred to the muffle furnace for ashing at 550°C for 4 hours. The crucible containing white or grey ash (free of carbonaceous material) was cooled in the desiccator and weighed to obtain W_2. The difference $W_1 - W_2$ gives the weight of fibre [6]. The percentage fibre was obtained by the formula:

$$\text{\% Fibre} = \frac{W_1 - W_2}{\text{Weight of sample}} \times 100$$

HCN determination

0.1gm of the biscuit samples were weighed into a flat bottom plastic bottle with a screw cap lid. 0.5 ml of 0.1m phosphate buffer at pH 6 was added in a pipette. A yellow picrate paper attached to a plastic strip immediately in the flat bottom plastic bottle containing samples and buffer, making sure that the picrate paper does not touch the liquid in the bottle. The bottle was immediately closed with the screw capped lid. A blank was also prepared as above into another screw capped bottle.

Linamarin standard stock solutions were also prepared using 10 mg linamarin in 10 ml, 0.1 ml phosphate buffer at pH 6. This was diluted to give concentrations of 25 ppm to 100ppm (ie 25, 50, 75, 100). This was used to standardize and calibrate the spectrophotometer. Linamarin paper of 50ppm concentration was treated as sample above and put in a separate screw capped plastic bottle containing phosphate buffer and linamarase enzyme and the bottle was closed immediately.

All bottles containing samples, blank and linamarine standard paper were allowed to stand for 16-24 hours at room temperature. At the end of 16-24 hours, the bottles were opened, plastic baking sheet if the picrate paper were removed and placed in a test tube. 5 ml of distilled water was pipette into the test tube containing the picrate paper and was allowed to stand for 30 minutes with occasional gentle stirring. The absorbance of all solutions in the test tube including linamarin standard solution were measured against blank on spectronic 20 spectrophotometer at a wavelength of 510 mm [7].

The total cyanide content (ppm) or mg/kg=396×absorbance.

$$\text{\% Total cyanide content} = \frac{\text{ppm cyanide}}{10,000}$$

Results and Discussion

Chemical composition of biscuit samples

Table 1 describes the chemical composition of biscuit samples. The hydrocyanic acid content (HCN) of biscuit processed solely from cassava flour was significantly higher than other biscuit samples. Although level of HCN (9.70 mgkg^{-1}) is within the safe level [8]. It was also observed that there was a consistent drop in the level of HCN in cassava-pigeon pea biscuit with increase in fortification levels. This trend is as a result of increased presence of pigeon pea flour due to substitution. Pigeon pea flour processed from parboiling is known to be deficient in HCN due to the fact that HCN is heat labile [9]. HCN of commercial biscuit and 30% cassava-pigeon pea biscuit are not significantly different from each other at P<0.05. The crude protein content of 100% cassava biscuit was significantly lower than other biscuit samples with 30% cassava-pigeon pea biscuit having a crude protein content of (7.97%). Cassava flour is known to be poor in protein; there was also a gradual increase in the level of % crude protein in cassava-pigeon pea biscuit with increase in fortification. It agreed with the findings of Ubbor and Akobundu [4] who asserted that protein content of cassava based composite flours could be elevated through the incorporation of legume flours. The ash content of 30% cassava-pigeon pea biscuit was higher than other biscuit samples at P<0.05. Ash content also increased with increase in level of fortification. Pigeon pea flour is known to be rich in minerals. Crude fat which is responsible for maintenance of human health and brain development [10] was significantly greater in commercial biscuit (control) than other samples. This could be due to greater content of animal fat used in the preparation of the commercial biscuit. It was also evident that percentage crude fat increased with increasing level of fortification. The crude fibre and moisture content of commercial biscuit (control) was higher than other biscuit samples at (P<0.05). Gradual increase in crude fibre content of cassava-pigeon pea biscuit with greater measure of pigeon pea flour was also observed. There was however no significant difference in the moisture content percentage in

	HCN (mgkg^{-1})	Crude Protein (%)	Ash (%)	Crude Fat (%)	Crude Fibre (%)	Moisture Content (%)	Dry Matter (%)
100% Cassava Biscuit	9.70^a	6.57^d	1.53^c	0.11^c	0.19^d	3.00^b	97.00^a
10% Cassava/pigeon pea biscuit	8.32^b	6.92^{cd}	1.45^d	0.14^c	0.14^c	3.50^b	96.50^a
20% Cassava/pigeon pea biscuit	2.58^c	7.53^{ab}	1.60^b	0.24^b	0.24^c	3.00^b	97.00^a
30% Cassava/pigeon pea biscuit	0.06^d	7.97^a	1.67^a	0.33^b	0.47^b	3.00^b	97.00^a
Commercial biscuit (Control)	0.02^d	7.09^{bc}	1.16^e	13.54^a	0.85^a	4.81^a	95.19^b

[a-d]Means along the same column followed by the same letter are not significantly different from each other (P<0.05).

Table 1: Chemical composition of biscuit samples.

	Colour	Taste	Texture	Flavour	General Acceptability
100% Cassava Biscuit	8.10[a]	8.20[a]	7.40[a]	7.40[a]	8.30[a]
10% Cassava/pigeon pea biscuit	6.00[b]	4.40[c]	5.20[c]	4.80[c]	4.70[c]
20% Cassava/pigeon pea biscuit	6.00[b]	6.60[b]	5.50[bc]	6.00[bc]	5.80[bc]
30% Cassava/pigeon pea biscuit	8.00[a]	7.40[ab]	6.80[ab]	6.80[ab]	7.00[ab]
Commercial biscuit (Control)	8.10[a]	8.10[a]	7.20[a]	8.10[a]	8.10[a]

[a-c]Means along the same column followed by the same letter are not significantly different from each other (P<0.05).

Table 2: Sensory evaluation of biscuit samples

cassava-based biscuits, which was also observed with respect to their dry matter content. The values observed with regards to their moisture content and crude fibre were within the acceptable levels [11-13].

Sensory evaluation of the biscuit samples

Table 2 shows the sensory evaluation of the biscuit samples. There was no significant difference in the colour, taste, texture and general acceptability of 100 percent cassava biscuit and commercial biscuit. This implies this implies that acceptable biscuits can be processed from 100% percent cassava flour. However, 30 percent cassava-pigeon pea biscuit was high in colour, taste, texture, flavor and general acceptability. It is noteworthy that 20 and 30 percent cassava-pigeon pea biscuit were acceptable. This result agreed with the findings of Taiwo [14] who observed that biscuits processed from cassava and legume flours come out with the good sensory attributes.

Conclusion

The result of this research reveals that acceptable biscuits can be processed from 100% cassava flour and 30% cassava- pigeon pea flour. This finding will serve as an alternative for biscuit confectionary industries in Nigeria and other tropical countries in Africa that depends solely on imported and expensive wheat flour for processing. The nutritive value of pigeon pea in the biscuit could improve the nutritional status of children who consumes biscuit as snack food.

References

1. Ajani EN, Onwubuya EA (2013) Analysis of use of improved cassava production technologies among farmers in Anambra state, Nigeria. Wood Pecker Journal of Agricultural Research.

2. Onabolu AO, Oluwole OS, Bokanga M, Roshing H (2001) Ecology variation of intake of cassava food and dietary cyanide load in Nigerian communities. Public Health Nutrition 4: 871-876.

3. Nwosu JN, Ojukwu M, Ogueke CC, Ahaotu I, Owuamanam CJ (2013) The Antinutritional properties and ease of Dehulling on the proximate composition of Pigeon Pea (Cajanus cajan) as affected by malting. International Journal of Life Sciences 2: 60-67.

4. Ubbor SC, Akobundu ENT (2009) Quality characteristics of cookies from composite flours of watermelon seed, cassava and wheat. Pakistan Journal of Nutrition 8: 1097-1102.

5. Eneche EH (1999) Biscuit making potential of millet/pigeon pea flour blends. Plant Foods for Human Nutrition 54: 21-27.

6. AOAC (1990) Official method of Analysis 15th Edition Association of Official Analytical Chemists, Washington D.C.

7. Bradbury MG, Egan SV, Bradbury JH (1999) Determination of all forms of cyanogens in cassava and cassava products using picrate paper kits. Journal Sci Food Agric 79: 593-601.

8. Ashaye OA, Obatolu VA, Amusa NA, Fasoyiro SB, Ayoola OT, et al. (2007). Tips and Strategies for Profitable Cassava Production, Processing and Export.

9. Onwuka GI (2006) Soaking, boiling and antinutritional factors in Pigeon Peas. (Cajanus Cajan) and Cowpeas (Vigna. Unguiculata). Journal of Food Processing and Preservation 30: 616-630.

10. Neha M, Ramesh C (2012) Development of functional biscuit from soyflour and rice bran. International Journal of Agricultureal and Food Science 2: 14-20.

11. Oluwamukemi MO, Oluwalana TB, Akinbowale OF (2011) Physicochemical and sensory properties of wheat-cassava composite biscuit enriched with soy flour 5: 50-56.

12. Ogunjobi MA, Ogunwolu SO (2010) Physicochemical and sensory properties of cassava flour biscuits supplemented with cashew Apple powder. Jouirnal of Food Technology 8: 24-29.

13. Camire ME, Douggherty MP, Briggs JL (2007) Functionality of fruit powders in extruded corn breakfast cereals. Food Chemistry 101: 765-770.

14. Taiwo KA (2006) Utilization potentials of Cassava in Nigeria: The Domestic and Industrial Products. Food Reviews International 22: 29-42.

11

Anti-Nutritional Factors of Green Leaves of *Cassia obtusifolia* and Kawal

Algadi MZ[1]* and Yousif NE[2]

[1]*Arab Center for Nutrition, Muharraq, Bahrain*
[2]*Food Science and Technology, Khartoum University, Sudan*

Abstract

Cassia obtusifolia (family leguminous) is a wild African plant found in wastelands in the rainy season. Its leaves can be fermented (named kawal) and is used by people from the eastern part of Chad and the western part of Sudan as meat replacer or meat extender. The role of kawal and the like is in providing the sauces which make these staples palatable. During famine years, kawal, a protein source, probably protected many children against kwashiorkor. Until a few years ago, kawal was little known to most Sudanese, for it was a product confined to the western provinces of the country, away from populated areas and centers of influence. Then as today, kawal was shunned by the elite who consider it unfit for modern social life because of its repugnant, fetid odor that lingers on the fingers for hours. The objectives of this study were to assess the effect of fermentation on anti-nutritional factors of *Cassia obtusifolia* leaves. The *in vitro* protein digestibility was significant ($P < 0.05$) increased from 49.43 to 61.87%. It is recommended to use fermentation to decrease anti-nutritional factors of *Cassia obtusifolia*.

Keywords: Fermentation; Kawal; Anti-nutrutional; Phytic; Tannin

Introduction

The fermentation process that an African woman employed was behind the dramatic improvement in the protein value of the food [1]. The fermentation of meals of cereal and legumes knows to increases the protein content [2]. The international community of food scientists has, in the past decades, shown a deep interest in three areas of food science and technology. The first of these is the area of indigenous fermented foods, where a preponderance of literature has revealed interesting facts including a substantial enhancement of the food as a result microbial growth in it [3]. The second field of interest is the areas of solid substrate fermentation in which the substance to be fermented albeit wet, is not fluid [4]. The third area is that of leaf protein [5]. Scientists, in their relentless quest for new protein sources to help feed an ever-increasing world population, found that the plant leaf can be a truly commendable candidate. *Cassia obtusifolia* (family leguminous) is a wild African plant found in wastelands in the rainy season. Its leaves can be fermented (named kawal) and is used by people from the eastern part of Chad and the western part of Sudan as meat replacer or meat extender [6]. The role of kawal and the like is in providing the sauces which make these staples palatable. During famine years, kawal, a protein source, probably protected many children against kwashiorkor. Until a few years ago, kawal was little known to most Sudanese, for it was a product confined to the western provinces of the country, away from populated areas and centers of influence. Then as today, kawal was shunned by the elite who consider it unfit for modern social life because of its repugnant, fetid odor that lingers on the fingers for hours. The objectives of this study was to assess the effect of fermentation on the anti-nutritional factors of *Cassia obtusifolia*.

Material and Methods

Kawal preparation method

In kawal fermentation according to [6], the green leaves are first freed of all extraneous matter, such as leaves of other plants, pods and flowers of the kawal plant itself, caterpillars and insect-damaged leaves. This process of sorting out the kawal leaves is strictly observed and in fact this part of the preparation procedure is the most tedious step as it takes hours of painstaking work. Green flower buds and delicate young pods may, however, be processed with the green leaves. The unwashed, healthy green leaves, now clean from all adulteration are beaten in a mortar-and -pestle to give a green paste. Pounding is done in such a way that the leaves are crushed without releasing their juice. In the final paste can be seen partially crushed leaves, twigs, mid-ribs and petioles. Meanwhile, a pit is dug in the ground in a shaded cool place. And earthenware pot (Burma) is fitted into the pit, leaving only the neck of the container above ground. The green paste is now packed into the pot by hand. Next, green sorghum leaves are folded onto the surface of the leaf paste in the Burma so that it is completely covered. Washed, dry stones are then placed on top of the sorghum leaves to weight them down. The mouth of the pot is then covered with some metal tray or dish and the whole sealed off with mud to prevent insect from entering. Every 3-4 days the jar is opened, the now yellow and dry sorghum leaves removed and the Burma thoroughly hand-mixed and repacked, this time a little loose. Fresh sorghum leaves are folded on the surface of the past and weighted down as before the Burma covered and sealed off again. The paste is next molded into small. Irregular balls or flattish cakes which are then sun dried for 3-4 days. The duration of the fermentation is about 25 days for the supply of an average family. *Cassia obtusifolia* leaves and kawal were obtained in dry form after been sun dried and freed from foreign materials and powdered by hummer mill with same mesh size and was kept in clean bottles at room temperature for further use.

Anti-nutritional factors determination

Determination of phytic acid content: Phytic acid content was determined by the method described by Wheeler and Ferrel [7]. Two grams of dried sample were weighted in 125 ml conical flask. The sample was extracted with 50 ml of 3% trichloroacetic acid (TCA) for 3 hr with mechanical shaking. The supernatant was centrifuged for 5 min. ten milliliters aliquot of the supernatant was transferred to a 40 ml tube

---

*Corresonding author: Algadi MZ, Arab Center for Nutrition, Muharraq, Kingdom of Bahrain, E-mail: mutasim.algadi@gmail.com

and 4 ml of $FeCl_3$ ($Fecl_3$ solution containing 2 mg Fe^{+3} ion/ml 3% TCA) were then added to the aliquot. The tube was heated in a boiling water bath for 45 min. One or two drops of 3% sodium sulphate (Na_2SO_4) in 3% TCA were added the tube was cooled and centrifuged for 10-15 min and the clear supernatant was decanted. The precipitate was washed by dispersing well in 25 ml 3% TCA, heated for 10-15 min in boiling water bath and then centrifuged again. Washing was repeated with distilled water , the washed precipitate was dispersed in few milliliters of distill water enriched with 3 ml of 1.5 N NaOH , and the volume completed to approximately 30 ml with distilled water . Heated on boiling water bath for 30 min and hot filtered using whatman No.2. The precipitate was washed with 60-70 ml hot water, and the washing was decanted. The precipitate from the filter paper was dissolved in 40 ml hot 3.2 N HNO_3 and placed in 100 ml volumetric flask. The paper was washed with hot distilled water and the washing was collected in the same flask then completed to volume. A 0.5 ml of aliquot was taken from the above solution and transferred into 10 ml volumetric flask. then 2 ml of 1.5 N KSCN(potassium) were added and completed to volume by water then immediately (with one min) read at 480 nm using (SP6 PyUnieam) spectrophotometer.

A standard curve of different Fe $(NO_3)^3$ concentrations was plotted to calculate the ferric ion concentration. The phytate phosphorus was calculated from the Iron concentration assuming 4:6 Iron to phosphorus molar ration.

$$\text{Phytate (mg/100 g)} = \frac{6/4\ A \times C \times 20 \times 10 \times 50 \times 100}{1000 \times S}$$

Where:

A = optical density

C = concentration corresponding to optical density

S = weight of sample

Determination of tannin content: Tannin content (TC) of *Cassia obtusifolia* leaves and kawal samples were estimated using modified vanillin-HCl in methanol as described by Price et al. [8]. About 0.2 g of ground sample was placed in 100 ml conical flask. 10 ml of 1% Hcl in methanol (v/v) were added, the contents were mechanical shaking for 20 min and centrifuged at 2500 rpm for 5 min. One ml of supernatant was pipettes into a test tube and 5 ml of vanillin-Hcl reagent (mixing equal volume of 8% concentrated Hcl in methanol and 1% vanillin in methanol) were added. The optical density was read using a colorimeter (Lab system Analyzer 9 filters, j, Mitra and Bros.Pvt .Ltd.) at 500 nm after 20 min incubation at 30ºC, a blank sample was carried out with each run of samples. A standard curved was repeated expressing the result of tannic acid, i.e. amount of tannic (mg per ml) which gives color intensity equivalent to that given by tannin after correcting for blank.

Calculation:

$$TC\ (\%) = \frac{C \times 10 \times 100}{200}$$

Where:

C = concentration corresponding to optical density

10 = volume of extract in ml

200 = sample weight in mg

Total polyphenol (TP) determination: Polyphenolic of each sample was estimated using Prussian blue assay, as described by Price and Butter [9]. About 60 mg of ground sample was extracted with 3 ml methanol in a 50 ml conical flask, and then poured into a filter paper. The tube was quickly rinsed with additional 3 ml methanol and the content poured once into the filter paper. the filtrate was diluted to 50 ml with distilled water , mixed with 3 ml 0.1 M $Fecl_3$ in 0.1 N Hcl for 3 min, followed by the time addition of 3 ml 0.008 M $K_3Fe(CN)_6$. The absorption was read after 10 min at 720 nm on spectrophotometer (corning, 259).

Standard curve preparation: Tannic acid standard curve was prepared by dissolving 100 mg tannic acid in distilled water in a 0.1 liter volumetric flask and made up to mark . This spread stock solution of 100 ppm. Various standard concentrations (0, 2, 4, 6, 8 and 10) were repeated. The Prussian blue assay described above was then employed to the standard solution. The standard curve was obtained by plotting concentration against the corresponding absorbance reading, which gave linear relationship.

Calculation:

$$\text{Total polyphenol (\%)} = \frac{C \times 56 \times 100}{60}$$

Where,

C = concentration corresponding to optical density

56 = total volume

60 = weight of sample in milligrams

Results and Discussion

Anti-nutritional factors of green leaves of *Cassia obtusifolia* and kawal is shown on Table 1. Fermentation was found to cause highly significant decrease (p>0.05) in phytic acid content. The phytic acid content was decreased from 649.13 to 340.92 mg/100 g. Generally fermentation is known to cause highly reduction in phytic acid content due to the low pH of fermented dough which considered to be optimum for the phytase activity. Fermentation was found to cause highly significant decrease (p>0.05) in tannin content. The tannin content was decreased from 2.39 to 2.24%. The values obtained in this study were in agreement with the value obtained by Babiker et al. [10] who reported that the tannin content of green leaves of *Cassia obtusifolia* 2.34%. But lower than the value obtained by Abdalla [11] who reported 2.50% for tannin content of green leaves of *Cassia Obtusifolia*. Fermentation was found to cause degradation of tannin content and this may be due to the action of enzymes. Fermentation was found to cause highly significant decrease (p>0.05) in total polyphenol content. The total polyphenol content was decreased from 4.77 to 3.80%. The total polyphenol of green leaves of *Cassia obtusifolia* in this study were in agreement with the value obtained by Ousman et al. [12] who was

sample	Tannin %	Phytic acid mg/100g	Polyphenol %
Cassia obtusifolia Leaves	2.39 ($\pm$0.012)[a]	649.13 ($\pm$7.137)[a]	4.77m ($\pm$0.252)[a]
Dry Kawal	2.24 ($\pm$0.021)[b]	340.92 ($\pm$5.952)[b]	3.80 ($\pm$0.200)[b]

- Each value in an average of three values expressed on dry weight basis.
- Values are means ($\pm$ standard deviation).
- Means not sharing a common letter in a column are significant at $p \geq 0.05$ as assessed by Duncan's multiple range tests.

Table 1: Anti-nutritional factorof green leaves of *Cassia obtusifolia* and kawal (as dry matter).

reported 4.8%. Reduction in polyphenols may be due to activation of polyphenol oxidase [13].

References

1. Dirar HA (1992) Traditional fermentation technologies and food policy in Africa. Appropriate technology 19: 21-23.

2. Tinay AHE, Gadir AHA, Hidai ME (1979) Sorghum fermented Kisra bread. 1: nutritive value of Kisra. Journal of Science Food and Agriculture 30: 859-563.

3. Steinkrause KH (1978) Contribution of Asian fermented foods to international food science and technology. Global Impacts in Applied Microbiology 5: 173-179.

4. Stanton WR (1978) Solid-substrate fermentation. Global impacts in applied microbiology 5: 180-189.

5. Pirie NW (1978) Leaf protein and other aspects of fodder fraction. Cambridge press, UK.

6. Dirar HA (1993) The Indigenous Fermented food of the Sudan. A study of African Food and Nutrition. CAB International, Walling ford.

7. Wheeler EL, Ferrel RE (1971) A Method for phytic acid determination in wheat and wheat fractions. American Association of Cereal chemists 28: 313-320.

8. Price ML, Scoyoc SV, Butler LG (1978) A critical evaluation of the vanillin reaction as an assay for tannin in sorghum grain. Agricultural Food Chemistry 26: 1214-1218.

9. Price ML, Butler LG (1977) Rapid visual estimation and spectrophotometric determination of tannin content of sorghum grain. Journal of Agriculture and Food Chemistry 25: 1268-1273.

10. Babiker A, Khalifa AEO, El Tinay AH (1998) A note on the effect of tannin content and the *in vitro* protein digestibility of some Sudanese flora. Journal of agricultural sciences 6: 157-161.

11. Abdalla ZED (1989) Studies on the chemical composition, Phenolic and polyphenolic constituents of some leafy vegetables consumed in Sudan. University of Khartoum, Sudan.

12. Ousman A, Ngassoum M, Kamga C (2005) Chemical composition of *Cassia Obtusifolia L.* leaves. Journal of Food Technology 3: 453-455.

13. Dhanker N, Chauhan BM (1987) Effect of temperature and fermentation time on phytic acid and polyphenol content of rabadi – A fermented pearl millet food. Journal of Food Science 52: 828-829.

12

Effect of Baked, Whipped and Fermentation on Antioxidant Activity in Red Raspberries

Darwish AZ[1]*, Bayomy H[2] and Rozan M[2]

[1]Dairy Science Department, Faculty of Agriculture, Assiut University, Egypt
[2]Food Science and Technology Department, Faculty of Agriculture, Damanhour University, Behira, Egypt

Abstract

Red raspberries (*Rubus idaeus*) are a good source of antioxidants and contains appreciable levels of phenolic compounds (TPC). Adding raspberry to the product are attributed the most significant health benefits of to the phenolic compounds. This study examined the three different manufacturing processes baked, whipped and fermentation on antioxidant activity in red raspberry. The phenolic compounds in red raspberry, sponge cake, whipping cream and yoghurt by red raspberry were determined by HPLC. Sensory evaluation found that the best proportions to add red raspberry to whipped cream and yogurt is 10% but in the sponge cake is 15%. The total phenols were 56%, 37% and 4%, 3% of red raspberry, red raspberry-yoghurt, red raspberry-whipped cream and red raspberry-sponge cake respectively. So the treatments were order in general to their effect of the TPC: fermentation > whipped > baked.

Keywords: Antioxidant; Baked; Fermentation; Red raspberry; Whipped

Introduction

Recently, Plants of the genus *Rubus* (family: Rosaceae) have been reported to exhibit several biological activities such as anti-diabetic, anti-oxidative, anti-inflammatory and anti-hyperlipidemic activities. These biological activities were due to their polyphenol components including anthocyanins present in some of the varieties [1]. Red raspberries (*Rubus idaeus*) are among the fruits containing the highest antioxidant levels. In addition to vitamin C, the antioxidant activity of red raspberries is primarily constituted by two classes of compounds: anthocyanins and ellagitannins. Ellagitannins, which are complex derivatives of ellagic acid [2], have been identified in tea, many medicinal plants, and several fruits, including raspberries [3,4]. In addition to their vasorelaxation properties [5], ellagitannins have been described to have general antioxidant effects [6]. Red raspberry could therefore be considered as a model fruit source for a variety of potentially healthy compounds [7]. Berries, fruits full of bioactive compounds, are also very delicious, have low energy [8]. To the bioactive compounds group in berries belong antioxidants such as phenolic compounds and fruit colorants (anthocyanins and carotenoids). Berries' phenolics represent a diverse group of compounds including phenolic acids, such as hydroxybenzoic and hydroxycinnamic acid conjugates; flavonoids, such as flavonols, flavanols, and anthocyanins. In addition, tannins, divided into condensed tannins (proanthocyanidins) and hydrolyzable tannins, are reported to be important bioactive compounds. These compounds are of great interest for nutritionists and food technologists due to the opportunity to use bioactive compounds as functional foods ingredients. Nutraceuticals and functional foods have become very popular for people due to the consumer demands for healthy nutraceutical foods that could possibly reduce some health risks and improve various health conditions [9]. The anthocyanins are stable at low temperatures and in the dark [10]. The highest temperature combined with a short baking time had the best effect on the preservation of polyphenols, in order to achieve the most favorable nutritional effect of baked products enriched with fruit pomace [11]. pH is closely related to microbial growth and the structural changes in phytochemicals during fermentation. For example, anthocyanin breakdown is dependent on the pH in the presence of oxygen, is also directly related to the level of pseudo base, and is inversely related to the cation concentration [12]. Several studies [13,14] have shown that anthocyanins are stable at low pH. Anthocyanins exhibit the highest stability, with the red flavylium cation stable around pH 1-2 [14]. The stability of anthocyanins is dependent on their structure; for instance, acylated anthocyanins are more stable than the non-acylated forms [15]. pH is a dominant factor in the radical scavenging capacity of wine anthocyanins, as an increase in pH often increases the capacity for radical scavenging [16]. Colour plays a very important role in the acceptability of some foods by many consumers [17]. In practice, most manufacturers tend to colour products which have dull colours and look unappealing to most consumers. Synthetic colourants have often been used in attempts to colour some foods and beverages [18]. However, the demand for foods with synthetic anthocyanins are a great interest as alternatives to synthetic colourants due to their bright colours and associated health benefits [19,20]. They are considered to be safe because they have been consumed for centuries in fruits, and vegetables without any health risks [21]. Whole fruit extracts containing non-acylated anthocyanins from *Berberis boliviana* L. showed improved colour and pigment stability when incorporated in yoghurt [22]. The anthocyanins are stable at low temperatures and in the dark [10] For that we believe that whole fruit juice extracts from red raspberries (*Rubus idaeus*) could serve as an appropriate colorant and nutraceutical in yoghurt. Raspberry fruits are rich in phenolic compounds contents such as phenolic acids [23,24] flavonoids [24,25] and anthocyanins [23]. The phenolic compounds in berries have been reported to have antioxidant, anticancer, antiinflammatory, and antineurodegenerative biological properties [26,27]. In recent years, red raspberry anthocyanins have in many occasions been applied in baked foods, and confectioneries [28]. In this study, we decided to investigate whether anthocyanins from Red raspberry (*Rubus idaeus*)

---

*Corresponding author: Darwish AZ, Dairy Science Department, Faculty of Agriculture, Assiut University, Egypt, E-mail: zahraadarwish@yahoo.com

could be used as potential colour additives in yoghurt since yoghurt has a low pH and it is stored under refrigerated conditions. Therefore, the current study was undertaken to use the red raspberries (*Rubus idaeus*) in industry of yoghurt, cake and the formation of cream as colorant. And study the effect of these processes on the antioxidant activity.

Methods

Materials

Whipping cream from the brand Almarai (Kingdom of Saudi Arabia) was purchased from the local supermarket. This is an ultrahigh temperature (UHT) product containing 33% milk fat, 1.9% protein and 3.5% carbohydrate. Cream was kept in fridge at 5°C or below during storage. Commercial wheat flour was purchased from Kuwait Flour mills & Bakeries Co. (Kuwait). Sunflower oil, sucrose, batter and fresh eggs were purchased from local market in Tabuk, Kingdom of Saudi Arabia.

Sponge cake preparation

The sponge cakes were prepared according to Chaiya and Pongsawatmanit [29]. The experiments used sponge cake batter formulations containing WF (50-100 g), 140 g liquid whole eggs, 10 g whole milk powder, 2 g baking powder, 120 g sugar, 80 g butter and 40 g water. In the cake batter preparation ~500 g), the liquid whole eggs, water, sugar were mixed in a using Kenwood-kitchen machine 1200 W (Chine) with machine speeds from 1 to 10 at speed 3 for 1 min and further mixed at speed 6 for 9 min. Then, dry ingredients (the flour blend of WF, whole milk powder and baking powder) were added simultaneously to the mixture at speed 1 for 1 min and further mixed at speed 3 for 2 min. The melted butter was added finally and mixed at speed 1 for 20 s. The batter was divided into four portions formulation each one 125 g (control and three with red raspberry puree as following: red raspberry spongy cake 10% (12.5 g), red raspberry spongy cake 15% (18.75 g) and red raspberry spongy cake 25% (31.25 g). Each one was placed in pan (8.5 × 16 × 5 cm) and baked in oven at 175°C for 20 min. After baking, the cakes were removed from the pans, cooled upside down on a wire rack for 30 min at room temperature and kept in plastic bags to prevent drying before being measured for sensory evaluation within 12 h.

Whipped cream preparing

Cream must be kept in fridge at least 24 h before tempering. In each experiment, 100 ml cream was mixed with different mount of raspberry puree (5, 10 and 15 g) using Kenwood-kitchen machine 1200 W (Chine).

Yoghurt preparing

Yoghurt was manufactured using lactobacillus delbrueckii supsp bulgaricus and streptococcus salivarious subsp thermophilus (1:1) commercial starter culture. Fresh cow's milk was heated to 90°C for 10 min, and cooled to 42°C. Milk was mixed with different percentage of red raspberry (5, 10 and 15%) after that adding 3% starter and incubated at same temperature.

Proximate analysis

Materials (red raspberry puree, cream, yoghurt and sponge cake) were dried at 70°C to a constant weight, moisture contents, ash, and crude fiber were determined by AOAC [30] methods. Total lipids from the samples were extracted with chloroform/methanol (2:1, v/v) and quantified gravimetrically [31]. Nitrogen content (N) of the sample was estimated by the method described by Kjeldahl [32] and crude protein

was calculated as N × 6.25 [33]. The amount of total carbohydrates was obtained by the difference between weight of the sample taken and sum of its moisture, ash, total lipid, protein, and fiber contents [34].

Analysis of total phenolic content

The total phenolic content was determined according to the Folin-Ciocalteu procedure [35]. Briefly, the extract (500 μl) was transferred into a test tube and oxidized with the addition of 250 μl of Folin-Ciocalteu reagent. After 5 min, the mixture was neutralized with 1.25 ml of 20% aqueous Na_2CO_3 solution. After 40 min, the absorbance was measured at 725 nm against the solvent blank. The total phenolic content was determined by means of a calibration curve prepared with gallic acid, and expressed as μg of gallic acid equivalent (GAE) per ml of sample.

Analysis of total flavonoid content

The total flavonoid content was determined according to Zilic et al. [35]. Briefly, 250 μl of 5% $NaNO_2$ was mixed with 500 μl of extract. After 6 min, 2.5 ml of a 10% $AlCl_3$ solution was added. After 7 min, 1.25 ml of 1 M NaOH was added, and the mixture was centrifuged at 5000 g for 10 min. Absorbance of the supernatant was measured at 510 nm against the solvent blank. The total flavonoid content was expressed as μg of catechin equivalent (CE) per ml of sample.

Determination of radical DPPH scavenging activity

Free radical scavenging capacity was determined using the stable 1,1-Diphenyl-2-picryl-hydrazyl (DPPH⋅) according to Hwang and Thi [36]. The final concentration was 50 μM for DPPH and the final reaction volume was 3.0 ml. The absorbance at 517 nm was measured against a blank of pure methanol at 60 min. Percent inhibition of the DPPH free radical was calculated by the following equation:

$$\text{Inhibition (\%)} = 100 \times (A_{control} - A_{sample})/A_{control}$$

Where

$A_{control}$ is the absorbance of the control reaction (containing all reagents except the test compound).

A_{sample} is the absorbance of the test compound. Also, the antioxidant activity was determined by means of a calibration curve prepared with Trolox, and expressed as mg of Trolox equivalent (TE) per unit (volume or weight) of sample.

Phenolic acids profile

Extraction of phenolic compounds: The sample was alkaline hydrolyzed according to Kim et al. [37]. Sample (1 g) was placed in quick fit conical flask and 20 ml of 2 M NaOH was added and the flasks were flushed with N_2 and the stopper was replaced. The samples were shacked for 4 h at room temperature. The pH was adjusted to 2 with 6 M HCl. The samples were centrifuged at 5000 rpm for 10 min and the supernatant was collected. Phenolic compounds were extracted twice with 50 ml ethyl ether and ethyl acetate 1:1. The organic phase was separated and evaporated at 45°C and the samples redissolved in 2 ml methanol.

Analysis of phenolic compounds by HPLC: HPLC analysis was carried out using Agilent Technologies 1100 series liquid chromatograph equipped with an auto sampler and a diode-array detector. The analytical column was an Eclipse XDB-C18 (150 × 4.6 μm; 5 μm) with a C18 guard column (Phenomenex, Torrance, CA). The mobile phase consisted of acetonitrile (solvent A) and 2% acetic acid in water (v/v) (solvent B). The flow rate was kept at 0.8 ml/min for a total run time of 70 min and the gradient programme was as follows: 100% B to 85% B in 30 min, 85% B to 50% B in 20 min, 50% B to 0% B in 5 min and 0% B to 100% B in 5 min. The injection volume was 50 µl and peaks were monitored simultaneously at 280 and 320 nm for the benzoic acid and cinnamic acid derivatives, respectively. All samples were filtered through a 0.45 µm Acrodisc syringe filter (Gelman Laboratory, MI) before injection. Peaks were identified by congruent retention times and UV spectra and compared with those of the standards.

Sensory evaluation

The hedonic test was used to determine the degree of overall liking for the whipped cream, sponge cake and yogurt. For this study, untrained consumers were recruited from the students, staff. All consumers were interested volunteers and informed that they would be evaluating whipped cream, sponge cake and yogurt. 15 consumers (7 males and 8 females, 19-55 years) received samples were asked to rate them based on degree of liking on a seven-point hedonic scale (1 = dislike extremely, 4 = neither like nor dislike, 7 = like extremely). Samples were placed on white plates and identified with random numbers. Panelists evaluated the samples in a testing area and were instructed to rinse their mouths with water between samples to minimize any residual effect [38]. Where the evaluation in terms of color, taste and smell and textures in addition to the overall acceptance.

Statistical analysis

Statistical analysis of experimental data was performed by analysis of variance (ANOVA) producers using SPSS version 9.0 program to examine statistical significance differences of sensory analysis means of experimental data. Results were considered statistically significant when p < 0.05. Mean ± standard deviation values were also presented.

Results and Discussion

Sensory evaluation

Results of sensory evaluation for reach to the best proportions to add red raspberry to whipped cream, sponge cake and yogurt are reported in Table 1. When evaluated by untrained consumers, statistically significant differences were detected in all of the sensory attributes evaluated (P ≤ 0.05). It is clear that the best proportions to add red raspberry to whipped cream and yogurt is 10% but in the sponge

cake is 15%. With regard to color, taste, smell texture and the overall acceptance 10% red raspberry whipped cream, 10% red raspberry yogurt and 15% red raspberry sponge cake were appreciated the most significantly higher preference scores than the other treatments (P ≤ 0.05).

Proximate composition

Table 2 describes the proximate composition of red raspberry puree and foods that have been selected from the sensory evaluation, which have ratios of red raspberry. These foods include whipped cream (with 10% red raspberry), yoghurt (with 10% red raspberry) and sponge cake (with 15% red raspberry).

Total phenolic content

The total phenolic content (TPC) for sponge cake, yoghurt and whipped cream (Table 3). Highest TPC in samples was found in sponge cake (0.709 ± 0.08 mg GEA/ml) which were hated at 175°C for 20 min. that may case decreasing percentage in TPC than red raspberry because though heat-treated lowered the antioxidant level, and adding ingredients such as sugar diluted the antioxidant concentration, products made from berries are high sources of antioxidants [39-41]. The highest value compare with yogurt and whipped cream may be due to the production of Maillard reaction products in the crust during thermal processing [42]. Similar observations have been made when baking rhubarb, whereby both TPC and FRAP AA were higher during the first 20 min and then decreased to low levels [43] and when baking chocolate cookies and chocolate cakes made with baking powder rather than baking soda [44]. TPC in red raspberry-whipped cream was 0.081 ± 0.004 mg GEA/ml decreasing percentage in TPC this may be due to whipped processes. Yogurt has recorded the lowest content of TPC 0.067 ± 0.001 mg GEA/ml. During fermentation process microbial yoghurt utilization of phenolic acids such as ferulic and p-coumaric acid and post acidification lead to the production of other phenolic acids such as vanillic and p-hydroxybenzoic acids before the aromatic ring structure is broken down [45]. Also the decreasing of TPC than the red raspberry there were increasing in the TPC red raspberry-yogurts than plain-yoghurt 0.008 mg GEA/ml that can be explained by the presence of indigenous phytochemical compounds in raspberry (e.g., flavonoids and phenolic compounds) [46]. The major TPC were 63.929 ± 3.000, 46.162 ± 5.100, 38.617 ± 5.000 and 11.320 ± 1.000 ug/g of gallic

	Whipped cream			Sponge cake			Yogurt		
	5%	10%	15%	5%	10%	15%	5%	10%	15%
Color	7.4 ± 0.06[c]	9.4 ± 0.02[a]	8.3 ± 0.03[b]	7.3 ± 0.04[b]	7.4 ± 0.05[b]	8.7 ± 0.04[a]	5.7 ± 0.06[c]	8.8 ± 0.03[a]	7.1 ± 0.01[b]
Taste	8.3 ± 0.02[b]	9.2 ± 0.02[a]	8.7 ± 0.05[b]	6.3 ± 0.07[a]	6.8 ± 0.07[a]	6.5 ± 0.07[a]	8.4 ± 0.03[b]	9.5 ± 0.02[a]	9.5 ± 0.02[a]
Smell	8.0 ± 0.04[b]	9.3 ± 0.03[a]	9.2 ± 0.03[a]	6.1 ± 0.05[b]	6.3 ± 0.03[b]	8.7 ± 0.06[a]	7.6 ± 0.04[b]	8.4 ± 0.04a[b]	8.7 ± 0.02[a]
Texture	7.9 ± 0.04[a]	7.4 ± 0.04[a]	6.3 ± 0.07[c]	8.9 ± 0.03[a]	8.3 ± 0.05[a]	8.2 ± 0.02[a]	9.2 ± 0.03[a]	9.4 ± 0.02[a]	8.3 ± 0.05[b]
Overall acceptance	8.5 ± 0.03[a]	9.2 ± 0.04[a]	8.6 ± 0.04[a]	7.8 ± 0.03[a]	7.8 ± 0.01[a]	8.4 ± 0.03[a]	7.6 ± 0.05[c]	9.1 ± 0.03[a]	8.5 ± 0.04[b]

Means within the same row without a common letter (a-c) are significantly different (P ≤ 0.05) for each type.

Table 1: Sensory evaluation for red raspberry-whipped cream, red raspberry-sponge cake and red raspberry-yogurt.

Proximate composition (g/100 g FW)	Red raspberry fruit	Whipped cream	Yoghurt	Sponge cake
Moisture	81.43 ± 2.06	61.53 ± 2.58	87.71 ± 2.73	35.82 ± 1.06
Ash	0.51 ± 0.08	0.14 ± 0.07	0.79 ± 0.08	1.22 ± 0.14
Lipid	0.57 ± 0.12	33.07 ± 1.07	3.26 ± 0.22	19.6 ± 0.77
Protein	1.49 ± 0.24	1.64 ± 0.32	3.42 ± 0.41	5.95 ± 0.30
Fiber	1.59 ± 0.68	0.23 ± 0.04	0.21 ± 0.09	0.87 ± 0.06
TC	14.41 ± 1.63	3.39 ± 0.24	4.69 ± 0.38	36.54 ± 1.21

FW: Fresh Weight; DW: Dry Weight; TC: Total Carbohydrates; Values (Mean ± SD)

Table 2: Proximate composition of red raspberry, whipped cream (with 10% red raspberry), yoghurt (with 10% red raspberry) and sponge cake (with 15% red raspberry).

Compound(ug/g)	Red raspberry	Sponge cake	Whipped cream	Yoghurt
Gallic acid	63.929 ± 3.000	35.497 ± 4.000	19.365 ± 2.000	16.996 ± 2.000
Protochatchuic acid	11.320 ± 1.000	4.814 ± 0.300	17.172 ± 5.300	28.343 ± 5.300
Gentisic acid	0.965 ± 0.070	ND	ND	ND
Catachine	46.162 ± 5.100	34.079 ± 6.000	37.875 ± 6.200	31.966 ± 4.000
Chlorgenic acid	ND	ND	ND	ND
Caffeic acid	4.073 ± 0.400	3.511 ± 0.200	2.458 ± 0.100	2.156 ± 0.300
Syrngic acid	ND	0.467 ± 0.030	ND	ND
Vanillic acid	1.832 ± 0.100	22.803 ± 0.300	2.938 ± 0.100	0.907 ± 0.400
Rutin	ND	ND	ND	ND
Coumarin	38.617 ± 5.000	29.571 ± 0.300	12.032 ± 1.00	10.680 ± 1.000
Ferulic acid	8.342 ± 0.300	1.061 ± 0.100	2.991 ± 0.002	2.058 ± 0.003
Sinapic acid	4.722 ± 0.006	3.094 ± 0.300	1.453 ± 0.001	1.542 ± 0.001
Rosmarinic acid	2.150 ± 0.003	0.751 ± 0.006	ND	0.569 ± 0.004
Cinnamic acid	1.274 ± 0.004	1.334 ± 0.400	0.530 ± 0.003	0.582 ± 0.002
Qurecetin	0.537 ± 0.003	ND	0.499 ± 0.003	0.169 ± 0.003
Kaempferol	0.709 ± 0.005	0.398 ± 0.002	2.717 ± 0.3	0.189 ± 0.001
Chyrsin	3.647 ± 0.004	3.075 ± 0.200	4.873 ± 0.300	2.090 ± 0.200

Mean ± standard deviation (n=3).

Table 3: Contents of main phenolic compounds of red raspberry, raspberry-sponge cake, raspberry-whipped cream and raspberry-yogurt by HPLC.

Sample	Total phenols (mg GAE/g)	Total flavonoids (mg CE/g)	DPPH (mg TE/g)
Red raspberry	1.088 ± 0.02	0.216 ± 0.003	2.270 ± 0.5
Yoghurt	0.067 ± 0.05	0.026 ± 0.001	0.289 ± 0.07
Whipped cream	0.081 ± 0.004	0.069 ± 0.001	0.353 ± 0.002
Sponge cake	0.709 ± 0.08	0.151 ± 0.004	0.591 ± 0.03

Mean ± standard deviation (n=3).

Table 4: Total phenols, total flavonoids and DPPH of red raspberry, raspberry-sponge cake, raspberry-whipped cream and raspberry-yogurt.

acid, catachine, coumarin and protochatchuic acid, respectively in the red raspberry. On the other hand, gallic acid, catachine, coumarin and vanillic acid were recorded 35.497 ± 4.000, 34.079 ± 6.000, 29.571 ± 0.300 and 22.803 ± 0.300ug/g respectively in the red raspberry- sponge cake. The higher TPC in the red raspberry-whipped cream were 37.875 ± 6.200, 19.365 ± 2.000 and 17.172 ± 5.300 and 12.032 ± 1.00 ug/g of catachine, gallic acid, protochatchuic acid and coumarin, respectively. The highest TPC found in red raspberry-yoghurt were 31.966 ± 4.000 28,343 ± 5.300, 16.996 ± 2.000 and 10.680 ± 1.000 ug/g of catachine, protochatchuic acid, gallic acid and coumarin, respectively. Adding raspberry to the product are attributed the most significant health benefits of to the phenolic compounds, such as flavonoids, phenolic acids [47]. So the treatments were order in general to their effect of the TPC: fermentation > whipped > baked.

DPPH scavenging activity

Red raspberry-sponge cake had higher antioxidant activity than red raspberry-whipped cream and raspberry-yogurt (Table 3). Highest DPPH in red raspberry-sponge cake than in red raspberry -whipped cream and red raspberry-yogurt were most likely contributed by individual phytochemical contents and as a result of microbial metabolic activities [48]. The lowest in yoghurt may be due to attributed to the metabolically active yogurt bacteria [49]. High antioxidant activities useful for protective cardiovascular effect [50]. The DPPH radical-scavenging highest with heat followed by whipped and fermentation. The DPPH radical-scavenging activity was 0.591, 0.353 and 0.289 mg TE/g of red raspberry-sponge cake, red raspberry-whipped cream and red raspberry-yogurt respectively (Table 4).

Total flavonoids

Table 2 shows that the baking treatment had higher total flavonoids than the whipping and fermentation. The mean values of red raspberry

were 0.216 ± 0.003 mg CE/g followed by red raspberry-sponge cake value 0.151 ± 0.004 mg CE/g and red raspberry-whipped cream 0.069 ± 0.001 mg CE/g, while yoghurt was recorded lowest value 0.026 ± 0.001 mg CE/g. flavonoid compounds of raspberry have significant antioxidant activities consumption may help prevent and/or moderate chronic diseases these antioxidant properties and health benefits and for tailored breeding for functional foods [51].

Conclusion

Red raspberry sponge cake could be further used as a source of natural antioxidants for application in the nutraceutical or functional food areas. Much higher losses of total phenols were found in baked, whipped and fermentation treatment. A strong effect of treatment on phenols content was found in fermentation treatment. The products produced with red raspberry contained 56%, 37% and 4%, 3% of raspberry total phenols of red raspberry, red raspberry -yoghurt, red raspberry -whipped cream and red raspberry-sponge cake.

References

1. Harauma A, Murayama T, Ikeyama K, Sano H, Arai K, et al. (2007) Mulberry leaf powder prevent atherosclerosis in apolipoprotein E-deficient mice. Biochem Biophys Res Commun 358: 751-756.

2. Quideau S, Feldman KS (1996) Ellagitannin chemistry. Chem Rev 96: 475-504.

3. Clifford MN, Scalbert AE (2000) Nature occurrence and dietary burden. J Sci Food Agric 80; 1118-1125.

4. Kahkonen MP, Hopia AI, Heinonen M, Berry S (2001) Phenolics and their antioxidant activity. J Agric Food Chem 49: 4076-4082.

5. Mullen W, McGinn J, Lean MEJ, MacLean MR, Gardner P, et al. (2002) Ellagitannins, flavonoids and other phenolics in red raspberries and their contribution to antioxidant capacity and vaso-relaxation properties. J Agric Food Chem 50: 5191-5196.

6. Okuda T, Yoshida T, Hatano T (1989) Ellagitannins as active constituents of medicinal plants. Planta Med 55: 117-122.

7. Beekwilder J, Jonker H, Meesters P, Hall R, Meer IM, et al. (2005) Antioxidants in raspberry on-line analysis links antioxidant activity to a diversity of individual metabolites. J Agric Food Chem 53: 3313-3320.

8. Namiesnik J, Vearasilp K, Nemirovski A, Leontowicz H, Leontowicz M, et al. (2014) In vitro studies on the relationship between the antioxidant activities of some berry extracts and their binding properties to serum albumin. Appl Biochem Biotechnol 172: 2849-2865.

9. Skrovankova S, Sumczynski D, Mlcek J, Jurikova T, Sochor J, et al. (2015) Bioactive compounds and antioxidant activity in different types of berries. Int J Mol Sci 16: 24673-24706.

10. Aramwit P, Bang N, Srichana T (2010) The properties and stability of anthocyanins in mulberry fruits. Food Res. Int 43: 1093-1097.

11. Paweł G, Karina JR, Vitalijs R, Inga P, Iveta P (2016) The impact of different baking conditions on the stability of the extractable polyphenols in muffins enriched by strawberry sour cherry raspberry or black currant pomace. LWT-Food Sci Technol 65: 946-953.

12. Su MS, Chien PJ (2007) Antioxidant activity anthocyanins and phenolics of rabbit-eye blueberry (Vaccinium ashei) fluid products as affected by fermentation. Food Chemistry 104: 182-187.

13. Cabrita L, Fossen T, Andersen OM (2000) Colour and stability of the six common anthocyanidin 3-glucosides in aqueous solutions. Food Chemistry 68: 101-107.

14. Nielsen ILF, Haren GR, Magnussen EL, Dragsted LO, Rasmussen SE, et al. (2003) Quantification of anthocyanins in commerce black currant juices by simple high-performance liquid chromatography. Investigation of their pH stability and antioxidative potency. J Agricultural and Food Chemistry 51: 5861-5866.

15. Devi PSSM, Mohandas S (2012) The effects of temperature and pH on stability of anthocyanins from red sorghum (Sorghum bicolor) bran. African J Food Sci 6: 567-573.

16. Borkowski T, Szymusiak H, Gliszczynska-Swiglo A, Rietjens IMCM, Tyrakowska B, et al. (2005) Radical scavenging capacity of wine anthocyanins is strongly pH-dependent. J Agricultural and Food Chemistry 53: 5526-5534.

17. Giusti MM, Wrolstad RE (2003) Acylated anthocyanins from edible sources and their applications in food systems. Biochemical Engineering Journal 14: 217-225.

18. Delgados-Vargas F, Paredes-López O (2003) Pigments as colourants. In natural colourants for food and nutraceutical uses. CRC Press: Boca Raton, London, New York, Washington DC.

19. Cevallos-Casals BA, Cisneros-Zevallos L (2004) Stability of anthocyanin-based aqueous extracts of Andean purple corn and red fleshed sweet potato compared to synthetic and natural colorants. Food Chem 86: 69-77.

20. Fan G, Han Y, Gu Z, Gu F (2008) Composition and colour stability of anthocyanins extracted from fermented purple sweet potato culture. LWT-Food Sci Tech 41: 1412-1416.

21. Bridle P, Timberlake CF (1997) Anthocyanins as natural food colors selected aspects. Food Chem 58: 103-109.

22. Wallace TC, Giusti MM (2008) Determination of color pigment and phenolic stability in yogurt systems colored with nonacylated anthocyanins from Berberis boliviana L as compared to other natural/synthetic colorants. J Food Sci 73: 241-248.

23. Rommel A, Wrolstad RE (1993) Influence of acid and base hydrolysis on the phenolic composition of red raspberry juice. J Agric Food Chem 41:1237-1241.

24. Turkben C, Sarıburun E, Demir C, Uylaser V (2010) Effect of freezing and frozen storage on phenolic compounds of raspberry and blackberry cultivars. Food Anal Methods. 3: 144–153.

25. Hakkinen SH, Karenlampi SO, Heinonen IM, Mykkanen HM, Torronen AR, et al. (1998) HPLC method for screening of flavonoids and phenolic acids in berries. J Agric Food Chem; 77: 543-551.

26. Seeram NP, Adams LS, Zhang Y, Lee R, Sand D, et al. (2006) Black-berry black raspberry blueberry cranberry red raspberry and strawberry extracts inhibit growth and stimulate apoptosis of human cancer cell in vitro. J Agric Food Chem 54: 329-339.

27. Seeram NP (2008) Berry fruits compositional elements biochemical activities, and the impact of their intake on human health performance and disease. J Agric Food Chem 56: 627-629.

28. Wu X, Liang L, Zou Y, Zhao T, Zhao J, et al. (2010) Aqueous two-phase extraction, identification and antioxidant activity of anthocyanins from mulberry (Morus atropurpurea Roxb.) Food Chem 2: 443-453.

29. Chaiya B, Pongsawatmanit R (2011) Quality of batter and sponge cake prepared from wheat-tapioca flour blends. Kasetsart J Nat Sci 45: 305-313.

30. AOAC (2000) Association of official analytical chemists. (17thedn). Official method of analysis Washington D.C., USA.

31. Christie VW (1983) Lipids. In: Aliphatic and related natural product chemistry. (2ndedn) Pergamon Press Oxford.

32. Kjeldahl J (1983) Determination of protein nitrogen in food products. Encyc Food Agric 28: 757-765.

33. Imran M, Khan H, Hassan SS, Khan R (2008) Physico-chemical characteristics of various milk samples available in Pakistan. J Zhejiang Univ Sci B 9: 546-551.

34. Muler HG, Tobin G (1980) Nutrition of food processing. Croom Helm Ltd, London.

35. Zilic S, Serpen A, Akıllıoglu G, Jankovic M, Gökmen V (2012) Distributions of phenolic compounds yellow pigments and oxidative enzymes in wheat grains and their relation to antioxidant capacity of bran and debranned flour. J of Cereal Sci 56: 652-658.

36. Hwang ES, Do-Thi N (2014) Effects of extraction and processing methods on antioxidant compound contents and radical scavenging activities of laver (Porphyra tenera). Prev Nutr Food Sci 19:40-48.

37. Kim KH, Tsao R, Yang R, Cui SW (2006) Phenolic acid profiles and antioxidant activities of wheat bran extracts and the effect of hydrolysis conditions. Food Chemistry 95: 466-473.

38. Stone H, Sidel JL (1993) Sensory evaluation practices (2ndedn), Academic press San Diego CA.

39. Patras A, Brunton NP, O'Donnell C, Tiwari BK (2010) Effect of thermal processing on anthocyanin stability in foods mechanisms and kinetics of degradation. Trends Food Sci Technol 21: 3-11.

40. Dinstel RR, Cascio J, Koukel S (2013) The antioxidant level of Alaska's wild berries high higher and highest. International J Circumpolar Health.

41. Rudy S, Dziki D, Krzykowski A, Gawlik-Dziki U, Polak R, et al. (2015) Influence of pre-treatments and freeze-drying temperature on the process kinetics and selected physico-chemical properties of cranberries (Vaccinium macrocarpon Ait.). LWT Food Sci Technol 63: 497-503.

42. Lindenmeier M, Hofmann T (2004) Influence of baking conditions and precursor supplementation on the amounts of the antioxidant pronyl-l-lysine in bakery products. J Agricultural and Food Chem 52: 350-354.

43. McDougall GJ, Dobson P, Jordan-Mahy N (2010) Effect of different cooking regimes on rhubarb polyphenols. Food chem 119: 758-764.

44. Miller KB, Stuart DA, Smith NL, Lee CY, Mchale NL, et al. (2006) Antioxidant activity and polyphenol and procyanidin contents of selected commercially available cocoa-containing and chocolate products in the USA. J Agricultural and Food Chem 54: 4062-4068.

45. Blum U (1998) Effects of microbial utilization of phenolic acids and their phenolic acid breakdown products on allelpathic interactions. J Chemical Ecology 24: 685-708.

46. Amirdivani S, Baba AS (2011) Changes in yogurt fermentation characteristics, and antioxidant potential and in vitro inhibition of angiotensin-1 converting enzyme upon the inclusion of peppermint dill and basi. Food Science and Technology 44: 1458-1464.

47. Paredes-López O, Cervantes-Ceja ML, Vigna-Pérez M, Hernández-Pérez T (2010) Berries improving human health and healthy aging and promoting quality life - A Review. Plant Foods for Human Nutrition 65: 299-308.

48. Thompson JL, Lopetcharat K, Drake MA (2007) Preferences for commercial strawberry drinkable yogurts among African American, Caucasian, and Hispanic consumers in the USA. J Dairy Sci 90: 4974-87.

49. Papadimitriou CG, Mastrojiannaki AV, Silva AV, Gomes AM, Malcata F (2007) Identification of peptides in traditional and probiotic sheep milk yoghurt with angiotensin I-converting enzyme (ACE)-inhibitory activity. Food Chemistry 105: 647-56.

50. Massey LK (2001) Dairy food consumption blood pressure and stroke. J Nutrition 131: 1875-1878.

51. Bradish CM, Perkins-Veazie P, Fernandez GE, Xie G, Jia W (2012) Comparison of flavonoid composition of red raspberries (Rubus idaeus L.) grown in the southern United States. J Agric Food Chem 60: 5779-5786.

13

Chemical and Nutritional Aspects of Some Safflower Seed Varieties

Al Surmi NY[1]*, El Dengawy RAH[1] and Khalifa AH[2]

[1]Food Industries Department, Faculty of Agriculture, Damietta University, Damietta, Egypt
[2]Food Science and Technology Department, Faculty of Agriculture, Assiut University, Assiut, Egypt

Abstract

In this study, three Safflower varieties (*Carthamus tinctorius* L.) Malawi, Giza1, Ethiopia were obtained from Egypt; of which the Ethiopian variety was analyzed for content of moisture, crude fiber, proteins, oils, carbohydrates, and ash. In addition, detailed studies were conducted on amino acid profile and minerals. Total phenols and fraction of phenolic compound were studied. The moisture content ranged from 5.24% to 6.23% and the protein content ranged from 14.70% to 16.21%, crude fiber 21.34% to 22.51 %,total lipid 32.47% to 35.12%, nitrogen free extract ranged from 22.47% to 26.11%, and ash 3.45% to 4.21% (on wet weight basis). Amino acid analyses revealed that Malawi, Giza1 and Ethiopian have higher level of arginine 5.28, 4.76, and 3.94 (g/100g) respectively. The total polyphenols content of defatted safflower meal was ranged from 452.52 mg to 677.27 mg (GAE /100g).

Keywords: *Carthamus tinctorius*; Safflower; Chemical composition; Amino acid; Minerals; Phenolic

Introduction

Safflower is a very ancient crop which originated in the Middle East [1]. Safflower (*Carthamus tinctorius* L.) is commonly known as khortom in Egypt, kusum in India and Pakistan and honghua (red flower) in China [2,3]. It is oil seed crops, contain about 80% oleic and linoleic acid, iodine value (148) and saponification value (190) [4], cultivated mostly for its high-quality oil, cut flowers, vegetables and medicinal plant. Safflower oils were used as a source of oil in the paint industry and edible oil for cooking, margarine production, and salad oil [2]. Therefore, it needed to be developed the oils produced from Safflower to be as a commercial product for edible oil, medicinal uses and pharmaceuticals, source of α-tocopherol, paint, varnishes and soap manufacturing industries. Safflower is a minor crop with a world production of about 834,000 tons in 2013 [5]. Several safflower seed products can be used as animal feeds: the seeds, the by-product of the oil extraction (safflower meal) and the hulls [6].

Materials and Methods

Safflower *Carthamus tinctorius L*. seeds of two Egyptian varieties namely: Malawi and Giza 1 were obtained from the farm of Agricultural Faculty, Cairo University during 2011/2012 season, while the imported variety from Ethiopian was obtained from local market of Cairo City during 2012 were used in this study.

Gross chemical composition

Moisture, crude oil, total nitrogen, ash content and crude fiber of safflower seed kernel were determined as outlined in AOAC [7]. Nitrogen free extract was calculated by difference.

Determination of minerals

Sodium and potassium were determined by Flame Photometer 410, whereas Calcium, magnesium, manganese, copper, iron and zinc were determined using Perkin-Elmer Atomic Absorption Spectrophotometer 2380, at Agricultural Research Center in Giza, as described in AOAC [8].

Amino acids composition

Amino acids were determined according to the method described by Pellett and Young [9] with some modifications, which could be summarized as follows: 200 mg of dried, defatted sample was hydrolyzed with 5 ml of 6 N HCl, in sealed tube at 110°C for 24 hours and the hydrolysate was filtered. The residue was washed with distilled water and the filtrate was evaporated on water bath at 50°C. The residue was dissolved in 5 ml/loading buffer (0.2 N sodium citrate buffer of pH 2.2). Amino acids were determined chromatographically using Beckmen Amino Acid Analyzer Model 119 CL, at National Research Center, Giza-Cairo.

Tryptophan determination: Tryptophan was determined calorimetrically using the method described by Sastry and Tummuru [10].

Determination of total phenolic compounds

The Folin-Denis method was used for total-phenolic (TP) analysis with tannic acid as a standard. Folin-Denis reagent is mixture of 10 g of sodium thungstate, 2 g of phosphomolybdic acid and 5 ml of phosphoric acid in 75 ml of distilled water that was refluxed for 2 h, cooled, and diluted to 100 ml with distilled water.

To use the tannic acid as a standard, the procedure described by Makkar [11] was applied: A sodium carbonate saturated-solution was obtained by adding 40 g of sodium carbonate to 150 ml of distilled water, then dissolved for 1h at dark and adjusted to 200 ml. A standard-solution of tannic acid was obtained by dissolving 50 mg of tannic acid in 100 ml of distilled water. Aliquots of 0, 20, 40, 60, 80 and 100 μl of the standard solution were dispensed into tubes containing 0.5 ml of Folin-Denis reagent and 2.5 ml of sodium carbonate saturated-solution. Finally, standards were diluted to 4 ml with distilled water and quickly shaken. Their absorbance was determined after 35 min in dark at 750 nm [8].

*Corresponding author: Al Surmi NY, Food Industries Department, Faculty of Agriculture, Damietta University, Damietta, Egypt
E-mail: nageeb_alsurmi@yahoo.com

Determination of total polyphenol was prepared by adding 0.5 ml of Folin-Denis and 2.5 ml sodium carbonate reagents to 1 ml of each safflower meal extracts. Using absorbance standard curve, TP content was estimated. Units of TP were expressed as μg of tannic acid equivalents per ml of safflower meal extracts. For safflower meal extracts, tannic acid equivalents were multiplied by 20, based on an extraction ratio of 1:20 (w/w).

Determination of phenolic compounds by high performance liquid chromatography (HPLC): Phenolic compounds of safflower cake were determined by using Hewlett Packared HPLC (Model 1100) at National Research Center, Cairo using a hypresil C18 reversed phase column (250 × 4.6 mm) with 5μm particle size-injection by means of Rheadyne injection value (Model 7125) with 50 μl fixed loop as used. A constant flow rate of 1 ml min-1 was used with two mobile phase (A) 0.5% acetic acid in distilled water at pH 2.65 and solvent(B) 0.5% acetic acid in 99.5% acetonitrile. The elution gradient was linear starting with (A) and ending with (B) over 35 min, using UV detector set at wavelength 245 nm. Phenolic compound of each sample were identified by comparing their relative retention time with those of the standards mixture chromatography. The concentration of an individual compounds was calculated based on peak area measurement, then converted to μ phenolic g-1dry weight.

Fatty acids composition

Preparation of methyl ester of fatty acids: The methyl esters of fatty acids were prepared from aliquots total lipids using 5 ml 3% H_2SO_4 in absolute methanol and 2 ml benzene as mentioned by Rossell et al. [12]. The contents were heated for methanolysis at 90°C for 90 minute. After cooling, phase separation was performed by addition of 2 drops distilled water and methyl esters were extracted with 3 aliquots of 2 ml hexane each. The organic phase was removed, and filtered through anhydrous sodium sulfate.

Gas liquid chromatographic of methyl esters of fatty acids: The methyl esters of fatty acids were separated using HP 6890 GC capillary column gas liquid chromatography with a dual flame ionization were carried out on (30 m × 0.32 mm × 0.25 μm) DB-225 capillary column, stationary phase (50% cyanopropyl phenyl + 50% dimethyl polysiloxane). Column temperature: initial temperature was 150°C, the temperature was programmed by increasing the temperature from 150-170°C at the rate of 10°C/minute, then increased from 170°C-192°C at the rate of 5°C/minute, holding for five minutes and then increased from 192-220°C during 10 minutes, holding three minutes. The injector and detector temperatures were 230°C and 250°C, respectively. Carrier gas: Hydrogen flow rate 40 ml/minute, nitrogen at the rate 3 ml/minute, and air flow rate was 450 ml/minute. Peak identifications were established by comparing the retention times obtained with standard methyl ester. The areas under the chromatographic peak were measured with electronic integrator. It was carried out in Agricultural Research Center in Giza–Cairo.

Results

Gross chemical composition

The proximate chemical composition of safflower seeds are presented in Table 1. Moisture content was ranged from 5.24 to 6.23% and found to be in agreement with the results of Bozan and Temelli [13], Vorpsi et al. [14], Ingale and Shrivastava [15] and Yu et al. [16] who stated that, the moisture content was 6.10, 6.26, 6.33 and 5.58%, respectively. However, the local Egyptian Malawi varieties recorded slight low moisture content compared to the imported Ethiopian one. In addition, the lower moisture content is an important factor for keeping quality during storage. The protein content of the studied safflower seeds was ranged from 14.70% to 16.21% (on dry weight basis). The obtained protein content values were in agreement with that reported by Zazueta [17], kim [18], Rahmatalla et al. [19] and Kim [20]. However Nagaraj [21] mentioned that, the protein content of safflower seeds was ranged from 14 to 19%. Oil content of safflower seeds is a very important economic trait for safflower varieties and considered one of the most important factors affecting the success of safflower insert in new areas. The crude oil content of the studied safflower seeds was ranged from 32.47 to 35.12% as shown in Table 1. The studied safflower varieties seed consider a good source of oil. Similar results were reported by Vorpsi et al. [14] and Mariod et al. [22]. Beside, among the Egyptian varieties, Giza 1 variety contained higher oil content (35.62%) compared by Malawi variety which contained (32.47%) while the imported Ethiopian variety contained intermediate level of oil (33.89%). Data in Table 1 revealed that Malawi safflower varieties seeds have the highest value of crude fiber (22.51%) followed by Giza 1 varieties (22.33%) then, Ethiopian varieties (21.31%). Similar research for crude fibers content in five varieties of safflower seeds (25.01% to 29.24%) was reported by Bardhi et al. [23]. On other hand, the data revealed that, the nitrogen free extract was ranged from 22.47 to 26.11% in the studied safflower seeds. The ash content was ranged from 3.45 to 4.21% as indicated in Table 2. Similar results for ash content of various varieties of safflower seeds were reported by Kim [18], Ingale and Shrivastava [15] and Yu et al. [16].

Minerals composition

Minerals analyses are essential to guarantee the quality of any food product. There are many minerals that are considered nutrients and are vital for the proper functioning of the body. Equally, there are a number of minerals that are toxic to the human body and interfere with its functioning and undermine health [22].

The results indicated that, safflower seeds are good source of phosphor which ranged from 663.00 to 770.40 mg/100g (on dry weight), while potassium (K) content was ranged from 156.15 to 203.60 mg/100g and calcium (Ca) was ranged from 59.00 to 101.50 mg/100g (dry weight). Similar results were recorded by Rahmatalla et al. [24] and Mckevith [25]. Potassium plays an important role in human physiology, and sufficient amounts of it reduce the risk of heart stroke,

Varieties	Moisture	Characteristics*				
		Crude protein	Crude oil	Fiber	Ash	Nitrogen free extract**
Malawi	5.24[a]	14.7[a]	32.47[a]	22.51[a]	4.21[c]	26.11[b]
Giza1	6.28[b]	16.21[a]	35.12[c]	22.33[a]	3.87[b]	22.47[a]
Ethiopian	5.67[a]	15.62[a]	33.89[b]	21.34[a]	3.45[a]	25.7[b]

[a,b,c]Values with different subscripts on the same column are significant (p<0.05)
Calculated by differences on dry weight basis (%)
**Calculated by differences

Table 1: Gross chemical composition of safflower seed varieties.

while calcium plays an important role in building stronger, denser bones early in life and keeping bones strong and healthy later in life [26]. However, the sodium content of safflower seeds was low and ranged from 12.22 to 39.32 mg /100g. With relatively high content in Giza 1 varieties compared to Malawi and Ethiopian varieties. Iron (Fe) the important element for blood building, content was ranged from 3.53 to 3.98 mg/100g and zinc (Zn) content was ranged from 1.49 to 2.06 mg/100g. However higher results for iron content (4.90 mg/100g) and zinc content (5.10mg/100g) was reported by Mckevith [27]. Magnesium (Mg) content of the studied safflower seeds was ranged from 30.55 to 61.10 mg/100g Moreover, Giza 1 varieties contained duplicate magnesium content compared with Malawi varieties, while, the imported Ethiopian variety contained intermediate level (40.84 mg/100g) as shown in Table 2. Copper (Cu) and Manganese (Mn) content of safflower seeds was similar to the results of Kim [27] and Yu et al. [13]. In addition, Cadmium (Cd) a toxic element was absent in the all studied safflower seeds

Amino acids composition of safflower seeds

Amino acids compositions of the three studied safflower seeds are presented in Table 3. The Data indicated that all essential amino acids were presented in safflower seeds and their content value were 14.03, 14.93 and 11.48 g /100g protein of Malawi, Giza 1 and Ethiopian safflower seeds; respectively. As it is well known, the human bodies do not synthesis the essential amino acid, but can be obtained by from food. Eight amino acids are generally regarded as essential for human; phenylalanine, valine, threonine, tryptophan, isoleucine, methionine, leucine and lysine [27]. Besides, the other, non-essential amino acids, except cysteine, which was not detected, constituted 28.65, 23.77 g/100g and 21.47 g/100g protein of Malawi, Giza 1 and Ethiopian safflower seeds; respectively. Moreover, among essential amino acids, leucine, phenylalanine and valine recorded the highest content in safflower seeds protein, while glutamic acid, arginine and aspartic acid recorded the highest values among the nonessential amino acids as shown in Table 3. However, the antioxidant activity of these amino acids suggests a disease preventive role as exemplifies by arginine which is beneficial for prevent of cardiovascular disease [28].

Arginine is a factor for maintaining the nitrogen balance in muscles and can enhance the lean tissue to fat tissue body fat ratio; a great factor for weight management [29]. Aspartic acid deficiency decrease cellular energy and may like be a factor in chronic fatigue. Adequate methionine prevents disorder of hair, skin and nail; reduce liver fat and protect the kidney. On other hand, essential amino acid contributes to good health and wellbeing. Deficiency of lycine leads to physical and mental handicap [30]. However, the essential amino acids represent 32.87, 38.61 and 34.84% of Malawi, Giza 1 and Ethiopian safflower

Amino acids	Safflower seed varieties		
(g/100 g protein)	Malawi	Giza1	Ethiopian
Essential amino acids			
Threonine	0.93	0.81	1.03
Valine	2.05	1.83	2.47
Methionine	0.2	1.08	0.47
Isoleucine	1.16	2.81	1.41
Leucine	2.93	1.43	3.92
Phenylalanine	2.67	2.05	3.15
Lysine	1.27	4.69	1.36
Tryptophan	0.27	0.23	0.22
Total essential amino acids	11.48	14.93	14.03
Non-essential amino acids			
Histidine	1.01	1.23	1.22
Arginine	3.94	4.76	5.28
Aspartic acid	2.59	2.59	3.23
Serine	1.19	1.3	1.61
Glutamic acid	6.35	6.73	8.82
Proline	1.32	1.92	2.95
Alanine	2.3	2.05	2.6
Cysteine	ND	ND	ND
Tyrosine	1.76	2.32	1.82
Glycine	1.01	0.87	1.12
Total non-essential amino acids	21.47	23.77	28.65
E.AA./non-E.AA. ratio	0.54	0.63	0.49

*ND: Not detected.

Table 3: Amino acids composition of safflower seed varieties.

seed protein; respectively. Similar results was reported by Mariod et al. [22] who found that, the essential amino acids represented 36% of total amino acids in safflower seeds protein. Moreover, the obtained amino acids composition of safflower seeds protein was found to be similar with reported values of Rahmatalla et al. [24], Ingale and Shrivastava [15] and Mariod et al. [22] for different varieties of safflower seeds.

Phenolic compounds of safflower seeds

There are many evidences that, natural products and their derivatives have efficient anti-oxidative characteristics, consequently linked to anti-cancer, hypolipidemic, anti-aging and anti–inflammatory activities. Phenols and polyphenols are stronger antioxidants than the vitamins. Several epidemiological studies showed a lower risk with increasing intakes plant foods and protection against DNA damage [31]. Anti-oxidative capacity of *Carthamus tincotorius* were evaluated by Koyama et al. [32] and they attributed their strong anti-oxidative activity to the major phenolic compounds serotonin and their glucoside derivatives.

Total polyphenols content of safflower seeds meal

Total polyphenols content of defatted safflower meal was tabulated in Table 4. The data revealed that the total polyphenols contents of the studied safflower seed meals were 452.52, 677.27 and 604.04 mg (GAE /100g) for Malawi, Giza 1 and Ethiopian varieties; respectively. The obtained results are in agreement with the results of Sreeramulu and Raghunath [33] who found that, total phenols in safflower seed was 599(mg GAE /100 g) but, it was lower than 1526 (mg /GAE 100g) which reported by Bozan and Temelli [13] and higher than 55.52 (mg GAE /100g) which reported by Yu et al. [16]. However as shown in Table 5 among the Egyptian safflower varieties, Giza 1 safflower seeds contained more total polyphenols compared with Malawi variety, but its phenols content was nearly similar to the imported Ethiopian variety.

Minerals	Safflower Varieties		
	Malawi	Giza 1	Ethiopian
K	194.95	203.60	156.15
P	770.40	673.80	663.00
Mg	30.55	61.10	40.84
Ca	101.50	99.00	59.00
Na	12.22	39.32	18.56
Fe	3.98	3.56	3.53
Zn	2.06	1.66	1.49
Cu	0.46	0.46	0.31
Mn	0.42	0.41	0.26
Cd	0.00	0.00	0.00

Table 2: Minerals content of safflower seeds (mg/100g on dry weight).

Varieties	Malawi	Giza1	Ethiopian
Total phenol(mg/100g of safflower defatted meal)	452.52[a]	677.27[c]	604.04[b]
[a,b,c] Values with different subscripts on the same row are significant (p<0.05)			

Table 4: Total phenol compound content in safflower meal (mg GAE/100g).

Phenolic compound	Safflower seed varieties		
	Malawi	Giza1	Ethiopian
N-(p-Coumaroyl) serotonin 7-O-β-D-glucoside	22.82	0.9	3.57
N-Feruloyl serotonin 7-O-β-D-glucoside	7.03	2.93	19.23
N-(p-Coumaroyl) serotonin	11.15	31.14	32.99
N-Feruloylserotonin	23.1	22.64	24.99
Matairesinoside	3.67	7.1	7.6
Acacetin 7-O-β-D-glucuronide	5.8	17.32	3.06
Acacetin	3.11	7.62	1.05
Luteolin	4.99	2.92	1.34
Unknown	3.1	0.65	2.55

Table 5: Phenolic compound content of safflower seed varieties (% of the total phenolic compound).

Determination of phenolic compounds of safflower seeds meal

By using of HPLC, phenolic compounds of safflower seeds were fractionated to eight fractions namely: N-(p-Coumaroyl) serotonin 7-O-β-D-glucoside, N-Feruloylserotonin 7-O-β-D-glucoside, N-(p-Coumaroyl) serotonin, N-Feruloylserotonin, Matairesinoside, Acacetin 7-O-β-D-glucuronide, Acacetin, Luteolin as well as unknown fraction as shown in Table 5. However, Kim et al. [34] isolated the polyphenols of safflower seed also, to nine phenolic compounds. The obtained results indicated that serotonin and its derivatives constituted the major content of polyphenols of safflower seeds. Similar observation was reported by Kim et al. [34], Katsuda et al. [35] Seo and Choi [36]. Moreover, as early reported by Zhang et al. [37] Serotonin derivatives in the safflower seeds are family of molecules containing seven to ten members featuring a serotonin moiety bound to a phenyl propanoid moiety via an amide bond. However, N-(p-Coumaroyl) serotonin and N-Feruloyl serotonin represented about 34% and 58% of total phenolic compound as indicated in Table 5. Beside, Malawi variety seeds contained more N-(p-Coumaroyl) serotonin 7-O-β-D-glucoside (22.82%) compared with Giza 1 variety (0.90%) and Ethiopian variety (3.57%)while Ethiopian variety contained more of N-Feruloylserotonin 7-O-β-D-glucoside (19.23%) compared with Malawi variety (7.03%) and Giza 1 variety (2.03%) of total phenolic compound. Moreover, Giza 1 variety contained more acacetin 7-O-β-D-glucuronide and acacetin compared with the other two varieties.

Fatty acids composition of safflower oils

As known, the fatty acid composition of vegetable oil is a main factor affecting its commercial uses and it influenced by a lot of factors such as genotype of the variety, environmental conditions, etc. [38]. The fatty acids composition of total lipids extracted from Ethiopian, Malawi and Giza1 safflower seeds are shown in Table 6 the results of analysis for fatty acid showed that the unsaturated fatty acid linoleic (74.60, 78.24 and 77.90%) and oleic (14.19, 11.22 and 11.39%) and the saturated fatty acid palmitic (6.03, 6.57 and 6.66%) and stearic (2.61, 2.01 and 2.06%) were the most abundant fatty acids in respecting decreasing order, which together composed about 97.43, 98.04 and 98.01% of the total fatty acids of Malawi, Giza 1 and Ethiopian varieties;

respectively. A negligible amount of linolenic acid was detected (0.07 - 0.08%) and minor amount of ecosenoic (C 20:1), palmitoleic (C16:1), arachidic (C 20:0) and behenic (C 22: 0) were present and the values of them did not exceed 0.89% of the total fatty acids. These results are comparable to data reported by Sabzalian et al. [39].

Besides compared to most other common edible oils, safflower oil contains the highest level of the linoleic acid, an essential fatty acid, which is make it as premium edible oil, because of its nutritional advantages and potential therapeutically properties in the prevention of coronary heart disease and cancer but the presence of the large amounts of linoleic acid makes the oil quite sensitive to oxidation [40,41]. Moreover, the essential fatty acid is not easily synthesized in the human system and must be supplied externally through the diet [42]. Smith indicated that, the importance of safflower seed oil is in its linoleic acid content, which is a required product with high polyunsaturated fatty acid clime. However, the total saturated fatty acid content of safflower oil was low (9.24% to 9.40%) of total fatty acids content while, the total unsaturated fatty acid was about 90% of total fatty acids content as shown in Table 6. Generally, high intakes of saturated fatty acids have been associated with raised blood cholesterol levels, one of the risk factors associated with heart diseases. In comparison mono unsaturated fatty acids decrease the bad cholesterol, (LDL-C), moreover polyunsaturated fatty acids also decrease LDL-C, intake of n-6 PUFA above 10% energy may have adverse effects on good cholesterol, (HDL- C) as mentioned by Clarke et al. [32]. On other hand, the nutritional quality index (linoleic/saturated fatty acids) is very high and ranged from 7.99 to 8.47 compared with that of groundnut oil which ranged from 1.8 to 2.4 as reported by Nagaraj [21]. In opposite the ratio of oleic to linoleic acid being a measure of oil keeping quality (oil stability index) was low and ranged from 0.14 to 0.19 as presented in Table 6. Besides, the all studied safflower seed oils contained long chain fatty acids (C_{20} - C_{24}) with minor mounts. The same observation was detected by Bozan and Temelli [13] when studied flax, safflower and poppy seed oils.

Conclusion

Total phenols and fraction of phenolic compound were studied. The moisture content ranged from 5.24 to 6.23% and the protein

Fatty acids	Carbon chain	Safflower seed varieties		
		Malawi	Giza1	Ethiopian
Palmetic	C16:0	6.03	6.57	6.66
Palmetolic	C16:1	0.06	0.09	0.09
Margaric acid	C17	0.02	0.03	0.02
Margaoliec acid	C17:1	0.00	0.01	0.01
Stearic	C18:0	2.61	2.01	2.06
Oleic	C18:1	14.19	11.22	11.39
Linoleic	C18:2	74.60	78.24	77.90
Linolenic	C18:3	0.07	0.08	0.08
Arachidic	C20:0	0.34	0.29	0.30
Ecosenoic	C20:1	0.19	0.17	0.17
Behenic	C22:0	0.24	0.24	0.25
Unknown	Unknown	1.51	0.91	0.93
Lignoceric acid	C24:0	0.10	0.10	0.11
Total saturated fatty acid		9.34	9.24	9.40
Total unsaturated fatty acid		89.11	89.81	89.64
Nutritional quality index (linoleic/ saturated fatty acid)		7.99	8.47	8.29
Oil stability index (oleic/linoleic)		0.19	0.15	0.15

Table 6: Fatty acids composition of total lipids of safflower seed varieties (% of total fatty acids).

content ranged from 14.70 to 16.21%, crude fiber 21.34 to 22.51%, total lipid 32.47% to 35.12%, nitrogen free extract ranged from 22.47% to 26.11%, and ash 3.45 to 4.21% (on wet weight basis). Amino acid analyses revealed that Malawi, Giza1 and Ethiopian have higher level of arginine 5.28, 4.76, and 3.94 (g/100g) respectively. The total polyphenols content of defatted safflower meal was ranged from 452.52 mg to 677.27 mg (GAE /100g).

References

1. Ashri A (1975) Evaluation of the germplasm collection of safflower. Distribution and regional divergence for morphological characters. Euphytica 24: 651-659.

2. Emonger V (2010) Safflower (Carthamus tinctorius L.) the underutilized and neglected crop: A review. Asian J Plant Sci.

3. Chavan VM (1961) Niger and Safflower. Indian Central Oilseeds Committee Publication, Hyderabad, India.

4. Rafiquzzaman M, Hossain MA, Hasan AM (2006) Studies on the characterization and glyceride composition of safflower (Carthamus tinctorius) seed oil. Bangladesh J Sci Ind Res 41: 235-238.

5. FAO (2014) FAOSTAT. Food and Agriculture Organization of the United Nations.

6. Oelke EA, Oplinger ES, Teynor TM, Putnam DH, Doll JD, et al. (1992) Safflower- Alternative field crop manual, University of Wisconsin-Extension, Cooperative Extension.

7. AOAC (1997) Official methods of analysis of AOAC International (16thedn). AOAC International, Suite 500, 481 North Frederick Avenue, Gaithersburg, Maryland, USA.

8. AOAC (1995) Official methods of analysis (16thedn). Association of Official Analytical chemists, Washington DC.

9. Pellett PL, Young VR (1980) Nutritional evaluation of protein foods. Food and Nutrition Bulletin, supplement published by the United Nation University.

10. Sastry CCP, Tummuru MK (1985) Spectrophotometric determination of tryptophan in protein. J Food Sci Technol 22: 146-147.

11. Makkar HPS (2000) Quantification of tannins in tree and shrub foliage. FAO/IAEA, IAEA/Vienna

12. Rossell JB, King B, Downes MJ (1983) Detection of adulteration. J Am Oil Chem Soc 60: 333.

13. Bozan B, Temelli F (2008) Chemical composition oxidative stability of flax, Safflower and poppy seed oils. Biores Technol 99: 6354-6359.

14. Vorpsi V, Harizaj F, Bardhi N, Vladi V (2010) Safflower seeds cultivated in Albania. Res j agri Sci 42: 26-331.

15. Ingale S, Shrivastava KS (2011) Chemical and bio-chemcal studies of new varieties of safflower (Carthamus tinctorius L.) PBNS-12 and PBNS-40 seeds. AAB Bioflux 3: 127-138.

16. Yu S, Lee Y, Kang YJ, Lee SN, Jang SK, et al. (2013) Analysis of food components of Carthamus tinctorius L. seed and its antimicrobial activity. Korean J Food Pres 20: 227-233.

17. Zazueta AJS (1986) Solubility and electrophoretic properties of processed safflower seed (Carthamus tinctorius) proteins. Department of Nutrition and Food Science, Arizona University.

18. Kim EO, Lee KT, Choi SW (2008) Chemical comparison of geriminated and ungerminated safflower (Carhtamus tinctorius) seeds. J Korean Soc Food Sci Nutr 37: 1162-1167.

19. Rahamatalla AB, Babiker EE, Krishna AG, El-Tinay AH (2001) Changes in fatty acids composition during seed growth and physicochemical characteristics of oil extracted from four safflower cultivars. Plant Foods Hum Nutri 56: 385-395.

20. Kim JH, Kim JK, Kang WW, Ha YS, Choi SW, et al. (2003) Chemical compositions and DPPH radical scavenging activity in different sections of safflower. J Korean Soc Food Sci Nutr 32: 733-738.

21. Nagaraj G (1995) Quality and utility of oil seeds, Directorate of Oilseeds Research (Indian Council of Agricultural Research) cultivars. 5th International Safflower Conference, USA.

22. Mariod AA, Ahmed SY, Abdelwahab SI, Cheng SF, Eltom AM, et al. (2012) Effects of roasting and boiling on the chemical composition, amino acids and oil stability of safflower seeds. Int J Food Sci Technol 47: 1737-1743.

23. Bardhi N, Susaj E, Dodona E, Kallço I, Mero G, et al. (2013) Productivity indicators of five safflower cultivars (Carthamus tinctorius L.) grown under Lushnja, Albania climatic conditions. Int Interdisci Res J.

24. Rahamatalla AB, Babiker EE, Krishna AG, El-Tinay AH (1998) Changes in chemical composition, minerals and amino acids during seed growth and development of four safflower cultivars. Plant Foods Hum Nutr 52: 161-170.

25. Mckevith B (2005) Nutritional aspects of oilseeds. British Nutrition Foundation Nutrition Bulletin 30: 13-26.

26. Kim SK, Cha JY, Jeong SJ, Chung CH, Choi YR, et al. (2000) Properties of the chemical composition of safflower (Carthamus tinctorius L.) sprout. Korean J Life Sci 10: 68-73.

27. Young VR (1994) Adult amino acid requirements: The case for a major revision in current recommendations. J Nutr 124: 1517-1523.

28. Balsubramanian SC, Ramasashtri BV, Gopalan C (1980) Nutritive values of Indian foods. National Institute of Nutrition, Indian council of Medical Research, Hyderabad 3: 28.

29. Medical Biochemistry (2005) Chemical nature of amino acids.

30. Papes F, Surili MJ, Langone F, Trigo JB, Arruda P (2001) Deficiency of lycine. FEBS Letter, 488: 34-38.

31. Vinson JA, Yong A, Xuelci S, Ligid Z, Bose P, et al. (2001) Phenol antioxidant and quantity and quality in foods. J Agric Food Chem 49: 5315-5322

32. Clarke R, Frost C, Collins R (1997) Dietary lipids and blood cholesterol: quantitative meta-analysis of metabolic ward studies. British Med J 314: 112-117.

33. Sreeramulu D, Raghunath M (2011) Antioxidant and phenolic content of nuts, oil seeds, milk and milk products commonly consumed in India. Food Nutri Sci 2: 422-427.

34. Kim EO, Lee JY, Choi SW (2006) Quantitative changes in phenolic compounds of safflower (Carthamus tinctorius L.) seeds during growth and processing. J Food Sci Nutr 11: 311-317.

35. Katsuda S, Suzuki K, Koyama N, Takahashi M, Miyake M, et al. (2009) Safflower seed polyphenols (N-(p-coumaroyl) serotonin and N-feruloylserotonin) ameliorate atherosclerosis and distensibility of the aortic wall in Kurosawa and Kusanagi-hypercholesterolemic (KHC) rabbits. Hypertens Res 32: 944-949.

36. Seo IH, Choi SW (2009) Preparation of high quality safflower (Carthamus tinctorius L.) seed extract by high-pressure extraction process. J Food Sci Nutr 14: 373-377.

37. Zhang HL, Nagatsu A, Watanabe T, Sakakibara J, Okuyama H, et al. (1997) Antioxidative compounds isolated from safflower (Carthamus tinctrius L.) oil cake. Chem Pharm Bulletin 45: 1910-1914.

38. Gecgel U, Demirci M, Esendal E, Tasan M (2007) Fatty acid composition of the oil from developing seeds of different varieties of safflower (Carthamus tinctorius L.). J Am Oil Chem Soc 84: 47-54

39. Sabzalian MR, Saeidi G, Mirloh A (2008) Oil content and fatty acid composition in seeds of three safflower species. J Am Oil Chem Soc 85: 717-721.

40. Oomah DB, Ladet S, Godfrey VD, Liang J, Giarard B, et al. (2000) Characteristics of raspberry(Rubus idaeusL.) seed oil. J Food Chem 69: 187-193.

41. Koyama N, Kuribayashi K, Seki T, Kobayashi K, Furuhata Y, et al. (2009) Serotonin derivatives, major safflower (Carthamus tinctorius L.) seed antioxidants, inhibit low-density lipoprotein (LDL) oxidation and atherosclerosis in apolipoprotein E-deficient mice. J Agric Food Chem 54: 4970-4976.

42. Mitra P, Ramaswamy HS, Chang KS (2009) Pumpkin (Curcubita maxima) seed oil extraction using super critical carbon dioxide and physicochemical properties of the oil. J of Food Eng 95: 208-213.

14

Benefit of Lactose Concentration between Goat's Milk and Commercialized Powder Milk

Nur Sofuwani ZA[1], Siti Aslina H[1*] and Siti Mazlina MK[2]

[1]Department of Chemical and Environmental Engineering, Faculty of Engineering, Universiti Putra Malaysia, Selangor, Malaysia
[2]Department of Food and Process Engineering, Faculty of Engineering, Universiti Putra Malaysia, Selangor, Malaysia

Abstract

Even though goat's milk naturally has lower lactose than cow's milk (~4.39% compared to 4.51%), when it's consumed in a large amount, those intolerant to lactose may suffer several inconvenient symptoms, such as bloating, nausea, and diarrhoea. Previous study had established that a high level of lactose removal from goat's milk could be attained by 10 KDa sized ultrafiltration (UF) membrane. Hence, the concentrated goat's milk obtained from the UF process and five local brands of commercial milk powder were compared in terms of nutrition facts. Lactose concentration as important nutrition is evaluated for the quality and the competitiveness between the products. While, proximate analysis was used as part of method to determine the chemical composition in the goat's milk, including moisture, protein, fat, ash, and carbohydrate. Then, the composition of the reconstituted concentrated powder milk and five others commercialized milk which homogenized with water was analysed by HPLC to determine the lactose concentration. As a finding, concentrated milk contained 5.63 g per 100 ml lactose concentration, which ranked at the second lowest concentration in the range of 2.81 to 7.91 g per 100 ml, proved that it is similar and comparable in standard as to commercial milk.

Keywords: Membrane; Ultrafiltration; Goat's milk; Proximate; Lactose concentration; Powder milk

Introduction

Ultrafiltration (UF) has molecular weight cut-off (MWCO) in the range of 1-500 KDa and corresponds to a nominal pore diameter of 1 nm to 100 nm [1,2]. The cut-off means molecular weight of the smallest molecule that cannot pass through the membrane. The pore size of the UF membrane is selected based on the size of the molecules being separated. In other words, larger molecules, such as proteins, fat, and carbohydrates, are fully or partially retained; depending on the pore size of the membrane used [3]. UF was mainly used for producing low lactose dairy products from cow's milk [4-6].

Hence, applying UF membrane with MWCO greater than 10 KDa would lead to increased transmission or loss of essential milk proteins, while using UF membrane with pore size smaller than 5 KDa may cause inefficiency in the UF process due to a significant reduction in lactose transmission. That is why the most common cut-off in dairy standard is 10 KDa. Meanwhile, membrane sizes within 6 to 9 KDa may have operated in a different pattern as UF membrane commonly has a definite or a diffuse separation limit. Membranes with a sharp cut-off separates lower molecular weight, while membranes with a diffuse cut-off allows permeation of some higher molecular weight solute and retains some lower molecular weight [3].

Fouling and concentration polarization (CP) on membrane surface during goat's milk processing which deteriorating the flux and gave negative impact on product yield are the major problems in the dairy industry. Until today, the issue concerning how to overcome fouling in cross-flow hollow fibre ultrafiltration unit is still debated due to the complex composition characteristic of milk that consists of proteins, minerals, lactose, and fat which contributes to the major foulants during the dairy UF process [7,8]. A recent study had proved that processing parameters condition of 0.18 L/min feed flow-rate and 0.55 bars trans-membrane pressure (TMP) gave the best condition in 10 KDa with UF membrane as it produced goat's milk with lowest lactose rejection and higher operating flux [9]. Hence, this paper focuses on evaluating the effects of this parameter on nutrition facts and lactose concentration.

The scope of work is that in order to ensure the comparability of low lactose goat's milk produced with other commercialized milk, all the milk samples were compared after atomization by spray-drying in terms of lactose percentage range and nutritional composition. Nonetheless, this research was not extended to determine the amount of protein content in concentrated milk.

Materials and Methodology

Conversion of liquid retentate milk into powder milk

The concentrated goat's milk from 10 KDa sized UF membrane had to undergo the spray-drying process, as portrayed in Figure 1, where it turned the liquid milk into milk powder. Lab-scale spray-dryer (Niro A/S, Mobile Minor, US) was used with air inlet and outlet temperatures at 150°C and 75°C, respectively. The optimum temperature for the growth of most of the bacteria was around 40°C, and hence, the experiments were carried out at above 40°C. The additive used was AAA maltodextrin (MDX) in the range of 15% w/w [10]. The spray-dried powder was stable during storage in vacuum sealed pack. The concentrated powder milk obtained from the fractionation was then reconstituted with warm water at 50°C. The standard formulation was used by adding and mixing 55 g of powder milk into 190 ml of warm water, 55 g/190 ml weight per volume or around one standard cupful [11]. The hot stirrer plate, along with the magnetic stirrer, played an important part in homogenizing the milk sample. Five other brands of commercialized milk powder were also homogenized.

*Corresponding author: Siti Aslina H, Department of Chemical and Environmental Engineering, Faculty of Engineering, Universiti Putra Malaysia, 43400 Serdang, Selangor, Malaysia, E-mail: aslina@upm.edu.my

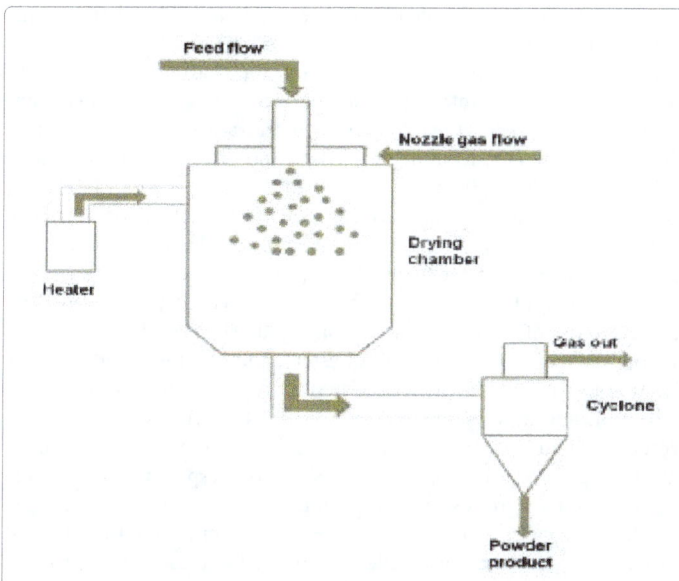

Figure 1: Schematic diagram of spray-drier (Niro A/S, Mobile Minor, US) (Food Engineering Laboratory, UPM) (T_i: 150°C, T_0: 75°C, 15% w/w MDX).

Proximate analysis

Proximate analysis was used in order to determine the chemical composition in the concentrated goat's milk, including moisture, protein, fat, ash, and carbohydrate [12] in Bioreactor Laboratory, Department of Process and Food Engineering, Universiti Putra Malaysia. The moisture content was measured in the laboratory using an infrared moisture analyzer (MX-50, A & D, Japan). Meanwhile, protein analysis was determined by using Kjeltec instruments (FOSS, Hillerod, Denmark), while fat analysis was determined via Soxtec extraction systems (Soxtec 2050, Foss Electric, Denmark) and ash content was indicated through the processes of ashing the powder milk sample by using furnace (Carbolite). Lastly, the carbohydrate content was measured by hundred minus total summations of other proximate.

Crude fat determination

The Soxtec method (Soxtec 2050, Foss Electric, Denmark) involved a direct solvent extraction. An empty solvent beaker was weighed. One gram (± 0.01) of pre-dried sample was weighed on a filter paper. The filter paper with sample was placed into a pre-dried extraction thimble, and was covered with cotton wool. The thimble was transferred into the Soxtec extraction system, and then, about 125 ml of petroleum ether was added in the solvent beaker. Mask and gloves were used as petroleum ether is a volatile substance. The Soxtec apparatus allowed the ether to soak into the sample. The sample was remained cool during extraction. The sample was held in porous thimbles, crucibles, filter paper, etc. The process was time-consuming. The unit was completed after 1.5 hours. The oven was pre-heat at 100°C for 10 min. After that, a towel was used to take the solvent beaker out. The solvent beaker with extracted fat was dried in an air oven at 100°C for 30 min, cooled in a desiccator for about 5 min, and then weighed.

Equation (1), (2) and (3) are the calculations used to calculate the fat content [11]:

Weight of fat = (beaker + fat) – beaker   (1)

$$\% \, fat \, wet \, basis = \left(\frac{weight \, of \, fat}{original \, sample \, weight} \right) \times 100 \quad (2)$$

$$\% \, fat \, dry \, basis = \left(\frac{weight \, of \, fat}{dried \, sample \, weight} \right) \times 100 \quad (3)$$

Protein determination

Kjeltec instrument (FOSS, Hillerod, Denmark) was used to determine the protein content in milk [11]. Protein determination in food solely depends on the amount of nitrogen. The three stages of the Kjeltec methods are digestion, distillation, and titration. Approximately 1 g of sample was weighed into the digestion tube followed by 2 tablets of Kjeltab (copper tablets with 3.5 g K_2SO_4 + 0.4 g $CuSO_4 5H_2O$). $CuSO_4$ was added during the digestion process to act as a catalyst and to increase the efficiency of sulfuric acid digestibility. 12 ml sulfuric acid (R&M; 98.08 g/mol) was added and the mixture was digested (FOSS Digestor Auto Lift) at 420°C for one hour. The amount of nitrogen was determined via sample digestion by using sulfuric acid (H_2SO_4) to form ammonium sulfate. The amount of ammonium to ammonia was measured by using NaOH solution via distillation. The ammonia released was captured through titration by the excess HCl solution to produce ammonium chloride while producing ammonium borate when captured using boric acid (weak acid). Then, crude protein content of the sample was obtained using Tecator Kjeltec protein analyser (FOSS, Hillerod, Denmark) by titration.

Moisture content determination

Moisture content of concentrated powder milk was determined by using infrared moisture analyzer (MX-50, A & D, Japan). Heating with infrared can be considered for sample of flour form or grind sample. Infrared moisture analyzer is an instrument that comprises of a balance and an infrared lamp that measure against the distance of sample to lamp. The sample was weighed 5 g and the temperature was set to 130°C. Moisture loss percentage was then directly determined.

Ash content determination

Ash content was indicated by gravimetric method according to AOAC [12] through the processes of ashing the powder milk sample by using a muffle furnace (Carbolite) at a very high temperature, 575°C for five hours. First, the weight of cool crucibles was recorded after pre-heating at 100°C. Two grams of the sample was weighed into the crucible and the weight was recorded. If the sample had been moist, it would be placed in an oven at 100°C and allowed to dry for an hour. Then, the sample was allowed to cool for a while before it was transferred to furnace at a temperature of 575°C for 5 h. After five hours, the furnace was turned off and it was opened after the temperature dropped to 50°C. The door was opened carefully to avoid losing ash that may be fluffy. Using a tong, the sample was placed into a desiccator and then, allowed to cool to room temperature. Lastly, the weight was recorded.

Equation (4) was used to calculate ash content [10]:

$$Total \, ash \, (on \, dry \, basis), \% \, by \, mass = \left(\frac{M_1 - M_0}{Sample \, wt} \right) \times 100 \quad (4)$$

Where,

M_1 = mass in g, of the crucible with ash

M_0 = mass in g, of the empty crucible

Carbohydrate content determination

Each gram of carbohydrate has four calories. Carbohydrate was calculated by 100 minus the sum of other proximates. The values for moisture, fat, protein, and ash contents were added and this value was subtracted from 100 to give the carbohydrate content. The calculation used to determine the total carbohydrate included lactose percentage by weight is as in equation (5) [10]:

$$\text{Total carbohydrate \% by wt.} = 100 - (P + F + M + A) \qquad (5)$$

Where,

P = Percent by mass of protein

F = Percent by mass of fat

M = Percent by mass of moisture

A = Percent by mass of ash

Results and Discussion

Production of concentrated powder milk

The concentrated milk from 10 KDa sized UF membrane was spray dried as shown in Figure 2 prior to proximate analysis and lactose concentration analysis. This helps in preservation, as powder milk has a longer shelf life than liquid milk, and does not need refrigeration due to low moisture content [13]. The powder retentate milk was then compared with other commercialized milk products.

To ensure the quality of goat's milk is maintained, the concentrated powder milk from 10 KDa membrane was then analysed in terms of nutritional composition and compared with five other reference samples from commercialized milk products labelled as A, B, C, D, and E. the proximate composition was defined by fat, protein, moisture, ash, and carbohydrate content. Table 1 represents the proximate composition

Figure 2: Goat's milk concentrated powder.

result. The proximate composition of retentate milk shows the means of total fat 8.12% ± 0.09, total protein 17.44% ± 0.13, total moisture 3.30% ± 0.04, total ash 26.90% ± 0.37, and total carbohydrate 44.24%. Based on Table 1, the total fat content in retentate milk obtained from fractionation process is the highest due to utilization of raw goat's milk, which is 8.12 ± 0.09. Moreover, the fat content in A, B, C, D, and E milk is 1.54 ± 0.07, 0.50 ± 0.14, 3.44 ± 0.20, 2.79 ± 0.14, and 1.05 ± 0.07, respectively. Lipids in milk are important in the aspects of nutrition, physical, and sensory characteristics of milk products [14]. The main lipid component of goat's milk is triacylglycerol's (TAG), which is about 97%, including a great number of esterified fatty acids [15]. The lipid component also contains simple lipids such as diacylglycerols, monoacylglycerols, and cholesterol esters, and complex lipids such as phospholipids [14]. The protein content of retentate milk is the second highest among other samples which is 17.44 ± 0.13. Meanwhile, the protein content for A, B, C, D, and E milk was 17.97 ± 0.14, 15.64 ± 0.06, 6.09 ± 0.06, 15.41 ± 0.02, and 13.58 ± 0.08, respectively as in Table 1. There are two groups of milk proteins which are casein and whey proteins. The caseins constitute 80% of the proteins and are classified as αs1, αs2, β and κ-caseins, while the major 20% refer to whey proteins namely β-lactoglobulin and α-lactalbumin [16].

The moisture content in retentate milk is second lowest which is 3.30 ± 0.04, while the moisture content percentage for A, B, C, D, and E was in the range of 2.60 to 4.87%, marked as 4.87± 0.04, 4.74± 0.13, 4.16± 0.02, 2.60 ± 0.02, and 4.67 ± 0.15, respectively. The ash content is generally recognized as a measure of quality for the assessment of the functional properties of foods [17]. Based on Table 1, the ash content of retentate milk is the second highest, which is 26.90 ± 0.37. While for A, B, C, D, and E, the ash content percentage was 22.74 ± 0.03, 21.78 ± 0.06, 23.98 ± 0.14, 29.46 ± 0.22, and 26.19 ± 0.21, respectively. This is because goat's milk has been reported to have higher content of potassium, chloride, calcium, phosphorus, selenium, zinc, and copper than cow's milk [18]. The main role of carbohydrates in diet is to produce energy. Each gram of carbohydrates provides us with 4.2 calories. Carbohydrates also act as a food store. Based on Table 1, the total carbohydrate of retentate milk is the lowest, which is 44.24%, while for sample A is 52.88%, B is 57.35%, C is 66.47%, D is 47.34%, and E is 54.52%. This is because lactose is a major carbohydrate in milk, and most importantly, retentate milk has lower lactose compared to other milk tested. Energy calculation was determined by multiplying the value of nutritional content of carbohydrate, protein, and fat with their conversion factor, which is 4 kcal for carbohydrate and protein, 9 kcal for fat, and that 1 Kcal is equal to 4.2 kJ. Serving size suggestion was 230 ml per serving. Table 1 represents the energy content in all

Nutritional Information							
Sample		Concentration	A	B	C	D	E
Carbohydrate	Mean ± SD(%)	44.24	52.88	57.35	66.47	47.34	54.52
	Per 100 ml	17.70 kcal	21.15 kcal	22.94 kcal	26.59 kcal	18.94 kcal	21.80624 kcal
Protein	Mean ± SD (%)	17.44 ± 0.13	17.97 ± 0.14	15.64 ± 0.06	6.09 ± 0.06	15.41 ± 0.02	13.58 ± 0.08
	Per 100 ml	6.98 kcal	7.19 kcal	6.25 kcal	2.43 kcal	6.16 kcal	5.4328 kcal
Fat	Mean ± SD (%)	8.12 ± 0.09	1.54 ± 0.07	0.50 ± 0.14	3.44 ± 0.20	2.79 ± 0.14	1.05 ± 0.07
	Per 100 ml	7.31 kcal	1.39 kcal	0.45 kcal	3.10 kcal	2.51 kcal	0.94221 kcal
Moisture	Mean ± SD (%)	3.30 ± 0.04	4.87 ± 0.04	4.74 ± 0.13	4.16 ± 0.02	2.60 ± 0.02	4.67 ± 0.15
Ash	Mean ± SD (%)	26.90 ± 0.37	22.74 ± 0.03	21.78 ± 0.06	23.98 ± 0.14	29.46 ± 0.22	26.19 ± 0.21
Energy (1 kcal = 4.2 kJ)	Mean ± SD (%)	31.98 kcal = 134.32 kJ	29.73 kcal = 124.85 kJ	29.64 kcal = 124.50 kJ	32.12 kcal = 134.90 kJ	27.61 kcal = 115.95 kJ	28.18 kcal = 118.36 kJ
Lactose	Per 100 ml	5.63	7.79	7.91	6.68	7.27	2.81

A and B: Lactose free cow's milk; C: Whole goat's milk; D: Lactose free soy milk; E: Almond milk

Table 1: Nutritional composition in different type of milk.

type of milk analysed per 100 ml. Thus, the energy content in retentate milk per serving is the second highest with 134.32 kJ, A is 124.85 kJ, B is 124.50 kJ, C is 134.90 kJ, D is 115.95 kJ, and E is 118.36 kJ.

The composition of reconstituted retentate powder milk and five others commercialized milk homogenized with water was analysed by HPLC to determine the lactose concentration. Table 1 states the range of lactose concentration in different brands of milk in Malaysia. From the data obtained, lactose concentration is in the range of 2.81 to 7.91 g per 100 ml, with concentrated milk is 5.63 g, A is 7.79 g, B is 7.91 g, C is 6.68 g, D is 7.27 g, and E is 2.81 g. Concentrated milk with a concentration of 5.63 g per 100 ml is in the range of lactose concentration given that it is in the rank of the second lowest amount of lactose concentration. Interestingly, sample A and B which claimed to be lactose fee cow's milk obviously had higher lactose content which is 7.79 g and 7.91 g per 100 ml, respectively compared to the other samples. Meanwhile, whole goat's milk in sample C contained higher lactose (6.68 g per 100 ml) than concentrated milk (5.63 g per 100 ml) obtained from this study. On the other hand, sample D contained quite a high proportion of lactose (7.27 g per 100 ml) which may be due to soy milk fortification that was subjected with sugar carbohydrate for the organoleptic importance. Finally, sample E from almond milk contained the lowest lactose concentration with 2.81 g per 100 ml due to the fact that lactose is a carbohydrate sugar exists in mammal milk only. This has proven that the concentrated milk obtained from this study is a low-lactose goat's milk and hence can be claimed as comparable standard or similar to other types of commercial milk.

Conclusion

Concentrated goat's milk from 10 KDa UF membrane size had been subjected to a comparison with other types of commercialized milk after atomization. Moreover, concentrated goat's milk with lactose concentration of 5.63 g per 100 ml was in the range of lactose concentration comparison with commercial milk. It can be concluded that a high degree of lactose removal from goat's milk could be achieved using MWCO 10 KDa cross-flow hollow fibre ultrafiltration system in producing low-lactose milk, which is as comparable as commercial milk.

Acknowledgement

Acknowledgement is due to Universiti Putra Malaysia for granting the financial support for my Masters studies in UPM, under Geran Universiti Putra Malaysia, GP-IPS/2013/9392300. Special thanks also extended to Department of Process and Food Engineering, Department of Chemical and Environmental Engineering, Department of Food Science and Technology, UPM, and Department of Veterinary Services, Alor Gajah, Melaka for providing facilities and equipment for this study.

References

1. Kulkarni SS, Funk EW, Li NM (2001) Ultrafiltration: Introduction and definitions. Membrane handbook, Norwell. Massachusetts, USA.

2. Zall R (1987) Accumulation and quantification of on-farm ultra-filtered milk: The California experience. Milchwissenschaft 42: 98-100.

3. Mulder J (1991) Basic principles of membrane technology. Kluwer academic publishers, The Netherlands.

4. Patel RS, Reuter H, Prokopek D, Sachdeva S (1991) Manufacture of low lactose powder using ultrafiltration technology. Food Sci Technol 24: 338-340.

5. Edelsten D, Meersohn M, Friis P, Nielsen EW, Sørensen KL, et al. (1983) Production of skim milk powder with lactose content reduced by ultrafiltration. Milchwissenschaft 38: 261-263.

6. Kosikowski F (1979) Low lactose yoghurts and milk beverages by ultrafiltration. J Dairy Sci 62: 41-46.

7. Cheryan M (1998) Ultrafiltration and microfiltration handbook. Technomic Publishing Company Inc, Lancaster, PA, USA.

8. Zeman LJ, Zydney AL (1996) Microfiltration and ultrafiltration: Principles and applications. Marcel Dekker Inc, New York.

9. Sofuwani N, Aslina S, Mazlina S (2016) Separation of lactose from raw goat's milk by cross-flow hollow fibre ultrafiltration membrane. Int Food Res J 23: 209-219.

10. Wang W, Zhou W (2010) Effect of maltodextrins on water adsorption and glass transition of spray dried soy sauce powders.

11. Savaiano DA, Levitt MD (1987) Milk intolerance and microbe-containing dairy foods. J Dairy Sci 70: 397-406.

12. AOAC (2000) Official methods of analysis. Association of official analytical chemists. Washington, DC, USA.

13. Westergaard V (2010) Milk powder technology. GEA Niro, Copenhagen, Denmark.

14. Park YW, Juarez M, Ramos M, Haenlein GFW (2007) Physico-chemical characteristics of goat and sheep milk. Small Rumin Res 68: 88-113.

15. Cerbulis J, Parks OW, Farrell HM (1982) Composition and distribution of lipids of goats′ milk. J Dairy Sci 65: 2301-2307.

16. Slacanac V, Bozanic R, Hardi J, Rezessyne J, Lucan M, et al. (2010) Nutritional and therapeutic value of fermented caprine milk. Int J Dairy Technol 63: 171-189.

17. Hofman PJ, Vuthapanich S, Whiley AW, Klieber A, Simons DH (2002) Tree yield and fruit minerals concentrations influence "Hass" avocado fruit quality. Sci Hort J 92: 113-123.

18. Saini AL, Gill RS (1991) Goat milk: An attractive alternate. Indian Dairyman 42: 562-564.

15

Development and Optimisation of Cassava Starch-Zinc-Nanocomposite Film for Potential Application in Food Packaging

Fadeyibi A[1]*, Osunde ZD[2], Agidi G[2], Idah PA[2] and Egwim EC[3]

[1]*Department of Agricultural and Biological Engineering, Kwara State University, Malete, Ilorin, Nigeria*
[2]*Department of Agricultural and Bioresources Engineering, Federal University of Technology, Minna, Nigeria*
[3]*Department of Biochemistry, Federal University of Technology, Minna, Nigeria*

Abstract

The improvement of biodegradable film used in the food packaging has been made possible through nanotechnology. This research was carried out to develop and optimize the cassava starch-zinc-nanocomposite films for potential applications in food packaging. The zinc nanoparticles were prepared by sol-gel method and established with the particle sizes ranging from 4 nm to 9 nm. The films were developed by casting the solutions of 24 g cassava starch, 0% to 2% (w/w) zinc nanoparticles and 45% to 55% (w/w) glycerol in plastic mould of 8, 10 and 12 mm depths. The average thickness of the films varied respectively with the depth as 15.14 ± 0.22, 16.21 ± 0.36 and 17.38 µm ± 0.13 µm. Permeability and stability of the films were determined at 27°C and 65% relative humidity and thermal range of 30°C to 950°C, respectively. Also, the mechanical properties were determined using the nano indentation technique. The films were optimised based on their characterized attributes using their desirability functions. The hardness, creep, elastic and plastic works, which determined the plasticity index of the films, decreased with thickness and zinc nanoparticles. The water vapour permeability increased with the concentrations of glycerol, zinc nanoparticles and thickness while the oxygen permeability decreased with the nanoparticles. The degradations of the Nanocomposites at 100°C were in the range of 2%-3%, which may indicate that the films are thermally stable. The optimum film whose desirability function is closer to the optimisation goal gave values of 49.29% glycerol, 17 µm thickness and 2% zinc nanoparticles for maximum thermal and mechanical properties. The low permeability, high thermal stability and low plastic work at higher concentration of zinc nancomposites may be essential in food packaging.

Keywords: Starch-zinc nanocomposite; Permeability; Thermal stability; Mechanical properties; Optimisation

Introduction

The study of the control of organic or inorganic matter at dimensions of roughly 1nm to 100 nm is termed nanotechnology. The addition of nanoparticles to renewable materials like starch and protein results in the formation of nanocomposite materials often transformed into flexible films [1-4]. The techniques involved in actualising this transformation are well understood, but their application in the food industry is greatly limited due to their poor service performance, especially in prolonging the shelf-life of food [5]. It is expected that the shelf-life of food should be improved by the addition of nanoparticles to flexible films, yet many nanocomposite materials are still unable to meet this packaging requirement [6,7] The addition of zinc nanoparticles as fillers to flexible films has proven quite promising due to their gas scavenging and antimicrobial activities, especially against gram-positive and gram-negative bacteria [8-10].

The thermal, mechanical and barrier properties of nanocomposite materials, which determined their suitability for packaging application, have been studied [6,11-16]. Zeng et al. [11], Taghizadeh and Sabouri [12] reported that the platelet nature of the nanoparticles usually hinders the diffusion of gases and specifically contributes to the improvement of their thermal and mechanical properties. Also, Huang and Yu [13] noted that introducing inorganic particles improved the thermal stability and mechanical resistance of starch-montmorilonite nanocomposite. The addition of chitosan in starch biofilms increased the thermal stability and ultraviolet absorption of the films [14-16]. Azeredo et al. [6], who carried out a study on the nanocomposite edible films from mango puree reinforced with cellulose nanofibre, revealed that the addition of the cellulose nanofibre significantly improved the thermal and mechanical properties of the films. However, most of the researches did not consider the antimicrobial and barrier scavenging

potentials of the nanocomposite materials, especially the roles of the nanoparticles in actualising these. Also, the methods presented in synthesising the nanoparticles are quite expensive. Besides, the materials are not environmentally friendly and are generally toxic to human cells, thus limiting their potential application in food packaging. Hence, there is the desire for an alternative nanocomposite packaging material which is cheap to synthesize, environmentally friendly and has strong potential for antimicrobial and barrier scavenging activities, higher mechanical strength and thermal stability.

An excellent example of nanocomposite material with such desired packaging requirements is the one blended with zinc nanoparticles [9,10]. Unfortunately, the studies on such material have not been reported in the literature. The effects of thickness of the material, zinc nanoparticles and glycerol concentration, which is often added as plasticizer, have not been studied. Also, the information on the optimisation of the experimental variables leading to the development of suitable flexible films for food packaging has not been reported. There is therefore the need to fill the existing knowledge gap by undertaking the present study. The objective of this research was to develop and

***Corresponding author:** Fadeyibi A, Department of Agricultural and Biological Engineering, Kwara State University, Malete, Ilorin, Nigeria
E-mail: adeshinaf601@gmail.com

optimize the cassava starch-zinc-nanocomposite film based on their barrier, thermal and mechanical properties for use in food packaging.

Materials and Methods

Preparation of cassava starch

Freshly harvested *gari* cultivar cassava tubers (5 kg), obtained from Kasuwan, Gwari Market in Minna, Niger State was used to prepare the starch. The produce was pealed and soaked in a clean bowl containing water for 24 hours after which it was ground into pastes. The ground paste was sieved using a muslin cloth and the resulting filtrate was left undisturbed for another 24 hours to allow the starch to settle at the bottom of the bowls to obtain starch slurry. The prepared starch slurry was dried under the sun for one week until a moisture content of 2% (wb) was achieved. The starch prepared from this cassava cultivar usually contained 16.3% amylase and 83.7% amylopectin [17].

Preparation and analyses of zinc nanoparticles

The sol-gel method, described by Fadeyibi et al. [18], was used to prepare the zinc nanoparticle at the Crop Processing and Storage Laboratory of the Department of Agricultural and Bioresources Engineering, Federal university of Technology, Minna, Nigeria. The method involves the mixture of two homogeneous solutions. The first solution was prepared by adding 30 mL of distilled water to 20 mL of triethanolamine with constant stirring while adding 2 mL (100 drops) of ethanol in drop wise. A 5.39 g of zinc acetate di-hydrate was added to 50 mL of distil water with continuous stirring, to obtain a 0.5 M solution, as the second homogeneous solution. The two solutions were mixed together in 500 mL beaker and a solution of ammonium hydroxide and 10 mL distilled water was added, in drop wise, with continuous heating and stirring for 20 minutes. Shortly after 30 min, a white bulky solution was formed. It was then washed 8-10 times with distilled water and filtered using a filter paper. The resulting residue was dried in the oven at 95°C for 8h.

The size distribution of the zinc particles was established using the zetasizer equipment (version 7.01) at the Centre of Genetic Engineering and Biotechnology of the Federal University of Technology Minna, Nigeria. The morphology of the zinc particles was studied using Siemens Bruker D5000 Multipurpose X-ray Diffractometer with the CuK α radiation of 1.54A°. This provides information on the intercalation and exfoliation processes of the short-range order of the zinc particles. The test was performed at the Chemistry Laboratory of Sheda Science and Technology Complex (SHESTCO), Abuja, Nigeria.

Development of cassava starch-zinc-nanocomposite film

The prepared starch from the cassava roots was mixed with the zinc nanoparticles and glycerol in definite proportions to form the different nanocomposites. The ingredients were homogenised with the help of an extruder, dried in an air circulated oven dryer (60°C and 80% RH) for 24 hours and ground to form the nanocomposites, as shown in Figure 1. This was followed by film casting in plastic moulds after adding 600 mL of distilled water to the nanocomposites and heated at 95°C.

The plastic mould, designed by determining the surface area using the expression in Equation (1), was used for casting the solution of the starch-glycerol-zinc nanocomposite. The surface area of the mould was 350 mm × 180 mm and the depth varied as 8, 10 and 12 mm. The thickness of the dried film at the four edges of the films was determined using a plastic micrometer, and the average and standard deviation were computed. The plastic mould with 8 mm depth gave an average

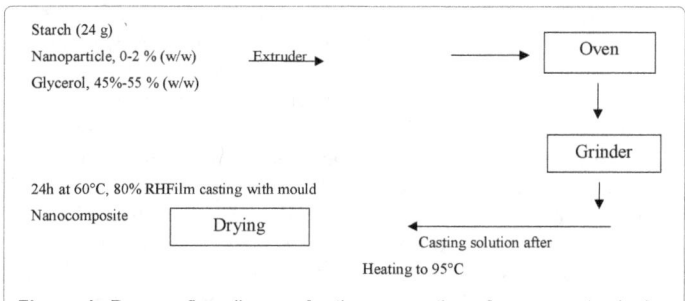

Figure 1: Process flow diagram for the preparation of cassava starch-zinc nanocomposite films.

dried thickness of 15.14 ± 0.22 μm, while those with 10 and 12mm depths were 16.21 ± 0.36 μm and 17.38 ± 0.13 μm, respectively.

$$TSA = Na + e \tag{1}$$

Where, N = number of equivalent biomaterials (500 g), a= surface area of the mould mm², e = allowance (assumed 600 mm²) and TSA = total surface area of the mould (mm²).

Characterisation of the film

Determination of thermal stability of the film: The thermal stability of the cassava starch-zinc nanocomposite films was determined using the Thermo Gravimetric Analyzer (TGA), at the Centre of Genetic Engineering and Biotechnology of the Federal University of Technology Minna, Nigeria. The chilling compartment of the TGA was initially set at a temperature of 15°C to maintain the equipment in the right cooling condition. The temperature scan was also set in the range of 30°C to 950°C and nitrogen gas was passed into the chambers at a flow rate of 20 mL/min and pressure of 2.5 bars for a period of 30 min. The weight of the crucible was zeroed after which a mass 2 mg of the sample was loaded in the TGA thermo-balance, with continuous weighing as heating progresses. Finally the programme was initiated and the weight loss (%) and time (min) data were measured and recorded as the temperature approached 950°C. The procedure was repeated thrice for each sample of the cassava starch-zinc-nanocomposite films and the values of the temperature were plotted against the weight loss of the film.

Determination of barrier properties of the film

Water vapour permeability: The water vapour permeability (WVP) of the nanocomposite films was determined according to the ASTM Standard [19]. The film sample was cut into circle of 4 cm diameter and placed on petri dish filled with 50 mL solution of sodium chloride to expose the film to 65% relative humidity at 25°C. The petri dish was allowed to equilibrate for two hours before taking the initial weight. The final weight was taken at the end of every one hour. The procedure was replicated three times; and the average of water vapour transmission rate and permeability were determined empirically as expressed in Equations (2) and (3).

$$WVTR = \frac{G}{tA} \tag{2}$$

$$WVT = \frac{WVTR}{\Delta p} \times L \tag{3}$$

Where, WVTR = Water vapour transmission rate (g/s. m²), WVP = water vapour permeability (g/m.Pa.s), G is the difference in mass (g), t is the time (h), and A is the area of the mouth of the petri dish (m²), L is the thickness of the test specimen (mm) and Δp is the partial pressure difference of water vapour across the film.

Oxygen permeability: The oxygen permeability (OP) of the cassava starch-zinc nanocomposite film was determined empirically using the column absorption method [20]. This was achieved by cutting 1.23 cm² sample of the film in the absorption column, immersed in a water bath and equipped with a hose from the cylinder containing oxygen gas. The sample was exposed to the oxygen gas, by this arrangement, for a period of one hour at 63% RH and 25°C. The oxygen concentration (%) in the column was subsequently measured, using an electronic balance (0.0001 g sensitivity) at the end of one hour. The data of properties shown in Table 1 contains the values of some parameters, including an established correlation for the computation of $P_{H_2O}{}^{sat}$, used for the calculation of the oxygen permeability (Table 1).

The difference in the partial pressure (ΔP) was computed using Eqs. (4) to (7).

$$[O_2] = \frac{m_{O_2}}{1000M} \tag{4}$$

$$[O_2]_i = \frac{RH \times P_{H_2O}{}^{sat}}{100RT} \tag{5}$$

$$[O_2]_t = \frac{[O_2]}{[O_2]_i} \times 100\% \tag{6}$$

Therefore, $\Delta P_{O_2} = \left([O_2]_t - [O_2]_i\right).P_{total}$ (7)

where, $[O_2]_t$ is the oxygen concentration inside the absorption column at time t, $[O_2]_i$ is the initial column oxygen content, m_{o2} is the mass of oxygen gas absorbed by the cassava starch nanocomposite film in the column, M is the molar mass of oxygen gas (16 g/mol), R is the molar gas constant (8.31×10^7 m³ Pa.mol⁻¹.K⁻¹) .

The total pressure (P_{total}) was calculated by taking into account the influence of the temperature (T) on the saturation pressure of water vapour $P_{H_2O}{}^{sat}$ and the relative humidity (RH) as shown in Equation (8). The value of OP was subsequently computed from Equation (9).

$$P_{total} = P_{atm} - P_{H_2O}{}^{sat}.RH \tag{8}$$

$$OP = \frac{\Delta m \Delta x}{\Delta t A \Delta P_{O_2}} \tag{9}$$

The procedure was repeated three times and the oxygen permeability was taken as the average of the oxygen permeability measured under the steady state condition.

Determination of mechanical properties of the film: The nanoindenter was used to determine the mechanical properties of the nanocomposite films. A typical profile of the load-displacement curve of the film, obtained from the nanoindenter, is shown in Figure 2.

The profile was used to compute the hardness, Young's modulus, elastic recovery and creep from the empirical relationships in Equations (10) to (13). The strain was computed as the ratio of the recovered depth to the original depth or thickness of the nanocomposite film. The stress was computed as the product of the strain and the Young's modulus of the material. The elastic and plastic works correspond to

Parameter	Value
Volume of the Column (V_v)	17.42×10^{-3} cm³
Density of Oxygen (ρ_{o2})	0.00133 g/m³
Atmospheric pressure (P_{atm})	101325 Pa
Saturated pressure of water vapour ($P_{H_2O}{}^{sat}$)	$P_{H_2O}{}^{sat}$ = 190.2T(°C)-1536.7(Pa)
Film area (A)	1.23×10^{-4}m²

Table 1: Data of properties used for the computation of oxygen permeability.

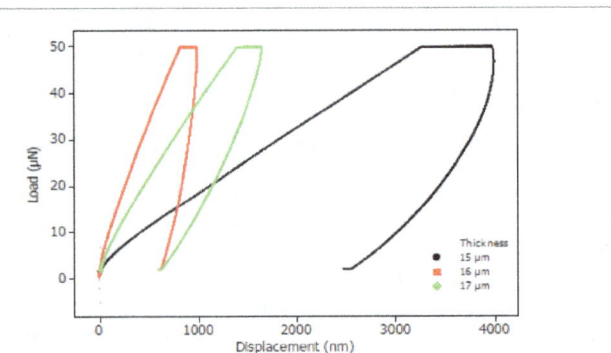

Figure 2: Load-displacement profiles of cassava starch-zinc nanocomposite films at 45% (w/w) glycerol and 2% (w/w) zinc nanoparticles and different thickness.

the areas under the loading and the unloading parts of the hysteresis loop [21-23].

$$H = \frac{P_{max}}{A_c h_c} \tag{10}$$

Where, P_{max} = maximum load, A_c = contact area (nm²), h_c = contact depth (nm), H = hardness of the nanocomposite film.

$$\frac{1}{E_r} = \frac{1-v^2}{E} + \frac{1-v_i{}^2}{E_i} \tag{11}$$

where, E_r = reduced modulus (MPa), v = poison's ratio of the nanocomposite film, which was obtained by assuming that the material is isotropic in nature with the elastic modulus evenly distributed in all crystallographic directions = 0.5, v_i = poison's ratio of the diamond indenter = 0.25, E_i = elastic modulus of the diamond indenter = 1140 GPa, E = Elastic modulus of the nanocomposite film.

$$ERP = \frac{h_e}{h_f} \tag{12}$$

Where, h_e = recovered depth (nm), h_f = depth of the unloading curve of the profile (nm), ERP = elastic recovery parameter.

$$\epsilon = \frac{1}{h_c} \frac{dh_c}{dt} \tag{13}$$

Where, h_c = depth at the holding region of the profile (nm), t = holding time (0.5 < t < 1), ϵ = strain rate or creep (nm/s).

Optimisation of the experimental variables of the film

Development of optimisation framework: The Response Surface Methodology (RSM) was applied to determine the optimum conditions of the experimental variables (concentration of glycerol, zinc nanoparticles and thickness) based on the characterised attributes of the films, namely mechanical, barrier and thermal properties. A Box-Behnken Design (BBD), which includes 17 experiments formed by 5 central points, was employed in the optimization process because of its suitability for a 3^3 full factorial experiment. The optimization framework of the BBD protocol is shown in Equation (14).

$$\forall \Phi = f(X_1, X_2, X_3) + e \text{, subject to the constraints:}$$

$$\begin{cases} 15 \leq X_1 \leq 17 \ \mu m \\ 45 \leq X_2 \leq 55 \ \% \\ 0 \leq X_3 \leq 2 \ \% \end{cases} \tag{14}$$

Where, Φ = characterised attributes of the film (barrier, thermal, mechanical properties), X_1 = thickness of the film, X_2 = concentration of glycerol, X_3 = concentration of zinc nanoparticles, e = error term of the optimisation function.

Desirability function: Desirability is an objective function that ranges from zero outside a limit to one at the goal. It is possible to alter the characteristics of a goal by either maximizing or minimizing the optimisation function. For several responses and factors, all goals can be combined into one desirability function [24]. Therefore, in this study the objective functions are to maximise the mechanical properties and minimise the permeability and thermal stability of the films. The reason for the choice of a particular objective function is centered on the need to select the most desirable film, whose characteristic feature is comparable to the conventional low density polyethylene material.

Results and Discussion

Establishment of zinc nanoparticles

The results of the size distribution of the zinc particles were shown in Figure 3. It can be seen that the size of the zinc particles are heterogeneous as it varied in the range of 4 nm-9 nm. The variation in the size of the zinc particles can be associated with the increase in their solubility in the water used as solvent [25]. Tang et al. [26] revealed that organic or inorganic particles having their particle size in the range of 1-100 nm can be regarded as nanoparticles, especially when used for food packaging application (Figure 3).

Moreover, Figure 4 showed the X-ray diffraction pattern of the zinc particles. The diffraction peaks obtained confirmed the presence of the heterogeneous structure of the zinc nanoparticles. It is possible that the galleries of the zinc nanoparticles are forced apart during the irradiation process thus leading to an increase in the gallery spacing (d-spacing) [27]. According to Bragg's law, this would cause a shift in the diffraction peak towards a lower angle. The peak, at the lower angle, becomes wider and finally broadens into the base line; thus revealing a complete heterogeneous structure. Similar behaviours have been observed for montmorillonite, which shows complete heterogeneous structure by the presence of distinct diffraction peaks [26,28]. The use of the zinc nanoparticles can help improve the performance of the film for food packaging application (Figure 4).

Thermal stability of the film

The thermal degradation curves of the cassava starch-zinc

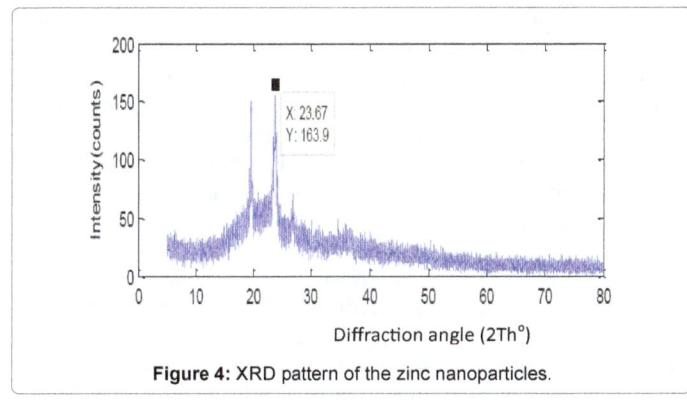

Figure 4: XRD pattern of the zinc nanoparticles.

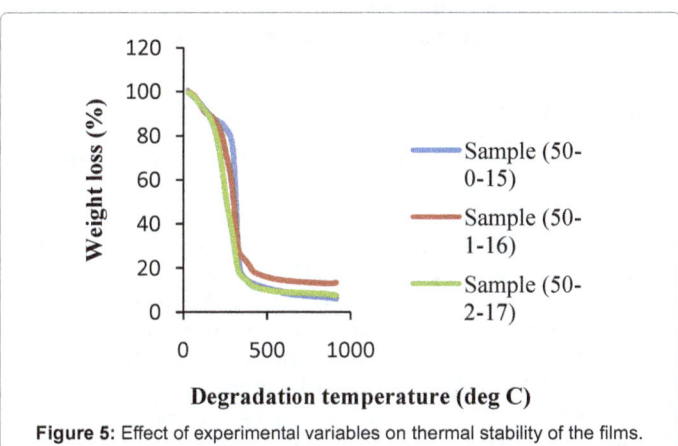

Figure 5: Effect of experimental variables on thermal stability of the films.

nanocomposite films are typically shown in Figure 5. It can be seen that the thermal decomposition of the films occurred in three main steps, which generally conforms to the three thermal degradation phenomena of most starch-based films reported in literature [29-31]. The initial stage of the thermal degradation of the films occurred at temperature less than 100°C. The weight loss of the cassava starch-zinc nanocomposite films at this stage are in the range of 2% to 3%, and can be associated with the evaporation or dehydration of loosely bound water molecules and low molecular weight compounds in the films. The cassava starch-zinc nanocomposite films show higher mass loss at 50% glycerol concentration and 15 μm thickness compared with the 17 μm counterparts at temperature lower than 100°C. This indicates that the films with 50% glycerol concentration and 0% zinc nanoparticles contain moisture than the ones with 50% glycerol concentration and 2% zinc nanoparticles. The second stage of the thermal degradation of the films ranges from 150°C to 260°C, which corresponds with the evaporation of glycerol compounds together with chemisorbed water molecules (Figure 5).

The degradation temperatures of the films agree with the findings of Zhong and Li [15] and Sanyang et al. [32] who reported the degradation temperature of glycerol-rich phase of kudzu starch-based edible films and biodegradable films based on sugar palm starch between 150°C to 280°C and 120°C to 290°C, respectively. Further heating beyond 260°C induced the highest thermal degradation rate which is reflected by the sharp weight reduction of the films. The onset of thermal decomposition of starches occurred at the residual weight, which is around 300°C for the films. According to Nescimonto et al. [33], this stage corresponds to the elimination of hydrogen groups, decomposition and depolymerisation of starch carbon chain. Generally, the residual weights represent almost complete decompositions of the films since more than 80% of the films

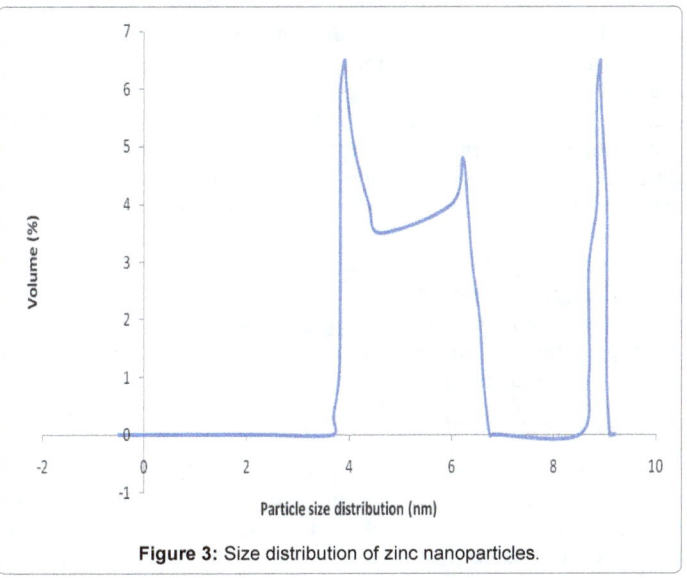

Figure 3: Size distribution of zinc nanoparticles.

have been degraded. Beyond this temperature range, the films become unstable and the weight loss of the films increased with an increase in temperature. Similar degradation responses with temperature change were obtained for other thicknesses and concentrations of glycerol and zinc nanoparticles, as was shown in Figure 5. It can be argued that the increase concentrations of the glycerol and the zinc nanoparticles are partly responsible for the slight difference in the three stages of decomposition of the films. The comparison of the behaviour of the cassava starch-zinc-nanocomposite film with that of the low density polyethylene material (LDPE) revealed that the thermal stability of the film is lower than the LDPE. According to Chatloff and Sircar [34], a single mass degradation step, occurring at 478°C, has often been observed for LDPE and other synthetic polymers. This is in sharp contrast to the thermal degradation behaviour of the cassava starch nanocomposite films, which revealed three mass degradation stapes. This study provides a fair knowledge about the thermal stability of the films for potential application in food packaging.

Permeability of the films

The influence of the experimental variables (thickness, concentration of glycerol and zinc nanoparticles) on the average values of the water vapour permeability is shown in Figure 6. It can be seen that the water vapour permeability increased generally with increase in the concentrations of the glycerol, zinc nanoparticles and thickness. This means that the experimental variables played significant roles in promoting the water vapour permeability of the resulting nanocomposite material. The presence of zinc nanoparticles together with increased thickness of the material help create a composite matrix with higher water absorption ability [30,34-36]. The reason for the increase in the water vapour permeability of the films with increasing concentration of the nanoparticles could be explained by structural modifications of the composite matrix, which may result from the high surface area to volume ratio of the nanoparticles. Thus, in order to maintain the quality of agricultural produce, it is necessary to package the food in a material with high water permeability (Figure 6).

The influence of the experimental variables on the oxygen permeability of the nanocomposite films is shown in Figure 7. The oxygen permeability decreased with increasing thickness and concentrations of glycerol and zinc nanoparticles. It is possible that the presence of the nanoparticles restricted the mobility of the gas molecules as they diffuse across the films. The mobility was altered

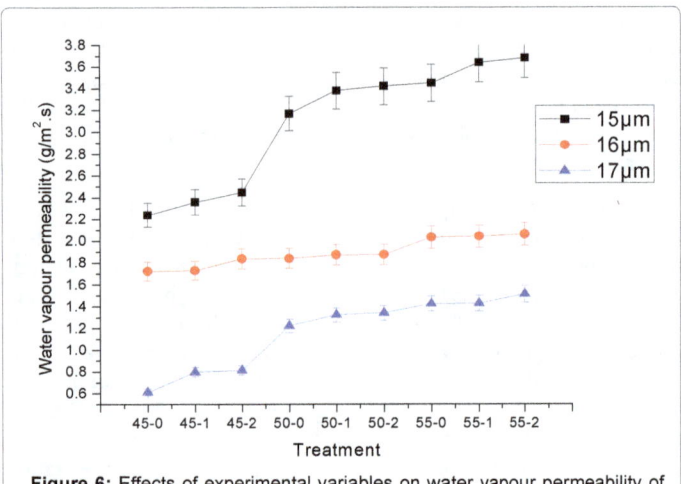

Figure 6: Effects of experimental variables on water vapour permeability of the films.

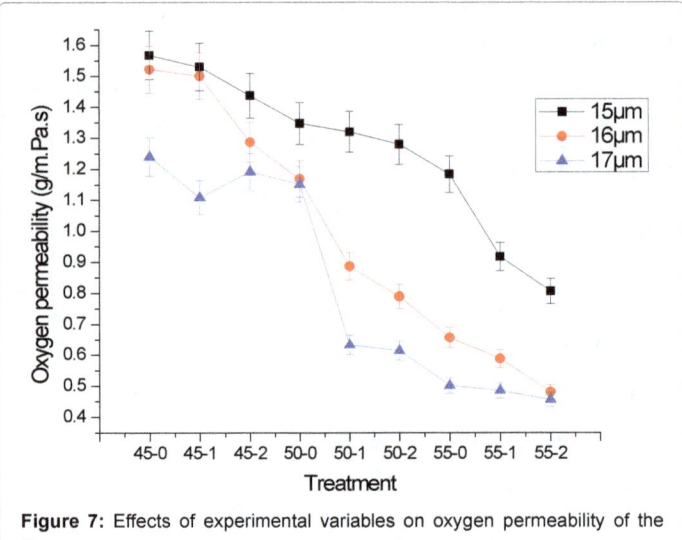

Figure 7: Effects of experimental variables on oxygen permeability of the films.

with the increase in the concentration of the nanoparticles, which is associated with their scavenging potential and the decreasing porosity of the composite matrix. It is important to note that, lower gas permeability of the films is an advantage in the nanocomposite technology, especially when they are to be used for packaging purposes. Therefore, by this action, the nanocomposite films can be regarded as an active or intelligent packaging material.

Mechanical properties of the films

The effects of thickness and zinc nanoparticles on the mechanical properties of the nanocomposite films are shown in Figures 8 and 9. The creep, which is essential in determining the strain rate sensitivity of the nanocomposite material, decreased with nanoparticles and thickness of the film. The hardness of the material decreased with thickness and zinc nanoparticles. The Young's modulus and elastic recovery parameter increased with nanoparticles but their behaviours were found to be inconsistent with thickness of the films, as shown in Figure 8. These behaviours can be attributed to the increased displacement of the film with a slight change in the applied load as the thickness increased.

The elastic and plastic works, which determine the plasticity index of the materials, generally decreased with thickness and zinc nanoparticles, as shown in Figure 9. This means that the higher the concentration of the nanoparticles and thickness, the lower the plasticity index and thus the viscoelastic behaviour of the resulting nanocomposite material. The strain and stress of the material decreased generally with thickness and nanoparticles. This may be associated with the increasing area of contact of the diamond indenter with the applied load. Jorge et al [36] corroborated our findings in their work on the mechanical properties of gelatin-montmorillonite nanocomposite film. The presence of montmorilonite increased the mechanical properties of the film. Hence, the nanoparticles be said to contribute in the improvement of the mechanical properties of the nanocomposite material due probably to their large surface area to volume ratio.

Optimally selected films

The optimum values of the permeability due to oxygen and water vapour of the nanocomposite films, which were determined by minimising the responses subject to the defined constraints of the experimental variables, is shown in Figure 10. The desirability function

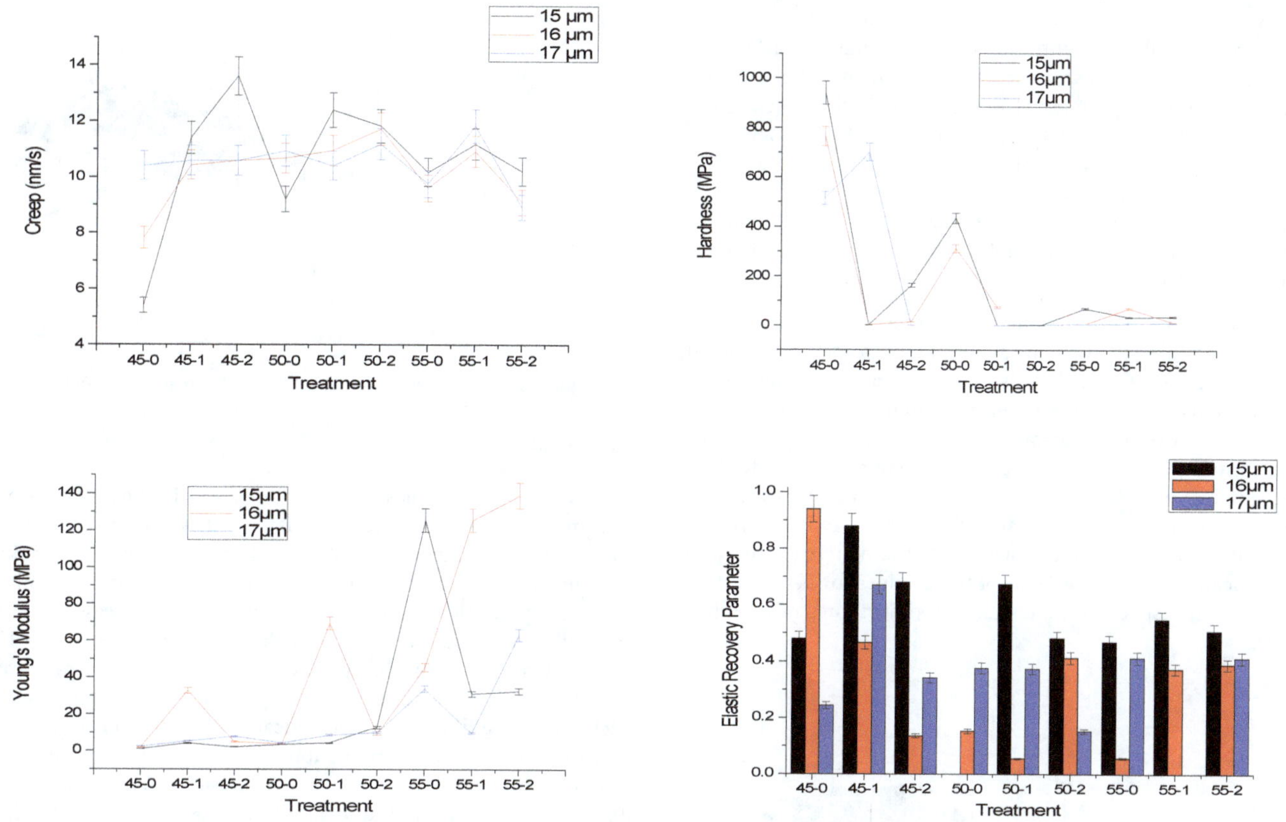

Figure 8: Effect of experimental variables on creep, hardness, Young's modulus, elastic recovery parameter of the films.

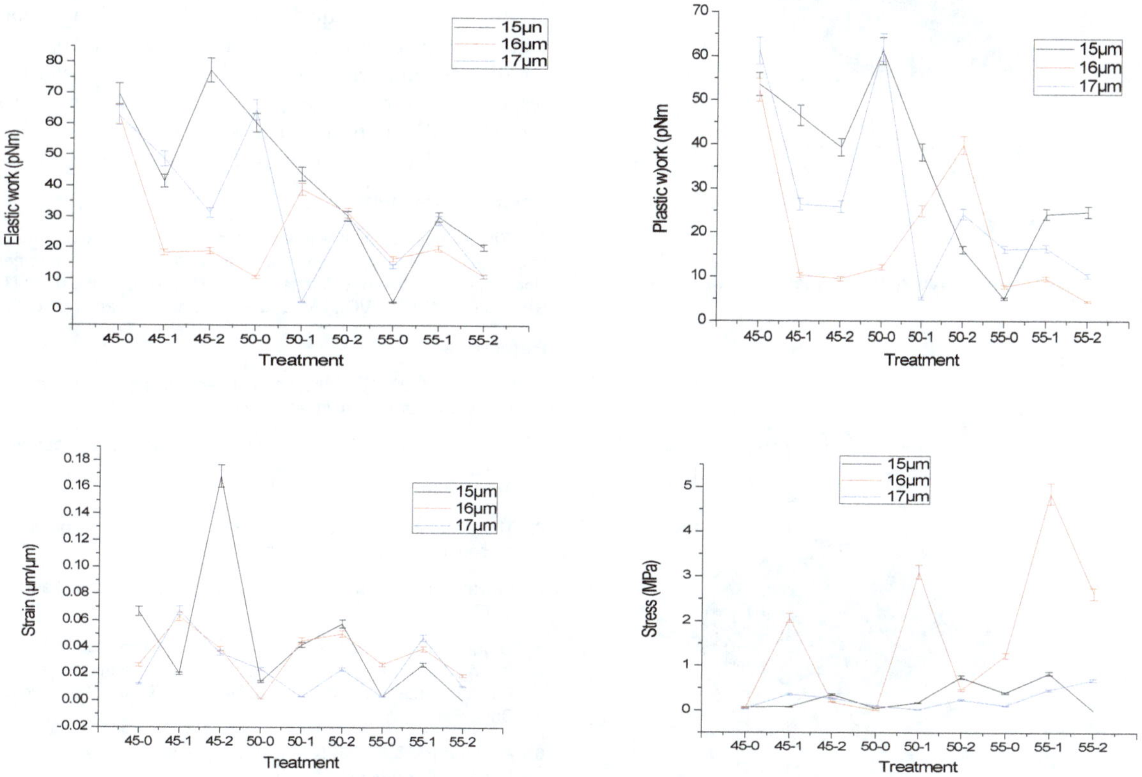

Figure 9: Effects of experimental variables on strain, stress, elastic and plastic works of the films.

of selecting the best combination of the experimental variables increased with thickness but decreased with the concentration of the glycerol regardless of the concentration of the zinc nanoparticles. The increased desirability may also be associated with the degree of the interaction among the experimental variables. Similarly, selecting the films based on their thermal stability also required higher degree of interaction among the experimental variables, as shown in Figure 11. It can be deduced from these observations that higher thickness and lower glycerol concentration will greatly influence the solution goals, which are to minimise the permeability and maximise the thermal stability (Figures 10 and 11).

Additionally, the optimum film was selected based on the desirability of obtaining the maximum mechanical response, as shown in Figure 12. The desirability, in this case, was mainly influenced by the effective contribution of the interaction among the experimental variables on the overall acceptability of the solution goal. The individual contributions of the variables were lesser compared to their associated interaction. Hence, the optimum film whose desirability function is closer to the solution goals gave values of 49.29% glycerol, 17 μm thickness and 2% zinc nanoparticles. The matrix of the ideal film, by implication, should be made of these values for better service performance with respect to the characterised attributes of the films (Figure 12).

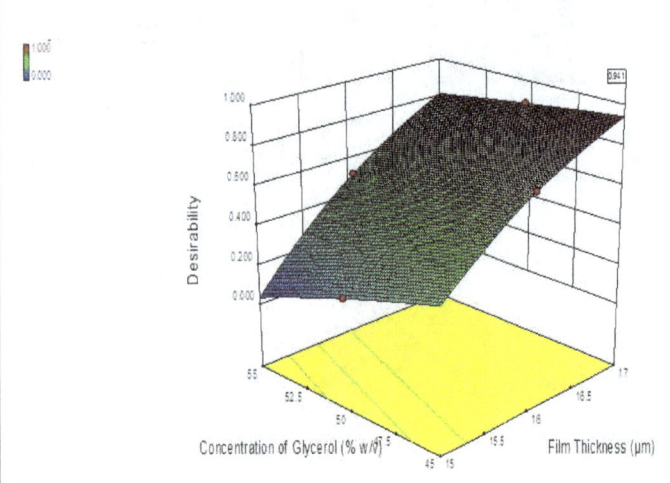

Figure 10: Surface response desirability for oxygen and water vapour permeability of the films.

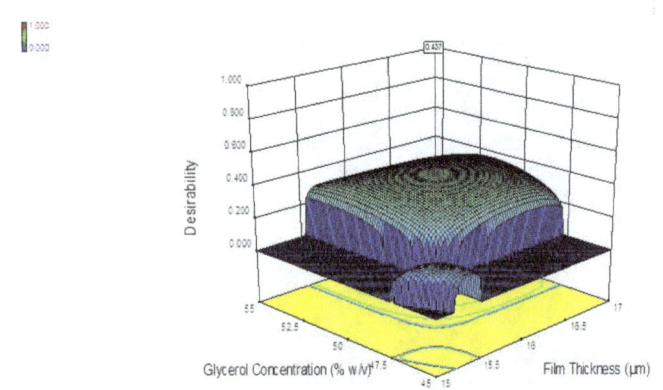

Figure 11: Surface response desirability for mechanical properties of the films.

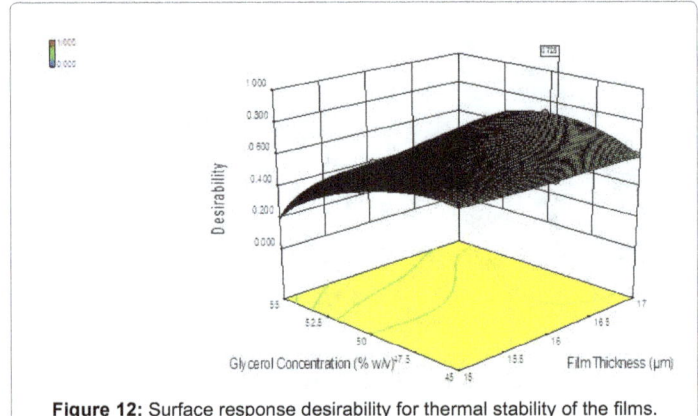

Figure 12: Surface response desirability for thermal stability of the films.

Conclusion

The food industry has seen great advances in the packaging sector with many innovations occurring lately. These advances have led to improved food quality and safety. This research was undertaken to develop and optimize the cassava starch-zinc-nanocomposite films for potential application in food packaging. The films were found to be thermally stable with 2% to 3% of their weights degraded below 100°C. The water vapour permeability increased while the oxygen permeability decreased with the concentration of glycerol, zinc nanoparticles and thickness of the films. The mechanical properties decreased generally with the thickness and zinc nanoparticles. The optimum film whose desirability function is closer the optimisation goal gave values of 49.29% glycerol, 17 μm thickness and 2% zinc nanoparticles for maximum thermal stability and mechanical properties. The lower gas permeability of cassava starch-zinc nanocomposite film can slow down the rate of respiration, just as the higher water vapour permeability can promote the keeping quality especially in packaged foods. It is likely that the oxygen scavenging effect of the zinc nanoparticles, which can retard ripening, senescence and consequent deterioration, may be responsible for the low gas permeability of the new packaging material. It can be deduced from the behaviour of the nanocomposite film that it has potential application in food packaging.

Acknowledgement

The authors would like to thank the Tertiary Education Trust Fund (TETFund) of the Federal Ministry of Education, Nigeria and the management of the Federal University of Technology, Minna, Nigeria for providing the grant (TETFUND/DESS/RP.DIS/FUT/MINNA/VOL.XV) and work space for carrying out this research.

References

1. Duncan TV (2011) Applications of nanotechnology in food packaging and food safety: Barrier materials, anti-microbials and sensors. J Coll Int Sci 12: 34-40.

2. Mehyar GF, Han JH (2004) Physical and mechanical properties of high-amylose rice and pea starch films as affected by relative humidity and plasticiser. J Food Sci 69: 449-459.

3. Tharanathan RN (2003) Biodegradable films and composite coatings: past, present and future. Trends Food Sci Tech 14: 71-78.

4. Kirwan MJ, Strawbridge JW (2003) Plastics in food packaging. Food Pack Tech 1: 174-240.

5. Lagaron JM, Sanchez-garcia M (2008) Thermoplastic nanobiocomposites for rigid and flexible food packaging applications, In: E. Chiellini (ed.), Environmentally friendly food packaging, FL pp. 62-89, Woodhead Publishers, Boca Raton.

6. Azeredo HMC (2009) Nanocomposites for food packaging applications. Food Res Int 42: 1240-1253.

7. Hernandez R, Selke SEM, Cultler J (2000) Plastics packaging: properties,

processing, applications, regulations. Hanser Gardner Publications, Munich, Germany.

8. Adams LK, Lyon DY, Alvarez PJJ (2006) Comparative eco-toxicity of nanoscale TiO_2, SiO_2, and ZnO water suspensions. Water Res 40: 3527-3532.

9. Jones N, Ray B, Ranjit KT, Manna AC (2008) Anti-bacterial activity of ZnO nanoparticle suspensions on a broad spectrum of microorganisms. FEMS Microbiol Lett 279: 71-76.

10. Emamifar A, Kadivar M, Shahedi M, Solaimanianzad S (2011) Effect of nanocomposite packaging containing Ag and ZnO on inactivation of Lactobacillu splantarum in orange juice. Food Control 22: 408- 413.

11. Zeng QH, Yu AB, Lu GQ, Paul DR (2005) Clay-based polymer nanocomposites: research and commercial development. J Nanosci Nanotech 5: 1574-1592.

12. Taghizadeh MT, Sabouri N (2013) Thermal degradation behaviour of polyvinyl alcohol/starch/carboxymethyl cellulose/ clay nanocomposites. Univer J Chem 1: 21-29.

13. Huang MF, Yu JG (2006) Structure and properties of nanocomposites and their characteristics, Carbohydrate composites. J App Poly Sci 99: 170-176.

14. Zhang L, Jiang Y, Ding Y (2008) ZnO nanofluids-A potential anti-bacterial agent. Prog Nat Sci 18: 939-944.

15. Zheng JP, Li P, Ma YL, Yao KD (2014) Gelatine/montmorillonite hybrid nanocomposite. I. Preparation and properties. J Appl Poly Sci 86: 1189-1194.

16. Dang KM, Yoksan R (2015) Development of thermoplastic starch blown by incorporating plasticized chitosan. Carbohydrate Polymers 115: 575-581.

17. Dakubu M, Bruce-smith SP (1979) Amylose content of starch from different varieties of cassava (*Manihot esculenta crantz*) in Ghana. Ghana J Agri Sci 12: 143-145.

18. Fadeyibi A, Osunde ZD, Agidi G, Evans EC (2016) Mixing index of a starch composite extruder for food packaging application. In: Inamuddin S (ed.) Green polymer composites technology: properties and applications. CRC press and Taylor and Francis Group, Asia Pacific.

19. Astm Standards (2005) E96-05 Standard test methods for water vapour transmission of materials. Philadelphia, PA.

20. Laksmana FL, Hartman Kok PJA, Frijlink HW, Vromans H, Voort Maarschalk K (2009) Gas permeation related to the moisture sorption in films of glassy hydrophilic polymers. J Appl Poly Sci 5: 1-10.

21. Syed VA, Nagamani H, Yeshwanth A, Sundaram S (2013) Nano-indentation behaviour of ultrathin polymeric films. Adv Aero Sci Appl 3: 235-238.

22. Tall PD, Ndiaye S, Beye AC, Zong Z, Soboyejo WO, et al. (2007) Nanoindentation of Ni-Ti thin films. Mat Manu Pro 22: 175-179.

23. Jian S, Chen G, Hsu W (2013) Mechanical properties of Cu_2O thin films by nanoindentation. Materials 6: 4505-4513.

24. Vargas-Lopez JM, Paredes-Lopez O, Espitia E (2006) Evaluation of lime heat treatment on physicochemical properties of amaranth by response surface methodology. Cereal Chem 67: 417- 421.

25. Makhluf S, Dror R, Nitzan Y, et al. (2005) Microwave-assisted synthesis of nanocrystalline MgO and its use as bacteriocide. Adv Funct Mat 15: 1708-1715.

26. Tang X, Alavi S, Herald TJ (2008) Effects of plasticizers on the structure and properties of starch-clay nanocomposite films. Carbohydrate Poly 74: 552-558.

27. Mcglashan SA, Halley PJ (2003) Preparation and characterization of biodegradable starch-based nanocomposite materials. Poly Int 52: 1767-1773.

28. Chung Y, Ansari S, Estevez L, Hayrapetyan S, Giannelis EP, et al. (2010) Preparation and properties of biodegradable starch-clay nanacomposites. Carbohydrate Poly 79: 391-339.

29. Dang KM, Yoksan R (2015) Development of thermoplastic starch blown by incorporating plasticized chitosan. Carbohydrate Poly 115: 575-581.

30. Sanyang ML, Sapuan SM, Jawaid M, Ishak MR, Sahari J, et al. (2015) Effect of plasticiser type and concentration on tensile, thermal and barrier properties of biodegradable films based on sugar palm (*Arenga pinnata*) starch. Polymers 7: 1106-1124.

31. Mehyar GF, Han JH (2014) Physical and mechanical properties of high-amylose rice and pea starch films as affected by relative humidity and plasticiser. J Food Sci 69: 449-459.

32. Nascimento TA, Calado V, Carvalho CWP (2012) Development and characterisation of flexible film based on starch and passion fruit mesocarp flour with nanoparticles. Food res int 49: 588-595.

33. Chatloff RP, Sircar AK (2007) Thermal analysis of polymers. Encyclopedia of polymer science and Technology, John Wiley and Sons.

34. Dai L, Qui C, Xiong L, Sun Q (2015) Characterisation of corn starch-based films reinforced with taro starch nanoparticles. Food Chem 174: 82-88.

35. Rojas-graü M, Tapia M, Martín-belloso O (2008) Using polysaccharide-based edible coatings to maintain quality of fresh-cut fuji apples. U-Technology 41: 139-147.

36. Jorge MFG, Vanin FM, de Carvalho RA, Moraes ICF, Bittante AMQB, et al. (2014) Mechanical properties of gelatin nanocomposite films prepared by spreading: effect of montmorillonite concentration.

16

Characterization and Classification of Different Tunisian Geographical Olive Oils using Voltammetric Electronic Tongue

Sana Mabrouk[1]*, Yosra Braham[1], Houcine Barhoumi[1] and Abderrazak Maaref[1]

[1]Laboratory of Interfaces and Advanced Materials (LIMA), University of Monastir, Tunisia

Abstract

In this work, we describe a sensor based on glassy carbon electrode, employed to discriminate between olive oils from different Tunisian regions. The characterization was made using three electrochemical techniques, cyclic voltammetry (CV), differential pulse voltammetry (DPV) and square wave voltammetry (SWV). Each type of oil provides a diversity of characteristic signals that can be used as an input variable of different statistics analysis, like principle component analysis (PCA), cluster analysis (CA) and discriminate factorial analysis (DFA).The results resulting from the electrochemical methods are compared. The obtained results show the reliability of the used methods on the discrimination between olive oil qualities obtained from different regions of Tunisia.

Keywords: Olive oils; Tocopherols; Electronic tongue; Voltammetry measurements; Principal components Analysis; Discriminate factorial analysis

Introduction

The quality of liquids is usually evaluated by several methods. Among these methods we can cite, chromatographic techniques, chemical analysis and mass spectrometry. In fact, none of the techniques mentioned above cannot be considered as a method of measurement completely characterizing the degree of freshness or bitterness of the food industry, where the need to search for new techniques to assess handle the reliable product quality in the food industry. In short, the food industry is looking for a technique that allows a practical, rapid, reliable and inexpensive to determine the state of freshness of products and analyze complex liquid samples. In consists an array of non-specific sensors coupled to a pattern recognition technique (E- tongue) [1-11]. The "electronic tongue" is a set (array) of partially selective chemical sensors with sensitivity to as wide a number of solution components as possible. Combined with multivariate analysis and/or pattern recognition programs such a system may provide quantitative information about multiple components of liquids and be used for recognition or identification of a liquid [12-24].

It is well- known that the virgin olive oils content different several classes of compounds having antioxidant proprieties such as polyphenols, tocopherols, carotenoids, sterols [25]. Such proprieties are the reason of the health benefits and protect oil from auto-oxidation [26, 27]. In this work, a novel method has been developed to discriminate between olive oils picked from different geographical regions and of different bitterness degrees. The objective of this work is to develop a novel electrochemical method to discriminate between the olive oils by means of an array of voltammetric sensors based on carbon paste. The global response of six olive oils has been recorded by three electrochemical techniques such as cyclic voltammetry (CV), differential pulse voltammetry (DPV) and square wave voltammetry (SWV)). The capability of discrimination of the olive oils has been analyzed by Principal Component Analysis (PCA), Cluster Analysis (CA) and Discriminate Factorial Analysis (DFA) of the obtained cyclic voltammograms.

Experimental Details

Reagents

Dichloromethane and tetrabutylammonium tetraphenylborate were purchased from Sigma-Aldrich. Six different geographical virgin olive oils (VOOs) varieties were picked from Tunisia regions.

Glassy carbon is an interesting electrode material because it is inexpensive and has a rich surface chemistry can be exploited to influence its reactivity. Also, its large range of potential provides a good survey of electrochemically. A glassy carbon electrode of 3 mm in diameter was used. The geometric surface area was 0.071 cm^2 and an Ag/AgCl/KCl (sat) as reference electrode.

Insulator glassy carbon electrode cleaning

The working electrode was treated by a polishing powder with alumina and rinsed between each polishing step by ultrapure water then sonicated for 10 min. After each measurement, electrodes were washed with dichloromethane. The glassy carbon electrode was activated electrochemically in solution of H$_2$SO$_4$ (0.5M) with oxygen. A series of 5 cycles from 1.8 to -0.5 V at scan rate of 100 mVs^{-1} vs.

Electrochemical characterizations and instrumentation

Voltammetric electronic tongue set up and measurements: Electrochemical experiments were performed using a potentiostat Autolab PG30 electrochemical analyzer. The software with a conventional three-electrode cell including an Ag/AgCl/KCl (Sat) as the reference electrode, a platinum electrode as the counter electrode and the glassy carbon as working electrode. The electrodes were immersed in the electrolytic solutions and the responses were measured for four times. After each measurement, electrodes were washed with dichloromethane. Electrochemical measurements were performed in 2 mL of olive oils solutions using 10 mL of dichloromethane and 0.02 g of tetrabutylammonium tetraphenylborate. Three electro analytical techniques were used: Voltammetry cyclic (VC), Differential pulse

*Corresonding author: Sana Mabrouk, Laboratory of Interfaces and Advanced Materials (LIMA), University of Monastir, Tunisia
E-mail: sanamabrouk@ymail.com

voltammetry (DPV) and Square wave voltammetry (SWV). Cyclic Voltammograms (VC) were registered from -1 to 3V at a scan rate of 100 mv/s. Differential pulse voltammetry (DPV) was performed at the same potential range by using a modulation time 0.02s, in interval time of 0.1s, step potential 0.0025 V and modulation amplitude 0.02 V. The square wave voltammetry (SWV) was performed at frequency of 15 Hz, amplitude of 0.09 V and potential step of 25 mV.

Pattern Recognition Methods

A pattern recognition statistical method was used for the evaluation and classification of the signals. The principal component analysis (PCA), differential factorial analysis (DFA) and cluster analysis (CA) were carried out using the software Spss 18. In this work we show a comparative study between PCA and DFA for the classification of the olive oils samples in the training set. PCA is an unsupervised learning technique that allows reduction of multidimensional data to a lower dimensional approximation, while simplifying the interpretation of the data by the first and second principal components (PC1 and PC2) in two dimensions and preserving most of the variance in the data. In addition, the samples can be classified without prior information on the samples. Conversely, DFA requires prior knowledge about the samples during the training. DFA is a supervised learning technique, which classifies the sample by developing a model and then identifies the unknown samples [28].

Results and Discussion

The voltammetric measurements

After cleansing glassy carbon electrode step and immersion in dichloromethane, the treated electrode was employed to classify the six olive oils by three electro-analytical techniques cyclic voltammetry (CV), differential pulse voltammetry (DPV) and square wave voltammetry (SWV). The cyclic voltammograms of several olive oils immersed in glassy carbon electrode showed a variety of responses. Figure 1 show three anodic peaks at -0.7 V, 0.6 V and 1.7 V. Figure 1 illustrates the cyclic voltammetry (CV) curves obtained using the glassy carbon sensors immersed in olive oils. Peaks associated to the polyphenolic content on the oils under study could be observed in the 0.1-0.8 V region (Table 1).

The main differences between curves consist in the position (potential), form and intensity of the peaks.

The Figure 2, of DPV voltammograms confirms the observations of the cyclic voltammetry. And shows the DPV responses recorded in six extra virgin olive oils coming from different Tunisian regions. For each oil sample, mixed with electrolyte support and dichloromethane as above, six replicates were collected. The patterns displayed in Figure 2 with full lines are average currents against potential plots, obtained from at least five replicates, while the shadow regions represent the signal variability range of each type of sample in both forward and backward scans of each measurement. It must be remarked that the hysteresis observed in each scan is due to the fact that, in these highly viscous media, planar diffusion affects to some extent the mass transport to the microelectrode surface [27]. As is evident from Figure 2, although the voltammograms obtained for the different olive oils have a similar shape, the associated current values differ significantly from one another, especially in the cathodic zone (Table 2). This is possibly due to a different composition of the oil matrix.

The peaks observed in the SWV voltammograms (Figure 3) translated the proprieties of the electro active compounds, antioxidants

present in the olive oils. Figure 3 shows that these oils were produced from the cultivars Kasserine, WEd Elil, Seliana, Mahdia, Ben Guerdne and Sidi Bouzid , respectively. In this case, the current circle in the anodic region, which was recorded reproducibly only in this oil, are a peculiar characteristic of this type of sample that allows distinguishing between all oils (Table 3).

The electronic tongue data

The characteristics of different samples are treated via pattern recognition techniques, such as the Principal Component Analysis (PCA) and Factorial Discriminant Analysis (FDA) [21].

PCA voltammetric electronic tongue: The discrimination capability of the method was evaluated by means of principal component analysis of the obtained signals. In order to reduce the high number of variables contained in a voltammetric curve to a few representative ones, a method was developed in our laboratory that uses mathematical functions that capture the information along the dynamic characteristics of the global response [29]. In this work we used a glassy carbon electrode for discrimination of vegetable oils. Figure 4 shows the PCA plot obtained (PC 2 vs. PC 1) of E-tongue results on Tunisian VOOs taste, and illustrates each measurement with the 20 normalized variables. As one can see, the variances explained by the first and the second principal components are 99.66%. However,

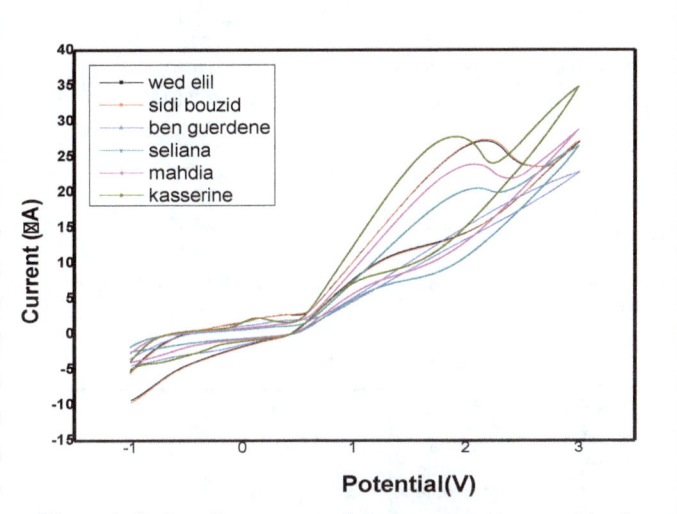

Figure 1: Cyclic voltammograms of glassy carbone immersed in six Tunisian olive oils, for scan range from -1V to 3V.

Echantillons	E(V)	I(µA)
Kasserine	0.11	2.29
	-0.64	0.047
	1.85v	27.76
Sidi bouzid	0.27	2.53
	-0.57	-0.41
	2.19	27.53
Ben Guerdene	0.32	1.82
	-0.44	0.24
Wed Ellil	-0.57	-0.4
	2.19	27.16
Mahdia	-0.72	-0.36
	2.07	23.68
Seliana	0.3	1.1
	-0.41	-0.76
	2.08	20.40

Table 1: The cyclic voltammetry parameters associated to the polyphenolic content on the oils.

Electronic tongue with differential pulse voltammetry data: Figure 5 shows the PCA plot obtained (PC 2 vs. PC 1) for E-tongue differential pulse voltammetry results on Tunisian VOOs taste and describes each measurement with the 20 normalized variables. We can see the variances explained by the first component (96.853%) and the second principal component 2.554%. But, in this plot there was a certain area of overlapping in which clear differentiation could be made on Oil_sidi bouzid, Oil_wed elil, Ben Guerdene and Kasserine. We note that the fingerprint oils from the region Sidi Bouzid, Mahdia and kasserine are situated in the right of the first component. That explains that these cities are located in the middle. The data clusters that belong to different simples were separated from each other. The first component tends to group the data according to their different regions of sample. Thus, the most olive oils are located on the right side, and the other finger print is localized on the left side. The second component seems to differentiate the different simple according to their area overlapping.

Electronic tongue with square wave voltammetry data: Figure 6 shows the PCA plot obtained (PC 2 vs. PC 1) for E-tongue. PC1 explains 99.673% variances and PC2 explains 0.283% variances. We can clearly see that two PC axes contribute potentially in the separation of the six Tunisian VOOs. Consequently, PCA results show a perfect classification between all the studied VOOs by square wave voltammetry.

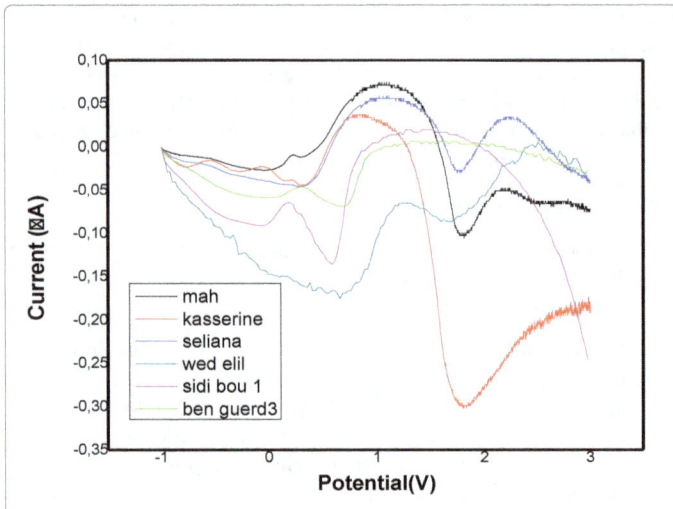

Figure 2: DPV voltammograms of glassy carbone immersed in six Tunisian olive oils for scan range from -1V to 3V.

Echantillons	E(V)	I(µA)
Kasserine	-0.52	0.015
	-0.033	-0.022
	0.91	0.034
Sidi bouzid	0.17	-0.064
	1.33	0.018
Ben Guerdene	0.28	-0.045
Wed Ellil	1.24	-0.0064
	1.18	0.001
Mahdia	0.21	0.0084
	1.07	-0.073
	2.17	-0.051
Seliana	1.08	0.056
	2.24	0.034

Table 2: DPV voltammograms parameters associated to the polyphenolic content on the oils.

Echantillons	E(V)	I(µA)
Kasserine	-0.52	0.0012
	0.25	0.21
	1.04	0.79
Sidi bouzid	-	-
Ben Guerdene	-	-
Wed Ellil	-	-
Mahdia	-0.59	-0.095
	0.24	-0.195
	1.18	0.61
	2.32	-0.12
Seliana	-0.59	-0.12
	0.227	-0.195
	1.2	0.73
	2.33	0.79

Table 3: SWV voltammograms parameters associated to the polyphenolic content on the oils.

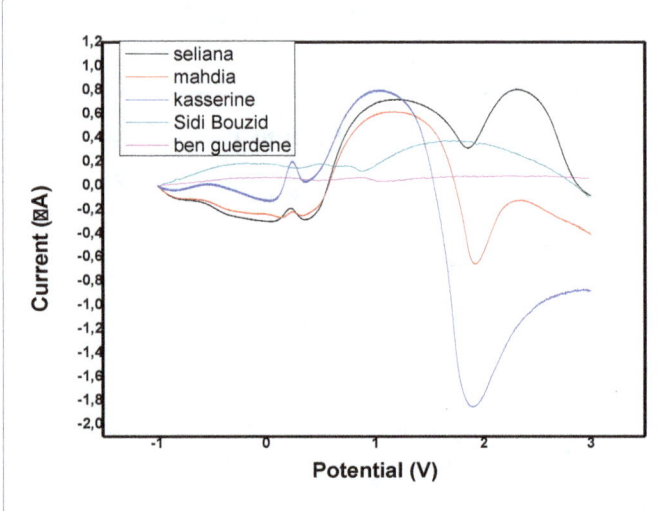

Figure 3: SWV voltammograms of glassy carbone immersed in six Tunisian olive oils for scan range from -1V to 3V.

in this plot, there was a certain area of overlapping in which no clear differentiation could be made on Oil sidi bouzid and Oil wed elil. Also, the analysis was extended until the third axis, but no great improvement can be revealed.

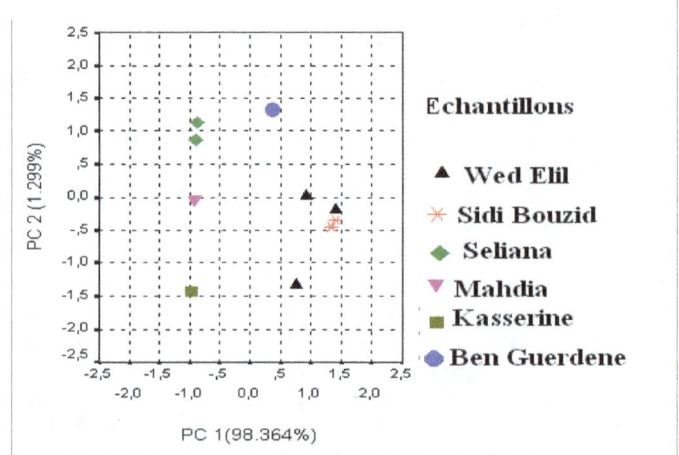

Figure 4: Two-dimensional PCA plot performed for six different geographical virgin olive oils (VOOs) varieties measured using the Electronic-tongue voltammetric.

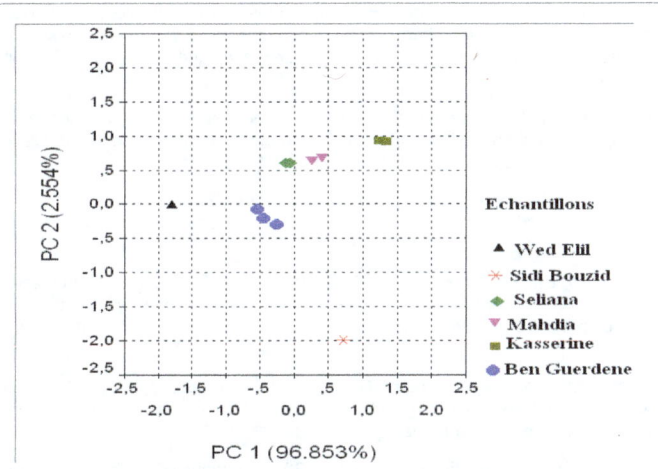

Figure 5: Two-dimensional PCA plot performed for Six VOOs varieties measured using the E-tongue differential pulse voltammetry data.

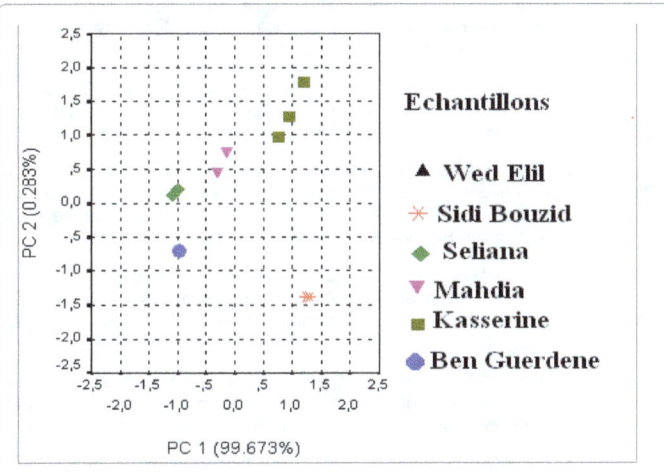

Figure 6: Two-dimensional PCA plot performed for Six VOOs varieties measured using the E-tongue square wave voltammetry data.

FDA plot of different varieties of olive oil by voltammetric electronic tongue: DFA is a multivariate technique for describing a mathematical function that will distinguish among predefined groups of samples. As an eigenvalue-eigenvector method, DFA has a strong connection to multiple regression and principal components analysis.

Applying FDA, a good separation between VOOs samples was obtained (Figure 7a). Function 1 and Function 2 appeared to contribute largely to discriminate mostly between all olive oil varieties. Through the DFA (Figure 7a), the first two factors explain 99.8% of the data variability, showing the distribution of the class separation between the dimensions, and a slight enhancement in the classification and separation among different categories is obtained. Although a clear differentiation was evident between the olive oils from different regions sample. However, after analyzing the same data set with DFA each group was clearly distinguishable (Figure 7b). The two discriminate functions accounted for 99.8% of the variance, indicating that the separated result is better with DFA (supervised method) than that with PCA (unsupervised method). We have observed a good separation between the Tunisian olive oil by DFA analysis, as results, 99.4% of the samples was successfully classified. However, after analyzing the same dataset with DFA each group was clearly distinguishable (Figure 7c). The two

discriminate functions accounted for 100% of the variance, indicating that the separated result is better with DFA (supervised method) than that with PCA (unsupervised method). Therefore, a better separation of the olive oils data was achieved by means of the DFA, demonstrating that DFA is better than PCA particularly when the number of samples is high producing a data overlapping in the clusters [28].

Conclusion

A novel method has been developed for the characterization of several olive oils cultivated from different region in Tunisia, by different voltammetric techniques. These methods are based on glassy carbon electrode. The features observed in the voltammograms are reflecting of properties of the electro-active compounds (antioxidants) present in the olive oils. The signal can be used as input variables in statistical studies. Principal component analysis using voltammetric signals fingerprint in multivariate data analysis allows a clear discrimination among oils pick from Tunisian geographic plot. The E-Tongue using square wave voltammetry are the best method to separate these olive oils. The characterization by different methods (VC, DPV and SWV) shows that the intensities of current differ from oil to another. The classification success rate obtained using a SWV (99.961%) is similar as CV (99.56%) and DPV (98.407%). A clearly obvious discrimination

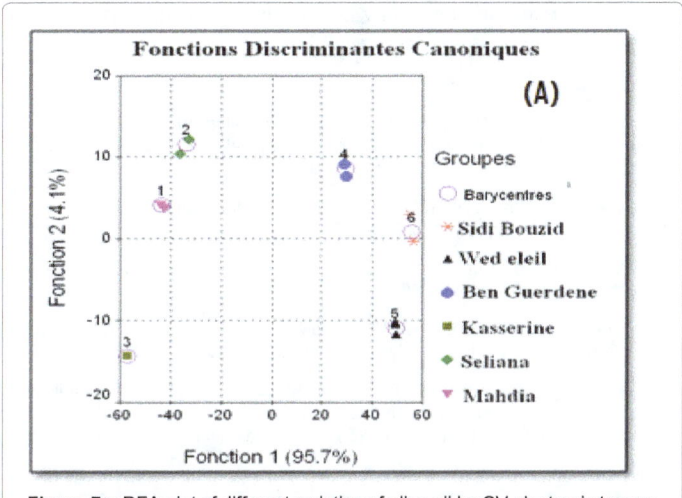

Figure 7a: DFA plot of different varieties of olive oil by CV electronic tongue

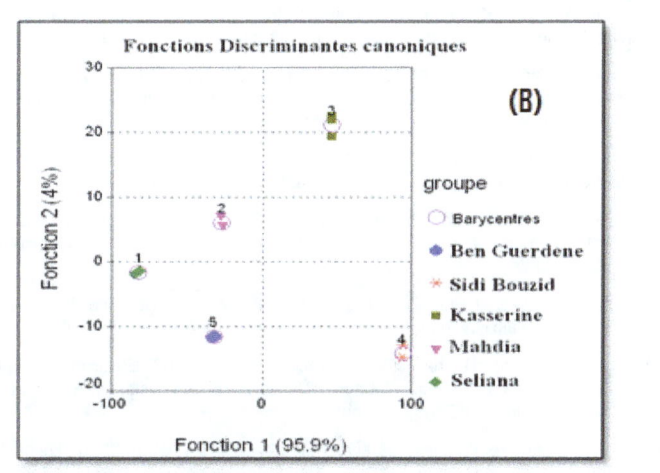

Figure 7b: DFA plot of different varieties of olive oil by CV by DPV electronic tongue.

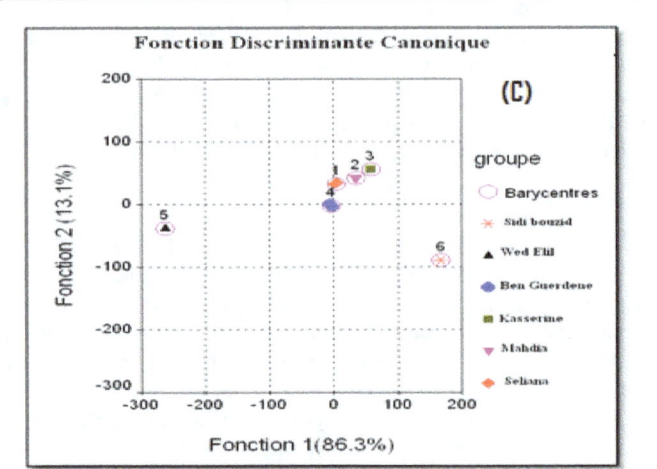

Figure 7c: DFA plot of different varieties of olive oil by SWV electronic tongue.

is found between all sources of olive oils by ACP. The voltammetry electronic tongue is a reliable tool to distinguish olive oil collected from different regions in Tunisia. We have observed good separations between the Tunisian olive oil by DFA analysis, as results, 100% of the samples were successfully classified.

References

1. Ciosek P, Janczyk M,WróblewskiW (2011) Classification of amino acids and oligopeptides with the use of multi-mode chemical images obtained with ion selective electrode array. Analytica Chimica Acta 699: 26-32.

2. Ciosek P,Wroblewski W (2007) Sensor arrays for liquid sensing-electronic tongue systems. Analyst 132: 963-978.

3. Valle D (2010) Electronic tongues employing electrochemical sensors. Electroanalysis 22: 1539-1555.

4. Escuder-Gilabert L, Peris M (2010) Review: Highlights in recent applications of electronic tongues in food analysis. Analytica Chimica Acta 665: 15-25.

5. Gay M, Apetrei C, Nevares I, Del Alamo M, Zurro J, et al. (2010) Application of an electronic tongue to study the effect of the use of pieces of wood and micro-oxygenation in the aging of red wine. Electrochimica Acta 55: 6782-6788.

6. Gutiérrez M, Llobera A, Ipatov A, Vila-Planas J, Mínguez S, et al. (2011) Application of an E-Tongue to the Analysis of Monovarietal and Blends of White Wines. Sensors (Basel) 11: 4840-4857.

7. Martins GF, Pereira AA, Straccalano BA, Antunes PA, Pasquini D, et al. (2008) Ultrathin ? lms of lignins as a potential transducer in sensing applications involving heavy metal ions. Sensors and Actuators 129: 525-530.

8. Mimendia A,Gutiérrez JM, Leija L, Hernández PR, Favari L, et al. (2010) A review of the use of the potentiometric electronic tongue in the monitoring of environmental systems. Environmental Modelling and Software 25: 1023-1030.

9. Riul A, Dantas CA, Miyazaki CM, Oliveira ON (2010) Recent advances in electronic tongues. Analyst 135: 2481-2495.

10. Vlasov YG, Ermolenko YE, Legin AV, Rudnitskaya AM, Kolodnikov V, et al. (2010) Chemical sensors and their systems. Journal of Analytical Chemistry 65: 880-890.

11. Winquist F,Olsson J, Eriksson M (2011) Multicomponent analysis of drinking water by a voltammetric electronic tongue. Analytica Chimica Acta 683: 192-197.

12. Legin A, Kirsanov D, Rudnitskaya Czechoslovak A (2006) Electronic Tongue An Array Of Non-Specific Chemical Sensors For Analysis Of Radioactive Solutions. Journal Of Physics Suppl.

13. Apetrei C, Apetrei IM, Nevares I,Del Alamo M, Parra V, et al. (2007) Using an e-tongue based on voltammetric electrodes to discriminate among red wines aged in oak barrels or aged using alternative methods. Electrochimica Acta 52: 2588-2594.

14. Busch JL, Hrncirik K, Bulukin E, Boucon C, Mascini M, et al. (2006) Biosensor measurements of polar phenolics for the assessment of the bitterness and pungency of virgin olive oil. Journal of Agriculture and Food Chemistry 54: 4371-4377.

15. Gutes A, Ibanez A, Cespedes F, Alegret S, Del Valle M (2005) Simultaneous determination of phenolic compounds by means of an automated voltammetric electronic tongue. Anal.Bioanal.Chem 382: 471-472.

16. Apetrei C, Rodríguez-Méndez ML, Saja DJA (2008) Evaluation of the polyphenolic content of extra virgin olive using an array of voltammetric sensors. Electrochimica Acta 53: 5867-5872.

17. Cosio MS, Ballabio D, Benedetti S, Gigliotti C (2007) Evaluation of different storage conditions of extra virgin olive oils with an innovative recognition tool built by means of electronic nose and electronic tongue. Food Chemistry 101: 485-491.

18. Rodriguez-Mendeza ML, Apetrei C, Nietoa M, Hernandez V, Lopez Navarrete JT, et al. (2009) Sensing properties of organised films based on a bithiophene derivative. Sensors and Actuators 141: 625-633.

19. Hong M, Caiwa Z, Ke Ning C, Donglin C (2013) Detection of Edible Oils Based on Voltammetric Electronic Tongue. Research Journal of Applied Sciences, Engineering and Technology 5: 1197-1202.

20. Lerma-García MJ, Simó-Alfonso EF, Bendini A, Cerretani L (2009) Metal oxide semiconductor sensors for monitoring of oxidative status evolution and sensory analysis of virgin olive oils with different phenolic content. Food Chemistry 117: 608-614.

21. Rodriguez-Mendez ML, Apetrei C, De Saja JA (2008) Evaluation of the polyphenolic content of extra virgin olive oils using an array of voltammetric sensors. Electrochimica Acta 53: 5867-5872.

22. Casilli S, De Luca M, Apetrei C, Parra V, Arrieta A, et al. (2005) Langmuir-Blodgett and Langmuir-Schaefer films of homoleptic and heteroleptic phthalocyanine complexes as voltammetric sensors: Applications to the study of antioxidants. Applied Surface Science 246: 304-312.

23. Apetrei C, Apetrei IM, Villanuev S, De Sajab JA, Gutierrez-Rosales F, et al. (2010) Combination of an e-nose, an e-tongue and an e-eye for the characterisation of olive oils with different degree of bitterness. Analytica Chimica Acta 663: 91-97.

24. Cosio S, Ballabio D, Benedetti S, Gigliotti C (2006) Geographical origin and authentication of extra virgin olive oils by an electronic nose in combination with artificial neural networks Maria. Analytica Chimica Acta 567: 202-210.

25. Apetrei C(2012) Novel method based on polypyrrole-modified sensors and emulsions for the evaluation of bitterness in extra virgin olive oils. Food Research international 48: 673-680.

26. Apetrei C, Gutierez F, Rodriguez-Mendez ML, De Saja JA (2007) Novel method based on carbone paste electrodes for the evaluation of bitterness in extra virgin olive oils. Sensors and Actuators, B121, 567-575.

27. Oliveri P, Antonietta Baldo M, Daniele S, Forina M (2009) Development of a voltammetric electronic tongue for discrimination of edible oils. Analytical and Bioanalytical Chemistry 395: 1135-1143.

28. Canoa M, Roalesa J, Castillero P, Mendoza P, Calerob AM, et al. (2011) Improving the training and data processing of an electronic olfactory system for the classification of virgin olive oil into quality categories. Sensors and Actuators B: Chemical 160: 916-922.

29. Apetrei C, Rodríguez-Méndez ML, De Saja JA (2005) Modified carbon paste electrodes for discrimination of vegetable oils. Sensors and Actuators B: Chemical 111: 403-409.

17

Effect of Different Drying Method on Volatile Flavor Compounds of *Lactarius deliciosus*

Qun Huang, Lei Chen, Hong-bo Song*, Feng-ping An, Hui Teng and Mei-yu Xu
College of Food Science, Fujian Agriculture and Forestry University, Fuzhou, China

Abstract

The effect of different drying methods on volatile flavor compounds of *Lactarius deliciosus* was investigated, such as hot-air drying, vacuum freeze drying and sunshine drying. By adopting the solid-phase micro-extraction method, volatile flavor compounds were extracted from lactarius deliciosus, then analyzed and identified through the application of gas chromatography-mass spectrometer (GC-MS). Results indicated that different drying methods could lead to large differences in volatile flavor compounds. Main volatile flavor compounds of fresh *Lactarius deliciosus* involved acids and aldehydes, which accounted for 86.31%; Hot-air dried *Lactarius deliciosus* mainly included acids and alkene, which accounted for 87.16%; *Lactarius deliciosus* dried with vacuum freezed *Lactarius deliciosus* was mostly composed of acids, esters and aromatic substance, which accounted for 94.74%; *Lactarius deliciosus* dried with sunshine was constituted by acid and aldehydes, which was as high as 90.67%.

Keywords: *Lactarius deliciosus*; Volatile flavor compounds; Drying method; Gas chromatography-mass spectrometer (GC-MS)

Introduction

The wild edible mushroom market is concentrated within a small group of species belonging to genera such as *Lactarius, Boletus, Cantharellus, Amanita, Pleurotus,* and *Calocybe* [1]. Among them, *Lactarius deliciosus* (i.e., *Agaricus deliciosus*), belonging to Russulaceae, is the most commercially important wild edible mushroom because of their high consumption by the rural population and their economic value in the markets of Southeast Asia [2].

As a popular natural and rare biological resource, *Lactarius deliciosus* have been used as food and food-flavoring material in soups and sauces for centuries, due to their unique and subtle flavour. Recently, they have become attractive as functional foods and as a source of physiologically beneficial medicines, while being devoid of undesirable side-effects. *Lactarius deliciosus* were also found to be medically active in several therapies, such as anti-tumor, antibacterial, antiviral and immune-modulating treatments [3,4]. However, like all wild edible mushrooms, *Lactarius deliciosus* are highly perishable with a short shelf-life of 1-3 days at room temperature, and tend to lose quality immediately after harvest due to high respiration rate and the lack of physical protection of cuticle [5,6]. Furthermore, *Lactarius deliciosus* lose marketability early during the storage period due to enzymatic activity, weight loss, shrinkage, browning, spore formation, bacteria and yeasts and moulds [7,8]. Because short shelf-life is an impediment to the distribution and marketing of the fresh product, its exploitation is mainly carried out by amateur personal, and the use of collected production is related to self-consumption or restaurants, intermediate links within the value chain being unregulated. Consumer interest for health and wellbeing has been reflected in a significant rise in the marketing and distribution of edible mushrooms in the latest years [9]. Thus, it's so very urgent to process and utilize this high quality wild resource in order to increase its added value.

It's a more perfect method to dry fresh *Lactarius deliciosus* for the sake of storage and development, but a series of biochemical reaction will react resulting in product quality deterioration through the change of colour, flavor and nutrition. To the best of our knowledge, there is no research focused on the effect of processing method on product quality of *Lactarius deliciosus*. For the first time, the present research investigated the effect of different drying methods, such as hot air, vacuum freeze and sunlight, on volatile flavor compounds of *Lactarius deliciosus* by extracting with head space solid phase micro extraction (SPME) technique and analyzing with gas chromatography-mass spectrometry (GC-MS). In addition, the change mechanism of volatile flavor was discussed in order to lay theoretical basis for the choice of drying methods and to provide reference data for production quality control.

Materials and Methods

Sample treatment

Fresh *Lactarius deliciosus* was purchased from Farmers. After washing raw materials, 4 samples of fresh *Lactarius deliciosus* were treated with different drying methods. Sample A was cut into cubes of 0.5 cm using knife and no longer processed, which served as the blank control, Sample B was dried with hot air, sample C was processed with vacuum freeze drying, sample D was treated with sunshine drying.

Hot air drying: 500 g of fresh *Lactarius deliciosus* was put in electric oven and dried at 45°C for 3-4 h. In the next step, the temperature rising was controlled at a rate of 1°C/h. *Lactarius deliciosus* was dried for 4-5 h when the temperature reached 50°C. And then, the temperature increased to 55°C at a rate of 1°C/h, and drying was performed for 8-9 h. Finally, the temperature rose to 65°C at the same rate, and drying was continued until the moisture content was about 13.0%.

Vacuum freeze drying: Fresh *Lactarius deliciosus* was frozen at -40°C for 9 h, and then dried at 65 Pa and -65°C for 8-9 h.

Sunshine drying: Fresh *Lactarius deliciosus* was exposed to the sunlight until the water mass fraction was less than 13.0%.

Volatile compounds extraction: Volatile flavor compounds

*Corresponding author: Hong-bo Song, College of Food Science, Fujian Agriculture and Forestry University, Fuzhou 350002, China
E-mail: sghgbode@163.com

were extracted using solid-phase microextraction (SPME). A SPME device (Supelco, Bellefonte, PA, USA) containing a fused-silica fibre (10 mm length) coated with a 50/30 μm layer of DVD/CAR/PDMS was used. The fibre was conditioned prior to analysis by heating it in a gas chromatograph injection port at 270°C for 60 min. Extraction was performed at 35°C for 30 min. Before extraction, samples were equilibrated for 15 min at the temperature used for extraction. Once sampling was finished, the fibre was withdrawn into the needle and transferred to the injection port of the gas chromatograph-mass spectrometer (GC-MS) system.

Gas chromatographic conditions: Agilent 19091s-433 (30 m × 250 μm × 125 μm) quartz capillary column was adopted. Temperature programming was following as: the temperature was maintained at 50°C for 3 minutes and increased to 80°C at a rate of 3°C/min. And then temperature was risen to 150°C at a rate of 5°C/min and maintained at 150°C for 4 minutes. Next, the temperature was speeded up to 240°C at a rate of 3°C/min and maintained at 240°C for 5 minutes. The carrier gas was high purity He, the inlet temperature was 270°C, the auxiliary heater temperature was 230°C, automatic injection volume was 1 μL and the split ratio was 1:1.

Mass spectrometric conditions: Advance sample temperature was 280°C, and EI ionization was employed. Ion source temperature was 230°C, and quadropole temperature was 150°C. Electron energy was 70 eV, and filament current was 34.6 μA. The scanning mass range was 50.0-550.0 aum, and the solvent delay time was 4 minutes.

Qualitative and quantitative analysis: Artificial analysis and computer retrieval (NIST) were performed on mass spectra, so as to determine chemical compositions corresponding to various spectral peaks. By adopting peak area normalization method, relative content (%) of each component was calculated. Assuming that the correction factor was 1 and using internal standard (phenyl ethyl acetate) method, the content of components to be determined was calculated as the equation below.

Results and Discussion

Total ion chromatogram and analysis of volatile flavor compounds

According to Figure 1, it could be known that volatile compounds of *Lactarius deliciosus* treated with different drying methods were quite different. Tables 1-3 revealed that totally 79 volatile compounds were identified in *Lactarius deliciosus* treated with different drying methods, including alcohols (5), alkenes (18), aldehydes (19), ketones (4), phenols (11), acids (5) and esters (17). The majority of these volatile substances were found naturally in lactarius deliciosus, while some were resulted from a series of chemical reactions during the drying process, such as Maillard reaction between reducing sugars and amino compounds, oxidation reaction of unsaturated fatty acid, decomposition reaction of ester oxide, which could generate new volatile flavor compounds [6,10].

Volatile flavor compounds of fresh *Lactarius deliciosus*

24 volatile flavor compounds were detected in fresh *Lactarius deliciosus*, including 1 alcohol (0.91%), 2 alkenes (1.16%), 9 aldehydes (13.50%), 1 ketone (0.87%), 1 phenol (0.65%), 3 acids (72.81%) and 7 esters (10.10%). Components with higher content mainly involved

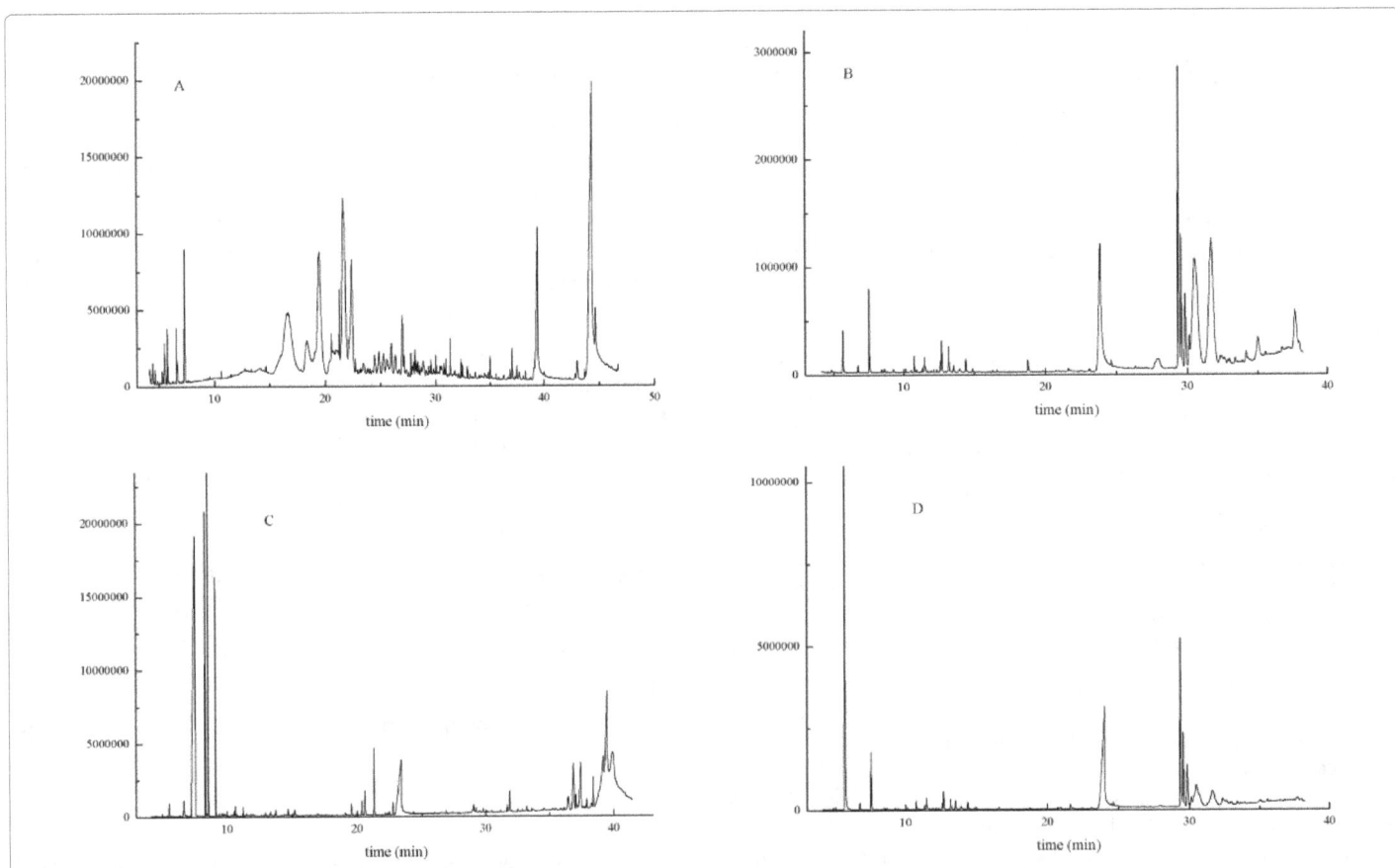

Figure 1: GC-MS total ion current of volatile compounds of lactarius deliciosus treated with different drying methods. (A: Fresh *Lactarius deliciosus*; B: Hot air dried *Lactarius deliciosus*; C: Vacuum freezed dried *Lactarius deliciosus*; D: Sunshine dried *Lactarius deliciosus*).

S.no.	Volatile compounds	Relative amount/%			
		A	B	C	D
1	Trimethyl butanol	-	0.26	-	-
2	N-butanol	-	-	-	0.29
3	Hexanal	1.31	0.57	0.79	0.88
4	Methoxy hexanal	-	6.77	-	-
5	Ethylbenzene	-	-	9.37	0.15
6	Para xylene	-	-	11.97	-
7	Benzene	-	0.26	-	-
8	O-xylene	-	-	-	0.21
9	1,3-dimethyl-benzene	-	-	6.13	-
10	Cyclohexanone	-	0.43	-	0.33
11	Heptanal	-	-	0.06	-
12	(1-methyl ethyl)-benzene	-	-	0.05	-
13	α-song pinene	-	0.28	-	-
14	1R-2,6,6-trimethoxy	-	-	-	0.22
15	Propyl benzene	-	-	0.14	-
16	Anti-2-heptenal	-	-	0.30	-
17	D-limonene	-	1.37	-	2.82
18	Limonene	-	3.29	-	1.18
19	Benzaldehyde	-	0.26	-	-
20	1-octene-3-alcohol	-	-	0.26	1.61
21	1-octene	-	1.36	-	-
22	Caprylic aldehyde	-	-	-	0.17
23	3-carene	-	0.25	-	0.21
24	2-carene	-	0.23	-	0.17
25	O- cymene	-	0.99	-	-
26	Isopropyl ortho toluene	-	-	-	0.87
27	Benzoic hexanal	-	2.78	0.15	1.73
28	γ-terpinene	-	0.80	-	-
29	Formic acid octyl ester	-	-	0.20	-
30	Anti-2-octene 1-alcohol	-	-	-	0.63
31	(+) -4- carene	-	1.48	-	-
32	Terpinolene	-	-	-	1.19
33	2-nonanone	-	-	-	0.25
34	Pelargonic aldehyde	-	0.44	0.23	0.34
35	Anti-2-nonene aldehyde	-	-	-	0.40
36	Anti-2-decene aldehyde	-	-	0.42	-
37	Anethole	-	-	0.55	0.35
38	E,E-2,2-dodecadienal	-	-	-	0.36
39	E,E-2,4-decedienal	4.44	0.43	1.09	1.10
40	2-undecenal	0.74	-	-	-
41	Tetradecanoic acid ethyl ester	-	-	5.13	-
42	Decanoic acid	-	59.14	-	69.73
43	9-oxo methyl nonanoate	3.46	-	-	-
44	Butylated hydroxytoluene	-	0.63	-	-
45	10-carbonyl methyl decanoate	2.03	-	-	-
46	Methoxy phenol	0.65	-	-	-
47	Anti-1,9-dodecadiene	0.74	-	-	-
48	Tetramethyl-1H-cyclopropene	-	17.97	-	14.81
49	Undecalactone	1.51	-	-	-
50	1,11-Dodecadiene	-	-	0.29	-
51	8-heptadecene	-	-	0.86	-
52	Triethylene cyclooctene	0.42	-	-	-
53	Cis-9-hexadecenal	1.28	-	-	-
54	2-pentadecanone	0.87	-	-	-
55	Pentadecanal	0.89	-	-	-
56	Methyl-12-methyl tridecanoate	-	-	0.28	-
57	3,3-dimethyl benzidine	-	-	0.09	-
58	Anti -9- eicosene	-	-	0.15	-
59	7-cis-10-cis-hexadecadienoic acid aldehyde	0.61	-	-	-

60	Trans-11-hexedecanal	1.69	-	-	-
61	Anti-2-tetradecene-1-alcohol acetate	0.43	-	-	-
62	Trans-9,12-Octadecadienoic acid methyl ester	-	-	1.32	-
63	Methyl cis-9-octadecenoate	-	-	4.36	-
64	9,12-Tetradecadien-1-acetate	0.70	-	-	-
65	Diisobutyl phthalate	-	-	0.63	-
66	13-tetradecenal	1.03	-	-	-
67	2-hydroxy-cyclopentadecanone	-	-	0.11	-
68	14-methyl-8-hexyne-1-alcohol	0.91	-	-	-
69	Methyl cis-9-hexadecenoate	-	-	0.35	-
70	Methyl stearate	-	-	0.40	-
71	Methyl 2-methyl-hexadecanoate	0.70	-	-	-
72	Palmitic acid methyl ester	-	-	1.23	-
73	Oleic acid	-	-	11.54	-
74	Tetradecanoic acid	31.41	-	-	-
75	Palmitic acid	1.48	-	18.95	-
76	Glyceryl monooleate	-	-	22.62	-
77	Methyl linoleate	0.44	-	-	-
78	Anti-methyl oleate	2.35	-	-	-
79	Cis-9, cis-12-octodecane dienoic acid	39.92	-	-	-

Note: Means not detected.

Table 1: GC-MS analysis of volatile flavor compounds of *Lactarius deliciosus* dried with different methods.

Compound	A	B	C	D
Alcohols	1	1	1	3
Alkenes	2	10	4	8
Aldehydes	9	6	7	7
Ketones	1	1	1	2
Phenols	1	2	6	3
Acids	3	1	2	1
Esters	7	0	10	0
Total	24	21	31	24

Table 2: Volatile flavor compound species of *Lactarius deliciosus* dried with different methods.

Compound	A	B	C	D
Alcohols	0.91	0.26	0.26	2.53
Alkenes	1.16	28.02	1.85	20.94
Aldehydes	13.5	11.26	3.04	4.99
Ketones	0.87	0.43	0.11	0.58
Phenols	0.65	0.89	27.74	1.23
Acids	72.81	59.14	30.48	69.73
Esters	10.1	0	36.52	0
Total	100	100	100	100

Table 3: Content of volatile flavor compounds of *Lactarius deliciosus* dried with different methods (%).

C-9,C-12-octodecane dienoic acid (39.92%), tetradecanoic acid (31.41%), E,E-2,4-Sebacic olefine aldehyde (4.44%), 9-oxo-methyl nonanoate (3.46%) and anti-methyl oleate (2.35%). The volatile composition of fresh *Lactarius deliciosus* was characterized by fresh fruity and slight green grass, dominating by acids and esters with the highest aldehydes content. Moreover, its acids and esters were dominated by unsaturated compounds, which were easy to produce peroxides under the action of enzymes during the drying process, then peroxide decomposition could generate aldehydes, ketones and alcohols compounds which play a harmonic, synergistic or complementary role in flavor resulting in different flavors [7].

Volatile flavor compounds of *Lactarius deliciosus* dried with hot air

21 volatile flavor compounds were detected in *Lactarius deliciosus* sample dried with hot air, including 1 alcohol (0.26%), 10 alkenes (28.02%), 6 aldehydes (11.26%), 1 ketone (0.43%), 2 phenols (0.89%) and 1 acid (59.14%). Components with higher content mainly involved decanoic acid (59.14%), methoxy hexanal (6.77%), limonene (3.29%), Benzene hexanal (2.78%). The volatile composition of *Lactarius deliciosus* dried with hot air was characterized by dominating by acids with higher content of aldehydes and alkenes, but the content of other volatile flavor compound was low. Compared with other drying methods, there were more unique volatile flavor compounds, such as γ-gamma terpinene (0.80%), 4-carene (0.99%), α-sobrerone (0.28%), O-cymene (1.48%) and 1-octene (1.36%). During the process of hot air drying, many small molecule volatile components were produced, and olefinic terpene was the most prominent which might be the outcome of decomposition reaction of unsaturated fatty acid and ester. As a result of these reactions and compounds formation, *Lactarius deliciosus* dried with hot air eventually formed the smell of bitter almonds, fried flavor and unique flavor of lactarius [5].

Volatile flavor compounds of *Lactarius deliciosus* dried with vacuum freeze

31 volatile flavor compounds were detected in *Lactarius deliciosus* dried with vacuum freeze, including 1 alcohol (0.26%), 4 alkenes (1.85%), 7 aldehydes (3.04%), 1 ketone (0.11%), 6 phenols (27.74%), 2 acids (30.48%) and 10 esters (36.52%). Components with higher content mainly involved glyceryl monooleate (22.62%), oleic acid (11.54%), limonene (3.29%), E,E-2,4-Decedienal (1.91%). The characteristic of volatile flavor compound in *Lactarius deliciosus* dried with vacuum freeze was more content of acids, esters and aromatic substances, but lower other compound. Compared with other drying methods, its unique volatile flavor compounds mainly included 8-Heptadecene (0.86%), n-Ethyl Myristate (5.13%), methyl cis-9-octadecenoate (4.36%), paraxylene (11.97%) and ethylbenzene (9.37%). Vacuum freeze is a desired method formed maximum volatile flavor compounds, it's deduced that the temperature was maintained at extremely low levels so as to original volatile constituents could be preserved perfectly.

Volatile flavor compounds of *Lactarius deliciosus* dried with sunshine

24 volatile flavor compounds were detected in *Lactarius deliciosus* dried with sunshine, including 3 alcohols (2.53%), 8 alkenes (20.94%), 7 aldehydes (4.99%), 2 ketones (0.58%), 3 phenols (1.23%) and 1 acid (69.73%). The mainly ingredients with higher content involved decanoic acid (69.73%), d-limonene (2.82%), benzene hexanal (1.73%), meaning that volatile flavor compounds of *Lactarius deliciosus* dried with sunshine were dominated by acids but lower content of other compound. Compared with other drying methods, its unique volatile substances mainly included Anti-2-octene-1-alcohol (0.63%), terpinolene (1.19%), isopropyl ortho toluene (0.87%), o-xylene (0.21%), (2E)-2-Nonenal (0.40%). Though the principle of sunshine drying was basically the same as hot air drying, its volatile composition is rather changeable and very different because of a circulation environment with a greatly fluctuated temperature. For example, sunshine drying produced more flavor substances of eight carbon alcohol compound (2.24%) which did not exist air drying.

Comparative analysis of major volatile flavor compounds

Alcohols: Alcohols is produced by fat oxidation or activity of part microorganism. The main alcohol in *Lactarius deliciosus* was 1-octene-3-alcohol, commonly known as mushroom essence, which respectively accounted for 0.26% and 1.61% of volatile flavor compounds of *Lactarius deliciosus* dried with vacuum freezed and sunshine. Alcohol is present richly in full mushroom flavour owing to its lower flavor threshold and existing in almost all kinds of mushrooms. It could be concluded that the main source of alcohols was the oxidation and decomposition products of unsaturated fatty acid because the precursor was mainly 16-22 carbon unsaturated fatty acid [11].

Carbonyl compounds: Carbonyl compounds mainly referred to some aldehydes and ketones, and the major contributor to flavor formation were micromolecular aldehydes and ketones and their precursors because of the low volatility of macromolecular carbonyl compounds. There are 2 kinds of carbonyl compounds in fresh Lactarius deliciosus, 3 in hot air drying, 7 in vacuum freeze drying and 5 in sunshine drying respectively. It could be known that carbonyl compound might be the intermediate product of fat oxidation and Maillard reaction [12,13].

Acid compounds: Acid compound, with fresh and sweet fruit flavor, is the leading volatile flavor compound in lactarius deliciosus, and it mainly included decanoic acid, myristic acid, oleic acid and palmitic acid. Decanoic acid, a important ingredient in acid compounds, accounted for 59.14% and 69.13% of volatile flavor compounds in hot air drying and sunshine drying respectively, but didn't exist in wet and vacuum freeze-dried lactarius, therefore it is the reaction product of during the heating process. Decanoic acid is allowed to be a food flavouring agent which is mainly used in the preparation of dairy, rum, brandy and coconut flavor for offering a special grease smell, thus the present experiment could provide a basis for further development research of *Lactarius deliciosus* flavor [14,15].

Ester compounds: Some mainly ester compounds, such as glyceryl monooleate, tetradecanoic acid ethyl ester, anti-methyl oleate and methyl cis-9-octadecenoate, were detected in the sample. The kinds of ester compounds of vacuum freeze-dried lactarius were 10 with the highest content of 36.52%, and wet lactarius 7 with 10.10%, but there is no ester detected in hot air dried and sunshine dried lactarius. Thus it can be seen that the esters of lactarius were unstable at relatively high temperature [15,16].

Acknowledgment

This research was supported by the Construction Projects of Top University at Fujian Agriculture and Forestry University of China (Grant No. 612014042).

References

1. Andres AI, Timon ML, Molina G, Gonzalez N, Petron MJ, et al. (2014) Effect of MAP storage on chemical, physical and sensory characteristics of *Lactarius deliciosus*. Food Pack. Shelf Life 1: 179-189.

2. Parlade J, Hortal S, Pera J, Galipienso L (2007) Quantitative detection of *Lactarius deliciosus* extraradical soil mycelium by real-time PCR and its application in the study of fungal persistence and interspecific competition. J Biotechnol 128: 14-23.

3. Isabel CFR, Paula B, Miguel VB, Lillian B (2007) Free-radical scavenging capacity and reducing power of wild edible mushrooms from northeast Portugal: Individual cap and stipe activity. Food Chem 100: 1511-1516.

4. Ding X, Hou YL, Ho WR (2012) Structure feature and antitumor activity of a novel polysaccharide isolated from *Lactarius deliciosus* Gray. Carbohyd Polym 2: 397-402.

5. Iqbal T, Rodrigues F, Mahajan P, Kerry JP (2009) Mathematical modelling of O_2 consumption and CO_2 production rates of whole mushrooms accounting for the effect of temperature and gas composition. Int J Food Sci Tech 7: 1408-1414.

6. Mahajan PV, Oliveira FA, Macedo I (2008) Effect of temperature and humidity on the transpiration rate of the whole mushrooms. J Food Eng 84: 281-288.

7. Fernandes A, Antonio AL, Barreira J, Botelho ML, Oliveira MN, et al. (2013) Effects of gamma irradiation on the chemical composition and antioxidant activity of *Lactarius deliciosus* L. Wild edible mushroom. Food Bioprocess Tech 6: 2895-2903.

8. Masson Y, Ainsworth P, Fuller D, Bozkurt HI, Banog LS, et al. (2002) Growth of *Pseudomonas fluorescens* and Candida sake in homogenized mushrooms under modified atmosphere. J Food Eng 54: 125-131.

9. Pettenella SL, Maso D (2007) NWFP&S marketing: Lessons learned and new development paths from case studies in some European countries. Small-scale For 4: 373-390.

10. Gu Q, Lu P, Huang WN (2015) Enzymatic hydrolysis of wheat protein for the production of precursors of bread flavor. J Food Sci Biot 4: 372-378.

11. Zhou ZY, Tan JW, Liu JK (2011) Two new polyols and a new phenylpropanoid glycoside from the basidiomycete *Lactarius deliciosus*. Fitoterapia 8: 1309-1312.

12. Angela F, Amilcar LA, João CM, Barreira M, Beatriz PPO, et al. (2012) Effects of gamma irradiation on physical parameters of *Lactarius deliciosus* wild edible mushrooms. Post-harvest Biol Tec 74: 79-84.

13. Goupry S, Rochut N, Robins RJ, Gentil E (2000) Evaluation of solid phase miroextraction for the isotopic analysis of volatile compounds produced during fermentation by lactic acid bacteria. J Agr Food Chem 6: 2222-2227.

14. Kim TH, Lee ML, Kim YS, Lee LHJ (2003) Aroma dilution method using GC injector split ratio for volatile compounds extracted by headspace solid phase microextraction. Food Chem 3: 221-228.

15. Alves GL, Franco MRB (2001) Headspace gas chromatography-mass spectrometry of volatile compounds in murici (*Byrsonimacrassifolia L. Rich*). J Chromatogr A 985: 297-301.

16. Carrapiso AI, Jurado A, Timón ML, García C (2002) Odor-activ compounds of Iberian hams with different aroma characteristics. J Agr Food Chem 22: 6453-6458.

18

Field Survey of Symptoms and Isolation of Fungi Associated with Post-harvest Rots of White Yam (*Dioscorea Rotundata* Poir.)

Ezeibekwe IO, Umeoka N* and Izuka CM

Department of Plant Science and Biotechnology, Faculty of Science, Imo State University Owerri, Imo State, Nigeria

Abstract

Investigations were carried out on field survey of symptoms and isolation of fungi associated with the Post-harvest rots of white yam (*Dioscorea rotundata* Poir.) at Orlu, Imo State. The results of disease incidence and severity showed that dry rot had the highest percentage incidence of 67.5%, followed by wet rot (47.5%) and soft rot 45.0%. Anthracnose recorded 37.5% and powdery mildew recorded 42.5%. The severity result also followed the same trend, with dry rot having the highest percentage of 26.8%, soft rot 23.7%, wet rot 23.2%, anthracnose 16.3% and powdery mildew recorded 15.1%. The fungi were isolated and identified as *Trichoderma viride* (Pers.), *Pythium aphanidermatum* (Edson), *Aspergillus fumigatus* (Fresenius), *Penicillium expansum* (Link.), *Geotrichum candidum* (Link.), *Fusarium oxysporum* (Link.), *Botryodiplodia theobromae* (Sac.) and *Aspergillus niger* (van Thieghem). Fungal organisms occurred consistently with soft rot, dry rot, wet rot and anthracnose of *D. rotundata* with *A. fumigatus* (Fresenius) occurring more frequently, 45.00%, followed by *T. viride* (Pers.) 20.00%, *P. aphanidermatum* (Edson) recorded 15.00%, *P. expansum* (Link.) and *G. candidum* (Link.) have 10.00% each.

Keywords: Symptoms; Isolation; Fungi; Post-harvest rots; *D. rotundata*

Introduction

Yams (*Dioscorea spp.*) are among the most important staple foods in the world, especially in some parts of the tropics and subtropics. The role played by yam in the food economy in most West African countries cannot be over-emphasized. It is one of the most important dietary sources of energy produced within the tropics [1]. Nigeria produces yams for local consumption and the export market. The country is a leading exporter of yam in the world (about 12,000 tons annually). Water yams (*D. alata*) are consumed when white yam becomes scarce or expensive [2]. One of the most pressing problems facing the countries of the third world is food scarcity. It is reported that nearly 1 billion people are challenged by severe hunger in these nations of which 10% die from hunger-related complications. A substantial part of this problem from hunger stems from inadequate agricultural storage and produce preservation from microbes-induced spoilages [3]. According to Arya [4], of all losses caused by plant diseases, those that occur after harvest are the costliest. Cassava, yam and sweet potato are important sources of food in the tropics. Others are cocoyam, rice, maize, wheat, sorghum, millet and various fruits, legumes and vegetables.

Yam tubers and plants are prone to several diseases. These diseases not only them unappealing but also reduce the quantity of yam produced. Viruses, bacteria, fungi, nematodes and many other factors are these diseases. Specifically, fungal infections have constituted a very limiting factor all over the world and to other tuber crops. A good number of pathogens are soil inhabiting, some gain their entrance into the plant through invasion of the roots causing harm to the plant [3]. Most rots of yam tubers are caused by pathogenic fungi such as *Aspergillus flavus*, *Aspergillus niger*, *Botryodiplodia theobromae*, *Fusarium oxysporum*, *Fusarium solani*, *Penicillium chrysogenum*, *Rhizoctonia spp.*, *Penicillium oxalicum*, *Trichoderma viride* and *Rhizopus nodosus* etc., [5]. The field diseases are those diseases that cause economic damage to yam in the field from the seedling stage to the point of harvest. Anthracnose disease of yam has a considerable impact on yam production world-wide [6]. This is caused mostly by the fungus *Colletotrichum gloeosporioides* [7]. IITA [8] reported that *Glomerela cingulata* (isolate number IMI W3725) was the yam anthracnose inducing pathogen in Southwestern Nigeria. *G. cingulata*

is the perfect state of *C. gloeosporioides*, the form that is usually found causing field anthracnose disease.

On susceptible yam cultivars, symptoms appeared at first as small dark brown or black lesion on the leaves, petioles and stems. The lesion is often surrounded by a chlorotic halo enlarged and coalesced, resulting in extensive necrosis of the leaves and die-back of the stem [9]. The withered leaves and stem die-back gave the plant a scorched appearance hence the name 'scorch' disease [8]. Previous work of Amusa [9] indicated that yam anthracnose is a disease complex, which has however been associated with the activities of *Colletotrichum gloeosporioides*, *Curvularia pallescens*, *Curvularia eragrostides*, *Pestalotia spp.* and *Rhizoctonia solani*. PANS [10] have reported that *Pestalotia spp.* mainly affects *D. esculenta* but appear to act as a secondary invader after infection by *C. gloeosporioides* on *D. alata* and *D. cayenensis*. There is need to carryout field survey of symptoms and isolation of fungi associated with the Post-harvest rots of white yam (*D. rotundata* poir.) to expand the frontiers of knowledge and hence advice the farmers accordingly. Considering the rate of tuber rots of yam and food scarcity as a result of damage caused by microorganisms especially those caused by fused, there is urgent need to determine the causative organisms and disease severity so as to offer a lasting solution and hence reduce food scarcity. The aim of the present study is to investigate the disease incidence and severity of tuber rots of *D. rotundata* at Orie-okporo market, Orlu L.G.A, Imo State. To isolate, identify and establish the pathogenicity of fungal organisms associated with tuber rots of *D. rotundata*.

Corresponding author: Umeoka N, Department of Plant Science and Biotechnology, Faculty of Science, Imo State University Owerri, Imo State, Nigeria
E-mail: izukachukwuemeka@gmail.com

Materials and Methods

Location of study

Samples were collected with clean polyethylene bags, labeled and taken to the laboratory at the Department of Plant Science and Biotechnology in Imo State University, Owerri were the practicals took place.

Sample collection

Yam tubers with symptoms of rot were obtained from Orie-okporo daily market in Isiala-okporo Autonomous Community (where many villages sell their products), Orlu L.G.A Imo State.

Disease survey

A survey of post-harvest rots of yam was carried out by obtaining 40 tubers of yam in each visit from Orie-okporo market in Isiala-okporo Autonomous Community, Orlu L.G.A Imo State on three (3) different days. Visual inspections of the yams were carried out by separating diseased yam tubers from apparently healthy yam tubers. The diseased yam tubers were further separated into groups based on symptoms. The field disease symptoms were observed and recorded. Disease incidence was calculated for each symptom by using the formula described by Ezeibekwe et al. [11].

$$Disease\ incidence = \frac{(Number\ of\ diseased\ yam\ tubers)}{(Total\ number\ of\ yam\ tubers\ sampled)} \times \frac{100\%}{1}$$

The severity of rot infection was assessed by using the following scale: -

0 = Healthy

1 = 1% to 25% slight infected.

2 = 26% to 50% moderate infected.

3 = 51% to 75% extensive infected.

4 = 76% to 100% completely rotted.

A percentage rot score per sample of the yam tubers was derived from the total rot scores as follows:

$$Severity = \frac{(Sum\ of\ numerical\ ratings)}{(Total\ number\ of\ observed\ yam\ tubers)} \times \frac{100\%}{Maximum\ diseased\ class\ (5)}$$

Medium preparation

The medium, Potato Dextrose Agar (PDA) which is a semi-synthetic medium was used for the experiment, and it was prepared following the manufacturer's directive. Thirty-nine grams (39 g) of the PDA media was dissolved in 1 litre of sterile distilled water and sterilized by autoclaving at 121°C and 15 psl for 15 minutes as instructed by the manufacturer. 0.5 gram of penicillin, an antibiotic was added to the autoclaved medium so as to inhibit any bacterial growth, and then it was shaken properly. It was allowed to cool to about 45°C and then sterile dispensed into sterile Petri-dishes. Slants were then prepared by dispensing the dissolved media into the McCartney bottles before autoclaving and then the bottles were slanted to cool. Sterility test was performed to the media by incubating them uninoculated for 24 hours at 37°C as described by Cheesbrough [12]. Only media that passed the sterility test were used.

Isolation of fungal species from rotten yam tubers

Pieces of diseased tissues cut from the periphery of rotten yam tubers with a sterilized knife were surface-sterilized in 5% sodium hypochlorite solution for 5 minutes. The surface sterilized diseased tissues were washed three times using sterile distilled water. The tissues were allowed to dry in a sterile Lamina flow chamber. The dried disease tissues were plated on a Potato Dextrose Agar (PDA) medium. Four to five days after incubation, mycelia that grew from the plated yam tissues were sub-cultured onto fresh PDA. Further sub-culturing was carried out until pure cultures of single specie isolates were obtained. From these pure cultures, inocula of the different fungal specie isolates were obtained for the pathogenicity tests. The percentage occurrence of the organisms isolated from Post-harvest rot of *D. rotundata* (white yam) tuber was recorded and calculated using the formula [13]:

$$Percentage\ occurance = \frac{Total\ number\ of\ fungal\ occurence}{Total\ number\ of\ plates} \times \frac{100}{1}$$

Identification of fungal isolates

Characteristics of fungal isolates from rotten yam tubers such as pigment production, pH, colony texture, spore or conidia-producing structures and spore shapes were observed and documented. The characteristics were observed from fungal tissues grown on PDA for one week or more, depending on the fungal species. Spore and mycelium characteristics were studied using the compound microscope. The microscopic examination was carried out by Lactophenol Cotton Blue (LPCB) wet mount [14] preparation which is the most widely used method of staining and observing fungi. The preparation has three components: phenol, which will kill any live organisms; lactic acid which preserves fungal structures, and cotton blue which stains the chitin in the fungal cell walls. LPCB mount was carried out to observe the structure of fungi. These characteristics were used in identifying the fungal organisms to the species level, following standards described by Barnett and Hunter [15].

Statistical analysis

The set up were arranged in Complete Randomized Design (CRD). The data collected were subjected to statistical analysis of variance (ANOVA) using SPSS 20.0 version (Statistical Package for the Social Sciences) (Dr. Mbagwu) to determine the means as expressed in Tables 1 and 2.

Results

Incidence and severity

The incidence and severity of surveyed Post-harvest rots of *D. rotundata* tubers obtained from Orie-okporo market, in Isiala-okporo Orlu, Imo State was as shown in Tables 1 and 2. Dry rot occurred

Day of Observation	Soft rot	Wet rot	Dry rot	Anthracnose	Powdery mildew
1st Day	15.0	17.5	25.0	15.0	12.5
2nd Day	17.5	15.0	20.0	10.0	15.0
3rd Day	12.5	15.0	22.5	12.5	15.0
Total % disease incidence	45.0%	47.5%	67.5%	37.5%	42.5%

Table 1: Incidence of post-harvest rots of *D. rotundata* tubers in Orlu, Imo State.

Day of Observation	Soft rot	Wet rot	Dry rot	Anthracnose	Powdery mildew
1st Day	10.0	8.8	8.1	7.5	6.3
2nd Day	5.6	6.9	8.1	5.0	5.0
3rd Day	8.1	7.5	10.6	3.8	3.8
Total % disease severity	23.7%	23.2%	26.8%	16.3%	15.1%

Table 2: Severity of post-harvest rots of *D. rotundata* tubers in Orlu, Imo State.

most in stored tubers with the percentage incidence of 67.5%. This was followed by wet rot (47.5%), soft rot have (45.0%), anthracnose (37.5%) and powdery mildew (42.5%). The severity results also followed the same trend as shown in Table 2 with dry rot being most severe (26.8%) and, soft rot 23.7%, wet rot 23.2%, anthracnose 16.3% and powdery mildew 15.1% (Figures 1-5).

Identified of fungal isolates

Macro and microscocopic characterization (Figures 1-5, Tables

Figure 1: Plate of Yam tuber infected with soft rot disease (Scale: 0.29).

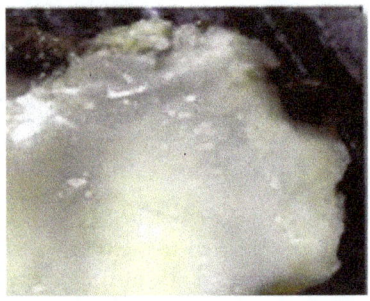

Figure 2: Plate of Yam tuber infected with wet rot disease (Scale: 0.29).

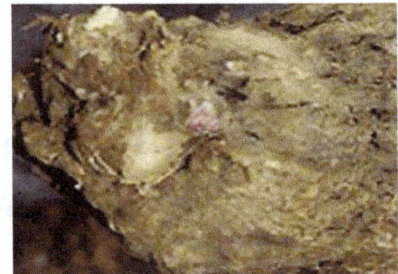

Figure 3: Plate of Yam tuber infected with dry rot disease (Scale: 0.29).

Figure 4: Plate of Yam tuber infected with anthracnose disease (Scale: 0.29).

Figure 5: Plate of Yam tuber infected with powdery mildew disease (Scale: 0.29).

Isolates	Morphological Features	Microscopic Features	Remark/Inference
A	White fluff colony on PDA and then became more compact and wooly. Later produced green patches due to formation of conidia. Reverse was light orange.	Septate hyphae; conidiophores were short, often branched, and flask-shaped at end. Conidia were rounding, single-celled, and clustered together at the end of each conidiophore.	Isolate identified as *Trichoderma viride* (Pers.)
B	Colony surface on PDA was white at first, and then became very powdery, blueish green, with a white border and reverse was white.	Septate hyphae with branched or unbranched conidiopores that had secondary branches known as metulae. On the metulae, arranged in whorls, were flask-shaped sterigmata that bore unbranched chains of round conidia. The entire structure formed the brush appearance.	Isolate identified as *Pythium aphanidermatum* (Edson)
C	Colony surface on PDA at first appeared white, then shade of black. Texture velvety and reverse was goldish.	Septate hyphae; unbranched conidiophore arising from a specialized foot cell. The conidiophore was enlarged at the tip, forming a rounded vesicle that was completely covered with flask-shaped stigmata that produce chains of round, conidia.	Isolate identified as *Aspergillus fumigatus* (Fresenius)
D	Colony on PDA appears light brown and beneath is mushy with earthly musty odour.	Conidia phialospores; phialides upright, brushlike conidiophores arising from the mycelium singly or less often in synnemata, branched near the apex, penicillate ending in a group of phialides, conidia hyaline, 1-celled, globose in basipetal chains.	Isolate identified as *Penicillium expansum* (Link.)
E	Colonies on PDA were white, moist, yeast-like and easily picked up at the early stage. Later submerged hyphae were seen at the periphery, giving the appearance of ground glass.	Coarse hyphae that segment into rectangular arthrospores varying in size and roundness of their ends.	Isolate identified as *Geotrichum candidum* (Link.)

F	Rapid growth on PDA at 25°C, produce woolly to cottony, flat, spreading colonies. From the front, the color of the colony was white. From the reverse, it was dark purple.	Phialides are cylindrical, with a component of a complex branching system. Macroconidia were produced from phialides on branched conidiophores. Macroconidia have a distinct basal foot cell and pointed distal ends. They tend to accumulate in rafts. Microconidia were formed on short simple conidiophores.	Isolate identified as *Fusarium oxysporum* (Mart.)
G	Black, ostiolate, erumpent, stromatic, confluent growth on PDA.	A compact mass of hyphae on which or in which conidia or fruit bodies are borne. Simple conidiosphores, conidia dark ovoid to elongate.	Isolate identified as *Botryodiplodia theobromae* (Sac.)
H	Initially white, quickly becoming black with conidial production. Reverse was pale yellow and growth in the PDA.	Hyphae were septate and hyaline. Conidia head were radiate initially, splitted into columns at maturity.	Isolate identified as *Aspergillus niger* (van Tieghem)

Table 3: Identification of organisms using morphological and microscopic features of fungi.

S/N	Fungi Isolates	No. of Occurrence	Occurrence (%)
1	*T. viride* (Pers.)	4	20
2	*P. aphanidermatum* (Edson)	3	15
3	*A. fumigatus* (Fresenius)	9	45
4	*P. expansum* (Link.)	2	10
5	*G. candidum* (Link.)	2	10

Table 4: Incidence of pathogen rotted tuber of *D. rotundata*.

1 and 2), of fungal isolates identified as the causative organisms of the *Dioscoria rotundata* rot implicated *Trichoderma viride* (Pers.), *Pythium aphanidermatum* (Edson), *Aspergillus fumigatus* (Fresenius), *Penicillium expansum* (Link.), *Geotrichum candidum* (Link.), *Fusarium oxysporum* (Link.), *Botryodiplodia theobromae* (Sac.) and *Aspergillus niger* (van Thieghem). Fungal organism occurred consistently associated with soft rot, dry rot, wet rot and anthracnose of *D. rotundata* (Poir) tuber with *A. fumigatus* (Fresenius) occurring often in various plates 45.00%, followed by *T. viride* (Pers.) 20.00%, *P. aphanidermatum* (Edson) recorded 15.00%, *P. expansum* (Link.) and *G. candidum* (Link.) have 10.00% each (Tables 3 and 4).

Discussion

The results on disease incidence and severity showed that there existed a high rate of post harvest rot of *D. rotundata* tuber obtained from Orie-okporo market, in Isiala-okporo Orlu, Imo State, with dry, soft and wet rot having the highest percentage value of disease incidence (67.5%, 45.0%, 47.5%) and severity and (26.8%, 23.7%, 23.2%). These findings are in agreement with the findings of Amusa and Baiyewa [16] who reported that soft rot disease is the most serious disease of yam tubers and it can also be known as wet breakdown. The dry rot is considered as the most devastating of all the storage diseases of yam. Dry rot alone causes a marked reduction in the quantity, marketable value and edible portions of tubers and those reductions are more severe in stored yams. The results of this study have shown that *T. viride, P. aphanidermatum, A. fumigatus, P. expansum, G. candidum, F. oxysporum, B. theobromea* and *A. niger* were the major casual

organisms that bring about the Post-harvest loses of *D. rotundata* tuber rot in Orlu, Imo State [5,11,13,17-20].

Conclusion

There are eight (8) fungi namely *T. viride, P. aphanidermatum, A. fumigatus, P. expansum, G. candidum, F. oxysporum, B. theobromae* and *A. niger* as casual agents of *D. rotundata* tuber rot at Orlu, Imo State. *A. fumigatus* recorded highest incidence in the disease occurrence while *T. viride* was most severe. Careful handling of the crops is recommended from harvest, through packaging, transportation and storage to minimize rot.

References

1. Okigbo RN, Ogbonnaya UO (2006) Antifungal effects of two tropical plant extracts (*Ocimum gratissimum* and *Aframomum melegueta*) on post-harvest yam rot. Africa J Biotechnol 5: 727-731.

2. FAO (2005) FAO Annual Report. Food and Agriculture Organization production year book. FAO, Rome.

3. Kana HA, Aliyu IA, Chammang HB (2012) Review on neglected and underutilized root and tuber crops as food security in achieving the millennium development goals in Nigeria. J Agri Vet Sci 4: 27-33.

4. Arya A (2010) Recent advances in the management of plant pathogens: Botanicals in the fungal pest management. Management of fungal plant pathogens, CAB International, UK.

5. Aidoo KA (2007) Identification of yam tuber rot fungi from storage systems at the Kumasi Central market. Imo state University, Nigeria.

6. Simon PW (1993) Plant pigments for colour and nutrition. Horti Sci 32: 12-13.

7. Nwankiti OA, Arene OB (1980) Diseases of yam in Nigeria. Pest articles and news summaries (PANS) 24: 468.

8. IITA (1993) Crops improvement division: Tuber root improvement program archival reports (1989-1993). Part III yam, Ibadan, Nigeria.

9. Amusa NA (1997) Fungi associated with anthracnose symptoms of yam (*Dioscorea spp*) in South-west Nigeria and their roles. Crop Res 13: 177-183.

10. PANS (1984) Pest control in tropical root crops: Pest articles and news summaries, Center for Overseas Pest Research, London 4: 147-162.

11. Ezeibekwe IO, Onuh Martin O, Anyaegbu PO (2008) Post-harvest rot diseases of fruits of *Asmina triloba* (*Carica papaya*) in some parts of Imo. Int Sci Res J 1: 107-112.

12. Cheesbrough M (2004) District laboratory practice in tropical district laboratory practice in tropical countries (Part-2). Cambridge University Press, Cambridge, UK.

13. Ezeibekwe IO, Opara MI, Mbagwu FN (2009) Antifungal effect of aloe vera gel on fungal organisms associated with Yam (*Dioscorea Rotundata* Poir) rot. J Mol Genetic 1: 11-17.

14. Thomas PA, Kuriakose T, Kirupashanker MP, Maharajan VS (1991) Use of lactophenol cotton blue mounts of corneal scrapings as an aid to the diagnosis of mycotic keratitis. Diagnostic Microbiol 14: 219-224.

15. Barnett HL, Hunter BB (1998) Illustrated genera of imperfect fungi (4thedn). Burges Publications Company, UK.

16. Amusa F, Baiyewu J (2003) Fungal toxic activity of extract of *Azadirachta indica* and *Senna alata*. Agri Res 11: 211-225.

17. Okigbo RN, Ikediugwu FEO (2001) Evaluation of water losses in different regions of yam (*Dioscorea spp.*) tuber in storage. Nigeria J Experiment Appl Biol 3: 320.

18. Okwu OE (2004) Phytochemical and vitamin content of indigenous spices of South Eastern Nigeria. J Sustain Agri Environ 6: 30-34.

19. Nwufo MI, Fajola AO (1984) Cultural studies on *Botryodiplodia theobromae* and *Sclerotium rolfsii* causing storage rots of cocoyam (*Colocasia esculenta*). Fitopatologia Brasileiva 11: 443-454.

20. Onuoha CI, Nwagbara EC (2011) Comparative studies of rot fungi of plantain and banana of different post-harvest ages. Global Res J Sci 1: 54-57.

19

Effects of Irrigations with Treated Municipal Wastewater on Phenological Parameters of Tetraploid *Cenchrus ciliaris* L.

Ben Said Ines[1]*, Adele Muscolo[2], Mezghani Imed[1] and Chaieb Mohamed[1]

[1]*University of Sfax, Department of Biology, Faculty of Sciences of Sfax, 3000, Sfax, Tunisia*

[2]*Department of Agriculture, Mediterranea University, Feo di Vito, 89124 Reggio Calabria, Italy*

Abstract

This study was conducted to investigate the use of treated municipal wastewater (TWW) in agriculture. Experiments have been carried out from July 2013 to July 2014, irrigating *Cenchrus ciliaris* with TWW or tap water (TW). The study, conducted under greenhouse conditions, compared the effect of TWW with the water normally used in irrigation, on the growth, phenological and phytomass production of *C. ciliaris* a species with high pastoral value. Firstly, our results evidenced that all the chemical parameters of TWW fell in the range of values permitted by Tunisian regulation except chloride. Additionally, TWW increased plant growth during the growth cycle, producing taller plant with respect to TW. All plants irrigated with TWW showed a better performance than plants irrigated with TW only. Similarly, TWW irrigations had positive impacts on flowering parameters during the reproductive cycle. Therefore, treated wastewater can be used as an alternative water resource in irrigation of annual fodder species, with the dual purpose of preserving fresh water and of increasing soil fertility as well as crop productivity.

Keywords: *Cenchrus ciliaris*; Pastoral species; Phenological and phytomass production; Waste water

Introduction

The volume of water used in the world increased more than twice the growth rate of the population and the growing number of regions reached a certain limit that made it impossible to provide reliable services and water supply for different uses FAO [1]. Population growth and economic development are placing unprecedented pressure on water resources, renewable but limited, particularly in arid regions. World total water resources is 1.4 billion m^3 DSİ [2] and only 1% of this amount is used as potable water [3]. Due to its arid and semi-arid climate, Tunisia is facing water scarcity problems, where the estimated available freshwater is only about 450 m^3 /citizen/year [4]. Since half of the world population lives in urban sections EC [5], the demand for fresh water is increasing every day and the production of municipal wastewater is increasing as well. Thus, the availability of good-quality water for irrigation is threatened Alobaidy, et al. [6] and irrigated agriculture faces the challenge of using less water, in many cases of poorer quality, to irrigate lands which provide food for an expanding population. The municipal wastewater has been recycled in agriculture for centuries as a means of disposal in cities such as Berlin, London, Milan and Paris AATSE [7]. However, in recent years wastewater has gained importance in water-scarce regions. In the most of these cases, the farmers irrigate with diluted, untreated, or partly treated wastewater. The lack of appropriate treatment and management of wastewater generated adverse health effects [8]. In this respect, it is necessary to adequately process wastewater before its use in the environment. Therefore, the usage of municipal treated wastewater for irrigation purpose, according to their composition and to the international standards of water irrigation quality, seems to be the most promising practice that may help to ensure safe and sustainable food crops in arid and semiarid regions. On the basis of the above statements, the aim of this paper is to evaluate the suitability of treated municipal wastewater to irrigate *Cenchrus ciliaris* L. (syn. *Pennisetum ciliare* L.) Link, Buffel grass). This species, native to dry areas of Africa, West Asia and India has been widely introduced in arid and semi-arid regions of the world for its high pastoral value [9-13]. Despite its importance as fodder, leaf of *C. ciliaris* contains compounds able to inhibit the bacterial/fungal growth much more than the standard drug used, representing an environmentally safe alternatives for plant disease control [14].

Additionally, *C. ciliaris,* is a hyper-root-accumulator of heavy metal and could be used for phytoremediation purpose [15]. *C. ciliaris,* has been used also in traditional medicines to relieve kidney pain, cure wounds, sores and tumors [16]. Due to its economic potentiality, in this study we used a tetraploid *C. ciliaris,* that are widely distributed in the most humid areas of Tunisia Kharrat-Souissi et al. [17], to evaluate the fertilizer potential of treated municipal wastewater.

The study of the effect of irrigation with treated wastewater on the growth, and production of plants is one of the promising aspects, under the projected climate change for the next time IPCC. Although, the ecophysiological aspects under natural rainfall conditions of *Cenchrus ciliaris* have been widely studied in Tunisia Visser et al. [18], it should be noted that no study of this species has irrigations was carried out in conditions with the treated wastewater. In this context, the present study conducted under greenhouse conditions compared the application of treated municipal wastewater with the ground water, normally used in irrigation, on the growth, phenological and phytomass production of *C. ciliaris*.

Materials and Methods

Municipal wastewater

Treated wastewater were sampled at the outlet of the Sfax wastewater treatment plant, where municipal wastewater was treated with the biological stabilization bonds, at different times and stored at 4°C before the chemical characterization. Effluent samples were analyzed for pH and electrical conductivity (ECw) using a pH meter (AFNOR

*Corresponding author: Ben Said Ines, Department of biology, Faculty of Sciences of Sfax, 3000 Sfax, Tunisia, E-mail: bensaidines55@yahoo.com

standard method N° NF T 90-008 AFNOR [19] and a conductimeter (AFNOR N° NF EN 27888 AFNOR [19]) respectively. Chemical oxygen demand (COD), suspended solids (SS), biochemical oxygen demand (BOD) and total phosphorus were measured according to the standard methods (AFNOR N° NF T 90-018, NF EN 872, NF T 90-103, NF EN 1189 AFNOR [19]. Heavy metal contents were measured following standard methods (APHA, 2005) Cations and anions were measured using ionic chromatography while carbonates and bicarbonates were estimated by titrating an aliquot of the effluent samples with HCl (AFNOR N° NF EN ISO 9963-2 AFNOR [19].

Plant material

4X *Cenchrus ciliaris* tetraploid (2n = 4x = 36), more adapted to wet areas in the extreme north of Tunisia Kharrat-Souissi et al. [17], were collected randomly from Morneg (south of the city of Tunis: latitude 36° 73 N, longitude 10° 24 E).

Experimental design

This experiment was carried out under a shelter greenhouse in the experimental field of the Olive Tree Institute of Sfax, (34° 43 N, 10° 41 E) in Central-Eastern Tunisia. The planting of tetraploid level, took place in August 2012 in pots under semi-controlled conditions. The pots were 20 L capacity, 30 cm in diameter and 30 cm in depth. The substrate used is a natural postural soil. Each pot contained one plant. Tap water was used during the installation containing 1.3 g/l of NaCl. The photon flux in the greenhouse varied between 163 to 389 µmol / m²/s the temperature ranged from 13.3 to 28.3°C with a photoperiod of 12-14 h. The relative humidity ranged from 43% to 83% and the evaporation ranged between 88.5 and 268.5 mm. One year after planting (in June 2013), that is to say as soon as we got adult plants, a cutting (3 cm) from above the soil surface was conducted for each plant to simulate the zero level of growth during the summer season. After the cutting procedure, two irrigation treatments were applied, after cutting, during July 2013-July 2014 with two growth cycles: the 1th cycle from July to November and the 2th one from March to July 2014. The frequency of irrigation was on a ten-day basis (1st, 10th, and 20th day of each month). The irrigation of plants was as follow:

T1: 800 mm tap water (TW).

T2: 800 mm treated wastewater (TWW).

At the end of the vegetative growth (6 months), the plant growth parameters in terms of height and tuft diameter, number of leaves, length of leaves, were detected. Additionally, reproductive parameters in terms of number of cobs per growth unit, total number of ears per individual, were measured monthly. At the end of the growth cycle (about 6 months), sections of 3 cm from above the soil surface including stems, leaves and cobs were collected from each plant. The plant material was dried in the oven at 80°C for 48 hours, and subsequently weighed to obtain phytomass.

Statistical analysis

Statistical analysis was conducted using the "SPSS 19" software, adopting an analysis of variance ANOVA, linear model generalized to two treatment factors. The mean values of all parameters were compared using the Dunnett test.

Results and Discussion

Climatic data and water characteristics

Note treated wastewater (Table 1), contained salts, in particular a high concentration of Cl- and Na+ that if added to the soil can increase salinity, soil osmotic potential inducing damage to cultivation. Additionally, the high total and fecal coliform content could affect crops and with consequence on human health. The biological treatment of wastewater improved their quality from a chemical point of view decreasing the concentration of Cl- and Na+, breaking down the polluting power. The physical and chemical characteristics of TWW and TW and the values admitted by Tunisian regulation are reported in Table 2. All the chemical parameters fall within the value permitted by Tunisian regulation except chloride. The pH of TWW and TW were 7.60 and 7.51, respectively, falling within the limits for *Cenchrus ciliaris* growth (7.0 to 8.0) [20]. The electrical conductivity (EC) was 6.80 dS m⁻¹ for TWW and 4.30 dS m⁻¹ for TW, indicating, a high and a moderate level of salinity respectively [21,22]. Cl- concentration in TWW was higher than the threshold values, as reported by Graham and Humphreys [23] in the guidelines for forage plants irrigation. As expected, the concentration of almost all elements was also higher in TWW than in TW, with the exception of Ca²⁺ and Mg²⁺ (Table 2), even if they were present in TWW at concentrations 9 times higher than that contained in the fertilizer normally used in agriculture. Both chemical and biological oxygen demands (COD and BOD5) of TWW were below the Tunisian thresholds for water reuse. According to the chemical parameters detected, the TWW represented a source of nutrients for crops. The content of heavy metals (Cd, Zn, Cr, and Pb) was lower than the toxicity limits (<0.004 mg/L) and it did not exceed the thresholds established by Tunisian regulation [24]. Neither coliforms nor fecal coliforms were detected in the irrigation water, resulting in an environmentally-friendly safe wastewater. The variations of climatic parameters over the experimental period are reported in Table 3. As expected the highest temperature was detected in august and the lowest one in February. The evaporation was the

Characteristics	Wastewater
pH	7.40 ± 0.2
EC	7.11 ± 1.9
TDS	2.20 ± 0.01
HCO_3^-	504.13 ± 0.5
SO_4^{2-}	398.66 ± 0.9
N total	108.03 ± 1.4
$N-NO_3^-$	18.96 ± 0.06
$N-NO_4^+$	72.3 ± 0.03
$N-NO_2^-$	97.99 ± 0.04
P total	26.22 ± 0.9
K^+	60.45 ± 0.1
Na^+	379.15 ± 0.04
Cl^-	2129 ± 0.08
Ca^{2+}	149 ± 0.03
Mg^{2+}	131 ± 0.01
Pb^{2+}	0.19 ± 0.01
Cd^{2+}	0.02 ± 0.00
Zn^{2+}	0.49 ± 0.01
Mn^{2+}	0.81 ± 0.01
SM	25.77 ± 0.03
COD	382 ± 0.7
BOD5	167 ± 0.2
Total coliforms	6.3 10⁶ ± 2.04 10⁴
Fecal coliforms	3.9 10⁵ ± 2.24 10⁴

Data represents mean values ± standard deviation. EC: electrical conductivity (mS/cm); TDS: total dissolved solids (g L⁻¹); SM: suspended matter (mg L⁻¹); COD (mg L⁻¹): chemical oxygen demand; BOD5: biological oxygen demand (mg L⁻¹); Total coliform and Fecal coliform (UFC/100 mL). anions, cations and total P and N are measured in (mg L⁻¹).

Table 1: Chemical characteristics of wastewater before treatment

highest in august because of the highest temperature and the lowest in December. Brightness, parameter linked to the length of the light cycle during the day, was higher in June and lowest in December. The relative humidity was the highest in august due to the high temperature, and the lowest in January, as a result of due to the scarcity of rain and the cold temperature. All these data reflect the climatic conditions of Mediterranean countries [25].

Growth and flowering parameters

The plant growth parameters: plant height, plant diameter, leaf

Characteristics	TWW	TW	Tunisian regulation
pH	7.60 ± 0.10	7.51 ± 0.11	6.50-8.50
EC	5.6 ± 0.02	4.30 ± 0.03	7
TDS	1.77 ± 0.02	0.93 ± 0.01	2
HCO_3^-	356.00 ± 0.3	223.30 ± 0.20	600
SO_4^{2-}	354.00 ± 0.7	67.50 ± 1.5	1000
N total	53.80 ± 1.20	-	30
$N-NO_3^-$	13.40 ± 0.01	0.97 ± 0.01	-
$N-NO_4^+$	35.6 ± 0.01	2.67 ± 0.04	-
$N-NO_2^-$	4.00 ± 0.02	0.04 ± 0.01	-
P total	9.44 ± 0.11	0.45 ± 0.02	0.05
K^+	33.80 ± 0.09	26.00 ± 0.05	50
Na^+	297 ± 0.01	430.00 ± 0.01	300
Cl^-	1767 ± 0.04	1340.00 ± 0.2	600
Ca^{2+}	98.50 ± 0.01	188.20 ± 0.02	-
Mg^{2+}	85.70 ± 0.01	126.20 ± 0.03	-
Pb^{2+}	< 0.004	0	0.1
Cd^{2+}	< 0.004	0	0.005
Zn^{2+}	0.33 ± 0.01	0.5 ± 0.01	5
Mn^{2+}	0.65 ± 0.01	0.13 ± 0.03	-
SM	12.20 ± 0.02	2.30 ± 0.02	
COD	74.00 ± 0.01	0	90
BOD5	20.00 ± 0.01	0	30
Total coliforms	nd	0	-
Fecal coliforms	nd	0	-

Data represents mean values ± standard deviation. EC: electrical conductivity (mS/cm); TDS: total dissolved solids (g L^{-1}); SM: suspended matter (mg L^{-1}); COD (mg L^{-1}): chemical oxygen demand; BOD5: biological oxygen demand (mg L^{-1}); Total coliform and Fecal coliform (UFC/100 mL). anions, cations and total P and N are measured in (mg L^{-1}); TW: tap water; TWW: treated wastewater; nd, Undetected.

Table 2: Chemical characteristics of the irrigation waters used in the experiment

length and number leaf, *Cenchrus ciliaris* tetraploids irrigated with tap water and the treated wastewater over the two growing cycles are shown in Table 4. Crops irrigated with treated wastewater showed a better growth during the two growth cycles than the plants irrigated with tap water. The quality of irrigation water did not affect significantly plant heights in the first cycle of growth. However, a significant increase in terms of height was observed in the second growth cycle, of plants treated with wastewater. The results evidenced that the plants irrigated with tap water were shorter than the plants irrigated with treated wastewater. Similar results were reported by Day et al. [26] who observed that wheat irrigated with wastewater produced taller plants, more heads per unit area, heavier seeds, higher grain yields than wheat grown with pump water alone. They attributed this increase to the nitrogen and phosphorus contained in the added wastewater. In contrast, Carter, et al. [27] for *Celosia argentea* and Grieve, et al. [28] for *Matthiola incana,* observed a regression in the height of these plants grown with wastewater. The diameter of plants irrigated with treated wastewater was larger than the diameter of the plants irrigated with tap water in both growth cycles. The largest diameter (52.60 cm) was observed in wastewater irrigated plants in July, the end of the second cycle. Except for April (p = 0.021), the quality of irrigation water did not cause significant differences in the diameter of plants. Regarding leaf length, in the first cycle, the irrigation with treated wastewater caused an increase in leaf length than irrigation with tap water. While no significant differences in leaf length were observed in the second cycle between the wastewater irrigated plants and the tap water irrigated ones. The highest leaf length (29.72 cm) was observed in treated wastewater irrigated plants in November. Nevertheless, the leaf numbers of plants irrigated with treated wastewater were greater than that of plants irrigated with tap water for both growing cycles. The greatest number of leaves was observed in wastewater treated plants in July. The irrigation with wastewater increased leaf number and the reproductive growth of *Cenchrus ciliaris* mainly during the second cycle. These results are in agreement with that of Oliveira-Marinho et al. [29] indicating an increase in the leaf number of *Rosa hybrida* 'Atmosphere' irrigated with wastewater with different salinity levels. An increase in leaf number has also been reported for *Arachis hypogaea* Saravanamoorthy and Kumari [30], *Sorghum bicolor* Khan et al. [31] and *Gossypium hirsutum.* Alikhasi et al. [32] when irrigated with biologically treated wastewater. TWW (containing high salt concentrations) increased not only the growth of the flowering

Cycle	1th Cycle					2th Cycle				
Months	July	Aug	Sept	Oct	Nov	Mar	Apr	May	June	July
Temperature(°C)	27.2	28.3	26.1	24.47	16.7	17.3	18.7	21.6	23.7	25.6
Evaporation (mm)	243.4	268.5	183.2	169	126.1	181	170.2	198.4	223.3	241.1
Brightness(µmol/m²)	380,3	319	245	225	174	232	248	301	389	379,9
Relative humidity (%)	82	83	71	67	60	73	75	78	77	81

Table 3: Mean temperature (°C), evaporation (mm), insulation (µmol/m/s²), and relative humidity (%) registered monthly in the greenhouse over the experimental period

Parameter	Irrigation water	1th cycle					2th cycle				
		July	August	September	October	November	March	April	May	June	July
Number of ears /UC	TW	0	0	0	0.1	0.1	0	1.4	2.5	3.2	3.2
	TWW	0	0	0.1	0.5	0.7	0	2.3	3.6	4.5	4.7
	Significance	.	.	0.331 n.s	0.054 n.s	0.004 n.s	.	0.001*	0.000**	0.000**	0.000**
Number of ears /plant	TW	0	0	0	0.3	0.3	2.8	9.6	16.7	23.8	23.8
	TWW	0	0.1	0.1	1.2	1.8	4.2	14.9	29.2	41.5	43.3
	Significance	.	0.331n.s	0.331 n.s	0.122 n.s	0.037 n.s	0.002 n.s	0.000**	0.000**	0.000**	0.000**

UC: unit of growth; n.s.: not-significant; **p < 0.001; *p < 0.01

Table 4: Effects of irrigation with tap and treated wastewater on growth parameters of tetraploid *Cenchrus ciliaris* L from July (2013) to July (2014).

power of *Cenchrus ciliarisbut* also intensified it (Table 5). All plants irrigated with treated wastewater showed a better performance than irrigated plants with tap water only during the second growth cycle even if the TWW contained a high concentration of chloride. Sun et al. [33] identified *Cenchrus ciliaris* as suitable plants to be utilized for bioremediation in surface saline soil or marine sediments, for its ability to grow in soil with (1-2% NaCl). The quality of irrigation water did not result in significant differences in the number of ears in the first cycle of treatment. The largest number of total ears per individual (43.03) was observed in July in irrigated plants with treated wastewater. Similarly, the highest value in the tap water irrigation was also observed in July (23.80). Regarding the number of ears per UC, the quality of irrigation water did not result in significant differences only in April, May, June and July during the second cycle of treatment which corresponded to the reproductive cycle of this species. The largest number of ears per UC (4.70) was observed in plants irrigated with treated wastewater in July. By taking into consideration all together these data, one can deduce that the treated wastewater increased the number of ears in tetraploid plants during the second cycle. Irrigation with treated wastewater had no negative effects on growth and flowering. These results were in agreement with those of Gerhart, et al. [34] on *Prosopis chilensis*, *Sophora secundiflora*, *Malephora* spp., *Cercidium* sp., *Leucophyllum* spp., *Rosmarinus officinalis*, *Acacia stenophylla*, *Caliandra californica* and *Dalea greggii*; and with those of Banon et al. [35] on *Lantana camara*.

Effect of municipal TWW on Biomass

Over the experimental time, the sheet dry matter of the plant was weighed at three different periods. The mean values for tetraploid *Cenchrus ciliaris* depended on the origin of irrigation water (Figure 1). Furthermore, the dry matter of tetraploid *Cenchrus ciliaris* irrigated with TWW was higher than those irrigated with TW. Statistical analysis showed significant differences between the average dry matter of the two treatments (TWW and TW) only at the end of the experiment. The irrigation with TWW showed a significant increase in dry mass over time. The irrigation with TW caused significant differences only between the first and second period. These results suggest that the application of TWW may add nutrients and bacteria to the soil, increasing biodiversity and abundance of soil organisms that are important to maintain agro-ecosystem services mainly in arid

and semiarid regions. This explanation is supported by data of del Mar Alguacil, et al. [36] showing that microbial activities were significantly higher in the soils irrigated with urban wastewater than in those irrigated with fresh water. Additionally, del Mar Alguacil, et al. [36] and Mousavi, et al. [37] showed that irrigation with treated municipal wastewater had a significant positive impact on the growth and quality of orange-tree and maize respectively, supporting the results of this study.

Conclusion

In short, we can conclude that the irrigation with treated wastewater increased plant growth and flowering with respect to tap water during the experimental period. Therefore, as no negative effects were observed on crop vitality and productivity, it seems that the treated wastewater can be used as an alternative source for irrigation of *Cenchrus ciliaris* tetraploid, with the dual purpose of not only saving fresh water for other uses, but also improving soil fertility and productivity in arid and semi-arid regions.

Acknowledgments

The authors gratefully acknowledge to the Ministry of Education and Science of Tunisia, CNRS (Centre National de la Recherche Scientifique) for funding this project.

References

1. FAO (2007) State resources for animal genetic food and agriculture the world. CGRFA Report, Rome, Italy.

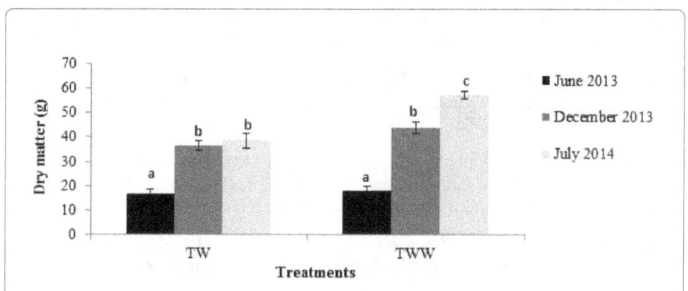

Figure 1: Effects of tap water (TW) and treated municipal wastewater (TWW) on growth of tetraploid *Cenchrus ciliaris*, during the experimental time (June 2013-July 2014). Growth of *Cenchrus ciliaris* is expressed as dry mass (g).

Parameter	Irrigation water	1th cycle					2th cycle				
		July	August	September	October	November	March	April	May	June	July
Plant height(cm)	TW	11.6	22.7	53.9	64.8	67.2	15.3	20.5	34.3	53.8	63.7
	TWW	12	24.3	55.1	66.6	68.9	18.9	25.4	43.4	64.9	69.1
	Significance	0.470n.s	0.021n.s	0.152n.s	0.021n.s	0.049n.s	0.000**	0.000**	0.000**	0.000**	0.001*
Plant diameter(cm)	TW	16.4	22	24.7	26.3	30.8	17.6	32.7	36.1	39.6	43.1
	TWW	15.6	27.4	33.5	37.2	44.2	22.7	34.3	42.3	51.7	52.6
	Significance	0.145n.s	0.000**	0.000**	0.000**	0.000**	0.000**	0.021n.s	0.000**	0.000**	0.000**
Leaf length (cm)	TW	10.69	14.02	17.12	21.38	23.17	4.98	13.97	22.28	25.58	27.27
	TWW	12.37	16.18	19.37	24.75	29.72	4.91	14.55	23.17	26.18	26.93
	Significance	0.011n.s	0.003*	0.001*	0.000**	0.000**	0.827n.s	0.336n.s	0.058n.s	0.266n.s	0.239n.s
Leaf number (n°)/ UC	TW	4.2	8	9,30	10.7	12.4	3.8	7.3	13	15.6	19
	TWW	5.3	11.3	14.5	15.8	18.2	3.7	8.6	15.3	20.6	24.9
	Significance	0.080n.s	0.000**	0.000**	0.000**	0.000**	0.801n.s	0.005*	0.000**	0.000**	0.000**

UC: unit of growth; n.s.: not-significant; **p < 0.001; *p < 0.01.

Table 5 Effects of irrigation with tap and treated wastewater on flowering parameters of tetraploid *Cenchrus ciliaris* L. from July (2013) to July (2014).

2. DSI (2012) Soil and water resources. Turkey Is Bank Cultural Publications, Istanbul, Turkey.

3. Cassaniti C, Romano D, Hop MM, Flowers TJ (2013) Growing floricultural crops with brakish water. Environmental and Experimental Botany 92: 165-175.

4. Louati M, Khanfir R, Alouini A, El Echi M, Frigui L, Marzouk A (2000) Drought Management Handbook in Tunisia. Internal report, Ministry of Agriculture, Tunisia.

5. EC (2012) Science for Environment Policy. DG Environment News Alert Service. European Commision.

6. Abdul HM, Alobaidy J, Mukheled A, Abass AJ, Kadhem MA, et al. (2010) Evaluation of Treated Municipal Wastewater Quality for Irrigation. Journal of Environmental Protection 1: 216-225.

7. AATSE (2004) Water recycling in Australia. Australian Academy of Technological Sciences and Engineering, Melbourne.

8. Angelakis AN, Bontoux L, Lazarova V (2003) Challenges and prospectives for water recycling and reuse in EU countries. Wat Sci Tech Wat Supply 3: 59-68.

9. Correll DS, Johnston MC (1970) Manual of the Vascular Plants of Texas. Texas Research Foundation, Renner, Texas, USA.

10. Clayton WD, Renvoize SA (1986) Genera Graminum: Grasses of the World. Kew Bulletin, Additional Series 13. Publisher: Her Majesty's Stationary Office, London.

11. Burquez-Montijo A, Miller M, Martinez-Yrizar A (2002) Mexican grasslands, thornshrub, and the transformation of the Sonoran Desert by invasive exotic buffelgrass (Pennisetum ciliare). In: Tellman B (ed.), Invasive Exotic Species in the Sonoran Region. University of Arizona Press, Tucson.

12. Mseddi K, Visser M, Neffati M, Reheul D, Chaïeb M,et al. (2002) Seed and spike traits from remnant populations of *Cenchrus ciliaris L.* in South Tunisia: high distinctiveness, no ecotypes. Journal of Arid Environments 50: 309-324.

13. Stieber MT, Wipff JK (2003) Cenchrus. Flora of North America north of Mexico. University Press, New York.

14. Singariya P, Kumar P, Mourya KK (2012) Evaluation Of Antibacterial Activity And Preliminary Phytochemical Studies On The Stem Of Cenchrus Ciliaris And Cenchrus Setigerus. Asian J Pharm Clin Res 5: 163-167.

15. Keeling SM, Werren G (2005) Phytoremediation: The Uptake of Metals and Metalloids by Rhodes Grass Grown on Metal-Contaminated Soil. REMEDIATION 2: 53-61.

16. Shahid, M, Rao NK (2011) Cenchrus Ciliaris: A Drought And Salt-Tolerant Grass For Arid Lands International Center for Biosaline Agriculture. Biosalinity News Newsletter of the International Center for Biosaline Agriculture 12: 1-8.

17. Kharrat-Souissi A, Siljak-Yakovlev S, Brown S, Chaieb M (2010) Cytogeography of Cenchrus ciliaris (Poaceae) in Tunisia. Folia Geobotanica 48: 95-113.

18. Visser M, Mseddi K, Chaieb M, Neffati M (2008) Assessing yield and yield stability of remnant populations of Cenchrus ciliaris L. in arid Tunisia: developing a blueprint for initiating native seed production. Grass Forage Sci 63: 301-311.

19. AFNOR (1997) Water quality and analysis methods.

20. Brzostowski HW (1962) Influence of pH and superphosphate on establishement of Cenchrus ciliaris from seed. Trop Agric Trinidad 39: 289-296.

21. Rhoades JD, Kandiah A, Mashali AM (1992) The use of saline waters for crop production. FAO Irrigation & Drainage Paper, FAO, Rome, Italy.

22. Wiesman Z, Itzhak D, Ben Dom N (2004) Optimization of saline water level for sustainable Barnea olive and oil production in desert conditions. Scientia Horticulturae 100: 257-266.

23. Graham TWG, Humphreys LR (1970) Salinity response of cultivars of buffel grass (Cenchrus ciliaris). Austral J Exp Agric Anim Husb 10: 725-728.

24. Ben-Hur M (2004) Sewage water treatments and reuse in Israel. Water in the Middle East and in North Africa, Springer-Verlag, Heidelberg, Germany.

25. Haarsma RJ, Selten F, Hurk B, Hazeleger W, Xueli WX, et al. (2009) Drier Mediterranean soils due to greenhouse warming bring easterly winds over summertime central Europe. Geophysical Research Letters 36: 1-7.

26. Day AD, Fadyen JA, Tucker TC, Cluff CB (1979) Commercial production of wheat grain irrigated with municipal waste water and pump water. J Environ Qual.

27. Carter CT, Grieve CM, Poss JA, Suarez DL (2005) Production and Ion Uptake of Celosia argentea Irrigated with Saline Wastewaters. Scientia Horticulture 106: 381-394.

28. Grieve CM, Poss JA, Amrhein C (2006) Response of Matthiola incana to irrigation with saline wastewaters. HortScience 41: 119-123.

29. Marinho LE, Tonetti AL, Stefanutti R, Coraucci Filho B (2013) Application of Reclaimed Wastewater in the Irrigation of Rosebushes. Water Air Soil Pollut 224: 1669-1674.

30. Saravanamoorthy MD, Ranjitha Kumari BD (2007) Effect of textile waste water on morphophysiology and yield on two varieties of peanut (Arachis hypogaea L.). Journal of Agricultural Technology 3: 335-343.

31. Khan FR, Bury NR, Hogstrand C (2010) Differential uptake and oxidative stress response in zebrafish fed a single dose of the principal copper and zinc enriched sub-cellular fractions of Gammarus pulex. Aquat Toxicol 99: 466-472.

32. Alikhasi M, Kouchakzadeh M, Baniani E (2012) The effect of treated municipal wastewater irrigation in non-Agricultural soil on cotton plant. Journal of Agriculture Science and Technology 14: 1357-1364.

33. Sun WH, Lo JB, Robert FM, Ray C, Tang CS, et al. (2004) Phytoremediation of petroleum hydrocarbons in tropical coastal soils. I. Selection of promising woody plants. Environ Sci Pollut Res Int 11: 260-266.

34. Gerhart VJ, Kaneb R, Glenn EP (2006) Recycling Industrial Saline Wastewater for Landscape Irrigation in a Desert Urban Area. Journal of Arid Environments 67: 473-486.

35. Banon S, Miralles J, Ochoa J, Franco JA, Sanchez MJ, et al. (2011) Effects of Diluted and Undiluted Wastewater on the Growth, Physiological Aspects and Visual Quality of Potted Lantana and Polygala Plants. Scientia Horticulturae 129: 869-876.

36. Alguacil MM, Torrecillas E, Torres P, García-Orenes F, Roldán A, et al. (2012) Long-term effects of irrigation with waste water on soil AM fungi diversity and microbial activities: the implications for agro-ecosystem resilience. PLoS One.

37. Mousavi SR, Galavi M, Eskandari H (2013) Effects of treated municipal wastewater on fluctuation trend of leaf area index and quality of maize (Zea mays). Water Sci Technol 67: 797-802.

20

Elephant Foot Yam (*Amorphophallus paeoniifolius*): Osmotic Dehydration and Modelling

Sangeeta* and Bahadur Singh Hathan

Department of Food Engineering andTechnology, Sant Longowal Institute of Engineering andTechnology, (SLIET), Sangrur, Punjab, India

Abstract

Osmotic dehydration of elephant foot yam was done in different concentration of sucrose solution at different temperature for regular interval of time. The osmotic solution concentrations used were 40, 50, 60°Bx, osmotic solution temperatures were 35, 45, 55°C and the process duration varied from 0 to 240 min. The fruit to solution ratio was kept constant i.e. 1:5 (w/w) during all the experiments. The experimental data of water loss and solute gain was fitted to different empirical kinetic models viz. Peleg, Penetration, Magee, and Azuara to know the best fitted model to the experimental data. Out of all the applied models, Magee model and Azuara model were the best fitted as compared to other models for water loss and solute gain of elephant foot yam, respectively.

Keywords: Elephant foot yam; Osmotic dehydration; Kinetics; Empirical models

Introduction

Elephant foot yam, *Amorphophallus paeoniifolius* is very much prevalent in Philippines, India, Malaysia, Indonesia, China, Sri Lanka and many other Southeast Asian countries [1]. The tubers of elephant foot yam are commonly used as a vegetable after cooking and in preparation of indigenous ayurvedic medicines [2]. The tubers are cheapest source of carbohydrates mainly starch and fibres, vitamins and minerals [3] and play a importantl role in food security and are the important staple or subsidiary food for a large group of population [4]. Tubers have a short shelf life because of their high moisture content. One of the best ways to preserve them may be by processing methods like drying, dehydration or by obtaining flour and/or starches. Due to the reduction of moisture content by various means the shelf life of corms can be increased. In recent years, for preservation of fruits and vegetables osmotic dehydration technique is gaining considerable amount of attention due to its potential to keep sensory and nutritional properties similar to the fresh fruits [5]. Osmotic dehydration is the process of water removal by immersion of water containing cellular solid in a concentrated aqueous solution of high osmotic pressure (hypertonic media) for a specified time and temperature. Water removal in osmotic dehydration is based on the natural and non-destructive phenomenon of osmosis across cell membranes. The driving force for water removal from cell is potential difference between osmotic pressure of fresh material and surrounding solution [6]. Osmotic dehydration is actually combination of simultaneous water and solute diffusion process [7] means mass transfer consists of two major simultaneous counter-current fluxes of water and solutes because complex cell wall structure is not perfectly selective [8]. Leaching of negligible amount of natural solutes from food into solution has considered as third minor flux [9]. This pre-treatment minimize color losses as well as reduce nutrient losses due to drying. The influence of the main process variables such as concentration and composition of osmotic solution, temperature, immersion time, pre-treatments, agitation, nature of food and its geometry, solution to sample ratio on the kinetics of mass transfer and product quality have been studied extensively [10,11]. Considerable effort has been made toward developing models to predict the mass transfer kinetics of osmotic dehydration process. In this regard, several equations based on Fick's second law have been proposed which are not useful practically because of unrealistic assumptions and complexity of the some equations. Some researchers like Peleg [12], Azuara

[13], Magee et al. [14] and Rahman [15] etc. recommended simpler empirical equations including parameters with physical meaning. These empirical equations have been used to model the rate of dehydration of different plant-based materials [16-21]. However, literature about the suitability of these equations to model the mass transfer kinetics of osmotically dehydrated elephant foot yam is very rare. So, the aim of present study was to evaluate the effect of temperature and sucrose solution concentration on mass transfer during osmotic dehydration process and to assess the predictive capacity of Peleg, Azuara, Magee and Rahman equations during osmotic dehydration of elephant foot yam cubes in sucrose solution.

Material and method

Osmotic dehydration of elephant foot yam cubes

Osmotic dehydration elephant foot yam (EFY) cubes having size 1 cm × 1 cm × 1 cm was done in osmotic solution of sucrose having different concentrations (40, 50, 60°Bx) and solution temperature (35, 45, 55 °C). Vegetable to solution ratio was kept 1:5 (w/w) [22] during osmotic dehydration for a regular interval of time period of (0-240 min). The temperature of the osmotic solution was maintained by hot water bath agitating@50 oscillations per minute. Agitation was given during osmosis for reducing the mass transfer resistance at the surface of the fruit and for good mixing and close temperature control in osmotic medium [23]. Stain less steel containers (of approximately 150 ml capacity) containing osmotic solution were kept in hot water bath. After attainment of desired temperature of the solution, known weight of EFY cubes was put in to the container. The EFY cubes from each container were removed at specified time and were immediately rinsed with running water to remove the solute adhered to fruit surface.

---

*Corresonding author: Sangeeta, Department of Food Engineering and Technology, Sant Longowal Institute of Engineering and Technology, (SLIET), Sangrur, Punjab, India, E-mail: ssaini.kataria@gmail.com

The cubes were then spread on muslin cloth to remove the free water from the outer surface of the EFY cubes. The cubes were then put in the pre-weighed petri-dish for determination of dry matter by oven method. During experimentation, it was assumed that the amount of solid (sugars, acids, minerals, vitamins) leaching out of product into the medium was considered quantitatively negligible [24]. The water loss and solute gain were calculated as given below:

Let, initial dry matter of fresh vegetable=Z%

Initial weight of vegetable taken for osmotic dehydration=W_0 (g)

∴Initial dry matter of vegetable= $\dfrac{W_o \times Z}{100}$ =S_o(say)

Let the weight of vegetable after osmotic dehydration for any time t=W_t(g)

And the dry matter of vegetable after osmotic dehydration for time t=S_t(g)

Then, Weight reduction, WR=W_o-W_t (g)

Solute gain after osmotic dehydration for time t, SG= S_t-S_o (g)

Water Loss, WL=WR+SG

Water loss in g/100 g fresh sample=$\dfrac{WL}{W_o} \times 100$ (1)

Solute gain in g/100 g fresh sample=$\dfrac{SG}{W_o} \times 100$ (2)

Validation of empirical models for osmotic dehydration of EFY cubes

The validity of empirical models for water loss and solute gain during osmotic dehydration (Table 1) was checked by non linear regression technique. Azuara et al. [13] developed a model from mass balance considerations to predict the kinetics and final equilibrium point of osmotic dehydration by using data obtained during relatively short period of osmosis. In Azuara model, the constant β_1 is related to the rates of water diffusion out from the sample (min $^{-1}$). For solute gain instead of β_1 and WL_∞, constant used are β_2 and, otherwise the formula used is same as that of water loss.

Adequacy of fit of empirical models

To fit the experimental data to the various empirical models, regression analysis has been carried out by statistical software STATSTICA 7.0 for windows (Statsoft, Inc Tulsa OK U.SA.). To select the best equation various statistical parameters, such as reduced χ^2 and root mean square error (RMSE) in addition to R^2, were also used as primary criterion [18]. For evaluating nonlinear mathematical models, these parameters are not a good criterion therefore, to select the best equation to account for variation in the drying curves of the dried samples, the percent mean relative deviation modulus ($E\%$) that indicate the deviation of the observed data from the predicted line was also used as recommended by several authors in their drying studies [25]. Therefore, the best model was chosen as one with the highest coefficient of correlation (R^2); and the least χ^2, RMSE, and mean relative deviation modulus ($E\%$).

R^2 is a measure of the amount of variation around the mean explained by model.

$$Chi\ Square = \chi^2 = \sum_{i=1}^{N}\left[\frac{(Experimental\ Value - predicted\ value)^2}{(N-n)}\right] (3)$$

Where, n=no. of unknown and

N=Data point measured

$$RMSE = Root\ mean\ square\ error = \sum_{i=1}^{N}\left[\sqrt{\frac{(Experimental\ value - predicted\ value)^2}{N}}\right] (4)$$

The mean relative deviation E (%) is an absolute value that was used because it gives a clear idea of the mean divergence of the estimated data from the measured data.

$$E(\%) = \frac{100}{N}\sum_{i=1}^{N}\left|\frac{Experimental\ Value - predicted\ value}{Experimental\ value}\right| (5)$$

The values of E less than 5.0 indicate an excellent fit, while values greater than 10 are indicative of a poor fit.

Results and Discussion

During the experiments on osmotic dehydration of EFY cubes an increase in water loss and solute gain has been observed with increase of osmotic solution concentration, process temperature and time. The rates of water loss and solute gain were higher in the initial stages and approached to zero in the later stages. The process variables have significant effect on the constants and exponents of the various empirical models fitted to the water loss and solute gain data obtained during osmotic dehydration. The validation of various models for water loss and solute gain during osmotic dehydration of EFY cubes has been discussed below.

Validation of Empirical Models for Water Loss

The values of statistical parameters, models constants and coefficients for water loss during osmotic dehydration are given in Tables 2 and 3. Out of the fitted models, the values of χ^2, RMSE and E% were lower for Magee model in comparison to the Peleg model and Azuara model. There was a very good adequacy between predicted and observed data with correlation coefficient 'R²' higher than 0.96 for water loss in case of Magee model. However, Azarpazhooh and Ramaswamy [26] reported that Peleg model was a best fit model for water loss in osmotic dehydration, but this model did not fit to the experimental data in the present study because of high value of E%, RMSE and χ^2.

The Azuara model (Table 3) indicates that the predicted values of equilibrium water loss were 40.231, 49.786, 59.324 g/100 g of sample at 35, 45, 55°C, respectively, for osmotic solution of 50°Bx concentration. Therefore, with increase of temperature of osmotic solution, the values of water loss at equilibrium have been increased. The predicted values of equilibrium water loss were 57.14, 59.32, 60.24 g/100 g of fresh fruit in 40, 50, 60°Bx, respectively, at 55°C of osmotic solution temperature. Therefore, with increase of concentration of osmotic solution, the values of water loss at equilibrium have been increased. The values of β_1 indicates that the rates of water loss were higher at higher concentrations and temperature in comparison to the low values of concentration and temperature may be due to the fact that increase in osmotic solution concentration increases the concentration gradient

Model Name	Model	Reference
Penetration model	WL or SG=K × $\sqrt{t}$	[15]
Peleg Model	WL or SG =K_1+K_2 × t	[12]
Magee Model	WL or SG=A+K × $t^{1/2}$	[14]
Azuara Model	$WL_t\ or\ SG_t = \dfrac{\beta_1 t(WL_\infty)}{1+\beta_1 t} = \dfrac{(WL_\infty)\ t}{\dfrac{1}{\beta_1}+t}$	[13]

Table 1: Selected osmotic dehydration models.

Conc (°Bx)	Temp. (°C)	Magee model (water loss)						Peleg model (water loss)					
		A	K	R^2	χ^2	E%	RMSE	K_1	K_2	R^2	χ^2	E%	RMSE
40	35	3.01677	2.5059	0.97	6.1022	11.1078	2.5723	14.78721	0.12359	0.91	16.9309	18.754	4.1147
40	45	6.2810	6.3820	0.98	1.6144	4.5476	1.3710	17.78264	0.12849	0.92	6.84894	11.119	2.61704
40	55	3.7845	3.8855	0.97	4.7640	6.4143	2.2828	20.80045	0.19062	0.90	22.1364	15.7564	4.70493
50	35	2.0695	2.0795	0.99	1.3906	4.3637	1.2793	15.54508	0.15215	0.92	10.15608	13.80795	3.1868
50	45	7.7284	7.7164	0.98	4.2946	6.6693	2.0825	21.47233	0.15069	0.88	16.8150	14.25272	4.10061
50	55	4.9950	4.8960	0.98	3.1760	4.9317	1.7724	23.45795	0.21256	0.92	20.9119	13.5888	4.5729
60	35	5.5333	5.4343	0.98	3.1847	5.6983	1.7943	20.37269	0.16658	0.90	15.7033	14.25569	3.9627
60	45	8.2063	8.1073	0.97	10.135	7.9162	3.998	25.21692	0.18684	0.85	31.79	15.0833	5.6382
60	55	4.0344	4.0234	0.99	3.1039	3.8266	1.7720	23.76347	0.23316	0.94	18.20551	11.78207	4.2667

Table 2: Various regression coefficient and statistical parameters of Magee and Peleg model for water loss.

Conc (°Bx)	Temp. (°C)	WL∞	β_1	R^2	χ^2	E%	RMSE
40	35	37.547	0.0114	0.99	3.0124	6.6984	0.3042
40	45	46.569	0.0135	0.98	3.2567	10.254	0.38547
40	55	57.142	0.0175	0.98	4.123	8.564	0.4587
50	35	40.231	0.0186	0.99	11.256	7.987	1.2354
50	45	49.786	0.0935	0.99	17.564	10.564	0.3154
50	55	59.324	0.0212	0.99	3.654	11.256	0.9574
60	35	42.214	0.0223	098	19.564	13.564	0.5604
60	45	52.321	0.0243	0.99	13.254	6.354	1.2635
60	55	60.245	0.0258	0.99	26.145	8.954	1.321

Table 3: Various regression coefficient and statistical parameters of Auara model for water loss.

and in turn the driving force for osmotic dehydration process [9] and increase in temperature decreases the viscosity of the osmotic solution, decreases the external resistance to mass transfer rate at product suface; and thus facilitate the outflow of water from cubes.

The comparative validity of the various models fitted to the water loss data can also be represented from the predicted curves of various models (Figure 1). The Figure indicates that the predicted values obtained from Magee model are very close to the experimental values.

Empirical Models for Solute Gain during Osmotic Dehydration

The solute gain during the process of osmotic dehydration at various concentrations and at various temperatures was observed at regular intervals of time. The penetration of solute goes on increasing with the passage of time and become almost constant at the end of process. There was a very good adequacy between predicted and observed data with correlation coefficient 'R^2' higher than 0.96 for solute gain (Tables 4 and 5) in case of Azuara model. The values for E%, RMSE and χ^2 are less as compared to other models and value of R^2 is high than other models, which is the criteria used for the adequacy of good fitting of Model. Adequacy of fitting of Azuara model is in good agreement with the results found by Mundada et al., [27] in case of osmotic dehydration of pomegranate arils.

The comparison of experimental and predicted values of various osmotic dehydration models for solute gain could be analyzed visually in the Figure 2. The predicted values of solute gain given by Azuara model were very close to the experimental values for solute gain during osmotic dehydration of EFY cubes.

According to Azuara model (Table 5), predicted values of equilibrium solute gain were 9.26, 9.64, 12.86 g/100 g of fresh sample at

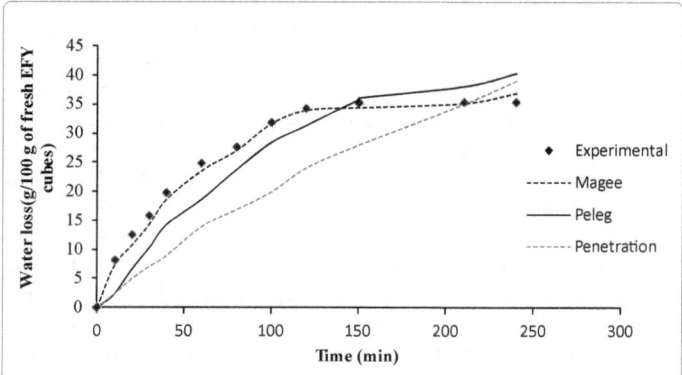

Figure 1: Plot for various predicted and experimental values for water loss with time at 40°Bx at 45°C.

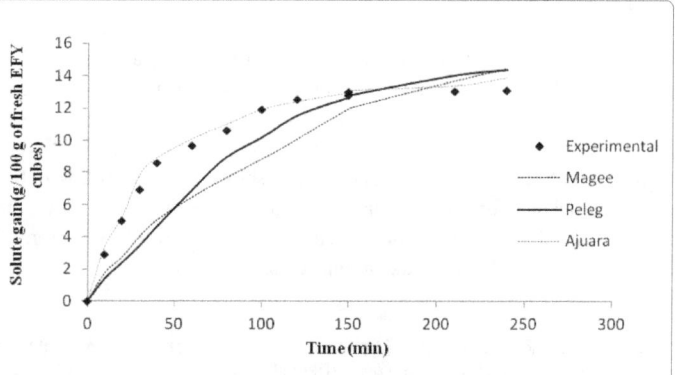

Figure 2: Plot for various predicted and experimental values for solute gain with time at 40°Bx at 45°C.

Conc.(°Bx)	Temp.(°C)	Magee model (solute gain)						Peleg model (solute gain)					
		A	K	R^2	χ^2	E%	RMSE	K_1	K_2	R^2	χ^2	E%	RMSE
40	35	1.6957	0.7824	0.98	0.2127	4.4275	0.3457	4.4329	0.03458	0.91	0.573	10.69	0.7573
40	45	2.9609	0.8472	0.99	0.0694	2.1855	0.2634	6.30406	0.04279	0.94	0.592	8.442	0.7695
40	55	3.5877	0.9240	0.99	0.1015	1.8886	0.3187	7.2927	0.04805	0.97	0.309	5.342	0.5559
50	35	2.2932	0.7861	0.98	0.1084	3.5912	0.3293	5.54218	0.04001	0.93	0.641	9.9769	0.8011
50	45	4.0481	0.8481	0.99	0.0529	1.9356	0.2400	7.46864	0.04399	0.97	0.282	5.4214	0.5312
50	55	4.5820	1.1402	0.99	0.1279	1.8948	0.3576	9.1689	0.05931	0.97	0.449	4.369	0.6707
60	35	2.6721	0.8614	0.99	0.0407	1.6916	0.2018	6.20413	0.04439	0.96	0.408	6.3006	0.6392
60	45	4.0604	1.0879	0.99	0.0562	1.7448	0.2573	8.4532	0.056201	0.96	0.586	5.5548	0.7656
60	55	4.6557	1.5219	0.99	0.2668	2.6868	0.5261	10.8119	0.07781	0.94	1.984	7.8928	1.4088

Table 4: Various regression coefficient and statistical parameters of Magee and Peleg model for solute gain.

Conc (°Bx)	Temp (C)	SG∞	β_2	R^2	χ^2	E%	RMSE
40	35	7.521	0.0348	0.98	0.0999	5.1326	0.0356
40	45	8.654	0.0254	0.99	0.0450	6.5478	0.0645
40	55	9.123	0.0088	0.98	0.0654	8.654	0.0795
50	35	9.2654	0.0045	0.99	0.0147	9.6479	0.0214
50	45	9.641	0.00145	0.99	0.1254	11.3255	0.0145
50	55	12.864	0.0013	0.99	0.3159	8.987	0.0478
60	35	10.764	0.0064	0.98	0.2647	4.679	0.0347
60	45	11.965	0.00564	0.99	0.1345	11.255	0.0614
60	55	12.954	0.00154	0.99	0.2359	8.789	0.0874

Table 5: Various regression coefficient and statistical parameters of Auara model for solute gain.

35, 45, 55°C, respectively, for osmotic solution of 50°Bx concentration. Therefore, with increase of temperature of osmotic solution, the values of solute gain at equilibrium have been increased. The values of equilibrium solute gain were 9.12, 12.86, 12.954 g/100 g of fresh sample in 40, 50, 60°Bx, respectively, at 55°C of osmotic solution temperature as predicted by Azuara model. Therefore, with increase of concentration of osmotic solution, the values of solute gain at equilibrium have been increased. The values of β_2 indicates that the rates of water loss were higher at higher concentrations and temperature in comprasion to the low values of concentration and temperature. It may be due to the fact that the low concentration of sugar syrup may get diluted and reach the near saturation point quickly. An increase in osmotic solution concentration increases the concentration gradient and in turn the driving force for osmotic dehydration process and high temperature decrease the resistance due to high viscosity by lowering down the viscosity of highly concentrated solution.

Conclusion

The osmotic solution concentration, temperature and time have significant effect on water loss and solute gain during osmotic dehydration of EFY cubes. The effect of process variables on water loss and solute gain can be represented by the model constants. Among different applied equations, Magee and Auara model showed the best fitting to the experimental data for water loss and solid gain, respectively. Therefore, the osmotic dehydration process of EFY cubes can be successfully represented by appropriate models for scale up purposes.

References

1. Ravi V, Ravindran CS, Suja G (2009) Growth and Productivity of Elephant Foot Yam (*Amorphophallus paeoniifolius* (Dennst. Nicolson): an Overview. J Root Crops 35: 131-142.

2. Mishra RS, Nedunchezhiyan M, Swam TMS, Edison S (2002) Mass

multiplication technique for producing quality planting material of *Amorphophallus paeoniifolius*. Trop Sci 34: 371-376.

3. Bradbury JH, Holloway WD (1988) Chemistry of Topical Root Crops: Significance for National and Agriculture in the Pacific. Australian Centre for International Agriculture Research 51-99.

4. Sreerag RS, Jayaprakas CA, Sajeev MS (2014) Physico-Chemical and Textural Changes in Elephant Foot Yam (*Amorphophallus paeoniifolius*) Tubers Infested by the Mealy Bug *Rhizoecus Amorphophalli* Betrem during Storage. J Post harvest Technol 02: 177-187.

5. García-Martínez E, Martínez-Monzo J, Camacho MM, Martínez-Navarrete N (2002) Osmotic Solution as Ingredient in New Product Formulation. Food Res Int 35: 307-312.

6. Corzo O, Bracho N (2005) Osmotic dehydration kinetics of sardine sheets using Zugarramurdi and Lupín model. J Food Eng 66: 51-56.

7. Rahman MS, Lamb J (1990) Osmotic dehydration of pineapple. J Food Sci Technol 27: 150-152.

8. Madamba PS (2003) Thin layer drying models for osmotically predried young coconut. Drying Technol 21: 1759-1780.

9. Rastogi NK, Raghavarao KSMS (2004) Mass transfer during osmotic dehydration of pineapple: Considering Fickian diffusion in cubical configuration. LWT Food Sci Technol 37: 43-47.

10. Rastogi NK, Raghavarao KSMS, Niranjan K, Knorr D (2002) Recent developments in osmotic dehydration: methods to enhance mass transfer. Trends Food Sci Technol 13: 58-69.

11. Panades G, Castro D, Chiralt A, Fito P, Nunez M, et al. (2008) Mass transfer mechanisms occurring in osmotic dehydration of guava. J Food Eng 87: 386-390

12. Peleg M (1988) An empirical model for the description of moisture sorption curves. J Food Sci 53: 1216-1219.

13. Azuara E, Beristain CI, Garcia HS (1992) Development of a mathematical model to predict kinetics of osmotic dehydration. J Food Sci Technol 29: 239-242.

14. Magee TRA, Murphy WR, Hassaballah AA (1983) Internal mass transfer during osmotic dehydration of apple slices in sugar solution. Irish J Food Sci Technol 7: 147-155.

15. Rahman MS (1992) Osmotic dehydration kinetics of food. Indian Food Ind 15: 20-24.

16. Kaymak-Ertekin F, Sultanoglu M (2001) Moisture sorption isotherm characteristics of peppers. J Food Eng 47: 225-231.

17. García-Pascual P, Sanjuan N, Melis R, Mulet A (2006) *Morchella esculenta* (morel) rehydration process modelling. J Food Eng 72: 346-353.

18. Singh B, KumarA, Gupta AK (2007) Study of mass transfer kinetics and effective diffusivity during osmotic dehydration of carrot cubes. J Food Eng 79: 471-480.

19. Khin MM, Zhou W, Perera CO (2006) A study of the mass transfer in osmotic dehydration of coated potato cubes. J Food Eng 77: 84-95.

20. Schmidt FC, Carciofi BAM, Laurindo JB (2009) Application of diffusive and empirical models to hydration dehydration and salt gain during osmotic treatment of chicken breast cuts. J Food Eng 91: 553-559.

21. Mercali GD, Marczak LDF, Tessaro IC, Norena CPZ (2010) Evaluation of water, sucrose and NaCl effective diffusivities during osmotic dehydration of banana (*Musa sapientum, shum.*). LWT - Food Sci Technol 2: 123-128.

22. Manivannan P, Rajasimman M (2008) Osmotic Dehydration of Beetroot in Salt Solution: Optimization of Parameters through Statistical Experimental Design. Int J Chem Biol Eng 1: 4.

23. Chopra CS (2001) Osmo-vacuum drying of carrots: Effect of salted syrup on drying behaviour and product quality. Beverage and Food World 15-17.

24. Singh S, Shivhare US, Ahmed J, Raghavan (1999) Osmotic concentration kinetics and quality of carrot preserve. Food Res Int 32: 509-514.

25. Azoubel PM, Murr FEX (2004) Mass transfer kinetics of osmotic dehydration of cherry tomato. J Food Eng 61: 291-295.

26. Azarpazhooh E, Ramaswamy HS (2010) Evaluation of diffusion and Azuara models for mass transfer kinetics during microwave-osmotic dehydration of apples under continuous flow medium-spray conditions. Drying Technol 28: 57-67.

27. Mundada M, Hathan BS, Maske S (2011) Mass transfer kinetics during osmotic dehydration of pomegranate arils. J Food Sci 76: 31-39.

21

Evaluating Physicochemical and Rheological Characteristics and Microbial Community Dynamics during the Natural Fermentation of Cassava Starch

Karine H Rebouças[1], Laidson P Gomes[1], Analy MO Leite[2], Thais M Uekane[1], Claudia M Rezende[1], Maria Ines B Tavares[3], Eveline L Almeida[4], Eduardo M Del Aguila[1] and Vania M Flosi Paschoalin[1*]

[1]Universidade Federal do Rio de Janeiro, Instituto de Química, Avenida Athos da Silveira Ramos, Cidade Universitária - Rio de Janeiro, Brazil

[2]Universidade Federal do Rio de Janeiro Campus Macaé. Rua Aloísio da Silva Gomes, Macaé-RJ, Brazil

[3]Universidade Federal do Rio de Janeiro, Instituto de Macromoléculas Professora Eloisa Mano, Brazil

[4]Universidade Federal do Rio de Janeiro, Escola de Química, Brazil

Abstract

The traditional fermentation of cassava starch was investigated by a polyphasic approach combining (i) microbial community identification using conventional and molecular techniques, (ii) analyses of organic acids, volatile compounds, fermentation products and spin-lattice relation time and (iii) evaluation of technological properties, such as pasting properties, water absorption and water solubility indexes. Cassava fermentation microbiota was dominated by bacteria and yeasts genera. Bacteria genera include *Lactobacillus*, *Leuconostoc*, *Lactococcus* and *Enterococcus*. *Lactobacillus* was the prevalent genera responsible for the acidification of cassava fermentation by the production of organic acids and also aromatic compounds. Yeast community was dynamically adjusted through the cassava fermentation *Pichia kudriavzevii* and *Issatchenkia orientalis* were succeeded by *Geotrichum candidum*, *Clavispora lusitaniae* and *Rhodotorula mucilaginosa*. *Candida rugosa*, *C. pararugosa*, *C. akabenensis*, *Cryptococcus albidus*, *Neurospora crassa* and *N. intermedia* were found exclusively in sour cassava. The acidification of sour cassava was due to the production of acetic, lactic and succinic acids. Volatile compounds, including aliphatic and aromatic hydrocarbons, esters and terpenes contribute to the aroma and correspond to 23% of compounds found after fermentation and sun-drying treatment. The acidification and fermentation process reduced the peak viscosity, paste viscosity, breakdown viscosity and set back viscosity in cassava starch. Solid-state NMR relaxometry measures were associated to the expansion ability and indicated that the fermented and sun-dried products were more inclined to expansion. Loaf expansion ability and pasting temperatures were increased in sour cassava (fermented and sun-dried). The results showed here should be useful to standardize the manufacturing of cassava starch in Brazil, providing homogeneous and high quality products.

Keywords: DNA sequencing; Headspace volatile analysis; Organic acids; Spin-lattice relaxation; Loaf expansion

Introduction

Cassava (*Manihot esculenta* Crantz), a wood scrub belonging to the *Euphorbiaceae* family (spurge), is considered an important source of food and dietary calories for large populations of tropical countries in Asia, Africa and Latin America [1]. Originally from Latin America, it is a shrubby plant, made up of a shoot and an underground portion. It is known as "tapioca" in Asian countries, as "mandioca", "aipim", "castelinha" and "macaxeira" in Brazil, as "yuca" in Spanish-speaking countries of Latin America, and as "manioc" in French-speaking countries in Africa [2]. In Brazil, cassava production is currently increasing and it is estimated that in the next 30 years the projected production should reach 106 million tons. Cassava starch has various applications in industry, such as in food, paper, and adhesives; however, only a small portion of starch is used in its native state, and mostly it is modified by chemical or physics agents. Cassava fermented and sun-dried starch or sour (sun-dried) cassava starch ("polvilho azedo" in Brazil or "almidón agrio" in Colombia) is used for the production of special types of gluten-free breads and biscuits that are very popular in some countries of South America [3]. Currently, a large number of individuals in many Western societies adopt a gluten-free diet, avoiding wheat, rye and barley. Although there are at least three clinical gluten-related conditions recognized-celiac disease, wheat allergy and non-celiac gluten sensitivity-most people change to a gluten-free diet even without any well-defined-medical reason [4]. The market for gluten-free products has been increasing speedily [5], opening opportunities for the development of new technologies using gluten-free ingredients as alternative for traditional manufacturing bakery products [6]. Sour (sun-dried) cassava starch can be used as an adjuvant for bread making or as the main ingredient for gluten-free breads.

Cassava starch fermentation is a common process conducted in small rural cassava starch factories to improve the textural qualities of the starch [7-9]. Natural fermentation is predominantly associated with the fermentative activities of bacteria and yeasts [10]. Cassava fermentation is carried out in tanks for a period of about 30-40 days. The wet acid starch is then sun-dried for a period, depending on the season, generating a non-uniform product. During fermentation considerable amounts of cyanide are removed and antimicrobial compounds are produced including bacteriocins, organic acids, hydrogen peroxide, and other active, low molecular weight metabolites [11].

Frequent variations occur in the quality of the final product from different producers and even from the same producer using raw material from the same origin. This occurs because there is no control parameters applied in the process. During manufacturing, sour cassava starch can be contaminated by unknown microorganisms, which may change the technological characteristics of the product [12]. The physicochemical properties determine the eating and cooking quality of sour cassava

*Corresponding author: Paschoalin F, Instituto de Química, Universidade Federal do Rio de Janeiro Av. Athos da Silveira Ramos 149 – Centro de Tecnologia – Bloco A - sala 545 Cidade Universitária - 21941-909, Rio de Janeiro, RJ, Brazil
E-mail: paschv@iq.ufrj.br

starch. A comprehensive analysis was performed using a polyphasic approach and multiple techniques that were simultaneously applied to describe dynamic changes in the physical, chemical, microbiological and rheological characteristics occurring during the natural fermentation of Brazilian cassava starch in the manufacturing of sour cassava starch. The dynamics of the microbial community involved in spontaneous fermentation was evaluated by conducting enumerations in specific culture media followed by the molecular identification of isolated microorganisms such as random amplified polymorphic DNA (RAPD) and partial DNA sequencing. Physical and chemical characteristics, including sugar, organic acid, volatile compound contents, fermentation product contents and the spin-lattice relaxation time were determined by high-performance liquid chromatography (HPLC), head space-solid phase micro-extraction coupled to a chromatograph quadrupole mass spectrometer (HS-SPME/GC-qMS) and low field NMR spectrometry (LF-NMR), respectively. Rheological properties, rapid visco analyzer (RVA) average parameters, water absorption and water solubility indexes were also determined. In view of the need for standardization and improvement of the Brazilian manufacturing process, knowledge and characterization of the manufacturing process and the final product can serve as a basis for planning and obtaining products with adequate physicochemical and pasting characteristics superior to the ones currently marketed in the country.

Materials and Methods

Sour cassava starch (fermented and sun-dried)

Cassava (*Manihot esculenta* Crantz) cultivated in the Paraná state, Southern Brazil, was processed into sour (sun-dried) cassava starch in accordance to the traditional small-scale processing of Northeastern Brazil. The cassava was washed, peeled and washed a second time for dirt removal. Subsequently, they were grated and the obtained mass was washed again and strained in fine mesh fabric until the water leaching from the cassava was transparent. The wet-extracted starch was sun-dried for 12h.

The sun-dried extracted starch was transferred to a sour cassava production plant located in the Bahia state, Northeastern Brazil. Natural fermentations were run from June to August 2013. The raw material was put in 1.63 m height polyethylene open tanks with 5,000 liters capacity and covered with a layer of running water of approximately 20 cm and allowed to naturally ferment for 30 days at ambient temperature (average temperature of 18°C). After fermentation, the cassava starch was laid on high density polyethylene (HDPE) black canvas and sun-dried for 12h. Cassava starch from several batches produced during June-August months in 2013 was sampled at ($t=1$) and ($t=30$) days alongside the fermentation process and from the final product, the sour (sun-dried) cassava starch, for further analyses.

Microorganism enumeration

Microorganism enumeration was carried out at the beginning of fermentation ($t=1$), after 30 days of fermentation ($t=30$) and in the final product, sour (sun-dried) cassava starch. For lactic acid bacteria enumeration and identification, M17 agar plates (HIMEDIA, Mumbai, Índia) were incubated at 30°C for 48 h, lactobacilli MRS agar plates (BD, Le Pont de Claix, France) at 37°C for 4 days under anaerobic jars using a Gaspak anaerobic generator (Becton Dickinson and Company, Franklin Lakes, USA) and Azide Blood Agar Base (Becton Dickinson and Company, Le Pont de Claix, France) at 37°C for 4 days. For yeast enumeration and isolation, malt extract agar (Himedia, Mumbai, India) and YPD 2% (2% peptone, 1% yeast extract and 2% glucose) were used,

incubated at 25°C for 7 days. Colonies with distinct morphologies were selected randomly and the cultures were stored in the corresponding isolation broth described above containing 20% glycerol, until further analyses.

Microorganism molecular identification

DNA templates from bacterial and yeast colonies were obtained as described by Sambrook and Russell (2001) [13] and quantified using the Qubit dsDNA HS kit (Invitrogen™, Grand Island, New York, USA). Partial amplification of the 16S rDNA, using the primer pair 27F (5′-AGAGTTTGATCCTGGCTCAG-3′) and 1512R (5′-ACGGCTACCTTGTTACGACT-3′) [14], was performed using Taq DNA polymerase (Invitrogen™, Grand Island, New York, USA) in a DNA thermocycler (MyCycler™, Bio-rad, Hercules, CA, USA).

The amplicons were digested with restriction enzymes *ApaI* and *XhoI* (Promega, Madison, USA) and *DdeI* (Fermentas, São Paulo, Brazil), following the manufacturer's instructions. The digestion profile was resolved by electrophoresis on 1.2% agarose gels, and electrophoresis was carried out in 1X TAE buffer for 70 min at 100 V and 200 mA. Gels were stained with GelRed (Biotium Inc., Hayward, CA, USA) diluted at 1:10,000 and documented under a MiniBis Pro UV light using the GelCapture software (DNR Bio-Imaging Systems, Hamisha, Israel). Representative profiles of each species observed in the amplified ribosomal DNA restriction analysis were selected for sequencing.

Yeast DNA templates were analyzed by RAPD-PCR using the primer EI1 (5'-CTG GCT TGG TGT ATG -3') [15]. The band profiles were resolved on 2% gels subsequently stained with GelRed (Biotium Inc., Hayward, CA) diluted at 1:10,000 and documented as described above.

The representative profiles were selected for sequencing and amplified with the primer pairs ITS1 (5′-TCCGTAGGTGAACCTGCGG-3′) and ITS4 (5′-TCCGTAGGTGAACCTGCGG-3′) [16]. PCR products from bacteria and yeast were purified by PCR DNA and the use of the Gel Band Purification kit (GE Healthcare Life Science Inc., Little Chalfont, Buckinghamshire, UK). The sequencing analysis of partial 16S rDNA gene and ITS region was accomplished with a 3130 sequencer (Applied Biosystems Inc., Tokyo, Japan) and subsequently used for identification of the bacteria and yeast, respectively. The identities of the sequences were determined by using the BLASTn algorithm at the GenBank database.

Physicochemical characterization of cassava starch

Moisture content evaluation: The moisture content of the analyzed starch was determined according to the protocol from the American Association of Cereal Chemists (AACC) International Approved Method 44-15.02.

Amylose and amylopectin content: The apparent amylose content was estimated in quadruplicate by iodine-based colorimetry according to the method number 66470 from the International Organization for Standardization. The absorbance was measured on a DU-730 spectrophotometer (Beckman Coulter, Fullerton, USA) at 620nm. The percentage content of amylopectin was estimated by difference (amylopectin percentage content = 100 - apparent amylose percentage content).

Determination of titratable acidity and pH: Titratable acidity and pH were determined according to the 016/IV e 017/IV methods, respectively, from the Adolfo Lutz Institute [17] using a calibrated potentiometer.

Organic acid content determination: The organic acid content of starch was determined as described by Leite et al. [18]. Briefly, 25 mL of H_2SO_4 45 mmol.L^{-1} were added to 5 g (dry weight) of starch and homogenized for 1 h on a rotatory shaker at 250 rpm. The supernatant resultant from of a centrifugation at 6,000 g was filtered through 0.45 μm filters (Millipore Corp, Billerica, USA). For carbohydrate analysis, the filtered samples were injected (50 μL) into an HPLC system (Shimadzu Corp., Tokyo, Japan) equipped with an HPX-87H Aminex fermentation monitoring column (150 × 7.8-mm i.d., Bio-Rad Laboratories Inc, Hercules, USA), hydrogen form, 9 μm particle size, 8% cross linkage, pH range 1-3, protected by a cation H$^+$ Micro-Guard cartridge (30 × 4.6-mm i.d.; Bio-Rad Laboratories Inc, Hercules, USA). The mobile phase (isocratic) was 3 mM H_2SO_4 at a flow rate of 0.7 mL.min^{-1} at 65°C. Organic acids (lactic, acetic, citric, succinic, butyric and propionic) were quantified by using a diode array detector model SPD-M20A (Shimadzu Corp, Tokyo, Japan), monitoring the absorbance at 210 nm [18]. Chromatograms from the HPLC and compound quantifications were obtained using the LC Solution software (Shimadzu Corp., Tokyo, Japan). Three independent samples each were collected at ($t=1$) and ($t=30$) days alongside the fermentation process. Standard curves based on peak area were calculated for the individual concentrations of the determined organic acids, covering a broad range of concentrations, by comparison with standard solutions. Standards (Supelco Analytical, Sigma, St Louis, MO, USA) were prepared in deionized water filtered through 0.45-μm filters (Millipore Corp.). The analyses were performed in triplicate.

Determination of volatile compounds: The volatile compounds present in the headspace samples were extracted using solid phase micro-extraction (SPME) for 30 min at room temperature using a three-phase fiber 50/30 μm DVB/CAR/PDMS into an Agilent 6890 gas chromatograph coupled to an Agilent 5973 N mass selective detector (GC/MS) and a DB-5 (30 m × 0.25 mm × 0.25 mm, J & W Scientific, Folsom, CA). Helium was used as carrier gas at a flow rate of 1.0 mL/min. The oven temperature was programmed from 50 to 250°C at 5°C/minute. Injector temperature was kept at 260°C. The mass detector was operated in an electronic ionization mode (70 eV) at 3.15 scan/s with a mass range from 30 μ to 550 μ. Transfer line was kept at 250°C, ion source at 230°C, and an analyzer at 150°C. The compounds were identified according to the Wiley mass spectrometer library (Enhanced data analysis software, Agilent, New York, USA) and the retention index data found in the literature [19]. Samples were analyzed in duplicate.

Determination of the spin-lattice relation time: The analyses were performed using a low field NMR spectrometer (LF-NMR) MARAN Ultra-23 (Oxford Instruments, Tokyo, Japan), operating at 23.4 MHz (for hydrogen) and equipped with an 18 mm variable temperature probe. Hydrogen spin-lattice relaxation times were determined directly by the traditional inversion-recovery pulse sequence (recycle delay - 180° - τ – 90°- acquisition). The 90° pulse, 4.6 μs, was calibrated automatically by the instrument software. The amplitude of the FID was sampled for twenty τ data point, ranging from 0.1 to 5,000 ms, with 4 scans each and 5s of recycle delay. The relaxation values (means of triplicate analysis) and relative intensities were obtained by fitting the exponential data with the aid of the WINFIT 2.4.0.0 software supplied from resonance. Distributed exponential fittings as plots of relaxation amplitude versus relaxation time were performed using the WINDXP software [20].

Technological properties of cassava starch and sour (fermented and sun-dried) cassava starch

Pasting properties: The pasting properties of starch were determined using a Rapid Visco Analyzer 4500 viscometer (Perten Instruments, Hägersten, Sweden) according to method No. 162 of the International Association for Cereal Science and Technology. Starch samples weighing 2.5 g were added to 25 mL distilled water (corrected volume considering 14% moisture content in flour) and the pasting temperature, peak viscosity, peak time, breakdown, minimum viscosity, final viscosity at 50°C and setback were evaluated. The analyses were performed in triplicate and mean values and standard deviations were calculated.

Water absorption index (WAI) and water solubility index (WSI) determination: The WAI and WSI analyses were performed in triplicate, following the method proposed by Anderson, Conway, Pfeifer, and Griffin [21].

Expansion power: The expansion power was determined as described by other researches with modifications [12]. Manual homogenization of 24 g of starch was carried out in 20 g of boiling water. Four portions of this dough, 7g each, were placed in aluminum containers (3.7 cm diameter and 6.8 cm height) and pre-heated in an electric oven at 150°C for 18 min. The expanded cassava starch samples were weighted and covered with Parafilm M®. The apparent volume was determined according to the AACC International Approved Method 10-05.01. The measurements were conducted in quadruplicate and the expansion was determined as the specific volume evaluated by the displacement method of millet seeds and expressed in mL.g^{-1} [22]. The specific volumes were used to classify the sour (fermented and sun-dried) cassava starch into small (<5.0 mL.g^{-1}), medium (5.0 ≤ x ≤ 10.0 mL.g^{-1}) and large (> 10.0 mL.g^{-1}) [12].

Statistical analyses

Data were expressed as means ± SD and the value significances were analyzed by the GraphPad Prism v.5 software package (San Diego, CA, EUA). Differences between means were compared by a one-way analysis of variance (ANOVA) with a Bonferroni post hoc test. A statistical significance level of 99.9% ($p<0.001$) was considered for all analyses.

Results and Discussion

Microbiological analyses of the spontaneous fermentation of cassava starch

At the initial ($t =1$) and final time points of the cassava starch fermentation ($t =30$), the presumptive count of lactobacilli in MRS agar showed values around 8 log units CFU.g^{-1}, whereas the final product, after sun-drying treatment, showed a reduction of 1.2 log units (Table 1). Previous studies have shown that the *Lactobacillus* genus is prevalent among the other genera belonging to lactic acid bacteria, being found throughout the manufacturing process of sour (sun-dried) cassava starch [23].

Regarding the presumptive count for streptococci on azide agar, values for cassava starch at the beginning and end of fermentation were also higher than in the sour (sun-dried) cassava starch, dropping from 7.8 ($t =30$) to 6.3 log units CFU.g^{-1} (Table 1). On the other hand, no *Lactococcus sp.* was detected in sour (fermented and sun-dried) cassava starch, probably because the number of microorganisms was below the limit of detection of the technique (10^2 CFU.g^{-1} log).

Sample (fermentation time)	Bacteria (Log CFU.g⁻¹)*			Yeast (Log CFU.g⁻¹)*	
	M17	MRS	Azide blood	YPD 2%	Malt extract
Cassava starch (*t=1*) (1st day)	7.6 ± 0.02[b]	8.0 ± 0.03[a]	7.7 ± 0.04[a]	7.9 ± 0.02[a]	7.8 ± 0.01[a]
Sour cassava starch (*t=30*) (30th day)	8.0 ± 0.02[a]	8.2 ± 0[a]	7.8 ± 0.04[a]	6.0 ± 0.02[c]	6.2 ± 0.02[b]
Sour (sun-dried) cassava starch	< 2[c]	6.8 ± 0.02[b]	6.3 ± 0.04[b]	7.5 ± 0.01[b]	7.6 ± 0.06[a]

*Values are expressed as mean ± standard deviation.
Bacteria were cultured on M17, MRS and azide blood agar plates. Yeast was cultured on YPD 2% and malt extract agar plates.
Samples were harvested at the 1st and 30th days of cassava fermentation and after the sun-dried treatment.
[a-c] Means with different superscript letters within a column are significantly different ($p < 0.001$), according to the Bonferroni post hoc test.

Table 1: Microbial enumeration (log CFU.g⁻¹) on selective media of lactic acid bacteria and yeast during cassava starch fermentation.

To identify bacteria, all microorganisms (n=131) isolated from the different batches and media (MRS, M17 and azide blood) were characterized by ARDRA (amplified ribosomal DNA restriction analysis). Eighteen distinct profiles were found and the isolates were identified by partial sequencing of the 16S rDNA. *Lactobacillus* sp. (42%), *Lactobacillus plantarum* (14.5%), *Leuconostoc citreum* (5.3%), *Lactococcus* sp. (12.2%), *Enterococcus* sp. (15.3%) and *Bacillus* sp. (10.7%) were identified based on homology identity (98-100%) searches at GenBank database.

Several studies have shown the involvement of LAB lactic acid bacteria in the spontaneous fermentation of sour cassava starch [23]. These bacteria contribute to the development of characteristic starch properties, such as taste, aroma, appearance, texture, shelf life and safety [10].

The involvement of various *Lactobacillus* species, mainly *L. plantarum* and *L. fermentum,* as the predominant species in sour (fermented and sun-dried) cassava starch manufacturing at two industrial plants in Southeastern Brazil has already been suggested by Lacerda et al. [10]. *L. plantarum* and other LAB are considered the prevalent microorganisms in the natural fermentation of cassava starch, responsible for the acidification of the product in the tanks during the fermentation process and the production of organic acids and aromatic compounds. Another group of LAB, the genus *Leuconostoc* has been identified since the beginning of the fermentation process (*t =1*). *Leuconostoc* sp has been usually isolated from fermented vegetables, including species that are capable of producing exopolysaccharides (EPS). This group has many applications in the food industry and pharmaceutical field [24].

Leuconostoc sp grow associatively with acid producing *Lactococcus* sp and can confer aroma and texture to sour (fermented) cassava starch. The associative growth between these two bacteria groups has been described as a synergistic functional relationship [24].

Bacteria belonging to the *Bacillus* genus were found only in samples collected at the end of the fermentation process (*t=30*). According to the European Food Safety Authority and ANVISA, RDC under number 263, 09/22/2005, Brazil's National Health Surveillance Agency, the presence of the *Bacillus* genus, Gram-positive, ubiquitous, characterized by spore-forming ability and usually present in soil, can be considered as a contamination indicator of fermentation since those opportunistic bacteria can grow due to inadequate hygiene conditions during the production process, which reinforces the need to include the concepts of good manufacturing practices in flour mills.

The yeast count, for both media used in the initial time point (*t=1*) of the cassava starch fermentation showed higher values (about 2 log units) than at the end of the fermentation process (*t =30*) (Table 1). On the other hand, in the sour (sun-dried) cassava starch showed an increase in yeast counts, reaching values of around 7.5 log units

CFU.g⁻¹ (Table 1). The presence of yeast species predominantly in the advanced stages of the process suggests a higher acid tolerance of these microorganisms [25], which can be considered a technological advantage.

A total of 157 yeasts isolated and analyzed by RAPD were grouped into 19 distinct profiles. The sequencing of the internal transcribed spacer of ribosomal DNA (rDNA ITS) of the 19 representative yeast RAPD profiles, presented homology 98% - 100% to the sequences of the GenBank. *Geotrichum candidum* (10.8%) *Pichia kudriavzevii* (33.1%), *Issatchenkia orientalis* (3.2%), *Clavispora lusitaniae* (6.4%), *Neurospora crassa* (1.9%), *Neurospora intermedia* (1.9%), *Rhodotorula mucilaginosa* (2.5%), *Cryptococcus albidus* (8.3%), *Candida akabenensis* (8.3%), *Candida pararugosa* (6.4%), *Candida rugosa* (4.5%) and *Geotrichum* sp. (12.7%) were identified during both the sour cassava fermentation and sun-drying treatment.

Microbial succession was more evident in the yeast community. Species such as *Pichia kudriavzevii* and *Issatchenkia orientalis* are part of natural cassava microbiota and were found only at the beginning of fermentation. *Geotrichum candidum, Clavispora lusitaniae* and *Rhodotorula mucilaginosa* were detected during cassava fermentation. *Candida rugosa, C. pararugosa, C. akabenensis, Cryptococcus albidus, Neurospora crassa* and *N. intermedia* were found in the final product, the sour cassava (fermented and sun-dried) starch.

It was found that the most common yeast species in sour (fermented and sun-dried) cassava starch when evaluating samples from distinct starches [10]. They identified *Galactomyces geotrichum* and a species of *Issatchenkia*, both of which were present throughout the process. Both species occurred at about 5.0 log CFU g⁻¹. The authors also commented that other species appeared only at certain times, especially in the initial phases of the process. However, none of the yeast isolated from the cassava fermentation were able to degrade starch.

In the present study, the identified yeast species have been shown to have high amylolytic activity, thus showing an ecological advantage in fermented sour and sun-dried cassava starch, since they can partially hydrolyze raw starch [26] to provide sugars such as glucose or maltose that can be used as an energy source by other microorganisms, as well as producing enzymes such as linamarase and polygalacturonase, and aldehydes and esters that impart a pleasant aroma to the final product [27].

Characterization and comparison of physical and chemical parameters of the fermentation musts

Moisture content: Cassava starch and sour (sun-dried) cassava starch presented significantly different mean moisture values ($p < 0.001$), 9.9% ± 0.02 to 12.9% ± 0.04, respectively.

According to the Brazilian Technical Regulation for starch products (ANVISA - RDC under number 263, 09/22/2005), samples of cassava starch and sour (sun-dried) cassava starch should not present significant differences regarding moisture content. The value recommended by Brazilian legislation, for both, is at most 18% w/w, in order to obtain good product preservation, stability, quality and appropriate composition of the final product. Although the 12-hour drying period was observed, the relative humidity in the area is 60%-90%, which may have hampered the drying of the final product. The moisture content of the products are within the limits recommended by the Brazilian legislation (ANVISA - RDC under number 263, 09/22/2005), similar to other commercial products available on the Brazilian market. Drying reduces moisture, volume and cyanide content of roots, thereby prolonging product shelf life. The higher moisture content would indicate that the process of drying the starch under the sun could not be completed within 12 h.

Apparent amylose and amylopectin content: No significant difference was observed between the apparent amylose and amylopectin content (p < 0.001) between the cassava starch and the sour cassava starch samples. Amylose content is an important parameter to study the changes that occur in starch pasting properties, which can affect their industrial applications [28]. Starch paste and thermal properties are negatively regulated by the amylose ratio produced in starchy vegetables [29]. The differences when comparing the results with other studies may be attributed to the use of starches from other sources, as well as the different methods used in the amylose analyses. A more careful and in-depth study demonstrated that amylose content can vary with cultivar conditions or crop planting time [3].

Determination of titratable acidity and pH of the flours during production: There was an increase in titratable acidity from 2.14 ± 0.01 mL of NaOH / 100g to 3.24 ± 0.24 mL NaOH / 100g and a reduction of pH from 4.8 ± 0.05 to 3.25 ± 0.03 for the initial timepoint (*t=1d*) and final (*t=30d*), respectively, while the final product, sour (sun-dried) cassava starch showed a titratable acidity of 5.25 ± 0.01 mL of NaOH / 100g and a pH of 3.63. According to the current legislation in Brazil (ANVISA – RDC under number 263, 09/22/2005), cassava starch should remain in the fermentation tank until the product reaches an acidity of about 5.0 mL NaOH / 100g of dry weight. However, in most starch flour in Brazil, the duration of the fermentation process is not controlled by the titratable acidity as a parameter to interrupt fermentation, being only the 30-day fermentation period observed. In the starch evaluated in the present study, there was no control of the endpoint of the fermentation process, with the sour cassava (fermented) starch submitted to sun-drying treatment, even with an acidity higher than recommended by the Brazilian regulatory agency.

Characterization of organic acids in cassava starch and sour (fermented and sun-dried) cassava starch: Among the organic acids, only acetic, lactic and succinic acids were produced during the fermentation and sun-drying of cassava starch. All of them were detected by HPLC (Figure 1), excepting acetic acid, which was evaluated by HS-SPMS/GC-qMS. The lactic acid concentrations found in the cassava starch and sour (sun-dried) cassava samples were of 0.15 g.L⁻¹ and 0.96 g L⁻¹, respectively. Succinic acid was also detected in the sour (sun-dried) cassava starch samples at a concentration of 0.084 g.L⁻¹. Lactic and succinic acid were detected by HPLC. Acetic acid comprised approximately 40% of the volatile compounds in the headspace of sour (sun-dried) cassava starch, as seen previously in Table 2 and Figure 2. The titratable acidity of sour (sun-dried) cassava starch could not be due to carboxyl groups resulting from residual acids due to degradation of amylose and amylopectin [30], since high molecular weight organic acids were not detected, neither were short-chain and long-chain fatty acids by HS-SPME/GC-qMS.

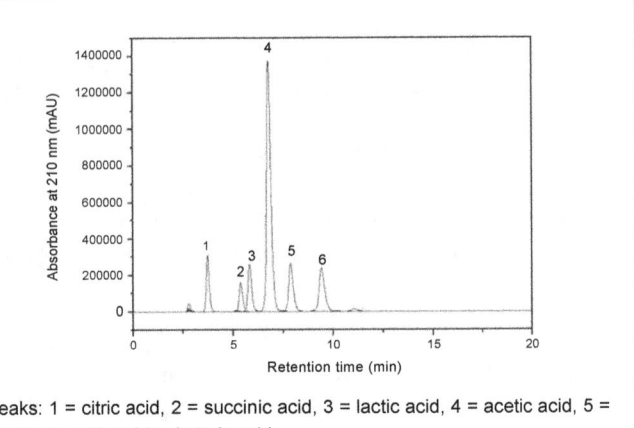

Peaks: 1 = citric acid, 2 = succinic acid, 3 = lactic acid, 4 = acetic acid, 5 = propionic acid and 6 = butyric acid.

Figure 1: Chromatogram from HPLC of organic acid standards.

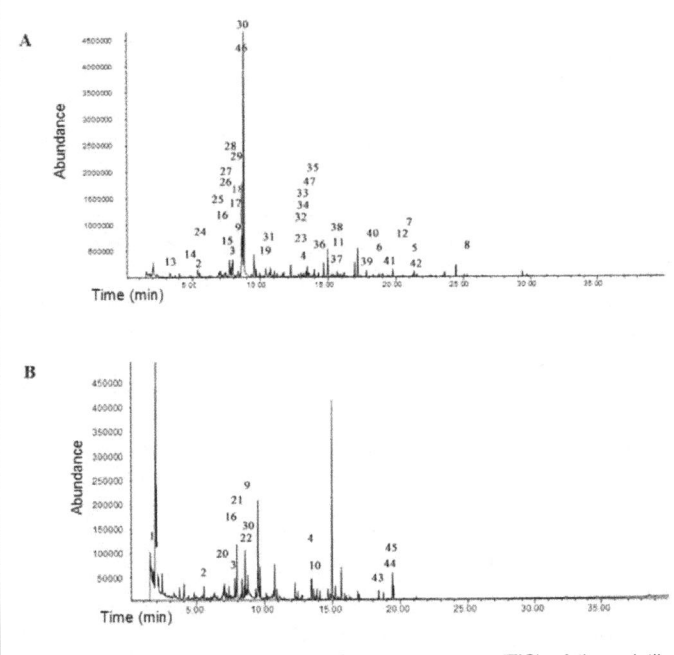

Figure 2: Typical HS-SPMS/GC-qMS chromatograms (TIC) of the volatile composition in the headspace of the cassava samples. A: Cassava starch; B:Sour (fermented and sun-dried) cassava starch.

Although in previous studies the increase in total acidity was ascribed to the production of organic acids, mainly lactic acid and substantial amounts of acetic and butyric acids [31]. The sour (fermented and sun-dried) cassava starch may have traces of propionic acid, without the butyric or propionic acid presence [32].

In regions with average temperatures around 18°C, fermentation is slow, with the predominance of lactic microbiota, mainly *Lactobacillus plantarum*, while in regions withaverage temperatures around 35°C, fermentation is quicker and butyric microbiota is predominant, mostly *Clostridium butyricum* [10].

Taken together, the data regarding pH, titratable acidity, acetic, lactic and succinic acid content and low room temperature indicate that fermentation was interrupted before completion during sour cassava starch manufacturing. Additionally, no *Clostridium butyricum* was

	Compound	Peak number	Cassava starch Relative area (%)	Sour (fermented and sun-dried) cassava starch Relative area (%)
Organic Acids	Acetic acid	1	-	51.2
Aliphatic hydrocarbons and ethers	n-Nonane	2	0.7	2.5
	n-Decane	3	3.5	9.3
	n-Dodecane	4	1.8	6.6
	n-Tridecane	5	0.1	-
	n-Tetradecane	6	0.5	-
	n-Pentadecane	7	0.3	-
	Di- N-octyl ether	8	0.3	-
	Nonane, 2,5-dimethyl-	9	0.6	2.0
	Octacosane	10	-	2.0
	Heptadecane, 8-methyl-	11	0.8	-
Ketone	2,6-ditert-butylcyclohexa-2,5-diene-1,4-dione	12	0.2	-
Aromatic hydrocarbons and aldehydes	Toluene	13	1.5	-
	m-Xylene	14	0.3	-
	1,3,5-Trimethylbenzene	15	1.5	-
	1,2,3-Trimethylbenzene	16	0.6	1.4
	1,2-Dichlorobenzene	17	0.3	-
	1-Methyl-3-(1-methylethyl)benzene	18	15.9	-
	1-Methyl-2-prop-1-en-2-ylbenzene	19	2.2	-
	1-Ethyl-3-methylbenzene	20	-	2.1
	1,2,4-Trimethylbenzene	21	-	3.4
	1-Ethyl-2,3-dimethylbenzene	22	-	0.7
	2-(4-Methylphenyl)propan-2-ol	23	0.7	-
	Styrene	24	1.1	-
	Benzaldehyde	25	0.9	-
Terpenes	Sabinene	26	0.4	-
	2-β-Pinene	27	1.0	-
	β-Myrcene	28	3.1	-
	δ-Carene	29	0.2	-
	Llimonene	30	44.6	8.6
	α-Terpinolene	31	1.1	-
	p-Mentha-1(7),8-dien-2-ol	32	0.4	-
	α-Terpineol	33	0.7	-
	Dihydrocarvone	34	0.8	-
	Trans-2-caren-4-ol	35	1.3	-
	Carvotanacetone	36	2.2	-
	p-Mentha-1,8-dien-7-al	37	0.3	-
	Carvacrol	38	0.3	-
	Neryl acetate	39	1.0	-
	Geranyl acetate	40	0.3	-
	α-Bergamotene	41	1.2	-
	β-Bisabolene	42	1.2	-
	β-Damascenone	43	-	1.5
	Caryophyllene	44	-	6.6
	α-Ionone	45	-	2.1
Alcohol	2-Ethyl-1-hexanol	46	5.1	-
Aldehyde	Decanal	47	1.0	-
TOTAL			100.0	100.0

The metabolites were all confirmed by comparison of retention times (Rfs) and mass spectra (MS) of authentic substances.
Arbitrary units (peak area) were used.
Mean values of duplicate identification.

Table 2: Volatile compounds identified in the "headspace" samples of cassava and sour (sun-dried) cassava starches by HS-SPME/GC-qMS (extraction temperature 25°C; extraction time, 30 min, using three-phase fiber 50/30μm DVB/CAR/PDMS).

found among the natural microbiota microorganisms. Succinic acid is a dicarboxylic acid produced as an intermediate of the tricarboxylic acid (TCA) cycle or as the major product of anaerobic fermentation by certain microorganisms [33]. Several yeast species found in the microbiota of sour cassava fermentation should contribute to succinic acid production during the oxidative metabolism of starch [34].

Characterization of volatile compounds and in cassava starch and sour (fermented and sun-dried) cassava starch: A higher number of distinct volatile compounds were found in the cassava starch in comparison to the sour (fermented and sun-dried) cassava starch (Figures 2A and 2B). This may be due to the sun-drying treatment, which may have caused the loss of some of the volatile compounds (Table 2). Aliphatic and aromatic hydrocarbons and terpenoidic

compounds were found in a higher amount in cassava starch than in the sour (fermented and sun-dried) cassava starch. Sour (fermented and sun-dried) cassava starch has acetic acid comprising 40% (Table 2) of the volatile compound, and seems to protect the natural fermentation against contamination by spoilage microorganisms. A reduction of the profile diversity of terpenes and aromatic hydrocarbons was observed.

Terpenoids are the main representative class in the volatile compounds in both starches (Table 2). They can be released by yeast α-glycosidases during the fermentation process, contributing to the aroma of the final product [35]. Limonene confers a fresh, citrus taste and odor, β-damascenone brings a sweet fruity smell and α-ionone, a tropical fruity and flowery smell which are all present in considerable amounts in the final product (http://theleafonline.

com/c/science/2014/09/terpene-profile-limonene), contributing to unique flavors, such as the pleasant aroma and taste of fermented cassava. Aromatic hydrocarbons had their relative content decreased in the sour (fermented and sun-dried) cassava starch and some of them like methylbenzene, 1-3-dimethylbenzene, 1,3,5-trimethylbenzene, 1,2-dichlorobenzene, 1-methyl-3-(1-methylethyl)benzene, 1-methyl-2-prop-1-en-2-ylbenzene, 2-(4-methylphenyl)propan-2-ol, styrene and benzaldehyde were removed by fermentation (Table 2).

Lactic acid bacteria and yeast present in cassava starch fermentation are able to produced volatile compounds formed via primary and secondary metabolism (MVOCs). Furthermore, moisture and temperature influence MVOC emission, and a prolonged growth phase due to a lower temperature may influence the production of certain compounds and extend the time for maximum production [36]. Other environmental factors such as substrate pH, light and CO_2 or O_2 levels can probably also influence the MVOC pattern. In the food industry, volatile organic acids are used as flavorings, preservatives and inhibitors of microbial growth.

H-NMR relaxometry information of cassava starch and sour (fermented and sun-dried) cassava starch: Solid-state NMR relaxometry allowed for the monitoring of the changes in cassava starch during fermentation and sun-drying processing, where the longitudinal relaxation time (T_1H) parameter provided information about the molecular dynamics of the starch before and after fermentation (Figure 3). The T_1H parameter showed a great variation in the proton spin-lattice relaxation times, from 83.3 to 69.3 ms, for cassava starch and sour (fermented and sun-dried) cassava starch, respectively. The shortest spin-lattice relaxation time T_1H ($p < 0.001$) indicates that the fermented and sun-dried product is more inclinable to expansion, since there is a population of hydrogen atoms in a lower confinement (or greater mobility) and greater heterogeneity in sour (fermented and sun-dried) cassava starch [20], as demonstrated by the line width and the peak intensity that changed during processing (Figure 3).

Technological properties of cassava starch and sour (sun-dried) cassava starch

Pasting properties: The sour (fermented and sun-dried) cassava starch presented a different pasting profile (p < 0.001) than cassava starch (Table 3). The sour cassava starch pastes presented lower viscosity at high temperature (lower peak viscosity), lower agitation stability (higher

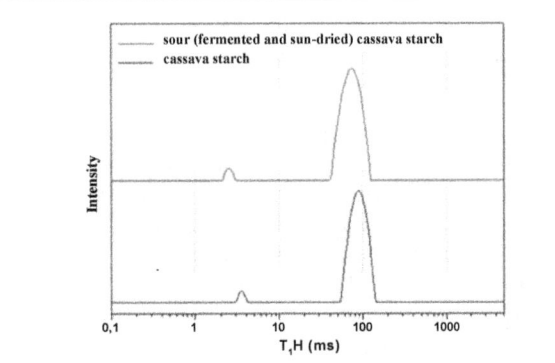

Figure 3: Distribution curves of domain relaxation times for cassava starch and sour (fermented and sun-dried) cassava starch obtained by LF-NMR. The relaxation values (means of triplicate analysis) and relative intensities were obtained by fitting the exponential data with the aid of the WINFIT 2.4.0.0 software supplied by resonance equipment.

Parameters	Cassava starch	Sour (sun-dried) cassava starch
Final viscosity at 50°C (RVU)*	172.5 ± 0.2[a]	55.8 ± 0.2[b]
Pasting temperature (°C)*	68.6 ± 0.1[ns]	70.2 ± 0.1[ns]
Peak viscosity (RVU)*	283.31 ± 3.1[a]	216.4 ± 2.4[b]
Peak time (min)*	3.7 ± 0[ns]	3.7 ± 0[ns]
Trough (RVU)*	123.3 ± 3.7[a]	37.3 ± 0.9[b]
Breakdown (RVU)*	160.0 ± 5.0[b]	179.2 ± 1.5[a]
Setback (RVU)*	49.2 ± 3.1[a]	18.5 ± 0.7[b]
WAI (g.g⁻¹)	2.0 ± 0.1[a]	2.1 ± 0[a]
WSI (%)	0.2 ± 0[a]	0.6 ± 0,1[a]
Expansion (mL.g⁻¹)	1.4 ± 0.4[b]	4.5 ± 0.4[a]
Amylose (%)	20.4 ± 0.4[a]	19.9 ± 0.3[a]
Amylopectin (%)	79.6 ± 0.4[a]	80.1 ± 0.3[a]

Means and standard deviation of triplicate measurements.
Different letters within a row are significantly different ($p < 0.001$), according to the Bonferroni post hoc test.
*RVA parameters
WAI: Water Absorption Index; WSI: Water Solubility Index; RVU: Rapid Visco Units.

Table 3: Technological characteristics of cassava starch and sour (sun-dried) cassava starch: pasting properties, water absorption, solubility indexes, amylose and amylopectin contents.

breakdown) and lower retrogradion tendency (lower setback) than the cassava starch. This cassava starch pasting profile alteration following fermentation and sun-drying treatment has also been observed in other studies [12,31,37]. Photochemical and enzymatic modifications occur during cassava starch manufacturing [37]. Starch molecules (amylose and amylopectin) in the amorphous regions of the granules were partially depolymerized by the amylolytic enzymes and organic acids produced by microorganisms from the natural environment and by UV irradiation (mainly UVB and UVC irradiation) during sun drying to the size-reduced starch molecules [37,38]. Besides depolymerization, starch molecules present carbonyl and carboxylate groups, which indicate that oxidation of the amylose and amylopectin hydroxyl groups is observed, in a mechanism involving free radicals [12]. The weakened granule organization caused by oxidative depolymerization during sour cassava starch fermentation [39] makes the starch granules show little swelling and, therefore, lower peak viscosity. The more weakened granule organization also causes the starch granules to readily disintegrate, which leads to higher breakdown. After being oxidatively depolymerized, amylose and amylopectin presented lower molecular weights, and a lower retrogradation tendency is observed (lower setback).

Water Absorption Index (WAI) and Water Solubility Index (WSI): Cassava starch and sour cassava starch (fermented and sun-dried) showed similar water absorption and water solubility index values, indicating that there were no effects following fermentation and sun drying (Table 3).

The water absorption index (WAI) is related to the degree of starch swelling or gelatinization. As the WAI was performed at 30°C, starches were not swollen because they did not reach the minimum energy required for the gelation process. Neither the cassava starch nor the sour cassava starch showed significant water uptake at this temperature (Table 3). The oxidative depolymerization of sour cassava starch was not efficient enough to prompt starch granules solubility in cold water, not assisting them to their pre-gelatinization state, since only pre-gelatinized starch granules can absorb water at ambient temperature, increasing starch viscosity [40]. The water solubility index (WSI) is a parameter that measures the total degree of degradation of the starch granule. Among the changes observed during the fermentation process of sour cassava starch (fermented and sun-dried), an increase in starch solubility is expected [37]. Herein, a significant increase in starch solubility was observed for sour cassava starch (fermented and sun-dried). Variations in the quality of the final product of a same producer are frequent due to the lack of control parameters in sour cassava starch processing. Therefore, the search for improvements in the technological process to obtain a final product with better quality and standardization is a major challenge for the sector. Generally, sour (sun-dried) cassava starch has a higher WSI value than its native starches. Studies have shown that cassava starch shows lower solubility than sour cassava starch, due to the presence of non-solubilized amylose in the crystalline region of the native granule, while the amylose of the fermented starch is already partially released [41].

Expansion properties: The bread making ability of both starches is represented by the loaf expansion values displayed in Table 3. There was a significant 3.2 fold increase in loaf expansion after starch fermentation and sun-drying treatment, although the production of organic acids during fermentation was discrete, as discussed previously, but still enough to promote physical changes in the granules, enhancing their ability to swell and solubilize in water [42].

The hydrolysis of the glycosidic bonds in the amorphous region of the granules by acids, enzymes and UV irradiation resulted in increased mobility and greater heterogeneity of the hydrogen molecules (relaxometry analysis), which probably results in the development of the expansion property. When comparing the specific volumes obtained to the loaf expansion indexes established [41], it can be observed that both starches showed low loaf expansion indexes, lower than 5 mL.g⁻¹ (Table 3). However, the loaf expansion values shown herein are similar to those observed for distinct genetic varieties of cassava [3]. A superior loaf expansion can be obtained, where a specific volume between 5 and 10 mL g⁻¹ can be reached after 83 days of fermentation [32].

The loaf expansion of cassava starch may be due to its high swelling capacity and solubility resulting from molecular degradation after acidification and irradiation [42]. Maximizing the expansion can depend on the degree of sour starch polymerization, the number of carboxyl and hydroxyl groups, pH, granule density, and other parameters that show significant correlation, whether positive or negative, with the expansion of the dough and its characteristics after cooking, as well as with its storage after cooking [37]. Although not fully established, the mechanism of sour cassava starch expansion may be similar to the one for extruded products, where the driving force would be water evaporation, and cell expansion would be governed mainly by dough-crust viscosity [41]. Partial depolymerization of sour cassava starch during fermentation and sun-drying provided small linear fragments and facilitated the development of an amorphous matrix structure of starch dough [38] reducing dough viscosity during expansion, aiding in the bubble expansion. However, other phenomena besides depolymerization could improve loaf expansion at different baking stages, including mass transfers, such as CO_2 or water displacement from the surrounding matrix to the expanding bubbles, inertia and surface tension [3].

Conclusions

Sour (sun-dried) cassava starch has great potential as a more economical and sustainable alternative to wheat flour in gluten-free bread production around the world. The results showed herein can be applied in order to supply the market with high-quality and homogeneous sour cassava starch. To control of several parameters in

the manufacturing process is necessary to increase the efficiency of the sour (sun-dried) cassava starch production process. The fermentation lifetime should be controlled by the physical and chemical parameters that can be an indicator of the quality of the final product. Formation of organic acids and volatile compounds as terpenoids and esters, and molecular relaxometry measures should be evaluated in order to achieve technological properties, which guarantee optimum quality of expansion of the final product and desirable flavor. Furthermore, changes in the ambient temperature and relative humidity should be minimized, since they may affect the efficiency of the fermentation process. The aim is to achieve ideal viscosity parameters, since this is the major quality technological importance that defines the acceptance and application of the product in the food industry, mainly associated with the production of cheese bread, where a mixture of cassava starch and sour cassava starch are associated with cured cheese powder, resulting in an appreciated and widely consumed product throughout the country.

Acknowledgement

The authors acknowledge the financial support from CNPq (Conselho Nacional de Desenvolvimento Científico e Tecnológico), CAPES (Coordenação de Aperfeiçoamento de Pessoal de Nível Superior), FAPERJ (Fundação Carlos Chagas Filho de Amparo à Pesquisa do Estado do Rio de Janeiro) and FAPESB (Fundação de Amparo à Pesquisa do Estado da Bahia). We also thank MSc Andrea Mattos and Dr Patrícia Ribeiro Pereira for fruitful discussions.

References

1. Olsen K, Schaal B (2001) Microsatellite variation in cassava (Manihot esculenta, Euphorbiaceae) and its wild relatives: further evidence for a southern Amazonian origin of domestication. Am J Bot 88: 131-142.

2. Pandey A, Soccol CR, Nigam P, Soccol VT (2000) Biotechnological potential of agro-industrial residues. II: Cassava bagasse. Bioresource Technol 74: 81-87.

3. Alvarado PM, Grosmaire L, Dufour D, Toro AG, Sánchez T, et al. (2013) Combined effect of fermentation, sun-drying and genotype on bread making ability of sour cassava starch. Carbohydr Polym 98: 1137-1146.

4. Lundin KE (2014) Non-celiac gluten sensitivity - why worry? BMC Med 12: 86.

5. Gallagher E, Gormley TR, Arendt EK (2004) Recent advances in the formulation of gluten-free cereal-based products. Trends Food Sci Tech 15: 143-152.

6. Schober TJ (2009) Manufacture of gluten-free speciality breads and confectionery products. In E. Gallagher (ed.), Gluten-free food science and technology (pp. 130-180). Oxford: Wiley-Blackwell.

7. Oluwasola O (2010) Stimulating rural employment and income for cassava (Manihot sp.) processing farming households in Oyo State, Nigeria through policy initiatives. J Dev Agric Econ 2: 18-25.

8. Lacerda IC, Miranda RL, Borelli BM, Nunes AC, Nardi RM, et al. (2005) Lactic acid bacteria and yeasts associated with spontaneous fermentations during the production of sour cassava starch in Brazil. Int J Food Microbiol 105: 213-219.

9. Holzapfel WH (2002) Appropriate starter culture technologies for small-scale fermentation in developing countries. Int J Food Microbiol 75: 197-212.

10. Demiate IM, Dupuy N, Huvenne JP, Cereda MP, Wosiacki G, et al. (2000) Relationship between baking behavior of modified cassava starches and starch chemical structure determined by FTIR spectroscopy. Carbohydr Polym 42: 149-158.

11. Sambrook J, Russel DW (2001) Molecular Cloning: A Laboratory Manual. (3rdedn), Cold Spring Harbor Laboratory Press, New York.

12. Wang X, Haruta S, Wang P, Ishii M, Igarashi Y, et al. (2006) Diversity of a stable enrichment culture which is useful for silage inoculant and its succession in alfalfa silage. FEMS Microbiol Ecol 57: 106-115.

13. de Barros Lopes M, Soden A, Henschke PA, Langridge P (1996) PCR differentiation of commercial yeast strains using intron splice site primers. Appl Environ Microbiol 62: 4514-4520.

14. Naumova ES, Ivannikova IuV, Naumov GI (2005) Genetic differentiation of the sherry yeasts Saccharomyces cerevisiae. Prikl Biokhim Mikrobiol 41: 656-661.

15. Adolfo Lutz Institute, (2008) Analytical Standards Institute Adolfo Lutz. (3rdedn), Sao Paulo.

16. Leite AM, Leite DC, Del Aguila EM, Alvares TS, Peixoto RS, et al. (2013) Microbiological and chemical characteristics of Brazilian kefir during fermentation and storage processes. J Dairy Sci 96: 4149-4159.

17. Adams RP (1995) Identification of essential oil components by gas chromatography mass spectroscopy. Illinois: Allured Publishing Corporation.

18. Brito LM, Tavares MI (2012) Evaluation of the influence of nanoparticles' shapes on the formation of poly (lactic acid) nanocomposites obtained employing the solution method. J Nanosci Nanotechnol 12: 4508-4513.

19. Anderson RA, Conway HF, Pfeifer VF, Griffin EL (1969) Gelatinization of corn grits by roll- and extrusion-cooking. Cereal Sci Today 14: 47.

20. Faubion JM, Hoseney RC (1982) High-temperature short time extrusioncooking of wheat starch and flour. I. Effect of moisture and flour type on extrudate properties. Cereal Chem 59: 529-533.

21. Ampe F, Sirvent A, Zakhia N (2001) Dynamics of the microbial community responsible for traditional sour cassava starch fermentation studied by denaturing gradient gel electrophoresis and quantitative rRNA hybridization. Int J Food Microbiol 65: 45-54.

22. Khue NT, Ngoc NH (2013) Exopolysaccharide in Lactobacillus rhamnosus Pn04 after co-culture with Leuconostoc mesenteroides Vtcc-B-643. JPAS 3: 14-17.

23. Halm M, Hornbaek T, Arneborg N, Sefa-Dedeh S, Jespersen L (2004) Lactic acid tolerance determined by measurement of intracellular pH of single cells of Candida krusei and Saccharomyces cerevisiae isolated from fermented maize dough. Int J Food Microbiol 94: 97-103

24. Rodriguez-Sanoja R, Morlon-Guyot J, Jore J, Pintado J, Juge N, et al. (2000) Comparative characterization of complete and truncated forms of Lactobacillus amylovorus a-amylase and role of the C-terminal direct repeats in raw-starch binding. Appl Environ Microbiol 66: 3350-3356.

25. Oyewole OB (2001) Characteristics and significance of yeasts' involvement in cassava fermentation for 'fufu' production. Int J Food Microbiol 65: 213-218.

26. Ascheri DPR, Boêno JA, Bassinello PZ, Ascheri JLR (2012) Correlation between grain nutritional content and pasting properties of pre-gelatinized red rice flour. Revista Ceres 59: 16-24.

27. Luo J, Jobling SA, Millar A, Morell MK, Li Z, et al. (2015) Allelic effects on starch structure and properties of six starch biosynthetic genes in a rice recombinant inbred line population. Rice (NY) 8: 15.

28. Silva GO, Takizawa FF, Pedroso RA, Franco CML, Leonel M, et al. (2006) Physicochemical characteristics of modified food starches commercialized in Brazil. Food Sci Technol 26: 188-197.

29. Adegunwa MO, Sanni LO, Maziya-Dixon B (2011) Effects of fermentation length and varieties on the pasting properties of sour cassava starch. Afr J Biotechnol 10: 8428-8433.

30. Aquino ACMS, Pereira JM, Watanable LB, Amante ER (2013) Standardization of the sour cassava starch reduces the processing time by fermentation water monitoring. Int J Food Sci Tech 48: 1892-1898.

31. Lee PC, Lee WG, Kwon S, Lee SY, Chang HN, et al. (2000) Batch and continuous cultivation of Anaerobiospirillum succiniciproducens for the production of succinic acid from whey. Appl Microbiol Biotechnol 54: 23-27.

32. Chaves-López C, Serio A, Grande-Tovar CD, Cuervo-Mulet R, Delgado-Ospina J, et al. (2014) Traditional fermented foods and beverages from a microbiological and nutritional perspective: The Colombian heritage. Compr Rev Food Sci F 13: 1031-1048.

33. Calleja A, Falqué E (2005) Volatile composition of Mencia wines. Food Chem 90: 357-363.

34. Sunesson AL, Nilsson CA, Carlson R, Blomquist G, Andersson B (1997) Influence of temperature, oxygen and carbon dioxide levels on the production of volatile metabolites from Streptomyces albidoflavus cultivated on gypsum board and tryptone glucose extract agar. Ann Occup Hyg 41: 393-413.

35. Marcon MJA, Vieira GCN, Simas KN, Santos K, Vieira MA, et al. (2007) Effect of the improved fermentation on physicochemical properties and sensorial acceptability of sour cassava starch. Braz Arch Biol Technol 50: 1073-1081.

36. Vatanasuchart N, Naivikul O, Charoenrein S, Srirot K (2005) Molecular properties of cassava starch modified with different UV irradiations to enhance baking expansion. Carbohyd Polym 61: 80-87.

37. Putri WDR, Marseno HDW, Cahyanto MNC (2012) Role of lactic acid bacteria on structural and physicochemical properties of sour cassava starch. APCBEE Procedia 2: 104-109.

38. Silva PA, Assis GT (2011) Development and characterization of an extruded breakfast cereal from cassava enriched with milk whey protein concentrate. Braz J Food Technol 14: 260-266.

39. Gomes AMM, Silva CEM, Ricardo NMPS (2005) Effects of annealing on the physicochemical properties of fermented cassava starch (polvilho azedo). Carbohyd Polym 60: 1-6.

40. Dias ARG, Zavareze ER, Elias MC, Helbig E, Silva DO, et al. (2011) Pasting, expansion and textural properties of fermented cassava starch oxidized with sodium hypochlorite. Carbohydr Polym 84: 268-275.

41. Nunes OLG, Cereda MP (1994) Effects of drying process on the development of expantion in Cassava starch hydrolyzed by lactic acid.

42. Bertolini AC, Mestres C, Lourdin D, Valle GD, Colonna P (2001) Relationship between thermomechanical properties and baking expansion of sour cassava starch (Polvilho Azedo). J Sci Food Agr 81: 429-435.

22

Effect of Process Variables on the Chemical Constituents and Sensory Characteristics of Nigerian Green Tea

Odunmbaku LA[1]*, Babajide JM[2], Shittu TA[2], Aroyeun SO[3] and Eromosele CO[4]

[1]*Food Technology Department, Moshood Abiola Polytechnic, Abeokuta, Nigeria*
[2]*Food Science and Technology Department, Federal University of Agriculture, Abeokuta, Nigeria*
[3]*Cocoa Research Institute of Nigeria, Ibadan, Nigeria*
[4]*Chemistry Department, Federal University of Agriculture, Abeokuta, Nigeria*

Abstract

Green tea possesses functional properties with attendant health benefits, Nigerian tea leaves has only been commercially processed into black tea. This study therefore evaluated the effect of Steaming Time (ST), Drying Temperature (DT) and Drying time (Dt) on the chemical and sensory properties of Nigerian Green Tea (NGT). Epical bud and two leaves from agronomic proven commercially viable clone were harvested from the Cocoa Research Institute of Nigeria experimental tea plots, Taraba State. Response surface methodology (Central Composite Design) was used to combine the three processing variables: ST (60, 90 and 120 s), DT (60, 65 and 70ºC) and Dt (90, 120 and 150 min). Epigallocatechin gallate (EGCG), Epigallocatechin (EGC), Epicatechin gallate (ECG) and Epicatechin (EC) contents of NGT were determined using High Performance Liquid Chromatography while descriptive sensory evaluation of NGT samples was carried out using semi trained panellist. Data generated were subjected to ANOVA and regression analysis. Results showed that NGT contain EGCG, EGC, EC and ECG contents that ranged from 46.90 to 178, 0.30 to 4.24, 1.03 to 8.83, and 8.05 to 33.96 (mg/g), respectively. Greenness, sweetness, bitterness, and astringency score of NGT extracts were 4.00-6.00, 1.00-2.23, 5.07-7.97 and 1.00-2.23 respectively on a 1-9 intensity scale. This study revealed that acceptable green tea can be obtained from Nigerian tea leaves in terms of chemical constituents, especially for the high EGCG content. The optimum process conditions for NGT were steaming for 60 s and drying at 70ºC for 150 min for high EGCG content and sensory acceptability.

Keywords: Camellia; Catechin; Optimization; Polyphenol; Tea

Introduction

Teas are classified into three major types depending on the manufacturing process: 'non-fermented' green tea; 'semi-fermented' oolong tea and 'fermented' black tea [1,2]. Green tea has long been exclusively consumed by the Chinese and Japanese, with cultural ties dating to the first millennium A.D. and, in particular, to tea ceremonies from the twelfth century. Gradually, green tea has received global acceptance due in part to the special health characteristics that have become more widely known through extensive scientific study.

Green tea has been considered a medicine and a healthful beverage since the olden days. The traditional Chinese medicine do recommend tea plant for headaches, body aches and pains, digestion, depression, detoxification, as an energizer and in general, longevity of life.

Several studies have established that, within each category of tea, differences in characteristics exist due to factors such as differences in the processing methods, stage of maturity of tea leaves at harvest, type of tree species, and the region where the tea was cultivated [3-5].

Tea importation to Nigeria has continued to increase steadily since the year 2003. The growth according to the Food and Agricultural Organisation was not unconnected with the perceived health benefit of tea consumption and as such, there is a need to develop the indigenous tea industry. It is equally important to translate the agronomical efforts on tea to more viable and economical ends. The objective of this study therefore is to determine the effect of steaming time, drying temperature and drying time on the chemical and sensory qualities of Nigerian Green tea.

Materials and Methods

Tea shoots sampling

Tea shoots comprising of apical bud and two leaves from agronomic proven commercially viable tea clone were harvest from the Cocoa Research Institute of Nigeria experimental tea plots, Taraba State, Nigeria. The plucked shoots were steamed, rolled and oven dried to make green tea.

Product optimization

Surface response methodology using Central Composite Design (CCD) at three levels, three variables was adopted in optimizing the process variables to obtain 15 experimental runs. The three independent variables experimented were; the steaming time, (60, 90 and 120 s), drying temperature (60, 65 and 70ºC) and drying time (90, 120 and 150 min). The first eight treatment combinations form a 2^3 factorial design. The next six treatment combinations are referred to as the axial runs, because they lie on the axes defined by the design variables. The last treatment combination represents the centre run and this arrangement of CCD as shown in Table 1 is in such a way that allows the development of the appropriate empirical equations (i.e., the second-order polynomial multiple regression equation) [6]. The model for predicting the quality of the green tea was expressed as:

$$Y_i = \beta_0 + \beta_1 X_1 + \beta_2 X_2 + \beta_3 X_3 + \beta_{12} X_{12} + \beta_{13} X_{13} + \beta_{23} X_{23} + \beta_1^2 X_1^2 + \beta_2^2 X_2^2 + \beta_3^2 X_3^2 + \varepsilon_i$$

*Corresonding author: Odunmbaku LA, Food Technology Department, Moshood Abiola Polytechnic, Abeokuta, Nigeria
E-mail: ollypo2000@yahoo.com

Chemical standard

Epigallocatechin gallate, epigallocatechin, epicatechin gallate and epicatechin standards were sourced from Sigma Aldrich Chemical Co. (USA); HPLC grade acetonitrile, ethyl acetate and methanol (Merck, Germany).

Equipment

Chromatography Acrodisc Syringe membrane Filters (0.45 μm, 30 mm diameter), Agilent (Germany); Water Distiller, AC-L4 Model, Optic Ivymen System, Europe; Vortex Mixer; KMC-1300 v Model, Vision scientific Co. Ltd., Korea; Syringes (1, 2 and 5 mL), Agary; Micro-pipette (200 and 1000 μL) Gilson, France; 5 mL Plain sample Bottles (Polypropylene); Porcelain Mortar and Pestle. Volumetric Flasks (10, 500 and 1000 mL), Borosilicate, Technico, England. Beakers (250, 500 and 1000 mL), Borosilicate, NAFCO, Nigeria; Digital Analytical Weighing Balance (Metler Toledo Instrument Company), North and South America; High Pressure Liquid Chromatography (HPLC), Agilent Technologies 1120 LC Compact series (Agilent), Germany and Japan. The system comprises a UV-Vis detector and an HP computer system. The HPLC column used was a ZORBAX SB C8

(75 × 4.6 mm, 3.5 μm) from (Hewlet Packard, HP). Data acquisition was done with Chemstation Software.

Determination of chemical constituents

High-Performance Liquid Chromatography (HPLC) methods was used to determine the amount of epigallocatechin gallate (EGCG), epigallocatechin (EGC), epicatechin gallate (ECG) and epicatechin (EC) present in the green tea [7,8]. Grounded green tea samples were extracted using Methanol and distilled water (95:5%) for 40 min at room temperature [9]. Solutions were filtered using a Millipore filter of 0.45 μm size. Filtered samples were filled into a vial bottles and 20 μl each programmed for injection twice per sample for HPLC auto sampler analysis. Detection was carried out by measurement of UV absorbance at 270 nm. Stock solutions of EGCG, EGC, ECG and EC were prepared by dissolving reference standards into mobile phase. Less concentrated solutions were prepared, as required, by dilution in the same mobile phase.

Sensory evaluation

Sensory attributes of the green tea infusion (Colour, Clearness, Dry Leaf Aroma, Green Tea Aroma, Sweetness, Bitterness, Green Tea Flavour and Astringency) were evaluated by semi trained panel (n=25) using descriptive analysis [10]. Attribute intensities were rated on a scale of 1-9 with 1 representing lowest intensity value and 9 representing highest intensity value. The samples were presented to the panellists in random order as coded samples out of tea cups covered by lids. This was prepared 30 min before evaluation. Water was provided for cleansing the palate between the samples.

Data analysis

The results were subjected to analysis of variance using SPSS 16b statistical package. The means were used to calculate linear regressions between individual Catechin and sensory evaluations responses using MATLAB R2012a.

Results

As presented in Table 2, the EGCG content of green tea samples ranged from 46.90-178 mg/g, EGC, 0.30-4.24 mg/g; EC, 1.03-8.83 mg/g and ECG, 8.05-33.96 mg/g. Table 3 showed the regression model result of the expected response against the observed responses. The

	ST (sec)	DT (°C)	Dt (min)
1	-1	-1	-1
2	+1	-1	-1
3	-1	+1	-1
4	+1	+1	-1
5	-1	-1	+1
6	+1	-1	+1
7	-1	+1	+1
8	+1	+1	+1
9	-1	0	0
10	+1	0	0
11	0	-1	0
12	0	+1	0
13	0	0	-1
14	0	0	+1
15	0	0	0

ST; steaming time, DT; Drying temperature, Dt; Drying time

Table 1: Experimental Runs of Optimized green tea sample.

Sample	ST time (Sec)	D T (°C)	D time (Min)	EGCG	EGC	EC	ECG
621	60	60	90	106.75 ± 0.00[d]	0.95 ± 0.00[de]	3.08 ± 0.40[ef]	11.29 ± 0.50[h]
738	60	70	90	112.45 ± 1.20[c]	0.62 ± 0.30[e]	2.21 ± 0.18[f]	10.64 ± 0.64[h]
926	60	60	150	145.67 ± 0.68[b]	0.30 ± 0.06[e]	2.10 ± 0.17[f]	8.60 ± 0.24[i]
194	60	70	150	178.06 ± 0.17[a]	1.03 ± 0.13[de]	1.03 ± 0.41[g]	8.05 ± 0.19[i]
531	60	65	120	112.05 ± 0.49[c]	0.79 ± 0.00[de]	2.61 ± 0.94[ef]	11.11 ± 0.16[h]
756	90	60	120	77.25 ± 0.24[i]	2.82 ± 0.93[bc]	6.10 ± 0.49[c]	17.99 ± 1.01[d]
573	90	70	120	91.94 ± 1.01[f]	1.88 ± 1.01[cd]	3.36 ± 0.07[e]	14.15 ± 0.31[f]
980	90	65	90	96.95 ± 0.82[e]	0.95 ±0.02[de]	3.27 ± 0.76[e]	12.77 ± 0.50[g]
292	90	65	150	90.74 ± 0.49[g]	2.46 ±0.36[c]	4.49 ± 0.31[d]	14.26 ± 0.76[f]
430	90	65	120	80.07 ± 0.14[h]	2.50 ±0.50[c]	4.71 ± 0.17[d]	16.13 ± 0.11[e]
658	120	60	90	46.90 ± 0.17[m]	3.88 ± 1.06[ab]	1.04 ± 0.05[a]	33.96 ± 0.09[a]
271	120	70	90	72.51 ± 0.59[j]	3.68 ±0.16[ab]	6.11 ± 0.68[c]	19.72 ± 1.26[c]
321	120	60	150	50.68 ± 0.00[l]	4.24 ±0.17[a]	8.83 ± 0.00[b]	24.32 ± 0.13[b]
564	120	70	150	72.22 ± 0.16[j]	3.72 ±0.24[ab]	6.39 ± 0.50[c]	23.55 ± 0.44[b]
250	120	65	120	70.05 ± 0.03[k]	3.99 ±0.38[a]	8.40 ± 0.33[b]	23.78 ± 0.34[b]

Values in the same column with same superscript are not significantly different at P≤ 0.05.

Table 2: Green tea Polyphenol content (mg/g).

ST (Sec)	DT (ºC)	Dt (min)	EGCG Measured	EGCG Predicted	EGC Measured	EGC Predicted	EC Measured	EC Predicted	ECG Measured	ECG Predicted	MSE
90	60	120	77.25	73.75	2.82	2.56	6.10	5.90	18.00	18.30	2.52
120	60	90	46.90	58.53	3.88	3.98	10.37	9.97	33.96	31.37	28.45
60	60	90	106.75	103.94	0.95	0.70	3.08	3.18	11.29	12.73	2.06
60	70	90	112.45	115.76	0.62	0.54	2.21	1.68	10.64	8.82	3.11
120	70	90	72.51	74.83	3.68	3.25	6.11	6.09	19.72	20.57	1.29
60	60	150	145.67	143.84	0.30	0.68	2.10	2.11	8.60	7.63	0.92
120	60	150	50.68	47.86	4.24	4.29	8.83	9.34	24.32	26.00	2.39
90	65	150	90.74	102.91	2.46	1.96	4.49	3.69	14.26	12.85	30.21
60	65	120	112.05	124.19	0.79	0.88	2.61	2.63	11.11	10.16	29.67
120	70	150	72.22	75.53	3.72	3.93	6.39	6.27	23.55	21.98	2.74
90	70	120	178.06	167.02	1.03	0.90	1.03	1.42	8.05	10.49	25.64
120	65	120	70.05	55.74	3.99	4.04	8.40	8.44	23.78	25.22	41.40
90	65	90	96.95	82.61	0.95	1.63	3.27	4.13	12.79	14.70	42.15
60	70	150	91.94	93.49	1.88	2.30	3.36	3.62	14.15	14.33	0.57
90	65	120	80.07	83.96	2.50	2.20	4.71	4.60	16.13	15.17	3.25

Table 3: Polyphenol Regression model (mg/g).

Sample Code	ST	DT	Dt	Greenness	Sweetness	Bitterness	Astringency	Clearness	DL Aroma	GT Aroma	GT Flavour
756	90	60	120	5.77 ± 0.57^{bc}	1.17 ± 0.38^{bc}	5.07 ± 0.87^{c}	2.20 ± 0.41^{b}	5.80 ± 0.48^{b}	2.83 ± 0.97^{d}	2.27 ± 0.45^{b}	6.87 ± 0.35^{b}
658	120	60	90	8.10 ± 0.48^{a}	1.00 ± 0.00^{c}	8.03 ± 0.46^{a}	1.00 ± 0.00^{a}	3.90 ± 0.55^{c}	1.87 ± 0.35^{e}	1.07 ± 0.25^{d}	8.07 ± 0.25^{a}
621	60	60	90	3.90 ± 0.40^{d}	2.10 ± 0.31^{a}	4.10 ± 0.40^{d}	4.10 ± 0.48^{c}	7.83 ± 0.53^{a}	5.97 ± 0.56^{a}	4.17 ± 0.38^{a}	4.77 ± 0.50^{c}
738	60	70	90	3.97 ± 0.41^{d}	1.03 ± 0.18^{c}	4.17 ± 0.46^{d}	4.00 ± 0.46^{a}	7.97 ± 0.49^{a}	2.03 ± 0.49^{e}	1.07 ± 0.25^{d}	8.13 ± 0.43^{a}
321	120	60	150	8.03 ± 0.41^{a}	1.00 ± 0.00^{c}	4.20 ± 0.41^{d}	1.20 ± 0.55^{a}	3.93 ± 0.52^{c}	2.13 ± 0.57^{e}	1.20 ± 0.41^{d}	8.13 ± 0.43^{a}
926	60	60	150	4.03 ± 0.49^{d}	1.20 ± 0.41^{b}	6.90 ± 0.31^{b}	4.20 ± 0.41^{b}	7.90 ± 0.61^{a}	1.93 ± 0.37^{e}	2.10 ± 0.31^{bc}	6.80 ± 0.48^{b}
271	120	70	90	8.03 ± 0.41^{a}	2.20 ± 0.41^{a}	8.00 ± 0.37^{a}	1.10 ± 0.31^{c}	3.93 ± 0.45^{c}	5.67 ± 0.71^{b}	4.17 ± 0.38^{a}	4.77 ± 0.50^{c}
292	90	65	150	6.00 ± 0.26^{b}	1.13 ± 0.35^{bc}	7.10 ± 0.31^{b}	2.10 ± 0.40^{b}	5.80 ± 0.55^{b}	2.77 ± 0.50^{d}	2.17 ± 0.38^{bc}	6.77 ± 0.43^{b}
531	60	65	120	3.97 ± 0.49^{d}	1.00 ± 0.00^{c}	4.17 ± 0.38^{d}	4.13 ± 0.43^{a}	7.83 ± 0.59^{a}	2.13 ± 0.43^{e}	1.07 ± 0.25^{d}	8.07 ± 0.37^{a}
564	120	70	150	8.03 ± 0.49^{a}	2.13 ± 0.35^{a}	8.00 ± 0.37^{a}	1.17 ± 0.46^{c}	3.90 ± 0.40^{c}	5.93 ± 0.37^{ab}	4.20 ± 0.48^{a}	4.93 ± 0.45^{c}
430	90	65	120	5.87 ± 0.43^{bc}	2.23 ± 0.43^{a}	7.97 ± 0.18^{a}	2.23 ± 0.68^{c}	6.00 ± 0.37^{b}	5.90 ± 0.40^{ab}	4.17 ± 0.46^{a}	4.83 ± 0.46^{c}
250	120	65	120	8.00 ± 0.52^{a}	1.00 ± 0.00^{c}	8.00 ± 0.41^{a}	1.27 ± 0.58^{a}	3.87 ± 0.51^{c}	2.13 ± 0.35^{e}	1.20 ± 0.48^{d}	8.27 ± 0.45^{a}
980	90	65	90	5.70 ± 0.47^{c}	1.13 ± 0.35^{bc}	6.87 ± 0.35^{b}	2.20 ± 0.41^{b}	5.77 ± 0.63^{b}	2.93 ± 0.52^{cd}	2.23 ± 0.57^{bc}	6.83 ± 0.38^{b}
194	60	70	150	4.00 ± 0.64^{d}	2.17 ± 0.38^{a}	4.13 ± 0.35^{d}	4.23 ± 0.43^{c}	7.97 ± 0.49^{a}	5.73 ± 0.52^{ab}	4.17 ± 0.38^{a}	4.77 ± 0.57^{c}
573	90	70	120	5.87 ± 0.43^{bc}	1.17 ± 0.38^{bc}	7.00 ± 0.37^{b}	2.37 ± 1.03^{b}	5.83 ± 0.53^{b}	3.13 ± 0.43^{c}	2.03 ± 0.18^{c}	6.87 ± 0.51^{b}

Values in the same column with same superscript are not significantly different at P≤ 0.05.

Table 4: Descriptive sensory evaluation mean scores of green tea samples.

regression coefficients (r^2) for EGCG, EGC, EC and ECG were 0.931, 0.939, 0.976 and 0.943 respectively. Descriptive sensory attributes of the green tea samples as affected by steaming time, drying temperature and drying time were presented in Tables 4 and 5. The result shows that green colour attribute ranged between 3.90 and 8.10, sweetness ranged between 1.00 and 2.23, bitterness ranged between 4.0 and 8.23 while astringency ranged between1.00 and 4.23. Clearness ranged between 3.87 and 7.97, dried leaf aroma ranged between 1.87 and 5.97 while GT aroma and flavour ranged from 1.07 to 4.20 and 4.77 to 8.27 respectively. Chromatogram of NGT chemical constituent and the effect of the steaming time, drying temperature and drying time on the EGCG content and sensory characteristics of optimized green teas were presented in Figures 1 and 2 respectively. The predicted models for the optimization of chemical constituents and sensory attributes of Nigerian green tea are presented in Tables 6 and 7.

Discussion

From the results, there were significant differences (P ≤ 0.05) in the chemical constituents and sensory attributes of the optimized Green tea samples. It was observed that, steaming for shorter period of time resulted in higher EGCG content compared with longer steaming duration. Xu and Chen [11] reported that gallated catechins usually convert into non-gallated catechins through hydrolysis under humid and heating conditions. Gulati [12] also reported that the chemical content and the composition of green tea catechins may vary with the conditions of processing. Reduction in the EGCG content at longer steaming duration therefore could be as a result of the release of EGC from EGCG.

The predictive model result for epigallocatechin gallate (EGCG) revealed that, increase in the steaming time, drying temperature, drying time and the product effect of the "steaming time and drying time" brings about a reduction in the amount of EGCG content. Increase in the product effect of the "steaming time and drying temperature", "drying temperature and drying time", quadratic effect of the steaming time as well as the quadratic effect of drying time positively enhances the EGCG content. The correlation coefficient R value was estimates to be 0.965. This indicates that there is a strong positive relationship between the EGCG and the independent variables.

The epigallocatechin (EGC) content increases with unit increase the in steaming time, drying time, product effect of the "steaming time and drying time", "drying temperature and drying time", quadratic effect of the steaming time and drying temperature while increase in

Parameters	Predicted model	RMSE	R^2	P-value
EGCG	$y_i = 193.61 - 1.1407x_1 + 0.8295x_2 - 1.9733x_3 + 0.0074685(x_1x_2) - 0.014046\beta_{13}(x_1x_3) + 0.018927(x_2x_3) + 0.006666x_1^2$	15.2	0.931	0.0198
EGC	$y_i = 33101 + 005159x_1 - 11996x_2 + 0065013x_3 - 000095174(x_1x_2) + 89843e^{(-05)}\,?_{13}(x_1x_3) + 000062478(x_2x_3) + 000028963x_1^2 + 00091122x_2^2 - 000045081x_3^2$	0.585	0.939	0.0148
EC	$y_i = 24593 + 015247x_1 - 0861x_2 + 0077131x_3 - 00039647(x_1x_2) + 0000123727\,?_{13}(x_1x_3) + 00013472(x_2x_3) + 00010407x_1^2 + 00063686x_2^2 - 000076346x_3^2$	0.719	0.976	0.0016
ECG	$y_i = 23657 + 050076x_1 - 66859x_2 - 0386611x_3 - 0011475(x_1x_2) - 72518e^{(-05)}\,?_{13}(x_1x_3) + 0011283(x_2x_3) + 00028049x_1^2 + 0045906x_2^2 - 00015459x_3^2$	2.87	0.943	0.0123
Colour	$y_i = -1.1278 + 0.046081x_1 + 0.053789x_2 - 0.00049034x_3 - 9.1084e^{-05}(x_1x_2) - 3.1885e^{-05}\beta_{13}(x_1x_3) - 2.5386e^{-05}(x_2x_3) + 0.00017436x_1^2 - 0.00031591x_2^2 + 2.6196e^{-05}x_3^2$	0.094	0.999	7.06e^{-07}
clearness	$y_i = 0.5076 + 0.063381x_1 - 0.049555x_2 + 0.017627x_3 - 0.000134(x_1x_2) - 1.3921e^{-05}\beta_{13}(x_1x_3) - 0.00019836(x_2x_3) + 7.5381e^{-05}x_1^2 + 0.00066464x_2^2 - 1.464e^{-05}x_3^2$	0.094	0.999	7.54e^{-07}
Dried Leaf Aroma	$y_i = 32.308 - 0.22121x_1 - 0.56524x_2 + 0.02394x_3 - 0.00035837(x_1x_2) - 5.1348e^{-05}\beta_{13}(x_1x_3) - 0.00024031(x_2x_3) + 0.0010424x_1^2 + 0.0048981x_2^2 - 1.5324e^{-05}x_3^2$	0.178	0.996	1.92e^{-05}
Green Tea Aroma	$y_i = 12.56 - 0.1406x_1 - 0.044293x_2 - 0.0024466x_3 - 2.526e^{-05}(x_1x_2) - 3.1978e^{-05}\beta_{13}(x_1x_3) - 2.526e^{-05}(x_2x_3) + 0.00052871x_1^2 + 0.00035264x_2^2 + 2.3157e^{-05}x_3^2$	0.078	0.999	1.03e^{-06}
Sweetness	$y_i = 9.125 - 0.11677x_1 - 0.055922x_2 + 0.0010414x_3 + 0.00019096(x_1x_2) + 4.0746e^{-06}\beta_{13}(x_1x_3) - 2.4558e^{-05}(x_2x_3) + 0.00046961x_1^2 + 0.0002948x_2^2 + 2.1918e^{-06}x_3^2$	0.032	0.999	1.01e^{-06}
Bitterness	$y_i = 4.1628 + 0.23717x_1 - 0.37469x_2 + 0.026199x_3 + 8.341e^{-05}(x_1x_2) + 2.2227e^{-05}\beta_{13}(x_1x_3) - 8.3168e^{-05}(x_2x_3) - 0.0010064x_1^2 + 0.0028732x_2^2 - 9.5125e^{-05}x_3^2$	0.567	0.961	0.00495
Green Tea Flavour	$y_i = 8.5731 + 0.14363x_1 - 0.32556x_2 - 0.0015971x_3 - 0.00015029(x_1x_2) - 1.3888e^{-05}\beta_{13}(x_1x_3) - 1.6506e^{-05}(x_2x_3) - 0.00042705x_1^2 + 0.0026267x_2^2 + 1.1808e^{-05}x_3^2$	0.075	0.999	5.76e^{-07}
Astringency	$y_i = 5.0908 - 0.15651x_1 + 0.16684x_2 + 0.015398x_3 + 0.00011728(x_1x_2) - 8.2575e^{-06}\beta_{13}(x_1x_3) - 4.8924e^{-07}(x_2x_3) + 0.00055648x_1^2 - 0.0013763x_2^2 - 5.4016e^{-05}x_3^2$	0.101	0.998	4.23e^{-06}

x_1, Steaming time; x_2, Drying temperature; x_3 Drying time.

Table 5: Regression Models of Polyphenols Content and Sensory Acceptance of Green tea.

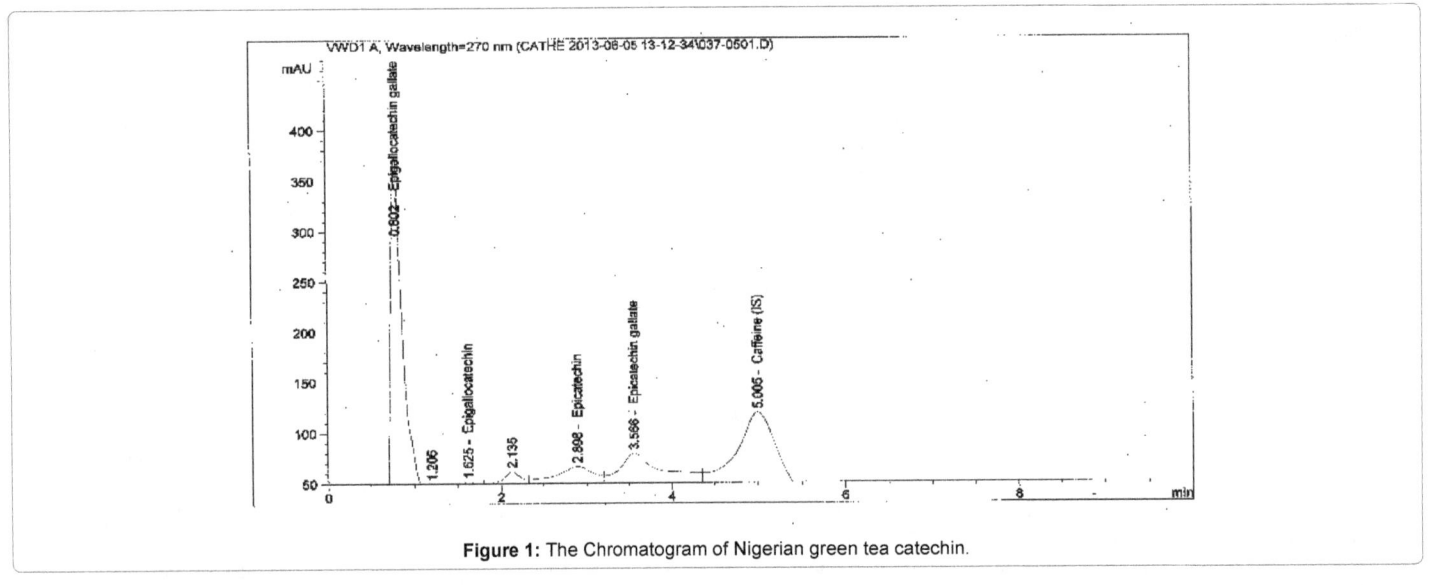

Figure 1: The Chromatogram of Nigerian green tea catechin.

drying temperature, product effect of the "steaming time and drying temperature" and quadratic effect of the drying time brings about a reduction in the amount of EGC content. The correlation coefficient value was estimated to be 0.969. This indicates that there is a strong positive relationship between EGC content and the independent variables.

The Epicatechin (EC) content, increases with unit increase in steaming time, drying time, product effect of the "steaming time and drying time", "drying temperature and drying time", quadratic effect of the steaming time as well as the quadratic effect of drying temperature. Increase in drying temperature, the product effect of the "steaming time and drying temperature" and quadratic effect of the drying time brings about a reduction in the amount of EC content. Strong positive relationship exists between the EC content and the independent variables.

Epicatechin gallate (ECG) increases with unit increase steaming time, product effect of the "drying temperature and drying time", quadratic effect of the steaming time and quadratic effect of the drying temperature while increase in drying temperature, drying time, product effect of the "steaming time and drying temperature", "steaming time and drying time" and quadratic effect of the drying time brings about a reduction in the amount of ECG content. The estimated correlation coefficient value of 0.971 indicates that there is a strong positive relationship between the amount of ECG content and the independent variables.

Somkiat [13] reported that colour is an important indicator for the quality of processed tea. The predictive model for the green colour intensity of green tea revealed that increase in steaming time and drying temperature brings about an increase in the green colour intensity of the tea extract while unit increase in drying time reduces the green colour intensity. The product effect of the "steaming time

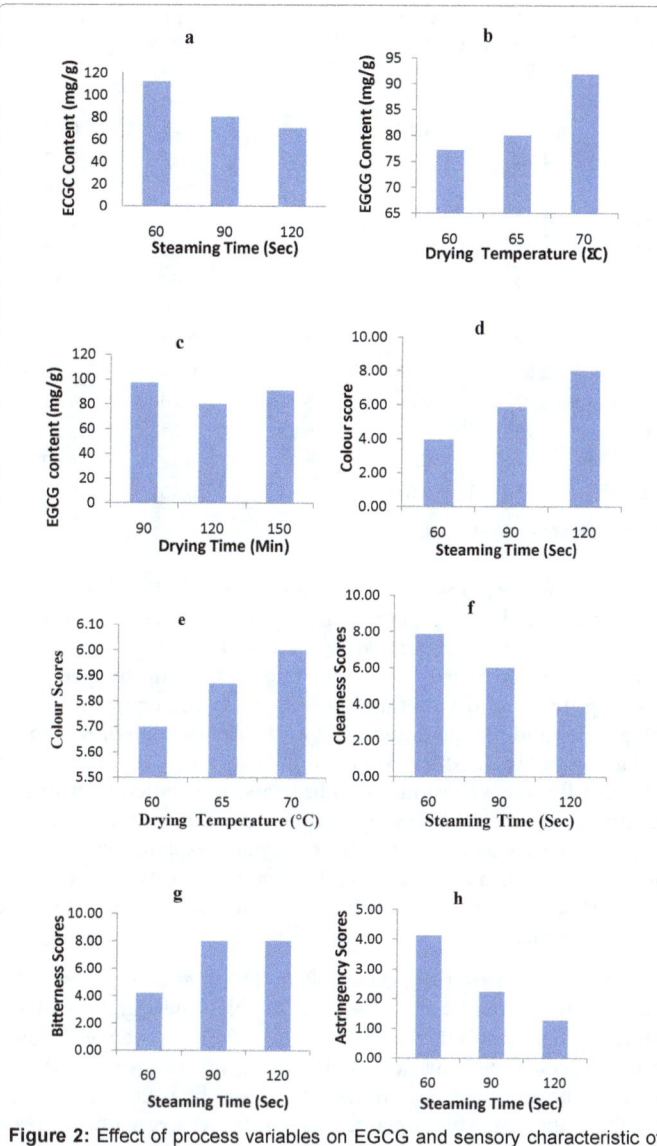

Figure 2: Effect of process variables on EGCG and sensory characteristic of NGT.

and drying temperature", "steaming time and drying time" and "drying temperature and drying time" also brings about a reduction in the green colour intensity. The quadratic effect of the steaming and drying time brings about an increase in the colour intensity of the green tea extract while the quadratic effect of the drying temperature leads to a reduction in the colour intensity. There exist a strong positive relationship between the green colour intensity and the independent variables. However, for the nine explanatory variables studied, only the quadratic effect of the steaming time exerts significant influence (r= 0.99, p=0.044629) on the green colour intensity of the tea extract.

Sample steamed for 120 sec, dried at 60°C for 90 min. was the most preferable based on green colour intensity. Sample 658 has the highest colour score (8.10) which was not significantly different (P ≤ 0.05) from samples 564 (8.03), 321 (8.03), 271 (8.03) and 250 (8.00). Green colour intensity of this samples were significant different (P ≤ 0.05) from those of all other samples, however, the least scores were recorded in samples 926 (4.03), 194 (4.00), 531 (3.97), 738 (3.97) and 621(3.90) in that order.

Unit increase in the steaming time, drying temperature as well as the product effect of "drying temperature and drying time" brings about a reduction in the sweetness of the green tea extract while increase in the drying duration as well as the product effect of the "steaming time and drying temperature", "steaming time and drying time", quadratic effect of the steaming time, drying temperature and drying time enhances the sweetness. A strong positive relationship exists between the sweetness of the green tea extract and the independent variables. However, for the nine explanatory variables studied, only the steaming time, product effect of the "steaming time and drying temperature" and the quadratic effect of the steaming time, exert significant influence on the sweetness (r=0.999, p=9.1011e^{-06}, 0.049795 and 4.0076e^{-06}). All samples taste attributes for sweetness were weak and therefore recorded low mean scores as recorded in sample 531, 250, 658 and 321 (1.00). Sample 430 had the highest sweetness mean score (2.23).

The predictive model for Bitterness indicates that, every unit increase in the steaming time as well as the drying duration, product effect of the "steaming time and drying temperature", "steaming time and drying time" and quadratic effect of the drying temperature brings about an increase in the bitterness of the tea extract while the drying temperature, product effect of the "drying temperature and drying time", quadratic effect of the steaming time and quadratic effect of

ST sec	DT °C	Dt min	Colour (Measured)	Colour (Predicted)	Sweetness (Measured)	Sweetness (Predicted)	Bitterness (Measured)	Bitterness (Predicted)	Astringency (Measured)	Astringency (Predicted)	MSE
90	60	120	5.77	6.12	1.17	1.01	4.13	5.94	2.20	1.86	0.91
120	65	120	8.10	7.98	1.00	1.02	7.97	8.01	1.00	1.13	0.01
60	60	90	3.90	3.91	2.20	2.22	4.20	3.85	4.10	4.15	0.03
60	65	120	3.97	4.32	2.13	2.00	4.17	4.73	4.00	3.69	0.16
120	60	150	8.03	8.39	1.00	0.84	8.00	8.20	1.10	0.75	0.10
60	70	90	4.03	3.60	2.23	2.42	4.17	3.82	4.20	4.60	0.15
120	60	90	8.03	7.60	1.00	1.21	8.03	7.10	1.20	1.62	0.36
60	60	150	6.00	5.70	1.20	1.30	6.90	6.28	2.10	2.34	0.16
60	70	150	3.97	4.33	2.17	1.99	4.13	4.89	4.13	3.76	0.25
120	70	150	8.03	7.97	1.03	1.04	8.00	8.20	1.17	1.17	0.01
90	70	120	5.87	5.75	1.17	1.22	7.00	5.93	2.13	2.29	0.32
120	70	90	8.00	8.24	1.00	0.93	8.00	8.46	1.27	1.08	0.10
90	65	150	5.70	5.35	1.13	1.35	6.87	6.35	2.20	2.68	0.20
90	65	90	4.00	4.59	2.10	1.75	4.10	5.26	4.23	3.55	0.63
90	65	120	5.87	5.42	1.13	1.34	7.10	5.88	2.23	2.58	0.47

Table 6: Regression results for the sensory attributes of Nigerian Green tea.

S T sec	DT °C	Dt min	Clearness (Measured)	Clearness (Predicted)	DLA (Measured)	DLA (Predicted)	GT A (Measured)	GT A (Predicted)	GT F (Measured)	GT F (Predicted)	M SE
90	60	120	5.80	6.16	5.93	4.10	2.27	1.86	6.87	7.15	0.91
120	65	120	7.90	7.86	2.13	2.27	1.20	1.26	8.27	8.14	0.01
60	60	90	3.87	3.84	5.67	5.98	4.17	4.19	4.77	4.75	0.03
60	65	120	3.93	4.24	5.93	5.24	4.20	3.90	4.93	5.23	0.16
120	60	150	7.97	8.30	2.13	1.83	1.20	0.84	8.13	8.53	0.10
60	70	90	3.90	3.50	5.90	6.34	4.17	4.59	4.83	4.39	0.15
120	60	90	7.97	7.54	2.13	3.11	1.07	1.56	8.07	7.63	0.36
60	60	150	5.80	5.56	1.93	2.68	2.10	2.38	6.80	6.57	0.16
60	70	150	3.90	4.25	5.73	4.91	4.17	3.75	4.77	5.16	0.25
120	70	150	7.83	7.79	2.03	1.85	1.07	1.11	8.13	8.11	0.01
90	70	120	5.83	5.74	3.13	4.29	2.03	2.20	6.87	6.76	0.32
120	70	90	7.83	8.00	1.87	1.27	1.07	0.85	8.07	8.26	0.10
90	65	150	5.77	5.38	2.93	3.48	2.23	2.68	6.83	6.31	0.20
90	65	90	3.93	4.62	5.97	4.84	4.17	3.46	4.77	5.48	0.63
90	65	120	6.00	5.45	3.20	4.31	2.17	2.64	6.77	6.43	0.47

Table 7: Regression for the sensory attributes of Nigerian Green tea.

the drying time, brings about a reduction in the bitterness. The result also indicated that there is a strong positive relationship between the bitter taste intensity and the independent variables. Sample 658 was rated highest in bitterness, followed by samples 250, 564, 271 and 430. Sample 621 (4.10) had the lowest score for bitterness. This could be as a result of the short steaming period.

From the predictive model result for astringency, every unit increase in the steaming time, drying temperature, drying time, product effects of the "steaming time and drying temperature", "drying temperature and drying time" as well as the quadratic effect of the drying temperature and drying time brings about a reduction in the astringency of the tea extract. Increase in the quadratic effect of the steaming time brings about an increase in the astringency of the green tea extract. The correlation coefficient value was estimated to be 0.999. This indicates that there is a strong positive relationship between astringency and the independent variables. However, for the nine explanatory variables studied, only the steaming time and quadratic effect of the steaming time exert significant influence on the astringency with p-Value of 0.0061525 for steaming time, and 0.00050398 for the quadratic effect of steaming time. Sample 194 (4.23) was the most preferred in astringency followed by 926 (4.20), 531 (4.13) in that order. This finding agrees with Wismer [14] that astringency is an important and often appealing characteristic of brewed tea. Sample 658 had the lowest scores for astringency which was significantly different (p ≤ 0.05) from those of all other samples.

Unit increase in the steaming time, quadratic effects of the steaming time as well as drying temperature and drying duration brings about an increase in the clearness of the green tea extract while increase in drying temperature, product effect of the "steaming time and drying temperature", "steaming time and drying time", "drying temperature and drying time" and the quadratic effect of the drying time reduces the clarity of the green tea extract. A strong positive relationship exists between the clarity of the tea extract and the independent variables. However, for the nine explanatory variables studied, only the steaming time exerts significant influence on the clearness (r = 0.999, p=0.0211). The highest clearness score was found in samples 738 (7.97), 194 (7.90), 531 (7.83) which had a better appearance attribute in terms of clarity. Steaming for a shorter period (60 s) could be attributed to the clarity of these samples. Sample 250 (3.87) was rate the least and significantly different (P ≤ 0.05) from other samples.

Increase in steaming time, Drying time, quadratic effects of the steaming time and drying temperature, brings about a rise in the dried leaf aroma of the tea extract while increase in drying temperature, the product effect of the "steaming time and drying temperature", "steaming time and drying time" as well as "drying temperature and drying time" and the quadratic effect of the drying time brings about a reduction in the dried leaf aroma of the tea extract. The correlation coefficient R value was estimated to be 0.998. This indicates that there is a strong positive relationship between dried leaf aroma and the independent variables. However, for the nine explanatory variables studied, only the steaming time and the quadratic effect of the steaming time exert significant influence (p=0.0017056 and 0.00038171) on the dried leaf aroma.

Steaming for a shorter period (60 s), drying at 60°C for 90 mins enhances the aroma of the green tea sample. The duration of the steaming process was said to be a key determinant in green tea flavour, aroma, and colour. Steamed leaves left at high temperature will lose their bright colour and their flavour and aroma will be negatively affected [15]. From the result for Green tea aroma (GTA), every unit increase in the steaming time, drying temperature, drying time, product effect of the "steaming time and drying temperature", "steaming time and drying time" as well as "drying temperature and drying time" brings about a reduction in the GTA while the quadratic effect of steaming time, drying temperature and time positively enhance the green tea aroma of the tea extract. Strong positive relationship exists between GTA and the independent variables however, for the nine explanatory variables studied, only the steaming time and the quadratic effect of the steaming time exerts significant influence (r=0.999, P=0.00029991, 0.00018735) on the GTA.

Unit increase in the steaming time while the drying temperature, drying time, product effect of the steaming time and drying temperature; steaming time and drying time; drying temperature and drying time, and the quadratic effect of the steaming time; drying temperature; and drying time are kept constant brings about an increase in the green tea flavour while increase in drying temperature brings about a reduction in the green tea flavour. There is equally a negative relationship between the drying time and the green tea flavour. This indicates that increase in drying time will bring about a reduction in the green tea flavour. Furthermore, a negative relationship exists between the product effect of the "steaming time and drying temperature" as well as "steaming time and drying time" and the green tea flavour; this indicates that increase in this variable brings about a reduction in the green tea flavour. Increase in the quadratic effect of drying temperature and drying time however brings about an increase in the flavour of the tea extract. The correlation coefficient R value for the green tea flavour was estimated to be 0.999. This indicates that there is a strong positive relationship exists between the green tea flavour and the independent variables, however, for the nine explanatory variables studied, only the steaming time and quadratic effect of the steaming time exerts significant influence ($p=0.00022153, 0.00042252$) on the green tea flavour.

Conclusion

There were significant differences ($P \leq 0.05$) among the optimized Green tea samples as influenced by the processing variables (steaming time, drying temperature and drying time). This study revealed that, steaming for shorter period of time resulted in higher EGCG content compared with longer steaming duration. Significant difference ($P \leq 0.05$) was also recorded in the EGC content of the optimized green tea samples. Steaming for longer period (120 s) positively enhanced the green colouration in green tea while shorter steaming regime (60 s) is desirable for clarity of green tea extract. Moderate steaming period (90 s) gives better aroma as the intensity of green tea aroma reduces with elongation of the steaming period. Astringency reduces with steaming time and this correlates with the Epigallocatechingalate (EGCG) content which is higher at minimum (60 s) steaming time.

Drying at higher temperature (70°C) enhances the colour intensity of green tea while an average drying temperature of 60°C gives better clarity of the extract. More EGCG is recorded in green tea dried at 70°C than at lower temperature, this could possibly be as a result of higher drying rate at higher temperature. The intensity of green tea flavour gets more pronounced as the drying temperature increases.

References

1. Willson KC (1999) Coffee, Cocoa and Tea. New York: CABI Publishing.

2. McKay DL, Blumberg JB (2002) The role of tea in human health: An update. J Am Coll Nutr 21: 1-13.

3. Jung DH (2004) Components and Effects of Tea (In Korean) Hongikjae, Seoul, Korea. 28-43.

4. Banga JR, Balsa-Canto E, Moles CG, Alonso AA (2003) Improving Food Processing Using Modern Optimization Methods. Trends in Food Science and Technology 14: 131-144.

5. Hakim IA, Weisgerber UM, Harris RB, Balentine D, van-Mierlo CAJ, et al. (2000) Preparation, composition and consumption patterns of tea-based beverages in Arizona. Nutrition Research 20: 1715-1724.

6. Mason RL, Gunst RF, Hess JJ (2003) Statistical Design and Analysis of Experiments-with Applications to Engineering and Science. John Wiley and Sons Inc, Hoboken, New Jersey, USA.

7. Rio DD, Stewart AJ, Mullen W, Burns J, Lean MEJ, et al. (2004) HPLC-MSn analysis of phenolic compounds and purine alkaloids in green and black tea. J Agric and Food Chemistry 52: 2807-2815.

8. Friedman M, Kim SY, Lee SJ, Han GP, Han JS, et al. (2005) Distribution of catechins, theaflavins, caffeine, and theobromine in 77 teas consumed in the United States. Journal of Food Chemistry and Toxicology 70: 550-559.

9. Agilent (2012) Extract from Green Tea. The Essential Chromatography and Spectroscopy Catalog 2011-2012 edition.

10. Stone H, Sidel JL (1993) Sensory Evaluation Practices. Academic Press, Inc, SanDiego.

11. Xu N, Chen Z (2002) Green tea, black tea and semi-fermented tea. In: Tea: Bioactivity and Therapeutic Potential, Boca Raton, FL, USA.

12. Gulati A, Rawat R, Singh B, Ravindranath SD (2003) Application of microwave energy in the manufacture of enhanced-quality green tea. Journal of Agric and Food Chem 51: 4764-4768.

13. Somkiat P, Paveena P, Somchart S (2004) Effective Diffusivity and Kinetics of Urease Inactivation And Color Change During Processing Of Soybeans With Superheated-Steam Fluidized Bed. Drying Technology 22: 2095-2118.

14. Wismer WV, Goonewardene LA (2004) Selection of an astringency reference standard for the evaluation of black tea. Journal of Sensory Studies 19: 119-132.

15. Processing of Sencha Green tea (2012) The crude tea manufacturing process for sencha.

23

Formulation of Nutritionally Superior and Low Cost Seaweed Based Soup Mix Powder

Jayasinghe PS[1]*, Pahalawattaarachchi V[1] and Ranaweera KKDS[2]

[1]*National Aquatic Resource Research and Development Agency, Crow Island, Colombo, Srilanka*

[2]*Faculty of Food science, University of Sri Jayewardenepura, Nugegoda, Srilanka*

Abstract

The demand for dried instant soups is increasing to their ease of use. In particular, the boom in healthy and functional foods, there is general preference for healthy soup prepared using vegetables, legumes, cereals and mushrooms. Dried instant soup has long shelf life because it is a dried food and frequently has its effective tasting time period set for a relatively long time. This study was conducted to develop a nutritious instant vegetable soup mixture incorporated with cereals, legumes and seaweed extracts such as agar or carrageenan use and was to replace of pectin. The agar and carrageenan area industrial food substitute enrich with protein, minerals, vitamins and amino acid which improve the overall nutritive values and viscosity of the soup. Organoleptic evaluation was conducted to select best combination of ingredients of vegetables, cereals, legumes and preservatives. In addition to that *Ulva* powder and seaweed extracts (agar or carrageenan) were incorporated in different percentages and was evaluated for the sensory attributes to optimize and compare physical chemical and sensory parameters with commercial vegetable soup mixture. The data rank some test revealed that best soup mixture were evaluated by sensory attributes one commercial available and two formulated soup mixtures with contain 80% vegetables, 10% grain, 3.5% legumes, 2.5% dried *Ulva* powder with 3% agar agar or 2% carrageenan and preservatives. The soup formulas have highest viscosity (698 cps, 766 cps, 951cps), water activity (0.618, 0.586, 0.437), crude protein (9.3%, 7.2%, 1.7%), carbohydrate (64.54%, 61.3%, 51.32%), iodine value and (0.35, 0.32, 0 mg/l) respectively in agaragar incorporated soup, carrageenan incorporated soup and commercial vegetable soup mixture. The total bacterial counts, total fungal counts and water activity were at the acceptable levels for consumption. The low levels of yeast and mould counts were detected in all the samples during the storage period. This two formulated products can be introduced to the commercial market as alternative medicinal heath food.

Keywords: Soup mixture; Agar; Carrageen an; Legumes; Dehydrated seaweeds.

Introduction

Lack of nutrition education leads to wrong choices of food and resultant malnutrition. The Interrelationship between diet, food habits and nutrient deficiency disease have made efforts to investigation of nutritive value of soup mixture and evaluate the properties of sensory, physical, chemical and nutrients. The nutrient content of most of soups available in the market is high in carbohydrate and low in protein. Deficiencies of nutrients are a major global health problem. More than 2 billion people in the world today are estimated to be deficient in key vitamins and minerals, particularly vitamin A, iodine, iron and zinc. Most of these people live in low income countries and are typically belong to poverty line. Micro nutrient deficiencies increase the general risk of infectious diseases because of compromised immune system as well as the risk [1]. Seaweed is known as rich source of protein, vitamins, amino acids and dietary iodine, macro and micro elements, dietary fiber [2]. The main two polysaccharides agar and carrageenan extracted from seaweeds are vicious hydrocolloids which have thickening properties of food. The growing demand in the role of macro micronutrients in optimizing the health in prevention or treatment of disease [3]. The iodine and other essential elements are rare in land plants which are highly abundance in seaweed. Previous studies have shown that iodine content in breast milk of lactating mothers has a strong correlation with the frequency and quantity of seaweed soup consumption [4]. Iodine is an essential macro and micro element required for thyroid hormone synthesis; impart some of its prevention of cardiovascular disease and cancer and other metabolic reactions. Seaweed soup is low cost nutritive sources and supply protein, minerals, vitamins and amino acids to health diet. The vegetable and cereal mixed soups are enriched with high protein [5]. The incorporation of seaweed extract to soup enhances the nutrition requirement of the diet. The combination of those macro and micro elements play important role in mounting immune response and substantially increase the risk of having poor immune responses to infections [6]. These compounds also help to influence adult and child survival as well as educational achievement, child survival and maternal health [7]. Method of preparation and storage of foods often leads to large loss of nutrients thus creating deficiency risk of nutrients. Therefore efforts have been geared towards the study of nutritional composition of market available and prepared soup. There are no scientific data available on nutritional values of Sri Lankan seaweed based soup mixtures. As the formulation and development of nutritious complementary foods from locally and readily available materials have received a lot of attentions, the present research work aimed to prepare and supplement dried vegetarian soup mixtures incorporated with agar or carrageenan (vicious polysaccharide) and evaluating their chemical, physical, rheological and sensory properties to assess the nutritional and technological quality of the resultant mixtures and soups.

Material and Methods

Preparation of seaweed extracts

The seaweed varieties *Kappaphycus alvarezii, Ulva lactuca,*

---

*Corresponding author: Jayasinghe PS, National Aquatic Resource Research and Development Agency, Crow Island, Colombo 15, Mattakkuliya, Srilanka
E-mail: pradee_jaya@yahoo.com

Gracilaria verrucosa were collected from Northeastern (Trincomalle), Southwestern (Unawatuna and Beruwala) coastal areas of Sri Lanka and transported to NARA laboratory in insulated boxes. *Ulva lactuca ,Gracilaria verrucosa* and *Kappaphycus alvarezii* species were washed in seawater initially to remove macroscopic epiphytes and sand particles and finally with fresh water to remove adhering salt. The fresh seaweeds were blanched in 95°C hot water in 10 minutes and washed in cold water. They were shade dry for four days followed by oven dry at 60°C for 12 hours. Then the materials were hand cursed and made as a coarse powder using mixer grinder. The dried grinded powder was stored in 0°C until it is used. The *Gracilaria verrucosa* (agar) and *Kappaphycus alvarezii* (carrageenan) were used to extract agar with acid treatments [8] and carrageenan with alkaline treatments [9] respectively. The dried grinded *Ulva* powder, agar and carrageenan were stored in 15°C until used.

Preparation of vegetables

The vegetables pumpkins, carrot, potato, mushrooms, tomato, B-onions, leaks, celery, carrot and cereals, red rice, soya bean, green gram, pepper, salt, corn starch were purchased from super market. Potato carrot and pumpkin samples were sorted, washed, peeled and sliced in cubic form and blanched in hot water at 95°C for 5 min, then washed in cold water then hot air flow drying were performed at 65°C in the first four hours and then reduced to 50°C till completely drying. Then milled and sieved (315 micron) into powdered form. Tomatoes were sliced then dried and milled as the above mentioned vegetables; the green gram and soya bean were subjected to some technological treatments as soaking in 1:2 w/v water ratios for 30 minutes, and then cooked for 15 minutes before dried and milled for soup formulation.

Formulation of the dried vegetarian soup mixtures

The prepared samples were seasoned with dried onion, garlic, coriander, black pepper, cumin salt and agar or carrageenan then mixed to formulate four dried vegetarian soup mixtures (three replicates for each formula) namely F_1 and F_2 as shown in Table 1. The obtained mixtures were packed into polyethylene bags and kept at -20°C for further analysis. The formula 1 were made using different percentages 4%, 3%, 2%, 1% of agar agar. The formula 2 were made out using same procedure for different carrageenan percentage 4%,

Ingredients	Formula 1 (g/100 g)	Formula 2 (g/100 g)
Agar	3g	
Carrageenan		2g
Ulva powder	2.5	2.5
Tomtato	5	5
Mush room	3	3
Pumpkin	10	10
Carrot	15	15
Green gram	15	15
Curry leaves	8	8
Soya bean	15	15
Red rice	15	15
Garlic	1	1
Salt	2.5	2.5
Black pepper	1	1
Onion	3	3
Citric acid	0.1	0.1
Sodium benzoate	0.3	0.3

Table 1: Formulas of dried vegetarian soup mixtures (g/100 g).

3%, 2%, 1%. Organoleptic parameters were measured to obtain of best percentages of agar or carrageenan.

Analysis of chemical parameters

The chemical, physical parameters were measured: macro and micro, proximate composition, viscosity (Brookfield viscosity meter), water activity (Spring water activity meter), dietary fiber [10], total Plate Count [11]. Standard deviation and variance of means (ANOVA) were calculated using statistical analysis package SPSS14 [12].

Rheological properties of the resultant soup samples

Rheological parameters (viscosity and shear rate) of dried vegetarian soup samples were measured according to Brookfield manual by using Brookfield viscometer. The sample was placed in a small sample adapter and a constant temperature water bath was used to maintain the desired temperature. The viscometer was operated between 10 and 60 rpm. Viscosity and shear rate data were obtained directly from the instruments; the SC4-21 spindle was selected for the measurement. Rheological measurements were made at the resultant soup samples (F_1 F_2 and F_3 commercial soup) at room temperature (25°C ± 1°C).

Organoleptic evaluation of the resultant soup samples

The resultant soup samples were organolyptically evaluated after dissolving in hot water (25 g dried vegetarian soup mixtures/250 ml water) for its sensory characteristics, taste, flavor, color, thickness and appearance, dissolution rate and overall acceptability. The evaluation was carried out by using ten panelists according to the method of [13].

Shelf life assessments

The shelf life were measured for prepared different soup formulas packed in air tight polystyrene packets and stored in three different temperatures, ambient temperature, 0°C and -18°C during six month period. Shelf life was determined by measuring of changes of overall acceptability, total bacterial count, total fungal count and water activity monthly over a six month period.

Results and Discussion

The best composition of selected from formula 1 was recorded for 3% agar 10% grains, 80% vegetables 2.5% *Ulva* powder and 10% legumes and other ingredients. The procedures followed in formula 2 which was formulated with carrageenan 2% are with other ingredients were selected as best composition. The viscosity of soup mixtures was lower than that of commercial available market soup sample. The viscosity level both the soup mixture was improved to market standards by incorporation of dried *Ulva* species powder and corn starch. This increased in the apparent viscosity of the formulated soup closer to the commercial sample.

Viscosity of the resultant soup samples

The physical properties of soup mixture were dramatically improved by incorporate of agar agar and Carrageenan. The highest apparent viscosity was found at commercial sample 951 cps followed by formula 2 and formula1. The significant variation could be observed among commercial sample and formula 1 and 2. Soup mixture commonly available market contains corn starch as thickening agent. The formulated soup shows the potential application of agar or carrageenan as a viscosity enhancer. Agar is generally water- soluble and very hydrophilic. The agar with low gelling property products provides the semi-solid body and high viscosity to the soup preparation. In high gel strength agar can be added limited amount as liquid thickener or viscose's media. In soup preparation, viscosity is an index of thickness [14]. Carrageenan or agar is used as thickening agents in food to provide the desirable viscosity to the soup mix [15]. ß-glucan has a lower or equal ability to increase viscosity as xanthan, guar gum, locust bean gum and Arabic gum. The good viscosity forming properties make ß-glucans potential alternatives as thickening agents in different food applications [16]. The functional properties of pulse proteins play an important role in food formulation and processing.

The relationship between shear rate (S) and values of the viscosity (cps) of the dried vegetarian soup supplemented with carrageenan agar and commercial sample given in Figure 1. It seems that the apparent viscosity (cps) of soup samples decreased as shear rate increased. This simply means that the three dried vegetarian soups F1 and F2 had a noticeable apparent viscosity pattern which could be characterized within the non-Newtonian pseudo plastic flow behavior [17]. The same Figure depicts that apparent viscosity pattern recorded the highest value compared to F1and F2, it decreased sharply while; apparent viscosity patterns in and F1 and F2 were much closer to each other and decreased gradually. The high viscosity pattern of F3 could be due to the higher proportion and the functionalities of corn flour starch and its ß-glucan. The reduction of apparent viscosity pattern of F1`and F2 may be due to the reduced proportion of agar or carrageenan during formulation.

There were no observed significant changes in water activity levels of three different formulas. The water activity content was less than the 0.6 and it range from 0.564 ± 0.4 to 0.597 ± 0.8. The low level of water activity content may have restricted the growth rate of microorganisms. Almost all microbial activity is inhibited below aw 0.6 [18]. while between most fungi are inhibited below a_w 0.7, most yeast are inhibited below aw 0.8 and most bacterial growth below aw 0.9 [18]. Very low values of aw are related to high lipid oxidation rates. A_w values of 0.2 and 0.4 lipids have been suggested to have optimal stability and oxidation rates increase with increasing aw [19]. According to the present study (Table 2) initial levels of microorganisms in tested samples were comparatively lower and in the range of 1×10^2 to 2×10^2. All the three samples were much lower compared to the maximum allowable limits. The product can be stored for further period of time.

Sensory attributes of different type of soup mixtures

Sensory evaluation considered to be a valuable tool in solving problems involving food acceptability. It is useful in product improvement, quality maintained and more important in new product development. It is desirable that the product to free from off flavors',

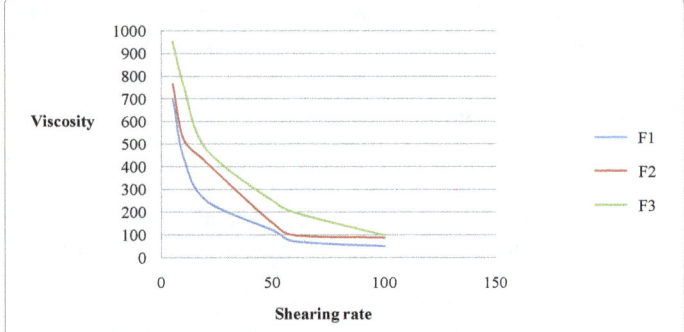

F1-agar incorporated soup mixture, F2-Carrageenan incorporated soup mixture, F3-Commercial product.

Figure 1: Viscosity of three different soup mixtures at different shearing rates.

Macro and Micro Elements mg/100g	Formula 1	Formula 11	Market soup
Cobalt (Co)	ND	ND	ND
Ni	ND	ND	ND
Cr	ND	0.08 ± 0.86	1.3 ± 0.7
Mn	ND	$1.97.0 \pm 0.45$	1.2 ± 0.31
K	53.39 ± 0.07	98 ± 0.42	5.6 ± 0.45
Mg	45.8 ± 0.98	54.9 ± 0.23	4.9 ± 0.32
Ca	250.56 ± 0.75	125.9 ± 0.21	3.8 ± 0.23
Na	115.76 ± 0.56	126.8 ± 0.32	5.3 ± 0.12
Iodine	0.62 ± 0.02	0.75 ± 0.02	0

Table 2: Comparison of mineral content of in different soup formula powder mixtures.

unacceptable aroma and faulty texture. The market sample was also observed to be in highest sensory quality. Formula 1 and 2 showed same points for sensory quality. Table 3 indicates the sensory quality attributes of resultant soup thickener with different soup mixtures. Incorporation of agar and carrageenan significantly affected thickness as well as dissolution rate. Sensory studies showed that all three samples were not significantly difference in appearance, color, texture, aroma, taste and overall acceptability. The commercial sample is used with corn starch as thickening agent in ß-glucane. Therefore agar or carrageenan can be successfully substituted to corn starch as a thickener with higher level of sensory acceptability.

Proximate composition of dry vegetable soup mixtures

The proximate composition of the different soup formulas were given in Table 4. The highest protein values were observed in the formula 1 followed by formula 2 and commercial vegetable soup sample. There were significant differences between protein content of commercial sample, F1 and F2 soup formulas. The protein content ranged from 1.7-11.3. There were no significant difference between F1 and F2. The soya bean, cow pea and added seaweed polysaccharides enhance the protein content in soup formula 1 and 2. The present data is similar to the observations of [20] who mention that legumes and seaweed are high in protein and especially rich in lysine and leucine, low in fat and excellent source of dietary fiber and complex carbohydrate.

The highest oil content 2.68% was observed at market soup samples followed by formula 1 and 2. There was no statistical difference among oil contents of soup formulas. The fat content ranged from 2.3%-2.68% at different soup formulas. Agar and carrageenan have low oil content [21]. The results of present study confirm the observations of Amal et al. [17] Observed similar oil contents in the diets from dried vegetable soup mixture. The highest carbohydrate content was also observed at formula 1 than in others soups. It can be concluded that formula 1 has the highest nutritional compounds than other two formulas. The soup mixture formula 1 supply highest carbohydrate content followed by formula 2 and then commercial soup sample. The carbohydrate content is over 60% in all seaweed polysaccharides. The application of these polysaccharides stabilized the food product against degradation. These polysaccharides enhance acceptability and shelf life [22]. The ash content is an indicator of minerals. Seaweed is rich source of minerals. The highest ash content was recorded in formula 2 followed by formula 1 and market soup. Dietary sources of essential elements are important for correct physiological functions of human body. A Deficient intake of certain minerals can produce diseases and lead to abnormal development [23].

The availability of macro and micro elements were observed in

Different soup Formulas	Taste (10) Mean ± SE	Colour (10) Mean ± SE	Flavour (10) Mean ± SE	Thickness Mean (10) ± SE	Dissolution rate (10) Mean ± SE	Overall acceptability (50) Mean ± SE	Colors
Formulas 1	8.3 ± 1.7	9.4 ± 1.23	8.6 ± 1.2	8.55 ± 1.27	8.45 ± 0.3	43.3 ± 1.14	Orange yellow
Formulas 2	8.4 ± .25	9.45 ± 1.54	8.4 ± 1.53	8.63 ± 1.6	8.9 ± 1.6	43.78 ± 1.8	Orange Yellow
Market available sample	9.10 ± 1.67	8.50 ± 1.32	9.1 ± 1.67	9.0 ± 1.9	9.1 ± 1.6	44.8 ± 1.63	Yellow

Table 3: Comparison of sensory attributes of different soup formulas (Mean ± SE).

Parameters	Formula 1	Formula 11	commercial soup sample
Protein (%)	11.3 ± 0.8	11.2 ± 0.6	1.7 ± 0.4
Oil (%)	2.4 ± 0.7	2.3 ± 0.6	2.68 ± 0.5
Carbohydrate (%)	64.54 ± 0.4	61.34 ± 0.5	51.32 ± 0.5
Moisture content (%)	12.6 ± 0.4	12.3 ± 0.6	14.5 ± 0.5
Ash (%)	9.5 ± 0.3	9.7 ± 0.3	9.3 ± 0.2
TPC (Total Plate Count)	$1 \times 10^2 ± 0.1$	$2 \times 10^2 ± 0.54$	$1 \times 10^3 ± 0.65$

Table 4: Comparison of proximate composition and TPC (Total Plate Count) in different soup formulas (dry weight basis, mean ± SE).

Parameters	Formula 1	Formula 2	Commercial soup sample
viscosity (cp)	698 ± 0.3	766 ± 0.5	951 ± 0.5
Overall acceptability	8 ± 0.2	8 ± 0.3	10 ± 0.6
water activity	0.564 ± 0.4	0.505 ± 0.6	0.597 ± 0.8
pH	6.3 ± 0.5	6.7 ± 0.8	6.9 ± 0.2

Table 5: Comparison of physical and sensory attributes of laboratory prepared soup and commercial soup samples (Mean ± SE).

Table 2 in three different soup Formulas. The availability of Ca, Na, Mg and K values were significantly very high in laboratory prepared two soup samples than in commercial samples.

The significantly (P<0.05) highest Ca content was found in formula 1 (250.056 mg/100 g) followed by formula 2 (125.9 mg/100 g). The lowest value was found in commercial sample (3.8 mg/100 g). The differences among the Ca content of samples were due to the high Ca availability of agar which was extracted from the red algae. These very high Ca content seaweed soup consumption may be useful in case of expectant mothers elderly and adolescents that all expose to risk of Ca deficiency [24]. Most of the vegetable soups in the market have low levels but not sufficient to meet adequate intake of calcium for adult 1000-1200 mg/day. Calcium deficiency is certainly a risk factor for osteoporosis in later life [25].

The Table 5 showed the Na content in dry soup formulas. The significantly highest value (126.8 mg/100 g) was found in formula 11 followed by formula 1(115.76 mg/100 g). The lowest value was observed in commercial market soup samples. The Na contents were ranged from 5.3 to 126.86 mg/100 g. Sodium is the major positive ion in the extracellular fluid and key factor in retaining body water. Under the FDA food-labeling rules, the daily value for sodium is 2400 mg. Two soups prepared in the laboratory were within the RDA (Recommended Daily intake for Adult) [26]. The highest macro and micro elements were observed in formula 2 where average K content was 98 ± 0.42 mg/100 g. The macro elements in formula 2 were three times higher than that of formula 1. The metals in both two soup formulas were significantly higher than that of market soup samples. This is due to highest availability of minerals in seaweeds than in land plants. Potassium plays a similar role with sodium in the biological system. But it is located in the intracellular fluids. Unlike sodium it is associated with lower levels rather than higher blood pressure values [6].

The Mg content was found significantly highest (45.8 ± 0.98) in formula 1 followed by formula 2 in 54.9 ± 0.23. The lowest Mg content was observed in market soup sample [27]. The very low Mg content found in market sample in the present study may be due to the unavailability of Mg found in land plants compared to Mg content of agar extracted from red seaweed. The variation between formula 1 and formulae 11 may be attributes to red seaweed species extracted agar and carrageenan and differences their habitats and metabolic preferences.

The whole samples were with free or low amount was observed in the toxic elements such as Co, Ni, and Cr. This formula 1 and 2 were observed as nutritionally rich source for healthy life of children's and pregnant mothers. The iodine content was found zero in commercial vegetable soup while it ranged 0.7-0.6 mg/100 g in a laboratory prepared soup mixtures. For people with lower organ functions and greater need of optimization of body iodine stores intakes up to 50 mg per day can be use safety [28]. Brown seaweeds are the main source of iodine treated for thyroid conversely, getting sufficient iodine seems to protect against this disease [29].

From the above nutritional data, it could be demonstrated that the dried vegetarian soup mixture had reasonable amount of required nutrients particularly protein, energy, fats, ash and micro elements.

Shelf life assessments

In (-18°C) deep freezer stored samples were found to have overall acceptability 9 (very much like) score from initially to until six month of storage [30-34]. The slight increase of microbial growth was observed after four months and finally it was 3 cfu/g. There were no significant difference observed in all the parameters during six month period storage at -18°C. In storage of 0°C (freezer) was found significant difference from 3rd month to six month in all three parameters analyzed were range 9-7, 1-90 cfu/g, 2-4 cfu/g respectively (Table 6). Overall acceptability, TBC and TFC. Dramatically increased in 27°C sample stored over months. The results show positive correlation of microbial growth and deterioration in sensory quality. The reduction of sensory quality may be due to the slight increase of microbial and fungal growth. The shelf life of (-18°C) deep freezer stored samples can be estimated more than six month. While freezer storage (0°C) samples can be kept five to six month without deterioration [35-37]. The samples stored in ambient temperature have short shelf life around three months.

Conclusion

The two seaweed extracts of carrageen and agar were identified as potential nutritive thickener and gelling agent in the soup industry. The small quantity of agar and carrageenan improved apparent viscosity

Parameters	-18°C storage (Deep freezer)			0°C storage (Freezer)			27°C storage (ambient temperature)		
	Initial	3rd month	Final	Initial	3rd Month	Final	Initial	3rd month	Final
Overall acceptability	9 ± 0.56	9 ± 0.56	9 ± 0.56	9 ± 0.56	8 ± 0.43	7 ± 0.26	9 ± 0.78	8 ± 0.45	6 ± 9.34
Total bacterial count (cfu/g)	1 ± 1.3	1 ± 1,8	3 ± 1,8	1 ± 1.3	3 ± 1..8	90 ± 12	1 ± 0.23	30 6.8 ±	900 ± 24.9
Total fungal count, (cfu/g)	2 ± 0.23	1 ± 0.2	2 ± 0.05	2 ± 0.05	4 ± 0.8	8 ± 1.7	2 ± 0.4	4 ± 0.3	8 ± 2
Water activity (A$_w$)	0.54 ± 0.066	0.54 ± 0.024	0.6 ± 0.034	0.54 ± 0.056	0.6 ± 0.075	0.67 ± 0.056	0.54 ± 0.075	0.73 ± 0.076	0.73 ± 0.087

Table 6: Effect on storage conditions six month duration on sensory and microbial quality of soup formula 1 (Mean ± SE).

value of the product considerably. But we can use limited amount of agar or Carrageenan for development of thickness of soup. The higher proportions of agar or carrageenan may induce to set into a gel in room temperature. It can be concluded that trace metal composition of improved seaweed base soup mixture is five times higher than that of commercially available vegetable soup mixture. The iodine values of the seaweed based soup formulas were higher than the commercial vegetable soup mixture in the available. The iodine level of the seaweed based soup mixtures supply the iodine requirement of thyroid. This soup can be recommended as therapeutic food for dietary iodine and mineral deficiencies. The experiments on two soup products stored in polystyrene packets was expired in three months duration in ambient temperature whereas more than six month in deep freezer and six months duration in freezer storage.

References

1. The World Health Report (2001) Reducing risks, promoting healthy life. World Health Organization, Geneva.

2. Matsuzaki S, Iwamura K (2001) Application of seaweeds to human nutrition and medicine. Nahrungausdem Meer, Food from the sea. Springer, New York.

3. Field CJ, Johnson IR, Schley PD (2002) Nutrients and their role in host resistance to infection. J Leukoc Biol 71: 16-32.

4. Zava TT, Zava DT (2011) Assessment of Japanese iodine intake based on seaweed consumption in Japan: A literature-based analysis. Thyroid Res 4: 14-19.

5. Tilakaratne BM (2013) Formulation of nutritionally superior and cost cereal based soup mix powder using source of protein. Institute of Post-harvest Technology, Anuradapura.

6. Walker A (2000) Micronutrients and infections: an African perspective. Nutrition 16: 1096-1097.

7. FSAU (2005) Micronutrients for healthy, happy families: Micronutrients in Somalia. FSAU/FAO, Kenya.

8. SOP/AP/001/(2003) Standard operation procedure, Determination of agar properties, agar yield, gel strength, ash content and clarity, gelation and melting points, sulphate content and moisture content. MSM Project Management Services Private Limited, Chennai, India.

9. Anon (2011) Researches urge awareness of dietary iodine intake in postpartum Korean American women who consume brown seaweed soup. Science News.

10. AOAC (1984) Official methods of analysis. Washington DC. Association of Official Analytical Chemists.

11. Busta F (1984) Colony count method. Compendium of method for microbiological examination of food.

12. SPSS 14 (2007) Bio-statistical analysis software package.

13. Kaplan BJ, Crawford SG, Field CJ, Simpson JS (2007) Vitamins, minerals, and mood. Psychol Bull 133: 747-760.

14. Ikegwu OJ, Oledinma NU, Nwobasi VN, Alaka IC (2009) Effect of Processing Time and Some Additiveson the Apparent Viscosity of "Achi" Brachystegiaeurycoma Flour. Journal of Food Technology 7: 34-37.

15. Abeysinghe CP, Illepruma CK (2006) Formulation of an MSG (Monosodium Glutamate) free instant vegetable soup mix. Journal of the National Science Foundation of Sri Lanka 34: 91-95.

16. Lyly M, Salmenkallio-Marttila M, Suortti T, Autio K, Poutanen K, et al. (2004) The sensory characteristics and rheological properties of soups containing oat and barley ß-Glucan before and after Freezing. LWT-Food Science and Technology 37: 749-761.

17. Amal MH, Abdel-Haleem, Azza AO (2014) Preparation of dried vegetarian soup supplemented with some legumes. Food and Nutrition Science, 2014, 5, 2274-2285.

18. Garcia JM, Chambers E, Matt Z, Clark M (2005) Viscosity measurement of nectar and honey thick liquid: product, liquid, and time comparisons. Communication Sciences & Disorders, School of Family Studies & Human Services, Kansas State University, Manhattan, Kansas, Dysphaga 20: 325-335.

19. Raitio R, Orlien V, Skibsted LH (2011) Storage stability of cauliflower soup powder: The effect of lipid oxidation and protein degradation reactions. Food Chem 128: 371-379.

20. Micale Donaldson (2011) Recent advance in iodine nutrition.

21. Luo YW, Xie WH, Cui QX (2010) Effects of phytase, cellulase, and dehulling treatments on iron and zinc in vitro solubility in faba bean (Vicia faba L.) Flour and Legume Fractions. J Agric Food Chem 58: 2483-2490.

22. Subba Rao PV, Vaibhave AM, Gunasena K (2011) Mineral composition of edible seaweeds Porphyra vietnamensis. Food Chemistry 102: 215-218.

23. Fellows P (2000) Food processing technology principles and practice. Wood head Publishing Limited and CRC Press LLC, Washington DC.

24. Burtin P (2003) Nutritional value of seaweeds. Elec J Environ Agric Food Chem.

25. Nan F (2011) Your thyroid may be hungry for some seaweed.

26. Greely A (1997) A pinch of controversy shakes up dietary salt. FDA consumer.

27. Wang R, Zhang M, Mujumdar AS, Sun JC (2009) Microwave Freeze-Drying Characteristics and SensoryQuality of Instant Vegetable Soup. Drying Technology 27: 962-968.

28. Istin Sri, Masao ohno, Hirozakusunose (1994) Methods of analysis for agar carrageenan and alginates in seaweeds. Bull Mar Sci Fish, Kochin University, India.

29. Jayasinghe PS, Pahalawattaarachchi V, Ranaweera KKDS (2011) Chemical composition of six edible seaweed species available in Sri Lanka. Proceedings of Annual sessions of National Aquatic Resource Research and Development Agency.

30. Hurtado AQ (2005) Socio-economic impact of epiphytes on the farming and trading of Kappaphycus and processing of Carrageenan (Terminal report).

31. Tadao T (1992) Seaweed, their chemistry and uses, Science of Processing Marine Food Products. Kanagawa International Fisheries Training Centre, Japan International Agency, Japan.

32. Fellows P (2000) Food processing technology principles and practice. Woodhead Publishing Limited and CRC Press LLC, Washington DC.

33. Wardlaw GN (1999) Perspective in nutrition. McGraw-Hills, Boston.

34. García-Pascual P, Sanjuán N, Melis R, Mulet A (2006) Morchella esculenta (Morel) Rehydration Process, Modeling. Journal of Food Engineering 72: 346-353.

35. Padua MD, Fontoura PSG, Mathias AB (2004) Chemical composition of Ulvaria oxysperma (kutzing) Bliding, Ulvalactuca (Linnaeus) and Ulvafascita (Delile). Barazilian Archives of B iology and Technology 47: 49-55.

36. Allen LH (2001) Micronutrients 2020 focus 5 - Health and nutrition emerging and reemerging issues in developing countries.

37. Adsule RN (1996) Food and feed from legumes and oil seeds. Food and feed from legumes and oil seeds, Chapman and Hall Publisher, London.

24

Estimation of Material Losses and the Effects of Cassava at Different Maturity Stages on Garification Index

Sobowale SS[1]*, Awonorin SO[2], Shittu TA[2], Oke MO[3] and Adebo OA[4]

[1]Department of Food Technology, Moshood Abiola Polytechnic, Abeokuta, Ogun State, Nigeria

[2]Department of Food Science and Technology, Federal University of Agriculture, Abeokuta, Ogun State, Nigeria

[3]Department of Food Science and Engineering, Ladoke Akintola University of Technology, Ogbomoso, Oyo State, Nigeria

[4]Department of Biotechnology and Food Technology, University of Johannesburg, Doornfontein 2028, South Africa

Abstract

Gari, a West African staple food was processed using fresh cassava tubers (TMS 30572 cultivar). Material losses and *garification* rate index as affected by the cassava ages of maturity, fermentation days and processing stages were determined. The material losses and yield of *gari* from 9, 12 and 15 months old cassava plants at different processing stages and fermentation days were evaluated. Results showed that the average peeling loss at different maturity ages ranged between 21 and 28.86% while percentage grating loss ranged from 3.71 to 5%. Likewise, estimated percentage dewatering/fermentation loss ranged from 25.55 to 30%, while sieving loss ranged from 4.24 to 5.14%. *Garification* losses ranged from 17.45 to 19.79% with an average *gari* yield between 19.86 and 23.68%. Cassava of 15 months maturity age generally produced higher yields of *gari* than those harvested early. The mean *garification* conversion rate achieved was 22% (0.22, wt/wt).

Keywords: Cassava; Material losses; Yield; Gari; Garification index

Introduction

Cassava (*Manihot esculenta crantz*) is one of the major root tubers produced in the forest zones of Nigeria, with yields as high as 38 million metric tons per annum [1]. It is a major source of staple food in West Africa, providing basic diet to millions of its teeming population across the different socio-economic classes. Due to its relative high moisture and cyanide content, cassava needs to be converted into suitable forms with longer shelf life and lower cyanide levels [2,3]. Different forms in which the processed root exists includes cassava bread, wet chips, *elubo*, *lafun*, cassava starch, roasted/boiled cassava, *fufu (akpu)* and *gari* [4,5].

In Nigeria, similar to other West African countries, over 70% of the cassava yield is processed into *gari*, a staple and convenient food [6]. It is the major form in which cassava is consumed in West Africa. Gari is a fermented roasted granule prepared from peeled, grated and fermented cassava root through a series of processing steps [7]. The conventional method of producing this involves peeling and grating the fresh roots. The grated pulp is then fermented for one to five days. Subsequently, the pulp is compressed to reduce its moisture content, after which it is sieved and fried in heated pans. A dash of palm oil may during grating or at the point of *garifying* to prevent burning, producing a yellowish-*gari* is distinct from white-*gari* in which no oil is added. This is either consumed immediately or further processed into various forms that combine diversity, convenience and nutritional value [8,9].

The traditional method of processing *gari* is an arduous, intricate and tedious operation. This requires a good understanding of combination factors which affects the quality and yield of the *gari* produced [10]. The inability to control moisture content as related to particle size and the exposure of processors to heat and fumes have been major mitigating factors reducing the efficiency of the traditional processing methods.

With the current food crisis in most part of the world and the potential of *gari* to substantially address calorie inadequacy of millions of people in Africa [11], it is imperative to maximize the yield of *gari* processing systems. This will ensure a comprehensive process analysis, to pinpoint factors causing material losses, as a guide in re-designing the *garification* process to minimize losses. This is vital for the development and assessment of new processing technologies [12]. Therefore, the purpose of this work was to establish a standard measure, to quantify the various losses and then determine the *garification* rate index associated with *gari* processing, using a locally fabricated electro-mechanical *garifier*.

Materials and Methods

Processing of cassava tubers into gari

The method described by Achinewhu, *et al.* [12] and Akingbala, *et al.* [13] was largely adopted. Fresh cassava tubers of the TMS 30572 cultivar were harvested at maturity ages of 9, 12 and 15 months. They were sorted according to size, weight and shape by visual assessment. 70 kg of between 300-400 pieces of each maturity age of fresh cassava tubers were selected. There were respectively peeled, washed, drained and grated in a mechanized commercial grater. The grated pulp (mash) was loaded into jute bags and tied with a string. The mash was left to naturally ferment for a period of 0, 1, 2, 3, 4, 5, and 6 days at ambient temperature before pressing out the juice with hydraulic press. The pressed out mash (cake) were manually crushed and sieved (1.50 mm mesh) to remove fibers. The starchy granules obtained were analyzed at different fermentation days and 1 kg from each of the samples was *garified* at a steady state in an electro-mechanical *gari* roaster developed locally with a capacity of 5 kg/h and functional efficiency of 80%. The machine (Figure 1) was operated at a constant stirring rate and allowed for gelatinization at 20 sec interval. This process is called *garification* as it dextrinizes the starch and dries the granules. The *gari* produced

*Corresponding author: Sobowale SS, Department of Food Technology, Moshood Abiola Polytechnic, Abeokuta, Ogun State, Nigeria
E-mail: sobowale.sam@gmail.com

Figure 1: Electro-mechanical *gari* roaster

was air dried, cooled and packaged in polyethylene bags for further processing and analysis.

Material losses

Material losses comprise of both heat and mass transfer, occurring at all the stages of processing. The losses at each processing stage were determined using the difference in the weight of material recovered before and after each stage [14]. This was then subsequently calculated as a percentage of the initial weight of the fresh cassava. Based on this, the individual percentage losses [Peeling (L_p), Grating (L_G), Dewatering/Fermentation (L_D), Sifting (L_s) and *Garifying* (L_R)] were calculated using Eq. (1)-(5).

$$L_p = (W_1 - W_2)/W_1 \times 100 \tag{1}$$

$$L_G = (W_2 - W_3)/W_1 \times 100 \tag{2}$$

$$L_D = (W_3 - W_4)/W_1 \times 100 \tag{3}$$

$$L_s = (W_4 - W_5)/W_1 \times 100 \tag{4}$$

$$L_R = (W_5 - W_6)/W_1 \times 100 \tag{5}$$

where W_1 is the initial weight of fresh cassava tubers (kg), W_2 is the weight of peeled tubers (kg), W_3 is the weight of wet mash collected from the grater (kg), W_4 is the weight of cassava cake after fermentation and dewatering (kg), W_5 is the weight of sifted granule (kg) and W_6 is the weight of the *garified* sample (kg).

Determination of the percentage yield of garification

The total loss for each age of maturity at different fermentation periods were obtained by totaling the losses in Eq. (1)-(5). Hence, the total amount of *gari* obtained, expressed as a percentage of the fresh cassava roots is calculated as follows;

% yield of gari = mass of gari/mass of cassava that produced the gari × 100   (6)

Mean values of the losses were also calculated. Therefore, from the obtained results, a relationship between the initial weight of fresh tubers (W_1) and *gari* yield was established thus:

$$W = W_1 K \tag{7}$$

Where W is the weight of yield of *gari* (kg) and K is a constant [*garification* conversion rate (wt/wt)].

Data analysis

The statistical analysis of the data was conducted using MATLAB commercial software package [Version 7.10.0 (R2010a), Neural Works Professional II/Plus, Neural Ware, Pittsburg, USA]. All experimental data obtained were subjected to analysis of variance (ANOVA) using nonlinear regression model (NLR) procedure of Statistical Analysis System Institute (SAS, 2003). Means were compared at 5% significance level using Duncan's multiple range test, DMRT [15].

Results and Discussion

The estimated material losses at different processing stages and the final yield of *gari* from cassava ages of maturity (9, 12 and 15 months) and days of fermentation (0-6 days) are shown in Figures 2-8. Likewise, the effects of processing variables on the moisture contents are presented in Figures 9-11. The average peeling losses for cassava of 9 months maturity age was 26.86%, the estimated percentage dewatering/fermentation loss ranged from 25.50 to 25.57%. The sieving loss ranged from 5.12 to 5.14%, while the value of *garification* losses ranged from 17.45 to 17.57%. The total loss was recorded to be between 79.93 and 80.14% with an average yield of *gari* ranging between 20.07 and 19.86%. The average peeling loss from 12 months cassava age of maturity was 22.43%. The percentage grating loss was 3.86%, while the percentage

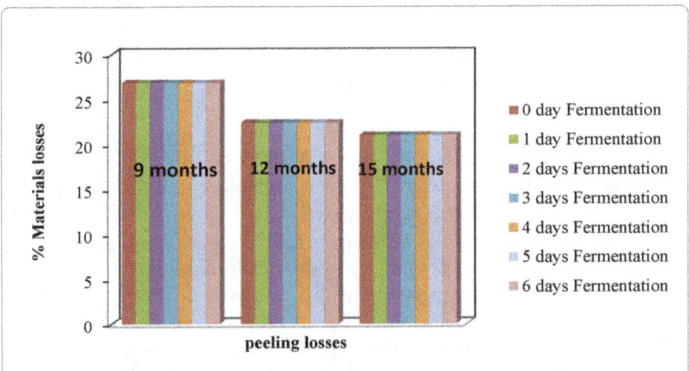

Figure 2: Estimated % peeling losses at different ages of cassava and fermentation days.

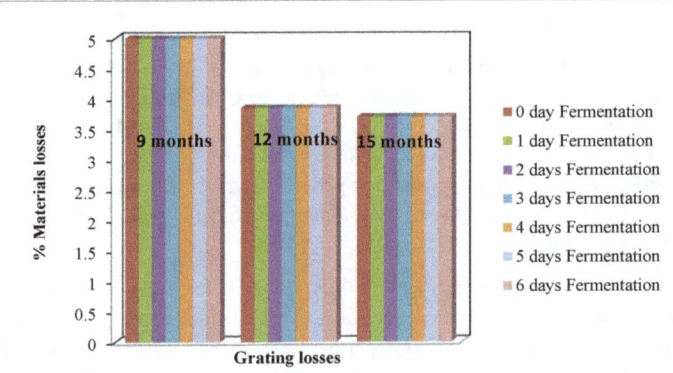

Figure 3: Estimated % grating losses at different ages of cassava and fermention days.

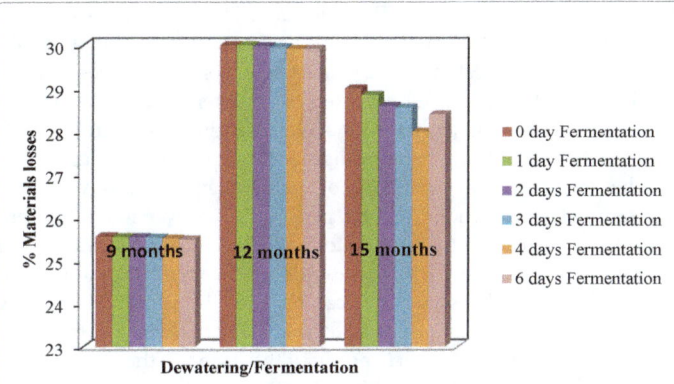

Figure 4: Estimated % dewatering/fermentation losses at different ages of cassava and fermentation days.

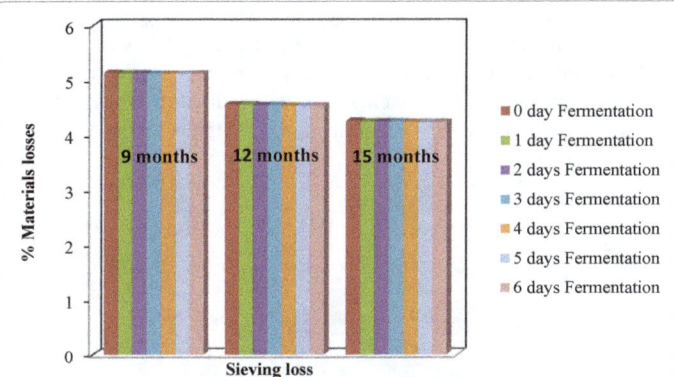

Figure 5: Estimated % sieving losses at different ages of cassava and fermentation days.

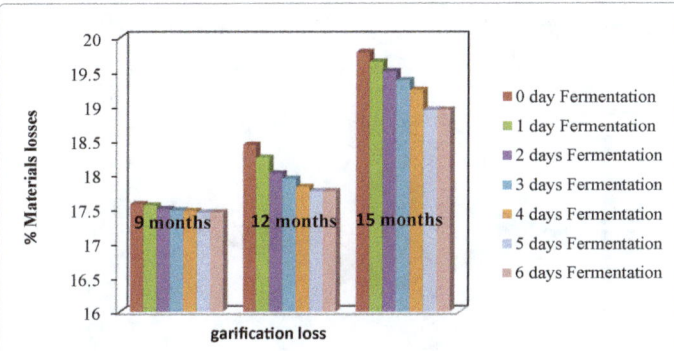

Figure 6: Estimated % *garification* losses at different ages of cassava and fermentation days.

The results from this study revealed that greater losses were recorded during the peeling, dewatering/fermentation and *garification* stages of processing, while grating and sieving had the lowest losses. According to Amoah, *et al.* [14], peeling requires the removal of the outer root cortex which contains the toxic cyanide, rotten portions and stumps. Alongside this, some considerable dry matter is also lost. The amount and quantity of these materials lost, however largely depends on cultivar type, peeling efficiency and quality of the cassava tubers. These peels make up about 10 - 15% of the tuber, but 25 - 30% losses is associated with hand peeling Opara [3] while its about 8.6% with a mechanized peeling machine [16] Losses observed with the mechanical grater used in this study ranged between 2 and 5%, which can be attributed to mash spillage from the hopper due to vibrations during cassava grating.

Losses at the dewatering stage are caused by the removal of the cyanide dissolved water and starch from the mash. The amount of expressed water is relative to the moisture content of the cassava tuber, which is a function of maturity ages. At this stage, lesser losses tend to correspond to cassava age of lesser moisture content. Cassava roots of 15 months age of maturity with higher dry matter accumulation and lower moisture content had lower dewatering losses compared to those harvested early. This result showed consistently higher dewatering losses for cassava harvested at 12 months age of maturity than corresponding 15 months cassava age, while 9 months age of cassava has the least. This may be due to duration of planting and variations in seasons of harvest. As noted by Hahn [17] during the rainy season, the dry matter content

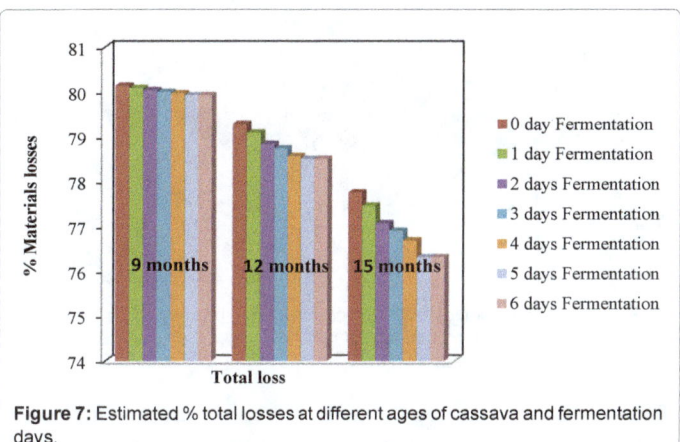

Figure 7: Estimated % total losses at different ages of cassava and fermentation days.

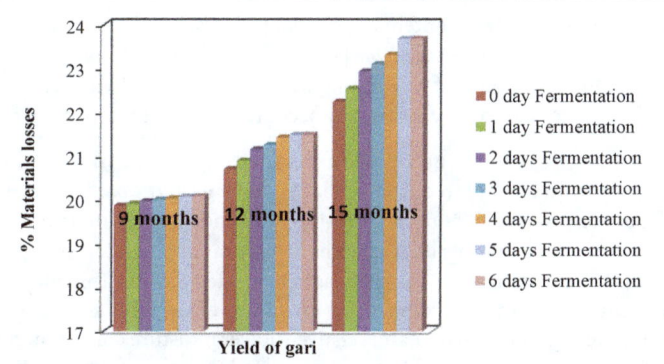

Figure 8: Estimated % yield of *gari* at different ages of cassava and fermentation days.

dewatering/fermentation loss ranged from 29.92 to 30%. The sieving loss ranged from 4.54 to 4.56%, while the *garification* losses ranged from 17.76 to 18.44%. The total loss was estimated to be between 78.51 and 79.29% with an average yield of *gari* ranging between 21.49 and 20.71%. The average peeling loss from 15 months cassava maturity age was 21%. The grating loss was 3.71%, while the estimated percentage dewatering/fermentation loss ranged from 28.42 to 29%. The sieving loss ranged from 4.24 to 4.26%, while the *garification* losses ranged from 18.95 to 19.79%. The estimated total loss was between 76.32 and 77.76% with an average yield of *gari* ranging between 23.68 and 22.24%.

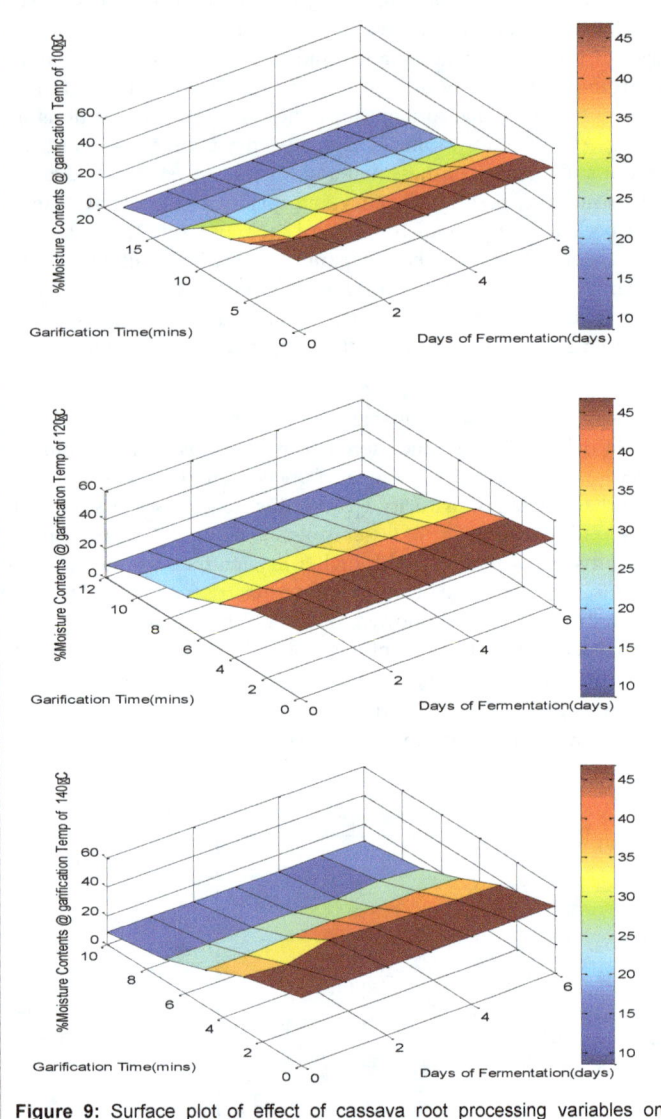

Figure 9: Surface plot of effect of cassava root processing variables on moisture content from 9 months age of maturity.

of roots is lower compared to the dry season. During growth season, both starch and roots production rapidly increases with moisture content to their maximum after which they decline afterwards [3]. The duration of fermentation showed slightly significant effect (p ≤ 0.05) on the moisture content at different processing stages of the *gari*. According to Achinewhu, *et al.* [12], fermentation could play a vital role in the processing of *gari* during which linamarin is hydrolyzed to hydrocyanic acid to a level that may become harmless after *garifying*.

Sifting losses are much more associated with the skill of the processor to ensure reduced spillage while *garification* losses are dependent on *garification* temperature and moisture content before *garifying*. Cassava roots of 15 months age of maturity produced higher yields of *gari* than other maturity ages harvested earlier. This may be accounted for as a result of vascular woodiness of the roots as observed from the higher dry matter and relatively lower moisture content in the older roots. The mean yield of *gari* associated with the *gari* processing system from the ages of cassava used in this study ranged from 19.86 to 23.68%, with average of about 21.77% of the initial weight of

fresh cassava tubers [17,18]. This translates into a mean *garification* conversion rate of 0.22 (wt/wt), which was achieved using a locally fabricated electro-mechanical *gari* roaster. A comparable *garification* rate index of 0.23 had been reported in the literature by Amoah, *et al.* [14]. Since, the cassava tubers harvested at different maturity ages were passed through the set of equipment and processing methods, it may be presumed that the effects of maturity ages of cassava tubers and fermentation periods on the *garification* conversion rate were neutralized by each other. Therefore, the average *garification* rate index of 0.22 can posited to represent the efficacy of the processing system adopted in this study. This can be applied as a numerical index and bench mark for comparison of the efficacy of diverse *garification* processing systems for *gari* production. Hence, the higher the *garification* index of a system, the higher the potential yield of *gari*.

Conclusions

Cassava tubers of 15 months maturity age generally produced higher yields of *gari* than those harvested earlier. The *garification* rate or index achieved was estimated to be averagely 0.22, which can be used as a numerical index and a bench mark for the determination of *garification* processing systems for *gari* production. The study also confirmed that the factors accounting for material losses are age of

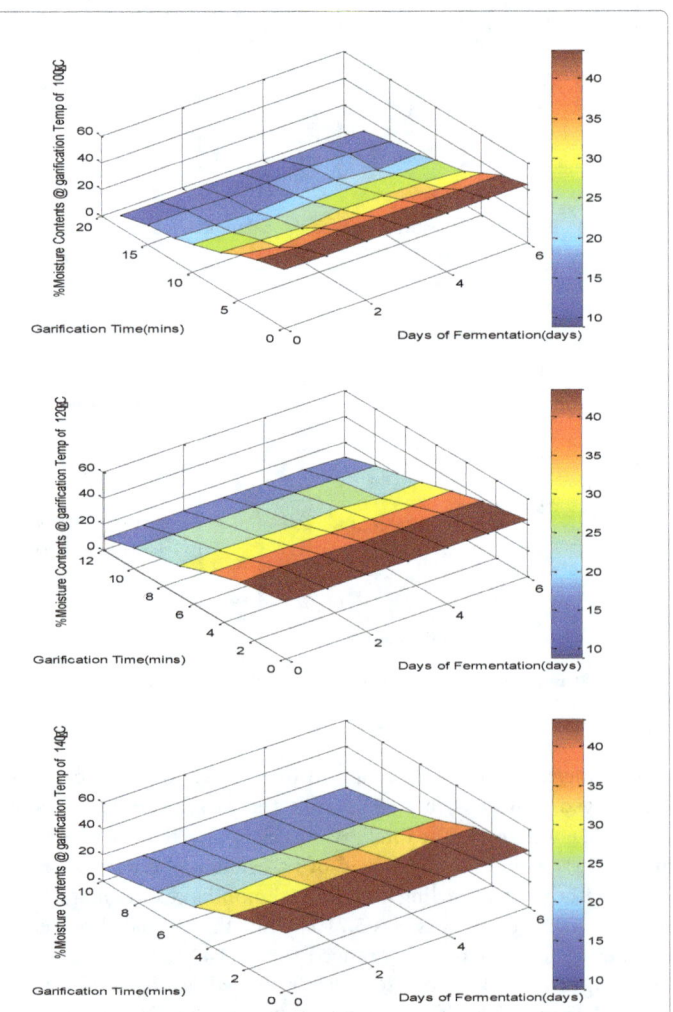

Figure 10: Surface plot of effect of cassava root processing variables on moisture content from 12 months age of maturity.

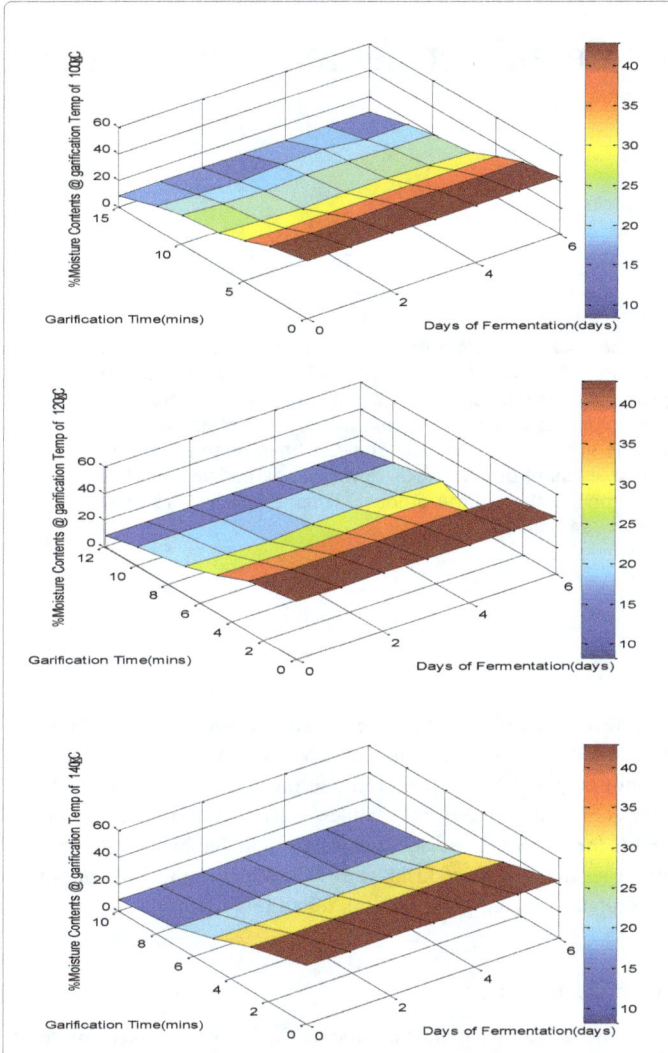

Figure 11: Surface plot of effect of cassava root processing variables on moisture content from 15 months age of maturity.

cassava at point of harvest and processing methods. The highest losses recorded were at the peeling, dewatering/fermentation and *garification* stages, while the least recorded losses were during the grating and sieving stages. Therefore, improved and increased *gari* yield should focus on the development and utilization of cassava tuber varieties with lesser peels and moisture content. These are important considerations to obtain an appreciable yield and quality *gari* for feasible commercial production.

References

1. Nwosu K (2006) Cassava Production, Proceedings of a workshop on electronics Application to information management, held at National Root Crops Research Institute (NRCRI), Umudike, Nigeria.

2. Asiedu JJ (1989) Processing tropical crops. A Technological Approach. Macmillian Press Ltd, London.

3. Opara LU (1999) Yam storage. In: CIGR Handbook of Agricultural Engineering. Agro Processing, The American Society of Agricultural Engineers, St. Joseph, MI, USA.

4. NRCRI (19870 Briefs on research extension and teaching. National Root Crops Research Institute, (NRCRI), Umudike, Nigeria.

5. FAO (2006) Bankable investment project profile. Cassava Production, Processing and Marketing Project. Food and Agriculture Organization of the United Nation Rome.

6. Achinewhu SC, Barber LI, Ijeoma IO (1998) Physicochemical properties and garification (gari yield) of selected cassava cultivars in Rivers State, Nigeria. Plant Foods Hum Nutr 52: 133-140.

7. Akingbala JO, Oyewole OB, Uzo-Peters PI, Karim RO, Baccus-Taylor GH, et al. (2005) Evaluating stored cassava quality in gari production. J Food Agric Environ 3: 75-80.

8. Amoah RS, Sam-Amoah LK, Adu BC, Duah F (2009) Estimation of the material losses and gari recovery rate during the processing of varieties and ages of cassava into gari. J Agric Res 5: 1-9.

9. Hahn SK (2006) An overview of traditional processing and utilization of Cassava in Africa.

10. Ingram JS, Humphries JRV (1972) Food crops of the lowland tropics. Oxford University Press, Oxford, London.

11. Jekayinfa SO, Olajide JO (2007) Analysis of energy usage in the production of three cassava-based foods in Nigeria. J Food Eng 82: 217-226.

12. Oduro I, Ellis WO (2000) Production of cassava and sweet potatoes based on snack food. Proposal Submitted to the Root and Tuber Improvement Programme, Ghana.

13. Owuamanam CI, Iwouno JO, Ihediohanma NC, Barber LI (2010) Cyanide reduction, functional and sensory quality of gari as affected by pH, temperature and fermentation time. Pakistan J Nutr 9: 980-986.

14. Sanni MO, Oluwabami AO (2003) The effect of cassava post-harvest and fermentation time on gari sensory qualities. Donald Danforth Plant Science Centre, Missouri, USA.

15. Sanni MO (1994) Garri processing in Ibadan metropolis: factors controlling quality at the small-scale level. Acta Hortic 380: 256-260.

16. Sobowale SS, Adebiyi JA, Adebo OA (2015) Design and performance evaluation of a melon sheller. J Food Proc Eng.

17. Steel RGD, Torrie JH (1980) Principle and procedure of statistics, Mc Grawhill, New York, USA.

18. Ukpabi UJ, Ndimele C (1990) Evaluation of the quality of gari produced in Imo state. Nig Food J 8: 105-109.

25

Improvement of the Nutritional Value of a Cereal Fermented Milk: 2-Dried Kishk Like

Nassar KS[1]*, Shamsia SM[1] and Attia IA[2]

[1]Department of Food and Dairy Science and Technology, Damanhour University, Egypt
[2]Department of Dairy Science and Technology, Alexandria University, Egypt

Abstract

The present study has been conducted to produce fermented milk fortified with different cereals like whole wheat, barley and freek (green wheat) burghul have been selected for their known nutrition benefits. The fermentation was occurred by using three types of cultures, yogurt starter, yogurt starter + Bio-yogurt or yogurt starter + *Lactobacillus plantarum*. All samples were stored at room temperature (25 ± 2°C) for three months and have been subjected to consumer sensory testing; dried kishk-like products were highly accepted by the tasting panel, furthermore fermented dairy products containing Freek gained the highest score of judging followed by wheat. Proximate composition, Colour, Organic acids and microbiologically analysis have been monitored in the fresh soft product and during storage. Nevertheless, the combined levels of organic acids, low pH, salt additive and low moisture content in the kishk samples were sufficient to ensure the microbial safety of the product. Thus, long shelf-life of all samples without changing in their chemical during the storage period has been noticed.

Keywords: Burghul; Probiotic bacteria; Dried kishk; Organic acids; Colour; Skim milk

Introduction

Fermentation is one of the oldest and most economical methods of producing and preserving food. Also, leads to a general improvement in the shelf life, texture, taste and aroma of the final product. Fermented foods are produced world-wide using various manufacturing techniques, raw materials and microorganisms [1]. Kishk is an extremely popular fermented food in many parts of the Middle East. In Egypt, Kishk is one of the traditional food products in Upper Egypt [2,3]. There are some other products similar to kishk such as tarhana (mixing yogurt, wheat flour, baker's yeast and variety of vegetables and spices in turkey), kushuk (milk- sour dough mixture with turnips in Iraq), atole (fermented cereal–milk porridge in Scotland and Greece) and tahonya/talkuna (fermented cereal mixture with vegetables in Finland and Hungary) [4]. Kishk is a natural, healthy, respect the environment and have great taste and cultural values that are increasingly attractive to the Egyptian consumers. Kishk made by mixing wheat with fermented milk (Laban zeer) and sun-drying the mixture to 8% to 12% moisture content [2,5] or made from different cereal products and fermented milk base by traditional methods of manufacture [6,7]. Kishk is usually reconstituted with water and served as a hot gruel, but with the incorporation of vegetables, spices, garlic or herbs, can form the base of savory and sweet dishes [8]. The aim of the present study was to focus some light on the chemical, biochemical, physical and sensory aspects of dried fermented milk made from various types of cereal and starter cultures.

Materials and Methods

In the previous paper, the preparation kishk-like from reconstituted skim milk and burghul from whole wheat, whole barley and fereek was described in Nassar et al. [9].

Dried Kishk-like manufacturing

Each type of burghul was mixed with reconstituted skim milk in a ratio of 1:4 (w/w) in addition to, 2% sodium chloride and then mixed thoroughly each 26%. Mixture was heated to 95°C for 10 seconds, and then rapidly cooled to 45°C, addition 3% of each Starter. The resultant paste was filled in polystyrene cups and covered then incubated at (43°C for W1, B1 and F1) and (37°C for W2, W3, B2, B3, F2 and F3) to 6 hours (Table 1). After that, the fermented paste was formed into nuggets (3-5 cm), placed into stainless steel trays and dried in air oven at 50°C for 15-18 h. the dried nuggets were milled by using a hammer mill. After that, the dried kishk were stored in airtight glass containers and kept at room temperature until tested (Figure 1).

Chemical analysis

In the previous paper, the same methods were adopted for cereal and dairy base analysis as described in Nassar et al. [9].

Dried Kishk-like analysis

Proximate composition: Total solids, protein, ash and fiber content were determined according to procedures described by AOAC [10]. The fat content, salt percentage, acidity and pH value were determined according to Ling [11]. Finally, carbohydrate was calculated as follows [12]:

Carbohydrate = total solids - (protein + fat + ash)

Colour of kishk-like samples: Colour of dried kishk samples was evaluated by Lovibond Schofield Tintometer (the Tintometer Ltd. Salisbury, England). Colours of samples were assessed. Reading obtained was further converted into C.I.E. (Commission International de L'E Clairage) units using the visual density graphs and direction booklet supplied with apparatus as described in AOCS [13].

*Corresponding author: Khaled S Nassar, Department of Food and Dairy Science and Technology, Damanhour University, Egypt
E-mail: Khalid.nassar@agr.dmu.edu.eg

Treatments	Cereals			Dairy base	Salt 2%
	Whole Wheat Burghul (W)	Whole Barley Burghul (B)	Freek Burghul (F)	Re-constituted Skim milk (15%)	
1	√			√	√
		√		√	√
			√	√	√
2	√			√	√
		√		√	√
			√	√	√
3	√			√	√
		√		√	√
			√	√	√

1: (3% Yoghurt starter)
2: (2% Yoghurt starter + 1% Bio-yoghurt starter)
3: (2% Yoghurt starter + 1% *Lactobacillus plantarum*)

Table 1: Experimental treatments.

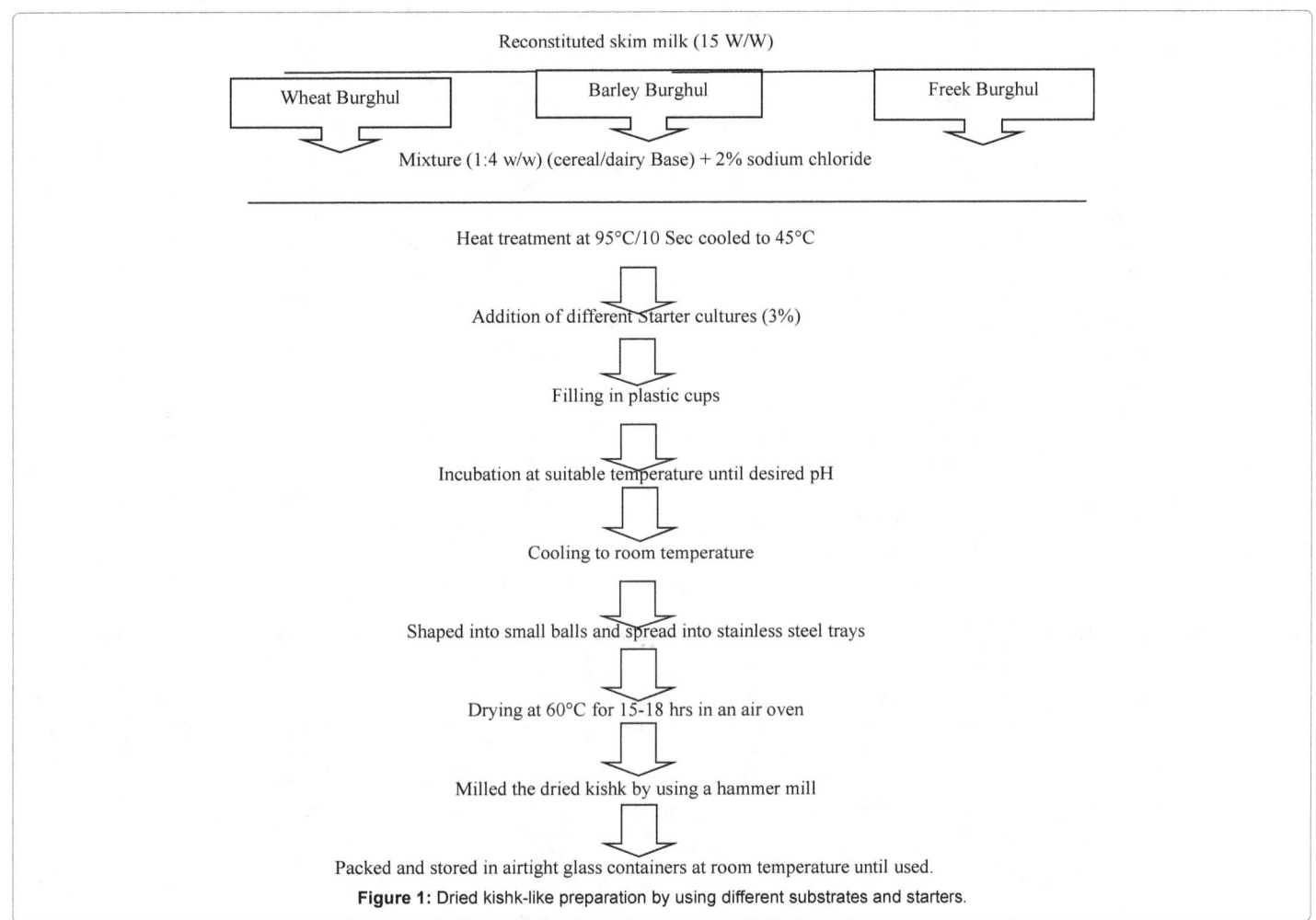

Figure 1: Dried kishk-like preparation by using different substrates and starters.

Organic acid determination: The concentrations of organic acids (Lactic, Propionic, acetic, and formic) in different dried Kishk samples were determined by HPLC (Spectra-Physics system, San Jose, CA, USA) method as described by Barrantes et al. [14]. Organic acids were extracted from dried Kishk (5 g) in a 50 mL beaker using 25 mL water-acetonitrile (1:4 v/v) (analytical grade, BDH Chemicals Ltd., Poole, UK). The extract after filtration through a Whatman No. 1 filter paper (Whatman Ltd., Maidstone, UK) was injected (20 µL) into the HPLC column. The flow rate of the solvent was 0.7 µL.min^{-1} at 65°C and the wave length of the detector was 220 nm [15].

Sensory evaluation

Organoleptic evaluation was carried out according to Abou-Donia et al. [16]. The samples were subjected to organoleptic analysis by 15 well-trained members of the Dairy Science and Technology Department (Fac. Agric. Alexandria Univ., Egypt). The sensory attributes evaluated were: The Flavor (1-45 points), Body and Texture (1-30 points), appearance and Colour (1-15 points) and acidity (1- 10 points). Soups were prepared by adding 20 g dried kishk to 170 mL of water and heating with gentle stirring to boiling, simmering for few min and cooling to 40°C. The samples (20 g) were placed in identical glass containers (8 cm diameter, 3.5 cm height) and served at 40°C [17].

Enumeration of microorganisms

The counts of lactic acid bacteria were enumerated as (CFU/g) using MRS agar according to De man et al. [18]. Proteolytic bacteria, coliforms, yeast and mould were enumerated respectively [19-21]. Whereas aerobic spore forming bacteria were enumerated according to the method described by Harrigan and McCance [22]. *Lactobacillus acidophilus* was enumerated according to the method described by Lapierra [23].

Statistical analysis

Statistical analysis was performed by applying three-way ANOVA and multiple comparisons of means of each treatment (cereals, starter cultures and storage time) using the Least Significant Difference (LSD) test at the confidence level of 95% [24].

Results and Discussion

Proximate composition

Proximate analysis of kishk-like products is presented in Tables 2-4. The results revealed that the effect of cereal type on the proximate analysis of the resultant dried kishk-like products was more pronounced ($P \leq 0.05$) than that of type of starter culture used.

There were significant differences ($P \leq 0.05$) in acidity and pH values between different cereals fermented milk products, depending on the type of cereal or starter culture. The fresh soft cereal fermented

Samples	Storage period (Days)	Acidity as lactic acid	pH	Total Solids%	Fat content%	Ash%	Crude Fiber%	Crude protein%	Carbohydrates%	Salt%
W1	Fresh soft pro.	0.300[PQR]	5.44[A]	27.42[RS]	0.280[L]	2.98[J]	0.36[E]	4.99[K]	18.80[NOP]	2.06[B]
	Fresh dried pro.	1.803[IJKL]	5.23[D]	93.27[A]	0.95[A]	10.15[A]	1.21[B]	16.97[ABCDEFG]	63.98[CDE]	7.00[A]
	30	1.880[HIJK]	5.17[EF]	92.65[AB]	0.95[A]	10.08[A]	1.206[B]	16.86[ABCDEFGH]	63.56[DEF]	6.96[A]
	60	2.080[CDEFG]	5.05[J]	92.25[BCDEFG]	0.94[B]	10.04[A]	1.20[B]	16.79[ABCDEFGH]	63.28[EFG]	6.93[A]
	90	2.136[BCDE]	5.02[K]	91.46[HIJKLM]	0.933[BC]	9.95[AB]	1.19[B]	16.64[BCDEFGHI]	62.74[FGH]	6.87[A]
W2	Fresh soft pro.	0.223[R]	5.23[D]	28.92[P]	0.290[K]	2.88[J]	0.36[E]	5.03[K]	20.36[M]	2.14[B]
	Fresh dried pro.	1.963[FGH]	5.14[GH]	92.35[BCDE]	0.94[B]	9.206[GHI]	1.15[B]	16.06[IJ]	64.99[ABC]	6.85[A]
	30	1.976[FGH]	5.12[H]	92.29[BCDEF]	0.94[B]	9.20[GHI]	1.15[B]	16.05[IJ]	64.95[ABC]	6.85[A]
	60	2.220[ABC]	5.01[K]	91.86[CDEFGHIJ]	0.933[BC]	9.16[GHI]	1.14[B]	15.97[J]	64.64[ABCD]	6.82[A]
	90	2.336[A]	4.93[M]	91.70[FGHIJKL]	0.930[C]	9.14[GHI]	1.14[B]	15.94[J]	64.53[ABCD]	6.80[A]
W3	Fresh soft pro.	0.226[QR]	5.32[C]	28.11[Q]	0.290[K]	2.85[J]	0.36[E]	5.02[K]	19.58[MN]	2.10[B]
	Fresh dried pro.	1.896[HIJ]	5.19[E]	92.28[BCDEF]	0.94[B]	9.36[EDFG]	1.18[B]	16.48[FGHIJ]	64.31[ABCDE]	6.91[A]
	30	1.926[GHI]	5.16[FG]	91.93[CDEFGHI]	0.94[B]	9.33[FGH]	1.17[B]	16.42[GHIJ]	64.07[CDE]	6.88[A]
	60	2.146[BCD]	5.08[I]	91.78[EFGHIJK]	0.94[B]	9.31[GH]	1.17[B]	16.39[GHIJ]	63.96[CDE]	6.87[A]
	90	2.226[AB]	4.92[M]	91.33[IJKLM]	0.933[BC]	9.27[GHI]	1.17[B]	16.31[HIJ]	63.65[DEF]	6.84[A]
SED		0.026	0.004	0.108	0.001	0.048	0.012	0.111	0.191	0.085
R-Square		0.996	0.999	0.999	0.998	0.999	0.9993	0.998	0.9997	0.996
Coeff. Var.		2.844	0.141	0.238	0.367	1.041	1.683	1.35	0.609	2.501

[A-M]: All the means were differentiated by a standard deviation (p<0.05).

Table 2: Chemical properties of cereal fermented milks using burghul from whole wheat.

Samples	Storage period (Days)	Acidity as lactic acid	pH	Total Solids%	Fat content%	Ash%	Crude Fiber%	Crude protein%	Carbohydrates%	Salt%
B1	Fresh soft pro.	0.423[P]	5.23[D]	27.96[QR]	0.266[MN]	2.86[J]	0.70[D]	5.10[K]	18.19[OP]	2.07[B]
	Fresh dried pro.	1.910[HIJ]	4.97[L]	91.98[CDEFGH]	0.900[DE]	9.71[BC]	2.38[A]	17.31[A]	61.68[HIJK]	7.01[A]
	30	1.986[EFGH]	4.93[M]	91.85[CDEFGHIJ]	0.900[DE]	9.70[BC]	2.38[A]	17.28[AB]	61.58[IJK]	7.00[A]
	60	2.233[ABC]	4.87[NO]	91.55[HIJKLM]	0.893[EF]	9.66[C]	2.37[A]	17.23[ABC]	61.39[IJKL]	6.98[A]
	90	2.310[A]	4.46[U]	91.43[HIJKLM]	0.890[F]	9.65[C]	2.37[A]	17.20[ABCD]	61.30[IJKL]	6.97[A]
B2	Fresh soft pro.	0.440[P]	4.94[M]	26.70[T]	0.260[N]	2.83[J]	0.70[D]	5.05[K]	17.85[P]	2.03[B]
	Fresh dried pro.	1.970[FGH]	4.87[NO]	91.09[LMN]	0.900[DE]	9.65[C]	2.40[A]	17.24[ABC]	60.90[JKL]	6.95[A]
	30	2.086[CDEF]	4.74[Q]	90.96[MN]	0.900[DE]	9.64[C]	2.39[A]	17.21[ABCD]	60.81[JKL]	6.93[A]
	60	2.250[AB]	4.66[S]	90.63[NO]	0.890[F]	9.62[CDE]	2.38[A]	17.15[ABCDE]	60.58[KL]	6.91[A]
	90	2.336[A]	4.52[T]	90.30[O]	0.890[F]	9.59[CDEF]	2.37[A]	17.09[ABCDEF]	60.35[L]	6.88[A]
B3	Fresh soft pro.	0.413[PQ]	5.43[AB]	27.29[ST]	0.270[M]	2.85[J]	0.70[D]	5.10[K]	17.94[P]	2.03[B]
	Fresh dried pro.	1.926[GHI]	5.22[D]	92.69[AB]	0.903[D]	9.70[BC]	2.38[A]	17.32[A]	62.38[GHI]	6.88[A]
	30	2.033[DEFGH]	5.07[IJ]	92.21[BCDEFG]	0.900[DE]	9.65[C]	2.37[A]	17.23[ABC]	62.05[HI]	6.85[A]
	60	2.190[ABC]	5.01[K]	91.84[DEFGHIJK]	0.896[DEF]	9.61[CDE]	2.36[A]	17.16[ABCDE]	61.80[HIJ]	6.82[A]
	90	2.266[AB]	4.89[N]	91.64[GHIJKL]	0.890[F]	9.62[CD]	2.35[A]	17.12[ABCDE]	61.64[HIJK]	6.81[A]
SED		0.026	0.004	0.108	0.001	0.048	0.012	0.111	0.191	0.085
R-Square		0.996	0.999	0.999	0.998	0.999	0.9993	0.998	0.9997	0.996
Coeff. Var.		2.844	0.141	0.238	0.367	1.041	1.683	1.35	0.609	2.501

[A-M]: All the means were differentiated by a standard deviation (p < 0.05)

Table 3: Chemical properties of cereal fermented milks using burghul from whole barely.

Samples	Storage period (Days)	Acidity as lactic acid	pH	Total Solids%	Fat content%	Ash%	Crude Fiber%	Crude protein%	Carbohydrates%	Salt%
F1	Fresh soft pro.	0.306[PQR]	5.41[B]	27.20[ST]	0.220[P]	2.72[J]	0.27[F]	4.93[K]	19.05[NO]	2.03[B]
	Fresh dried pro.	1.490[O]	5.24[D]	92.46[BCD]	0.733[I]	9.24[GHI]	0.93[C]	16.77[ABCDEFGH]	64.77[ABC]	6.90[A]
	30	1.596[NO]	5.09[I]	91.67[FGHIJKL]	0.730[I]	9.166[GHI]	0.92[C]	16.63[CDEFGHI]	64.22[BCDE]	6.85[A]
	60	1.690[LMN]	5.01[K]	91.36[HIJKLM]	0.730[I]	9.136[GHI]	0.92[C]	16.58[DEFGHIJ]	64.00[CDE]	6.82[A]
	90	1.790[IJKL]	4.92[M]	91.22[KLMN]	0.720[J]	9.12[GHI]	0.92[C]	16.55[EFGHIJ]	63.90[CDE]	6.80[A]
F2	Fresh soft pro.	0.296[PQR]	5.33[C]	27.93[QR]	0.230[O]	2.74[J]	0.27[F]	4.94[K]	19.74[MN]	2.05[B]
	Fresh dried pro.	1.593[NO]	5.14[GH]	92.47[BC]	0.750[H]	9.09[GHI]	0.91[C]	16.37[GHIJ]	65.33[A]	6.81[A]
	30	1.740[KLMN]	5.08[I]	92.28[BCDEF]	0.750[H]	9.07[HI]	0.91[C]	16.34[GHIJ]	65.20[AB]	6.79[A]
	60	1.806[IJKL]	4.82[P]	91.89[CDEFGHI]	0.750[H]	9.13[GHI]	0.90[C]	16.27[HIJ]	64.82[ABC]	6.76[A]
	90	1.940[FGHI]	4.70[R]	91.72[EFGHIJK]	0.750[H]	9.02[I]	0.90[C]	16.24[HIJ]	64.80[ABC]	6.66[A]
F3	Fresh soft pro.	0.310[PQR]	5.41[B]	27.12[ST]	0.290[O]	2.86[J]	0.27[F]	4.92[K]	19.67[MN]	2.05[B]
	Fresh dried pro.	1.500[O]	5.05[J]	91.44[HIJKLM]	0.766[G]	9.35[EDFGH]	0.89[C]	16.11[IJ]	64.31[ABCDE]	6.71[A]
	30	1.516[O]	5.02[K]	91.34[IJKLM]	0.763[G]	9.34[EDFGH]	0.88[C]	16.09[IJ]	64.25[ABCDE]	6.71[A]
	60	1.623[MNO]	4.88[N]	91.32[IJKLM]	0.760[G]	9.34[EDFGH]	0.88[C]	16.09[IJ]	64.24[ABCDE]	6.70[A]
	90	1.766[JKLM]	4.85[O]	91.25[JKLMN]	0.760[G]	9.333[FGH]	0.88[C]	16.08[IJ]	64.18[BCDE]	6.67[A]
SED		0.026	0.004	0.108	0.001	0.048	0.012	0.111	0.191	0.085
R-Square		0.996	0.999	0.999	0.998	0.999	0.9993	0.998	0.9997	0.996
Coeff. Var.		2.844	0.141	0.238	0.367	1.041	1.683	1.35	0.609	2.501

[A-T]: All the means were differentiated by a standard deviation (p < 0.05)

Table 4: Chemical properties of cereal fermented milks using burghul from freek (green wheat).

milk products containing Barely (B2, B1 and B3, Respectively) were characterized by higher acidity as compared with their containing of freek (F3, F1 and F2) and wheat (W1, W3 and W2, Respectively). These results are in agreement with [9,16,25-28]. There is no significant difference ($P \leq 0.05$) between the values of ash, salt, protein and carbohydrate of kishk-like before draying. While, ranged as follow: (2.72-2.98), (2.03-2.14), (4.92-5.10) and (17.85-20.36), respectively. The crude fibers content was significant ($P \leq 0.05$) depending on the type of cereal used. Whereas, the barley kishk products had higher its values (0.70%) than other samples, this is due to the higher content of crude fiber which reached 2.51% [9,25]. As expected the type of starter culture used in the fermentation did not effect on the total solids and fat contents but the variation could be attributed to the fat content of kishk components and blends. These results were in agreement with those obtained by [2,9,25,29].

There was a significant ($P \leq 0.05$) increase in pH and acidity percentages for all fresh samples incorporation with the corresponding values of dry products (Tables 2-4). However, the samples of dried cereal fermented milk products which fermented with Yoghurt and Bio-Yoghurt starter (2) were characterized with higher acidity rates than either fermented with mixed cultures contained only Yoghurt starter culture (1) or fermented by mixed cultures of Yoghurt starter culture and *Lactobacillus plantarum* (3), respectively. During storage at (25 ± 2°C) for 90 days, significant decrease ($P \leq 0.05$) were recorded in pH of different dried kishk-like products. Moreover, gradual decrease in pH could be observed in all samples, with extending the storage period (3 months), that due to limit growth of various bacterial starter cultures and the slow fermentation of residual lactose [9,25,29-31].

Dried kishk-like samples had slightly decrease in total solids, fat, Ash, crude fibers, protein, carbohydrate and salt contents until the end of storage period [16,31]. The ranges of previous contents were as follow: (90.30% to 93.27%), (0.720% to 0.950%), (9.02% to 10.15%), (0.89% to 2.40%), (15.94% to 17.31%), (60.35% to 65.33%) and (6.66% to 7%) respectively. These results are in agreement with previous studies [2,16,31-35].

Colour of kishk samples

Data presented in Table 3 show the colour of dried kishk samples as obtained using Lovibond Tintometer. The values of the primary Colours (Red, yellow and blue) showed that all dried kishk samples as ranged (3 - 5), (6 - 6.9) and (1.9 - 4.9), respectively. The data for primary colours reflects the values of X, Y and Z coordinates. The X value for (B1) treatment had significantly higher than among of samples (0.395) while the (F3) sample had the highest value of Y coordinate (0.544). Whereas, the Z coordinate showed the opposite figure. As a result of saturation and visual density, the brightness of (W1) sample was the highest percentage (66.07%). on contrary, (F3) treatment had lower brightness (42.66%) than other samples. These results are in agreements with Toufeli et al. [17], Bilgicli and Ibanoglu [36].

Organic acids determination

The concentrations (*ppm*) of organic acids (Lactic acid, Propionic acid, Acetic acid and formic acid) in the dried Kishk samples after 90 days of storage made with wheat, barley and Freek Burghul Skim milk is shown in Table 4. Lactic, propionic and acetic acid formed during lactic fermentation. The (W1) treatment had the highest levels of Propionic, Acetic and Lactic acids respectively, except formic acid comparing to among of samples [6,15]. Furthermore, propionic acid had the lowest or not detected in kishk samples made with barley, and these results are in agreement with Tamime et al. [37].

Microbiological analysis

Microbiological composition of kishk-like products are shown in Table 5. To produce healthy and safety of kishk that it's based on the critical control points during the making of kishk were cooking, fermentation, drying and storage [35,38]. The fermented fresh soft products had 2.30-2.48 and 1.60-2.44 c.f.u × 10^{-3}/ gm. for lactic acid bacteria and *Lactobacillus acidophilus*, respectively. However, the drying treatment gets rid of the bacterial starter, and these results agreement with [2,16]. Furthermore, all the samples did not contain any growth in 0.1 gm on SDA, VRBA, NA and MSA media in either fresh or dried products through the storage period. These results are revealed

Samples	Red	Yellow	Blue	X	Y	Z	Visual density	Saturation%	Brightness%
W1	3	6	1.9	0.355	0.415	0.23	0.18	38.64	66.07
W2	4	6	2.6	0.34	0.425	0.235	0.23	45.45	58.88
W3	4	6	2.7	0.355	0.475	0.19	0.235	59.57	58.21
B1	4	6	2.7	0.395	0.415	0.19	0.235	59.57	58.21
B2	4.6	6.9	3.3	0.385	0.465	0.15	0.28	66.67	52.48
B3	4.4	6.9	3.3	0.35	0.53	0.12	0.25	70.21	56.23
F1	5	6.9	3.9	0.36	0.53	0.11	0.33	79.07	46.77
F2	4.6	6.9	3.9	0.36	0.525	0.115	0.35	70.21	44.76
F3	4.9	6.9	4.9	0.356	0.544	0.1	0.37	83.72	42.66

B: Dried kishk like manufactured from fermented whole barley burghul skim milk.
F: Dried kishk like manufactured from fermented freek burghul skim milk.
W: Dried kishk-like manufactured from fermented whole wheat burghul skim milk.
Starters: 1: (3% Yoghurt starter); 2: (2% Yoghurt starter + 1% Bio-yoghurt starter); 3: (2% Yoghurt starter + 1% *Lactobacillus plantarum*).

Table 5: Colour of dried kishk-like samples as measured by Lovibond and C.I.E system.

Samples	Organic acids			
	Lactic acid *ppm*	Propionic acid *ppm*	Acetic acid *ppm*	Formic acid *ppm*
W1	1134.57	3522.16	2930.35	< 50
W2	908.94	3341.66	2728.21	< 50
W3	901.86	3451.5	973.285	< 50
B1	903.9	< 50	2147.14	< 50
B2	889. 16	< 50	< 50	320.05
B3	717.5	< 50	2153.64	366.248
F1	601.25	562.33	< 50	738.954
F2	691.52	562.7	481.357	< 50
F3	638.33	598.72	< 50	< 50

B: Dried kishk-like manufactured from fermented whole barley burghul skim milk.
F: Dried kishk-like manufactured from fermented freek burghul skim milk.
W: Dried kishk-like manufactured from fermented whole wheat burghul skim milk.
Starter: 1: (3% Yoghurt starter), 2: (2% Yoghurt starter + 1% Bio-yoghurt starter) and 3: (2% Yoghurt starter + 1% *Lactobacillus plantarum*).

Table 6: Organic acids concentrations of dried cereal fermented milks.

Samples	Storage Period (Days)	MRS	MRS+L-Cysteine	NA	SDA	VRBA	MSA
W1	Fresh soft pro.	2.3					
	Fresh dried pro. 90	< 0.001	----				
W2	Fresh soft pro.	2.56	1.6				
	Fresh dried pro. 90	< 0.001	< 0.001				
W3	Fresh soft pro.	2.84					
	Fresh dried pro. 90	< 0.001	----				
B1	Fresh soft pro.	2.32					
	Fresh dried pro. 90	< 0.001	----				
B2	Fresh soft pro.	2.31	1.8				
	Fresh dried pro. 90	< 0.001	< 0.001	Not detected in 0.1 gm.			
B3	Fresh soft pro.	2.76					
	Fresh dried pro. 90	< 0.001	----				
F1	Fresh soft pro.	2.48					
	Fresh dried pro. 90	< 0.001	----				
F2	Fresh soft pro.	2.68	2.44				
	Fresh dried pro. 90	< 0.001	< 0.001				
F3	Fresh soft pro.	2.64					
	Fresh dried pro. 90	< 0.001	----				

W: Soft kishk-like manufactured from fermented whole wheat burghul skim milk.
B: Soft kishk-like manufactured from fermented whole barley burghul skim milk.
F: Soft kishk-like manufactured from fermented freek burghul skim milk.
Starter: 1: (3% Yoghurt starter), 2: (2% Yoghurt starter + 1% Bio-yoghurt starter) and 3: (2% Yoghurt starter + 1% *Lactobacillus plantarum*). (--): Not determined.
NA: nutrient agar; MSA: manitol salt agar; SDA: Sabouraud dextrose agar; MRS: Man rogosa sharpe agar; MRS: Man rogosa sharpe agar.
VRBA: Violet Red Bile Agar.

Table 7: Changes in viable microbial counts (c.f.u × 10^{-3}/ gm.) in dried cereal fermented milks.

the good hygiene sanitation during manufacture different products. The low pH (4.46-5.23), release of organic acids due to fermentation, salt additive and low moisture content (6.73% to 9.70%) lead to a harsh environment (bacteriostatic effect) for pathogenic microorganisms, in which food spoilage may not occur and shelf life increases [35].

Sensory evaluation of recombined dried kishk like

Sensorial evaluation of recombined dried kishk samples is given in Tables 6-8. There were no significant differences ($P \leq 0.05$) between the soup samples in flavour and body and texture scores. However, significant difference existed with regard to appearance and color and acidity, with barley skimmed milk Kishk soup having the lowest score. It was clear that, the addition of wheat or freek burghul didn't affect the

Sample	Flavour -45	Body / texture -30	Appearance and colour -15	Acidity -10	Total -100
W1	38AB	25A	13A	7BC	83A
W2	39AB	25A	12A	7BC	83A
W3	40A	26A	11A	6C	82A
B1	33C	24A	8B	6C	71B
B2	36BC	24A	8B	5D	70B
B3	34C	24A	9B	6C	74B
F1	40A	25A	13A	9A	87A
F2	40A	24A	12A	8AB	84A
F3	40A	24A	13A	8AB	85A
SED	0.555	0.368	0.43	0.248	0.881
R-Square	0.897	0.53	0.928	0.883	0.95
Coeff. Var.	2.534	2.583	6.663	6.246	1.891

W: Dried kishk-like manufactured from fermented whole wheat burghul skim milk.
B: Dried kishk-like manufactured from fermented whole barley burghul skim milk.
F: Dried kishk-like manufactured from fermented freek burghul skim milk.
Starter: 1: (3% Yoghurt starter); 2: (2% Yoghurt starter + 1% Bio-yoghurt starter); 3: (2% Yoghurt starter + 1% *Lactobacillus plantarum*)
SED: Standard Error of Difference.

Table 8: Organoleptic properties of recombined mixtures.

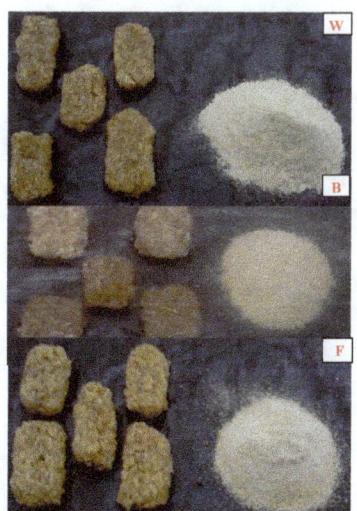

Figure 2: The pictures of dried Kishk-like products.
W: Dried kishk-like manufactured from fermented whole wheat burghul Skim milk with different types of starter cultures.
B: Dried kishk-like manufactured from fermented whole barley burghul Skim milk with different types of starter cultures.
F: Dried kishk-like manufactured from fermented freek burghul Skim milk with different types of starter cultures.

general acceptability of them, whereas the addition of barley burghul had lowering the total score acceptability (Figure 2) [16,17,37]. These products could be used to feed infants to 6 months as a complementary diet, children and elderly persons who need special care instead of the commercial extracts because the nutritive value of cereal fermented milks higher than of cereal alone. On the other hand the therapeutic effect of crude fibers and wheat bran in diets [16] (Tables 7 and 8).

Conclusion

The cereal fermented skim milk shown long shelf-life without changing in either chemical or microbial characteristics during the storage period at room temperature. This phenomenon is accepted as result of those mixtures.

References

1. Blandino A, Al-Aseeri ME, Pandiella SS, Cantero D, Webb C (2003) Cereal-based fermented foods and beverages. Food Res Int 36: 527-543.

2. Attia IA, Khattab AA (1985) Microbiological and chemical studies on Kishk. J Alex Sci Exch 6: 63-71.

3. Abou-Zeid NA (2016) Review of Egyptian cereal-based fermented product (Kishk). Int J Agri Innov Res 4: 600-609.

4. Tamime AY, Muir DD, Khaskheli M, Barclay MNI (2000) Effect of processing conditions and raw materials on the properties of Kishk: 1. Compositional and microbiological qualities. Ebensmittel-Wissenschaft Technol 33: 444-451.

5. El-Gindy SM (1983) Fermented foods of Egypt and Middle East. J Food Protection 46: 358-567.

6. Tamime AY, O'connor TP (1995) Kishk-A dried fermented milk/cereal mixture. Int Dairy J 5: 109-128.

7. Tamime AY, Muir DD, Barclay MNI, Khaskheli M, Mcnulty D (1997) Laboratory-made Kishk from wheat, oat and barley: 1. Production and comparison of chemical and nutritional composition of Burghol. Food Res Int 30: 311-317.

8. Kurmann JA, Rasic JL, Kroger M (1992) Encyclopedia of fermented fresh milk products. Van Nostrand Reinhold, New York.

9. Nassar KS, Shamsia SM, Attia IA (2016) Improvement of the nutritional value of cereal fermented milk: 1- Soft Kishk like. J Food Process Technol 7: 1.

10. AOAC (2007) Official Methods of Analysis (18thedn). Association of Official Analytical Chemists, Washington DC, USA.

11. Ling ER (1963) Text book of Dairy chemistry. Chapman and Hall.

12. FAO (2003) Food energy-methods of analysis and conversion factors. FAO Food and Nutrition Paper 77. Food and Agriculture Organization of the United Nations, Rome.

13. AOCS (1990) American Oil Chemist's Society. Official methods and recommended practice of the American oil chemist's society.

14. Barrantes E, Tamime AY, Sword AM, Muir DD, Kalab M (1996) The manufacturer of set type natural yogurt containing different oils: 1. Compositional quality, microbiological evaluation and organoleptic properties. Int Dairy J 6: 811-826.

15. Tamime AY, Barclay MNI, Amarowicz R, Mcnulty D (1999) Kishk-a dried fermented milk /cereal mixture: 1. Composition of gross components, carbohydrates, organic acids and fatty acids. Lait 79: 317-330.

16. Abou-Donia SA, Attia IA, Khattab AA, EL-Shenawi Z (1991) Formulation of dried cereal fermented milks with prolonged storage life. Egyptian J Dairy Sci 19: 283-299.

17. Toufeili I, Melki C, Shadarevian S, Robinson RK (1999) Some nutritional and sensory properties of bulgur and whole wheat meal kishk (a fermented milk-wheat mixture). J Food Qual Pref 10: 9-15.

18. De Man JC, Rogosa M, Sharp EM (1960) A medium for the cultivation of lactobacilli. J Appl Bacteriol 22: 130-134.

19. APHA (1992) American Public Health Association, standard method for the examination of dairy products (16thedn). Washington DC, USA.

20. IDF (1985) Milk and milk products-enumeration of coliform, colony count technique and most probable number technique at 30°C, standard 73A. International Dairy Federation, Brussels.

21. IDF (1990) Milk and milk products-Enumeration of yeasts and moulds colony count at 25°C, standard 94B. International Dairy Federation, Brussels.

22. Harrigan WF, McCance ME (1960) Laboratory methods in food and microbiology (Revised edition). Academic Press, London.

23. Lapierra LP, Undeland P, Cox LJ (1992) Lithium chloride-sodium propionate agar for the enumeration of Bifidobacteria in fermented dairy products. J Dairy Sci 75: 1192.

24. SAS (2013) Statistical analysis system user guide, Version 9.3. SAS Institute Inc, Cary, NC, USA.

25. Hussein GAM (2011) Production and properties of some cereal-based functional fermented dairy products. Egyptian J Dairy Sci 39: 89-100.

26. Mehanna AS, Hefnawy SA (1990) A study to follow the chemical changes during processing and storage of zabadi. Egypt J Dairy Sci 18: 425-434.

27. Mehanna AS (1991) An attempt to improve some properties of zabadi by applying low temperature long incubation period in the manufacturing process. Egypt Dairy Sci 19: 221-229.

28. Kailasapathy K, Rybka S (1997) *Lactobacillus acidophilus* and *Bifidobacterium* spp. Their therapeutic potential and survival in yoghurt. Australian J Dairy Tech 52: 28.

29. Barrantes E, Tammime AY, Muir DD, Swoed AM (1994) The effect of substitution of fat by microparticulated whey protein on the quality of set-type natural yoghurt. J Soci Dairy Technol 47: 61.

30. Aklain AS (1996) L (+), D (-) Lactic acid content and aroma profile in bioghurt, bifigurt, biograde in comparison with yogurt. Egypt J Dairy Sci 24: 227.

31. El-Nawawy MA, Ibrahim R, Al-Bonayan AM, El-Beialy AR (2012) Development of functional food products. Int J Diary Sci Res 1: 12-17

32. Tamime YA, Barclay MNI, Amarowicz R, McNulty D (1999) Kishk-a dried fermented milk/cereal mixture.1. Composition of gross components, carbohydrates, organic acids and fatty acids. Lait 79: 317-330.

33. Tamime AY, Barclay MNI, Law AJR, Leaver G, Anifantakis EM, et al. (1999) Kishk-a dried fermented milk/cereal mixture: 2. Assessment of a variety of protein analytical techniques for determining adulteration and proteolysis. Lait 79: 331-339.

34. Tamime YA, Muir DD, Kaskheli M, Barclay MNI (2000) Effect of processing conditions and raw materials on the properties of Kishk: 1. Compositional and microbiological qualities. Lebensm Wiss Technol 33: 444-451.

35. Mashak Z, Hamidreza S, Banafsheh M, Shahram N (2014) Chemical and microbial properties of two Iranian traditional fermented cereal-dairy based foods: Kashk-e-Zard and Tarkhineh. Int J Biosci 4: 124-133.

36. Bilgicli N, Ibanoglu S (2007) Effect of wheat germ and wheat bran on the fermentation activity, Phytic acid content and colour of tarhana, a wheat flour-yoghurt mixture. J Food Eng 78: 681-686.

37. Tamime AY, Muir DD, Barclay MNI, Khaskheli M, McNulty D (1997) Laboratory-made Kishk from wheat, oat and barley: 2. Compositional quality and sensory properties. Food Res Int 30: 319-326.

38. Dalgıc AC, Belibagli KB (2008) Hazard analysis critical control points implementation in traditional foods: a case study of Tarhana processing. Int J Food Sci Technol 43: 1352-1360.

26

Effects of Supplementation with Different Forms of Barley on Feed Intake, Digestibility, Live Weight Change and Carcass Characteristics of Hararghe Highland Sheep Fed Natural Pasture

Sefa Salo[1], Mengistu Urge[2]* and Getachew Animut[2]

[1]Department of Animal Science, Wachemo University, Hosanna, Ethiopia

[2]Department of Animal Sciences, Haramaya University, Dire Dawa, Ethiopia

Abstract

This study was conducted using 24 yearling intact male Hararghe highland sheep with initial body weight (BW) of 15.7 ± 2.3 kg (Mean ± SD), to determine effects of supplementing different forms of barley grain to natural pasture hay basal diet on feed intake, digestibility, average daily BW gain (ADG) and carcass parameters. Animals were grouped into 6 blocks of 4 animals based on initial BW and were randomly assigned to the four treatments. Treatments were feeding hay *ad libitum* alone (T_1) or supplemented with 300 g dry matter (DM) of raw barley (RB, T_2), malted barley (MB, T_3) or cracked barley (CB, T_4). All animals received 50 g DM supplemental noug seed cake (NSC) and had a free access to water and mineral block. The experiment consisted 90 days of feeding and 7 days digestibility trials and carcass evaluation at the end. The crude protein (CP) content of hay, NSC, RB, MB and CB were 6.6, 35.7, 11.7, 12.5 and 11.6%, respectively. Hay DM intake was higher for T1 (523 g/day) than other treatments (360-425 g/day). Total DM intake (573, 710, 723 and 775 g/day (SEM = 29.5)) and CP intake (52, 77, 77 and 83 g/day (SEM = 2.0) for T_1, T_2, T_3 and T_4, respectively) was lower for T_1 than supplemented groups, with no difference (P > 0.05) among the supplemented treatments. Digestibility of CP (55.8, 71.1, 69.0 and 70.0% for T_1, T_2, T_3 and T_4, respectively (SEM = 1.93)) were higher (P < 0.05) in supplemented sheep than T_1. ADG of 13, 73, 87 and 83 g/day for T_1, T_2, T_3 and T_4, respectively (SEM = 6.0), was also greater (P < 0.05) for the supplemented groups than T_1. Barley supplementation resulted in a higher (P < 0.05) hot carcass weight than T_1 (6.0, 10.0, 10.7 and 10.5 kg for T_1, T_2, T_3 and T_4, respectively (SEM = 0.56). The results of this study highlighted that treatment of barley as in malting and cracking do not alter the performance of sheep as compared to the untreated barley. Thus, supplementation with raw barley is recommended. In general, supplementing animals with energy dense diet has proven to improve animal performance and profitability.

Keywords: Barley processing; Digestibility; Intake; Live weight; Carcass parameter

Introduction

Barley (*Hordeum vulgare* L.) is one of the cereal founder crops, domesticated about 10,000 years ago in the Fertile Crescent [1]. In Ethiopia, the long history of cultivation and the diverse agro-ecological and cultural practices have resulted in a wide range of barley diversity. Barley is the predominant cereal in the high altitudes and cultivated in some regions in two distinct seasons: Belg that relies on the short rainfall period from March to April and Meher that relies on the long rainfall period from June to September [2]. In the highland areas, barley is the major source of food, homemade drinks, animal feed and cash [3].

Barley is a very important grain in the world today. It is very versatile, and has been well adapted through its evolution. Barley grain is one of the most common feed grains used in diets for ruminant livestock species [4]. It is very well suited for sheep and cattle as a source of energy [5]. Barley contains a large proportion of starch and can be a good source of energy supplement to ruminants [6]. Barley as feed has similar nutritive value as corn. It has been found to contain more protein and a better amino acid balance than corn, and as a result, barley-based diets require less protein supplementation [7]. Processing methods for barley were originally developed to improve starch and protein digestibility of the grain [8]. Nicholson et al., [9] found that organic matter digestibility was decreased when whole barley was fed. Dry rolled barley increased organic matter digestibility by 42% and starch digestibility by 100% [10]. Likewise, Morgan and Campling [11] reported a 49% decrease in starch digestibility with whole barley, and Mathison [12] found 37% average decrease in starch digestibility when whole barley was compared to rolled barley.

Different types of barley grain processing (namely cracking, dry rolling and to a lesser extent grinding) are recommended to farmers in order to improve their feeding values in ruminants. Malts are also supposed to have higher nutritive values than seeds. A number of workers have reported increased levels of nutrients from malt [13]. However, the benefit if any, from these processing methods over the grain fed intact for sheep is not well quantified. Therefore, the objectives of this study were to evaluate the effect of supplementation of different forms of barely grain on intake, digestibility, live weight gain, and carcass characteristics of Hararghe highland sheep fed natural pasture hay.

Materials and Methods

The experiment was carried out at Haramaya University goat farm, Ethiopia, which is located 515 km east of Addis Ababa. The site is located at an altitude of 1950 masl at latitude 9° 26° N and longitude 42° 3° E. The mean annual rainfall and temperature of the study area is 790 mm and 16°C, respectively.

*Corresponding author: Urge M, Department of Animal Sciences, Haramaya University, P.O. Box: 138, Dire Dawa, Ethiopia, E-mail: urgeletta@yahoo.com

Experimental animals and management

Twenty-four intact yearling male Hararghe highland lambs weighing 15.7 ± 2.3 kg (Mean ± SD) were purchased from *Kulubi* and *Chelenko* markets and used for the experiment. The age of the animals was determined by their dentition and information obtained from the owner. The animals were quarantined for 21 days and during this period; they were drenched with albendazol against internal parasites and sprayed with acaricide against external parasites. All animals were vaccinated against common disease in the area and penned individually.

Feed preparation and feeding

Natural pasture hay was purchased from *Babile*, stored under shade, and used as basal diet throughout the experimental period. Barley grain sufficient for the study was purchased from *Weter* market. The purchased barley was then divided in to three equal portions. One part remains untreated, the second one was malted and the third portion was cracked. Malted barley was prepared by steeping barley in water for 24 hours and then water was drained out. After that, it was allowed to sprout/germinate under room temperature for 3-4 days. Finally, it was partially sun dried under shade and stored properly. Cracked barley was prepared by crashing the grain at the size normally used for soup making with hammer mill in locally available food grinder. Noug seed cake was purchased from Dire dawa oil factory and ground with hammer mill before provided to experimental animals.

Experimental design and treatments

To conduct the experiment a completely randomized block design with six blocks and four treatments was used. At the end of the quarantine period, the experimental animals were blocked in to six blocks of four animal based on their initial body weight (BW) and animals within a block were randomly assigned to one of the four treatment diets. The initial BW was determined as a mean of two consecutive weighing after withholding feed and water overnight. The four experimental treatments are as shown in Table 1. Hay was fed *ad libitum* to animals to allow 20% refusal. Noug seed cake 50 grams per day per animal was supplemented to all animals to prevent loss of body weight for animals not supplemented with barley. The noug seed cake was provided to all animals at 08:00 hour in a separate container, where as the supplemental barley was offered to the animals at 08:00 and 16:00 hours in two equal portions. All animals had free access to water and mineralized salt block. The mineralized salt block contain a mixture of 18.8% calcium, 2.8% phosphorus, 31.2% sodium chloride, 0.09% magnesium, 0.084% manganese, 0.17% zinc, 0.0068% iodine, 0.055% copper, 0.115% iron, 0.0009% cobalt and 0.0004% selenium.

Feeding trial

The feeding trial was conducted for 90 days following an acclimatization period of 15 days to make animalsadapted to the experimental diets and pens. The amount of feed offered and that of refused was weighed and recorded daily. Daily feed intake was calculated as the difference between quantity of feed offered and refused. Sample was taken from batches of feed offer, thoroughly mixed, and sub-sampled for chemical analysis. Feed refusal samples were daily taken per animal, pooled on treatment basis, mixed thoroughly and sub-sampled for chemical analysis. The body weight (BW) of the animals was measured initially at the beginning of the feeding trial and at ten days interval after overnight with holding of feed. Daily BW change was calculated as the difference between final BW and initial BW divided by the number of feeding days. The feed conversion efficiency was calculated as a proportion of daily BW gain to daily feed intake.

Digestibility trial

The digestibility trial was conducted after the feeding trial. The digestibility trial was undertaken for 7 days. Feed offered and refusal was recorded daily. Total fecal collection for the digestibility trial was done using fecal bags. Each animal was fitted with fecal collection bags (harness) for three days of adaptation period. This was followed by a 7 days fecal collection period. The fecal output per animal was collected and weighed each morning before offering the morning meal. After weighing the daily total feces voided by each animal, the feces were thoroughly mixed, and a sub sample of 20% was taken to form a single weekly composite fecal sample for each animal. Composite samples per animal were stored in airtight plastic bags in deep freezer at -20°C. The composite fecal samples were thawed and thoroughly mixed for each animal and a sub-sample was taken for chemical analysis. A grab of fed samples from each feed and refusal for each animal was taken daily to make a weekly composite sample. Refusal samples were then bulked per treatment.

$$\text{Apparent disgestibility coefficient} = \frac{\text{Nutrient intake - Nutrient in feces}}{\text{Nutrient intake}}$$

Carcass parameters

At the end of the digestibility trial, all experimental sheep were fasted for about 12 hours and slaughtered for carcass analysis. Animals were weighed immediately before slaughter. The sheep were killed by severing the jaguar vein and the carotid artery with a knife. The blood was drained in to bucket and its weight was recorded. The esophagus was tied off close to the head to avoid leaking of gut contents. The animal was then suspended head down over a container to collect any remaining blood droppings. The head was detached from the body after blood flow ceases. The skin was flayed, fore and hind legs were trimmed off at the carpal and the tarsal joints and weighed. The entire alimentary tract (stomach, small and large intestines) with its contents was removed and weighed. Then the internal content of the gut was emptied and weight of the empty gut was recorded. Empty body weight was then determined following the procedure of Ashbrook [14] as slaughtered weight less gut content. The remaining internal organs were removed and weighed. After dressing and evisceration, hot carcass weight was recorded to assess dressing percentage on slaughter weight and empty body weight basis. The hot carcass weight was estimated after removing weight of the head, thorax, abdominal and pelvic cavity contents, as well as legs below the hock and knee joints. Rib eye muscle area was traced on transparency paper and measured by using square paper after cutting the vertebrae between the 12th and 13th ribs.

Chemical analysis

The feed, refusal and fecal samples were partially dried in an oven at 60°C for 72 hours and ground to pass 1 mm screen. The sample of feed offered, refused, and feces were analyzed for DM, ash and nitrogen (N) according to the procedures of AOAC. The CP content was estimated as N × 6.25. Neutral detergent fiber (NDF), acid detergent fiber (ADF)

	Supplement (g DM/day/animal)				
Treatment	Hay	NSC	RB	MB	CB
T 1	*ad libitum*	50	-	-	-
T 2	*ad libitum*	50	300	-	-
T 3	*ad libitum*	50	-	300	-
T 4	*ad libitum*	50	-	-	300
NSC: Noug Seed Cake; RB: Raw Barley; MB: Malted Barley; CB: Cracked Barley					

Table 1: Experimental treatments.

and acid detergent lignin (ADL) were analyzed by the method of Van Soest and Robertson [15]. The ADL was analyzed for feed samples only. Organic matter was calculated by subtracting the ash content from 100.

Statistical analysis

The data was subjected to analysis of variance (ANOVA) in a randomized complete block design using the general linear model procedure of SAS [16]. The treatment means were separated using list significant difference.

Result and Discussion

Chemical composition of feed

The chemical composition of the treatment feeds is given in Table 2. The CP (5.35%) content of hay used in this study was lower than the CP value of 8.9 reported by Yoseph. The low CP content of hay in this experiment may be due to the over maturity of the grass at the time of harvest. McDonald et al., [17] suggested that as plants mature the percentages of the CP normally decreases. The CP content of NSC in this study was slightly higher than the CP content of NSC as reported by Wondwosen [18] which was 28.9%. The difference in CP content of NSC used in this study and others might be due to the oil extraction method, laboratory procedures and the difference in variety of the noug seed used [19]. The ADF and NDF contents in NSC used in this study were comparable to the ADF content of 30.57% and NDF content of 36.27% for NSC reported by Abebaw [20], but lower than the ADF content of 33.12% and NDF content of 40.18% reported by Almaz [21]. The ADL content of NSC was lower than ADL content of 12 to 13% reported by others [20,21].

Barley grain used in this study had DM values comparable to the DM value of 90% reported by NRC [22]. The ADF and NDF content of barley grain in this study were higher than ADF value that ranges from 5.7 to 7.2% and NDF values that ranges from 22.3 to 25.6% reported by Fairbairn et al. [23]. The ADL content of barley used in this study was greater than the 2.3% reported by Hatfield et al. [24]. The variation in nutrient content is generally attributed to soil fertility and climatic conditions [8].

The CP content of barley grain used in this experiment was comparable to the value of 11.0% reported by NRC [22]. The malted barley contained 12.5% CP that was relatively comparable to the values of 13.0 reported by Peer and Leeson [25]. In agreement with this result, Morgan et al., [26] noted that changes in the protein contents for malted barley occur rapidly from day four corresponding with the extension of the root.

Dry matter and nutrient intake

The mean daily hay and total DM intake and nutrient intake of Hararghe highland sheep is presented in Table 3. The daily hay DM

Variables	Hay offer	Hay refusal							
		T1	T2	T3	T4	NSC	RB	MB	CB
DM (%)	90.1	90	90.3	90.4	90	91.8	89.7	88.7	89.9
OM (% DM)	91.3	90.3	89.9	89.7	91.3	92.4	94.2	96	93.1
CP (% DM)	6.6	5.3	5.1	5.4	5.6	35.7	11.7	12.5	11.6
NDF (% DM)	80.1	81.9	82.8	80.1	80.9	36.5	30.6	33.7	31
ADF (% DM)	52.7	53.6	50.4	48.6	51.1	31.2	15.6	13.5	16
ADL (% DM)	11.2	-	-	-	-	7.9	4.9	4.1	4.1

DM: Dry Matter; OM: Organic Matter; CP: Crude Protein; NDF: Neutral Detergent Fiber; ADF: Acid Detergent Fiber; ADL: Acid Detergent Lignin; NSC: Noug Seed Cake; RB: Raw Barley; MB: Malted Barley; CR: Cracked Barley

Table 2: Chemical composition of treatment feeds and hay refusal.

intake of the non-supplemented sheep was higher (P < 0.05) than that of the supplemented group. Supplementation through substitution may cause reductions in forage intake by grazing and pen-fed ruminants. However, supplementation for energy and/or protein can be very desirable at times, based on factors of forage quantity and quality and production demands [27].

The animals consumed almost the entire offered supplement. Consequently, total DM intake of the supplemented groups were greater (P < 0.05) than the non-supplemented sheep. Nevertheless, supplemented animals had similar DM intake (P > 0.05). Lower total DM intake in T_1 may be due to the quality of the basal diet used in this study. The higher structural carbohydrate content of the hay compared to the supplements might have limited intake of the animals due to their effect on ruminal fill.

The positive effects of supplementation on feed intake could be a reflection of the increase in the intake of essential nutrients such as energy, vitamins, and minerals and in particular nitrogen [28]. Similarly, Inoue et al., [29] reported that concentrate supplementation to low quality feeds increase feed intake because the supplements stimulate the rumen microbial function and thereby reduce digesta retention time.

Digestibility of dry matter and nutrients

The apparent digestibility of DM and nutrients for Hararghe highland sheep fed hay as a basal diet and supplemented with different forms of barley grain are given in Table 4. The DM and OM digestibility in supplemented animals were significantly higher (P < 0.05) than in control treatment. An animal in T_3 also had higher (P < 0.05) DM digestibility relative to animals in T_4. Increases in DM and OM digestibility were likely a result of the supplement being more digestible than the grass hay [30]. In parallel to this result, Fonseca et al., [31] confirmed that supplementation with maize increased DM and OM digestibility of wheat straw.

The apparent digestibility of CP was higher (P < 0.05) in supplemented sheep than the control ones, with no difference (P > 0.05) among the supplemented groups. The lack of differences in the digestibility of DM and nutrients in the supplemented group in this study indicates similar level of nutrient supply from the different forms of barley grain. This is in agreement with the results of Rainey [32] who found similar nitrogen digestibility in rolled and whole barley. Similarly, Bengochea et al., [30] reported that digestibility of CP were not affected by degree of barley processing. In the present result, there was no significant (P > 0.05) difference in NDF and ADF digestibility. Feeding grain based commercial supplements may reduce ruminal pH and decrease fiber digestibility. Moore et al. [33], which may be the cause for similar fiber digestibility in the supplemented and non-supplemented treatments. The ability of the microbial population within the rumen to digest fiber decreases when the amount and proportion of readily fermentable carbohydrate digested in the rumen increases [34].

Body weight change and feed conversion efficiency

The supplemented sheep had significantly (P < 0.05) higher final weight and daily live weight gain (ADG) compared with the sheep on the control treatment (Table 5). This might be due to enhanced DM intake and nutrient supply for production through improved intake and digestibility of nutrients as the result of supplemental barley. Animals in the supplemented group had similar performance (P > 0.05) in body weight gain and final weight. The results indicated that

Intake (g/day)	Treatment				
	T₁	T₂	T₃	T₄	SEM
Hay DM	522.5ᵃ	360.2ᵇ	375.2ᵇ	424.6ᵇ	28.8
Total DM	572.5ᵇ	710.2ᵃ	722.6ᵃ	774.6ᵃ	29.2
Total OM	523.1ᵇ	657.4ᵃ	674.0ᵃ	713.1ᵃ	26.7
Total CP	52.4ᵇ	76.7ᵃ	77.3ᵃ	83.4ᵃ	2
Total NDF	436.5ᵃᵇ	398.4ᶜ	418.9ᵇᶜ	451.1ᵃ	23.2
Total ADF	290.8ᵃ	252.1ᵇ	253.3ᵇ	287.3ᵃ	15.2

ᵃ⁻ᶜmeans within a row not bearing a common superscript are significantly different (P < 0.05); SEM: Standard Error of Mean; DM: Dry Matter; OM: Organic Matter; CP: Crude Protein; ADF: Acid Detergent Fiber; NDF: Neutral Detergent Fiber; NSC: Noug Seed Cake; RB: Raw Barley; MB: Malted Barley; CR: Cracked Barley

Table 3: Dry matter and nutrient intake of Hararghe highland sheep fed a basal diet of natural pasture hay and supplemented with different forms of barley grain.

Digestibility (%)	Treatments				
	T₁	T₂	T₃	T₄	SEM
DM	70.0ᶜ	71.3ᵃᵇ	71.9ᵃ	70.7ᵇ	0.43
OM	61.4ᵇ	73.4ᵃ	73.8ᵃ	70.6ᵃ	1.70
CP	55.8ᵇ	71.1ᵃ	69.0ᵃ	70.0ᵃ	1.93
NDF	60.6	62.7	62.9	59.8	2.76
ADF	62.5	64.5	64.8	61.4	2.56

ᵃ⁻ᶜmeans within a row not bearing a similar superscript letter significantly differ (P < 0.05); DM: Dry Matter; OM: Organic Matter; CP: Crude Protein; NDF: Neutral Detergent Fiber; ADF: Acid Detergent Fiber; SEM: Standard Error of Means; NSC: Noug Seed Cake; RB: Raw Barley; MB: Malted Barley; CR: Cracked Barley

Table 4: Apparent dry matter and nutrient digestibility of Hararghe highland sheep fed a basal diet of natural pasture hay and supplemented with different forms of barley grain.

Parameter	Treatments				
	T₁	T₂	T₃	T₄	SEM
IBW (kg)	15.75ᵃ	15.75ᵃ	15.58ᵃ	15.75ᵃ	0.93
FBW (kg)	17.0ᵇ	22.4ᵃ	23.5ᵃ	23.3ᵃ	0.68
ADG (g/day)	13.3ᵇ	73.2ᵃ	87.3ᵃ	83.1ᵃ	6.02
FCE (g gain/ g feed)	0.025ᶜ	0.105ᵇ	0.123ᵃ	0.110ᵃᵇ	0.01

ᵃ⁻ᶜmeans within a row not bearing a similar superscript letter significantly differ (P < 0.05); IBW: Initial Body Weight; FBW: Final Body Weight; ADG: Average Daily Gain; FCE: Feed Conversion Efficiency; NSC: Noug Seed Cake; RB: Raw Barley; MB: Malted Barley; CR: Cracked Barley

Table 5: Body weight change of Hararghe highland sheep fed a basal diet of natural pasture hay and supplemented with different forms of barley grain.

the supplements were comparable in their potential to supply nutrients for improving the weight gains of sheep. L'estrange [35] also reported presence of non-significant difference in sheep supplemented with whole and ground barley to hay and wheat straw based basal diet.

The level of ADG observed in this study for sheep was within the range of 69.04-104.11 g/day found by Abebe [36] from supplementation to low quality feed positively impact animal performance [37,38]. Similarly, study performed by Yagoub and Babiker [39] to determine effect of energy supplementation on performance of goats indicated that animals supplemented with high energy diets performed better than animals supplemented with low energy containing diets. This indicates that supplementation of animals with energy source improve their performance. Thus, energy appeared to be an important factor for growing sheep [40]. The synthesis of microbial protein in the rumen requires the presence of digestible carbohydrate to provide energy Rids [41] amount of feed nutrient converted in to a gram of body weight gain in supplemented sheep was higher than non-supplemented sheep. An animal in T3 also had higher (P < 0.05) feed conversion efficiency

Parameter	Treatments				
	T₁	T₂	T₃	T₄	SEM
Slaughter weight (kg)	16.0ᵇ	21.7ᵃ	22.33ᵃ	21.7ᵃ	1.23
Empty body weight (kg)	12.2ᵇ	17.8ᵃ	18.2ᵃ	17.8ᵃ	1.10
Hot carcass weight (kg)	6.0ᵇ	10.0ᵃ	10.7ᵃ	10.5ᵃ	0.56
Dressing percentage (%)					
Slaughter weight basis	37.5ᶜ	46.1ᵃ	48.0ᵃ	48.49ᵃ	3.33
Empty weight basis	49.5ᵇ	56.2ᵃ	58.8ᵃ	59.0ᵃ	4.90
REA (cm²)	5.6ᵇ	8.4ᵃ	8.2ᵃ	8.6ᵃ	0.60

ᵃ⁻ᶜmeans within a row not bearing a similar superscript letter significantly differ (P < 0.05); REA: Rib-Eye Muscle Area; SEM: Standard Error of Mean; NSC: Noug Seed Cake; RB: Raw Barley; MB: Malted Barley; CR: Cracked Barley

Table 6: Carcass parameters of Hararghe highland sheep fed a basal diet of natural pasture hay and supplemented with different forms of barley grain.

relative to animals in T2. Diets that promote a high rate of gain will usually result in a greater efficiency than diets that do not allow rapid gain [42].

Carcass parameters

Effect of treatment on slaughter weight and empty body weight were significantly higher (P < 0.05; Table 6). Both slaughter weight and empty body weight were greater (P < 0.05) for the supplemented than non-supplemented sheep. Hot carcass weight, dressing percentage and rib eye muscle area were also higher (P < 0.05) for the supplemented than non-supplemented animals. In almost all carcass parameter values for the supplemented treatments were generally similar (P > 0.05). Dressing percentage is both yielding and value determining factor and is therefore, an important parameter in assessing performance of meat producing animals. According to Alexandre et al., [43], the carcass weight increased with the increasing live weight at slaughter. Thus, greater carcass yield for the supplemented animals is consistent with enhanced ADG presumably due to the supplemental barley. The lower dressing percentage in animals not supplemented with barley may be due to retarded muscle growth of animals possibly due to lack of sufficient and balanced nutrients [44]. Reduced dressing percentage in non-supplemented animals' shows reduced in carcass yield, and body fat and increase in gut fill due to more hay intake by the control group [39]. Similarly, Sayed [37] reported that dressing percentage increased with increase of fatness associated with feeding high dietary energy. Rib-eye muscle area is an indicator of the amount of lean muscle associated with a carcass since these two parameters arepositively correlated [45,46]. Greater rib-eye muscle area for the supplemented sheep in this study suggests that supplementation with energy dense diets might induce the development of muscle in growing sheep. The rib-eye muscle area of 8.2 to 8.6 cm² was within the range of 7.10 to 10.7 cm² reported by Takele and Getachew when Horro sheep fed on vetch haulm basal diet was supplemented with wheat bran and/or Acacia albida leaf meal.

Conclusion

The above results indicate that barley processing for sheep was not given additional benefit over feeding intact barley. However, malted barley contains higher CP (12.5%) than intact barley. It can be concluded that processing barley for sheep is might not necessary to improve performance.

References

1. Lev-Yadun S, Gopher A, Abbo S (2000) Archaeology. The cradle of agriculture. Science 288: 1602-1603.

2. Grando S, Bekele B, Alemayehu F, Lakew B (2005) Food barley: Importance, uses and local knowledge. H ICARDA Aleppo 53- 82.

3. Eticha F, Woldeyesus S, Grausgruber H (2010) On-farm diversity and characterization of barley (Hordeum vulgare L.) landraces in the highlands of West Shewa. Ethno-botany research and applications: 25-34.

4. Taghizadeh A, Zabihollah N (2008) Degradability characteristics of treated and untreated barley grain using in situ technique. Amer J Anim Vet Sci 3: 53-56.

5. Paris RL (2000) Potential of hulless winter barley as an improved feed crop. Virginia Polytechnic Institute and State University Blacksburg Virginia.

6. Neuman CW, McGuire CF (1985) Barley. In: Rasmusson C (ed.) Am Soc Agron Crop Sci Soc and Soil Sci Soc, pp: 403-456.

7. Bhatty RS (1993) No malting uses of barley. In: MacGregor AW, Bhatty RS (eds.) Barley: chemistry and technology. American Association of Cereal Chemists Inc, St.Paul, MN, pp: 355-418.

8. Boyles SL, Anderson PL, Koch PL (2000) Feeding barley to cattle. The Beef InfoBase version 1.2 Jan. 2001/Feeding and Nutrition. ADDS Inc. USDA Coop State Res Ext Ed Ser NCBA.

9. Nicholson JWG, Gorrill ADL, Burgess PL (1971) Loss in digestible nutrients when ensiled barley is fed whole. Can J Anim Sci 51: 697-700.

10. Toland PC (1976) The digestibility of wheat barley or oat grain fed either whole or rolled at restricted levels with hay to steers. Aust J Exp Agric Anim Husb 16: 71-75.

11. Morgan CA, Campling RC (1978) Digestibility of whole barley and oat grains by cattle of different ages. Anim Prod 27: 323-329.

12. Mathison GW (1996) Effects of processing on the utilization of grain by cattle. Anim Feed Sci and Tech 58: 113-125.

13. Anderson VL, Bock EJ (2000) Malting barley co-products or wheat midds in corn based diets for growing yearling steers. Carrington research extension center beef production field day proceedings 23: 20-21.

14. Ashbrook FG (1955) Butchering processing and preservation of meat. D van Nostrand compony Inc, London.

15. Van Soest PJ, Robertson BJ (1985) Analysis of forage and fibrous feeds. A laboratory manual for animal science. J Dairy Sci 74: 3583-3597.

16. SAS (1998) SAS/STAT Guide to Personal Computers Version 9. Prentice Hall, London.

17. McDonald P, Edwards RA, Greenhalgh JFD, Morgan CA (2002) Animal nutrition. Prentice Hall, London.

18. Alemu W (2008) Effect of supplementing hay from natural pasture with oil seed cakes on feed intake digestibility and live weight change of Sidama goats. School of Graduate Studies of Haramaya University, Ethiopia.

19. Mogus S (1992) The Effect of processing method of oilseed cakes in Ethiopia on their nutritive value: in vitro N-degradability in growing sheep fed a basal diet of maize stover. Bonn University Germany.

20. Abebaw Nega (2007) Effects of supplementation with rice bran, noug seed (guizotia abyssinica) cake and their mixtures on feed utilization and live weight change of Farta sheep. School of Graduate Studies of Haramaya University, Ethiopia.

21. Ayenew A (2008) Supplementation of dried atella noug seed (guizotia abyssinica) cake and their mixtures on feed intake digestibility and live weight change of local sheep fed finger millet (eleusine coracana) straw basal diet. School of Graduate Studies of Haramaya University, Ethiopia.

22. NRC (National Research Council) (1985) Nutrient requirements of sheep National Academic Perss. Washigton DC.

23. Fairbairn SL, JF Patience, HL Classes, RT Zijlstra (1999) The energy content of barley fed to growing pigs: Characterizing the nature of its variability and developing prediction equations for its estimation. J. Anim. Sci 77: 1502-1512.

24. Hatfield PG, Walker JW, Glimp HA, Adams DC (1991) Effect of level of intake and supplemental barley on marker estimates of fecal output using an intra-ruminal continuous-release chromic oxide bolus. J Anim Sci 69: 1788-1794.

25. Peer DJ, S Lesson (1985) Feeding value of hydroponically sprouted barley for poultry and pigs. Anim Feed Sci and Tech 13: 183-190.

26. Morgan EK, Gibson ML, Nelson ML, Males JR (1991) Utilization of whole or steam rolled barley fed with forages to wethers and cattle. Anim Feed Sci and Tech 33: 59-78.

27. Caton JS, Dhuyvetter DV (1997) Influence of energy supplementation on grazing ruminants: requirements and responses. J Anim Sci 75: 533-542.

28. Sisay Z, Shenkoru T, Tegegne A, Woldeamanuel Y (2006) Feed intake water balance and water economy in highland sheep fed teff (eragrostis teff) straw and supplemented with graded levels of leucaena leucocephal. Eth J Anim Prod 6: 67-82.

29. Inoue T, IM Brookes, A John ES. Kolver, TN Barry (1994) Effects of leaf shear breaking load on the feeding value of perennial rye grass (Lolium perenne) for sheep 2 Effects on feed intake particle breakdown rumen digesta outflow and animal performance. J Agri Sci 123: 137-147.

30. Bengochea WL, GP Lardy, ML Bauer, Soto-Navarro SA (2005) Effect of grain processing degree on intake digestion ruminal fermentation and performance characteristics of steers fed medium-concentrate growing diets. J Anim Sci 83: 2815-2825.

31. Fonseca AJM, daSilva D, Lourenço ALG (2001) Effects of maize and citrus-pulp supplementation of urea-treated wheat straw on intake and productivity in female lambs. Bra Soc Anim Sci 73: 123-136.

32. Rainey BM (2004) Effect of beef cattle age, gender and barley grain processing method on rate and efficiency of gain and nutrient digestibilities. Montana State University Montana.

33. Moore JA, Poore MH, Luginbuhl JM (2002) By-product feeds for meat goats: effects on digestibility, ruminal environment, and carcass characteristics. J Anim Sci 80: 1752-1758.

34. Hoover WH (1986) Chemical factors involved in ruminal fiber digestion. J Dairy Sci 69: 2755-2766.

35. L'estrange JL (1977) The performance and carcass fat characteristics of lambs fattened indoors on concentrate diets. Ir Agric Res 16: 221-232.

36. Hailu A (2008) Supplementation of graded levels of concentrate mix on feed intake digestibility live weight change and carcass characteristics of Washera sheep fed urea treated rice straw. School of Graduate Studies of Haramaya University, Ethiopia.

37. Abdel-Baset NS (2009) Effect of different dietary energy levels on the performance and nutrient digestibility of lambs. Vet. World 2: 418-420.

38. Bowman JGP, Sanson DW (1996) Starch or fiber based energy supplements for grazing ruminants. In: proceedings grazing livestock. Nutrition Conf Rapid City USA, pp: 118-135.

39. Yagoub YM, Babiker SA (2008) Effect of dietary energy level on growth and carcass characteristics of female goats in Sudan. LRRD.

40. Abdias M, Geleti D, Gizachew L, Diriba T, Hirpa A (2001) The effect of improved forages and/or concentrate supplementation on live weight of horro lambs and growing bulls. Proceedings of the 9th annual conference of the Ethiopian society of animal production (ESAP), Bako Research Center Addis Ababa, Ethiopia.

41. Rids PM (1983) The pools of tissue constituents and products. Elsevier Science Publishers, Amsterdam, the Netherlands.

42. Matiwos Solomon (2007) The effect of different levels of cottonseed meal supplementation on feed intake digestibility live weight changes and carcass parameters of Sidama goats. School of Graduate Studies of Haramaya University Ethiopia 119: 137-144.

43. Alexandre G, Coppry O, Bocage, Fleury J, Archimède H (2008) Effect of live weight at slaughter on the carcass characteristics of intensively fattened Martinik sheep fed sugar cane supplemented with pea flour. LRRD.

44. Woldemariam B (2008) Supplementation with dried foliages of selected indigenous browse: effects on feed intake body weight gain carcass characteristics of Abergalle goats offered hay. School of Graduate Studies of Haramaya University, Ethiopia.

45. Yirga H (2008) Supplementation of concentrate mix to Hararghe highland sheep fed a basal diet of urea-treated maize Stover: effect on feed utilization live weight change and carcass characteristics. School of Graduate Studies of Haramaya University, Ethiopia.

46. Mekasha Y, Urge M, Kurtu MY, Bayissa M (2001) Effect of strategic supplementation with different proportion of agro-industrial by-products and grass hay on body weight change and carcass characteristics of tropical Ogaden bulls (Bos indicus) grazing native pasture Afri J Agri Res 6: 825-833.

27

Effect of Peanut Addition on the Fatty Acid Profile and Rheological Properties of Processed Cheese

Rafiq SM* and Ghosh BC

ICAR (Indian Council of Agricultural Research), National Dairy Research Institute (SRS), Bengaluru, India

Abstract

The aim of the present study was to increase the unsaturated fatty acid content of the processed cheese by adding peanuts. Processed cheese was prepared from a blend of young and old Cheddar cheese. Partially roasted peanuts were added in form of paste at a level of 5%, 10% and 15%. Fatty acid analysis by gas chromatography showed that peanut added cheeses were characterized by a reduced level of saturated fatty acids (SFA=57.37% to 64.19%) and increased level of polyunsaturated fatty acids (PUFA=5.64% to 10.71%) compared to the control processed cheese (SFA=67.18%; PUFA=2.73%). Addition of peanuts upto 10% had no significant effect (p>0.05) on the rheological properties, meltability and sensory properties of processed cheese however at 15% addition all these properties were significantly (p<0.05) influenced. Our results suggest that processed cheese fatty acid profile can be improved by adding peanuts up to a level of 10% without affecting its quality attributes.

Keywords: Processed cheese; Fatty acid profile; Peanuts; Gas chromatography; Texture profile analysis; Meltability

Introduction

Processed cheese (PC) is a rich source of fat containing about 40% on dry matter basis and most of the fat is saturated. This saturated fat content in cheese and other dairy products has been associated with cardiovascular diseases. Milk and dairy products with a healthier fatty acid profile and/or more functional or nutraceutical properties has prompted researchers worldwide to evaluate different methods for increasing the level of polyunsaturated fatty acids (PUFAs) in the milk and dairy products of various ruminant species [1]. Several scientists have worked on enhancing the fatty acid profile of milk or milk products by supplementation of unsaturated fatty acids through feed and fortification of dairy products with oils high in PUFAs [2].

Peanuts (*Arachis hypogaea*) containing high unsaturated fatty acid have drawn the attention of manufacturers to be used as an ingredient in various foods as they are inexpensive and nutritionally powerful [3]. Peanuts are rich in protein, oil and fibers [4]. All these components are present in their most beneficial forms. This plant protein, unsaturated fat and complex fiber in peanut are proved to be very good for human nutrition. Peanut fat profile contains about 50% monounsaturated fatty acids (MUFAs), 33% Polyunsaturated fatty acids (PUFAs) and 14% saturated fatty acids which is a heart friendly combination of fatty acids [5]. Consumption of peanuts elicits several biological effects such as weight-loss [6], prevention of cardiovascular diseases by lowering blood pressure and blood cholesterol levels [7], anti-inflammatory effects [8], and inhibition of cancer [9]. Peanut milk has been used in the preparation of yogurt [10] and cheese spread [11] but peanut addition in processed cheese to enhance the fatty acid profile has not been reported. Therefore, an attempt has been made to add peanuts into the processed cheese for a healthier fatty acid profile and to improve its functional value. Addition of peanuts might alter the rheological and sensory properties of processed cheese. Hence, the objective of this study was to identify the optimum level of peanut addition into the processed cheese without affecting its overall quality and to report how best the fatty acid profile of the processed cheese can be improved.

Materials and Methods

Materials

Cheddar cheese, both young (1 to 2 months ripened) and old varieties (6 months ripened) were obtained from the dairy plant of NDRI Bengaluru. Peanuts and common salt were purchased from the local market of Bengaluru. Tri-sodium citrate was used as emulsifying salt for PC preparation.

Manufacture of processed cheese

A blend of Cheddar cheese (75% young and 25% old) was used for PC preparation. The young and old Cheddar cheeses were cleaned, quartered, and milled by using milling machine (WFC and Company USA: Model no 32). Double jacketed steam kettle with a capacity of 10 Kg (Milkmax Engineers, Bengaluru) was used for manufacture of PC. Milled cheese was taken into cheese processing kettle. Calculated amount of salt (1%) and tri-sodium citrate (3%) were dissolved into required amount of hot water and added into the cheese intermittently 2 to 3 times while heating. When the mass became homogeneous, peanut paste was added and mixed thoroughly. Peanut paste was prepared by roasting, peeling and grinding it in a mixer/grinder and the ground peanuts (0%, 5%, 10% and 15%) were mixed with 30% to 40% water to make paste. The mixture of cheese and peanuts were heated to 75°C to 85°C for 5 to 6 minutes with continuous stirring and scraping with a steel ladle. Thereafter, the heating was stopped and the hot product was transferred into moulds, cooled for 2 to 3 hrs at room temperature, packed and stored at refrigeration temperature till its further use.

Compositional analysis

Moisture content of the PC was determined by gravimetric method [12]. Fat content, titrable acidity and ash content were determined as per the method described in AOAC [13]. Total protein was determined by micro kjeldhal method according to AOAC with some modification. The pH of cheese was measured as described by Awad et al. [14].

*Corresponding author: Rafiq SM, ICAR (Indian Council of Agricultural Research), National Dairy Research Institute (SRS), Adugodi, Bengaluru-560030, India, E-mail: mansharafiq@gmail.com

Carbohydrate content was reported by difference method i.e. hundred minus total content of protein, ash, moisture and fat.

Fatty acid analysis

Fatty acids were analyzed using gas chromatography (GC) as per the procedure of O'Fallon et al. [15] with some modifications.

Sample preparation (FAME Synthesis)

Samples (500 mg) were placed into a 16 mm × 125 mm screw-cap Pyrex culture tube (capacity 15 ml) to which 1 ml of the C13: 0 (tridecanoic acid) as internal standard (0.5 mg of C13: 0/ml of methanol), 0.7 ml of 10 N KOH in water, and 5.3 ml of methanol were added. The tube was incubated at 55°C in water bath for 1.5 h with intermittent vigorous shaking for 5 secs at every 20-min interval to properly permeate, dissolve, and hydrolyze the sample. After cooling below room temperature under cold tap water, 0.58 ml of 24 N H_2SO_4 was added to it. The tube was mixed by inversion and incubated again in water bath at 55°C for 1.5 h with intermittent vigorous shaking for 5 secs at every 20-min interval. Thus, fatty acid methyl ester (FAME) synthesis in the tube was cooled in a cold tap water bath. Three milliliters of hexane were added, and the tube was vortex-mixed for 5 min on a multi tube vortex. The tube was centrifuged for 5 min in a tabletop centrifuge, and the hexane layer, containing the FAME was used for fatty acid analysis in GC.

Gas chromatography

The fatty acid composition of the FAME was determined using GC unit (M/s Agilent Technologies, USA, serial No-7890A) equipped with flame ionization detector and a HP-88 capillary column (100 m × 0.25 mm × 0.20 μm). The carrier gas was hydrogen (1.7 ml/min) and the sample volume was 1 μl. The split ratio was 50:1. The injector and the detector were set at 280°C and 250°C respectively. Identification of fatty acids was achieved by comparing the retention times of the peaks with previously run pure standard compounds (Supelco 37 component FAME Mix). Fatty acids were expressed as % of total fatty acids identified in the samples.

Texture profile analysis

Textural properties of the processed cheese were assessed using a TA.XT.plus texture analyzer (Stable Micro Systems Ltd., Godalming, UK). The samples were cut (2.0 cm × 2.0 cm × 2.0 cm) using a stainless-steel knife and measurements were performed by two sequential compression events using a 75-mm compression platen probe at a speed of 2 mm/s and a trigger force of 2 g. Each sample was subjected to 50% compression after tempering the samples for 45 minutes at 25°C ± 2°C. The TPA hardness, cohesiveness, gumminess, adhesiveness, chewiness and springiness were measured [16].

Meltability

Meltability was measured using a modified Schreiber test [17]. Cheese discs (16 mm diameter and 10 mm thickness) were placed in centre of covered glass Petri dishes and kept in a pre-heated oven at 150°C for 5 minutes, after which they were removed and cooled for 30 minutes. Spread area was measured using Image J software and meltability was expressed as a ratio of melted area and initial area.

Sensory analysis

Sensory analysis was performed according to the methodology described by Meyer [18] on a 20-point score card using 8 trained panellists. The sensory panel was selected based on the normal sensitivity for basic color/appearance, body/texture and flavor of cheese. Sensory panelists were well trained on evaluation of processed cheese and were made familiar with the test methods used. Duo-trio tests were used to determine a candidate's ability to detect differences among similar products with different ingredients. Cheese slices (about 2 cm to 3 cm) of 10 g to 15 g of each were presented in a covered glass petri plate in a random order with coded numbers. During each session of evaluation, three samples were presented at a time along with one reference (control) sample. The judged parameters were: appearance (4), body and texture (8) and flavour (8).

Statistical analysis

The statistical analysis was executed using the statistical software SPSS 16.0 (Stat Soft Polska Sp. z o. o., Kraków, Poland). One-way analysis of variance (ANOVA) was performed and significant differences among samples were reported according to Duncan's test at p<0.05.

Results and Discussion

Compositional analysis

It was observed that with the addition of peanut into the processed cheese, moisture content increased however, the increase in moisture content was nonsignificant (p>0.05) as compared to the control (Table 1). A significant increase in the fat content was observed at 10% and 15% peanut addition whereas at 5% addition no significant difference (p>0.05) was found as compared to control. The increase in fat content may be due to the higher level of peanut addition as the oil content in peanut is about 48% to 50% [4]. A significant difference (p>0.05) was found in the protein content at 15% addition. Ash content also decreased and was significantly lower in all the cheeses (p<0.05) than the control. Carbohydrate content increased with peanut addition and it was significantly higher (p<0.05) at 10 and 15% peanut addition. The pH value increased and the acidity decreased with the increased level of peanut addition but no significant difference (p>0.05) was found as compared to control.

Fatty acid composition

Fatty acid profile chromatograms of the peanuts, control processed cheese and peanut added processed cheese are shown in Figure 1. The fatty acids were identified based on retention times obtained using Supelco 37 component FAME mix as standard and tridecanoic acid (C13: 0) as an internal standard. The retention time of the standard and fatty acid compositions of peanut, processed cheese and processed cheese with added peanuts are presented in Figure 2. The saturated fatty acids butyric (C4: 0), caproic (C6: 0), caprylic (C8: 0), capric (C10: 0) and lauric acid (C12: 0) were not detected in peanuts whereas in processed cheese the level of these fatty acids ranged from 1.22% to 3.57%. The

Attributes	Control PC	Peanut added PC		
		5%	10%	15%
Moisture (%)	40.95 ± 1.75ᵃ	40.94 ± 1.48ᵃ	42.29 ± 0.22ᵃ	42.50 ± 1.63ᵃ
Fat (%)	30.26 ± 0.42ᵇ	31.35 ± 0.49ᵃᵇ	31.82 ± 0.83ᵃ	32.21 ± 0.60ᵃ
Protein (%)	21.82 ± 0.64ᵃ	19.79 ± 0.68ᵃᵇ	19.46 ± 01.08ᵃᵇ	18.89 ± 0.98ᵇ
Ash (%)	4.03 ± 0.05ᵃ	3.78 ± 0.03ᵇ	3.74 ± 0.03ᵇ	3.87 ± 0.06ᶜ
Carbohydrate (%)	0.84 ± 0.35ᵇ	1.32 ± 0.31ᵃᵇ	1.58 ± 0.42ᵃ	1.90 ± 0.24ᵃ
pH	5.34 ± 0.10ᵃ	5.38 ± 0.11ᵃ	5.41 ± 0.10ᵃ	5.44 ± 0.10ᵃ
Acidity (% lactic acid)	1.24 ± 0.11ᵃ	1.03 ± 0.29ᵃ	0.96 ± 0.25ᵃ	0.87 ± 0.26ᵃ

Results are expressed as Mean ± S.D; means with different superscripts in a row differ significantly (p<0.05) (n=3).

Table 1: Proximate composition of control and peanut added processed cheeses.

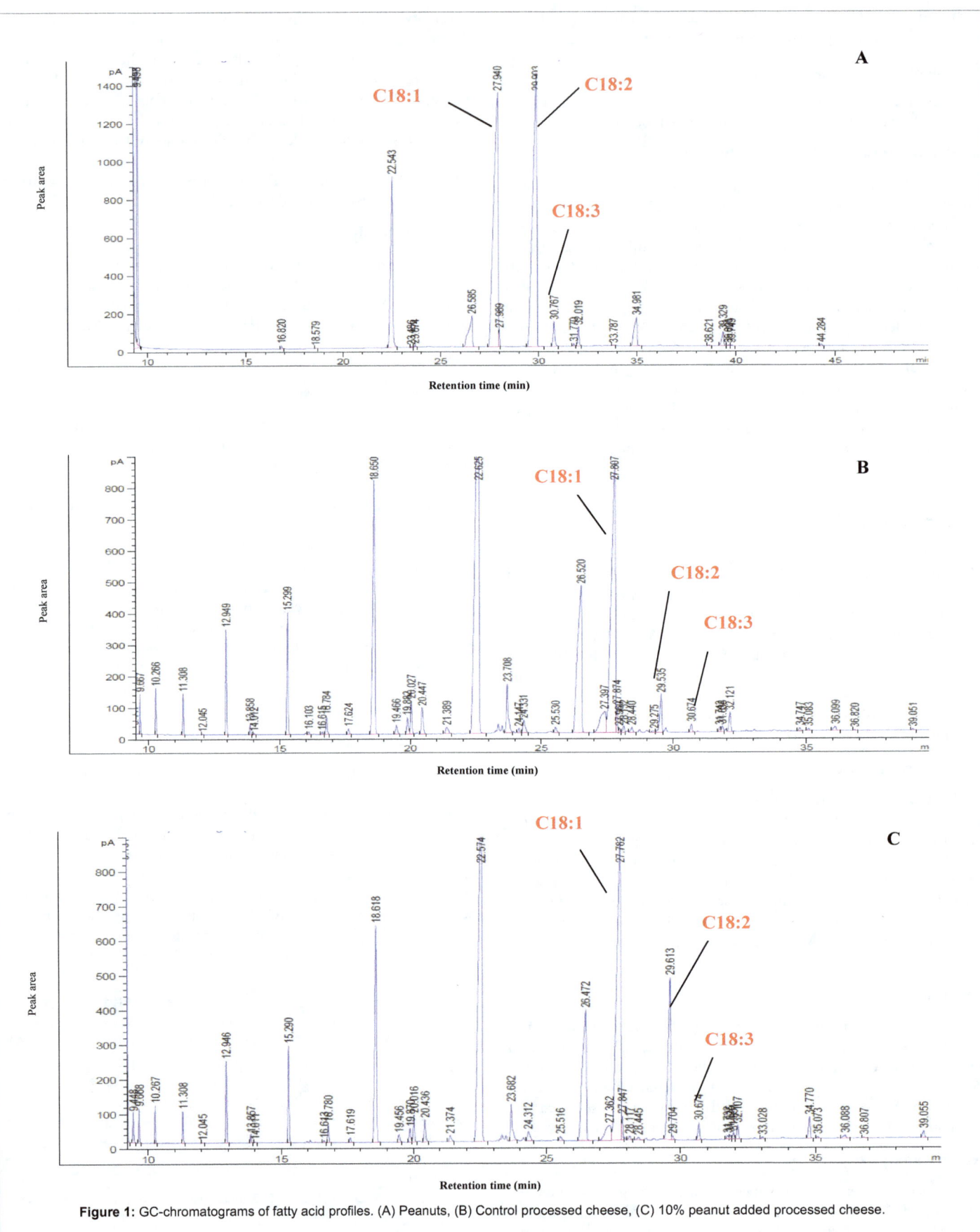

Figure 1: GC-chromatograms of fatty acid profiles. (A) Peanuts, (B) Control processed cheese, (C) 10% peanut added processed cheese.

Fatty acids (%)		Retention time of standard (min)		Control PC	Peanuts	Peanut added PC		
						5%	10%	15%
Butyric acid	C4:0	9.665	mean	1.41[a]	nd	1.40[a]	1.14[b]	1.05[b]
			SD	0.27		0.06	0.01	0.07
Caproic acid	C6:0	10.265	mean	1.55[a]	nd	1.29[ab]	1.13[b]	1.10[b]
			SD	0.42		0.06	0.05	0.17
Caprylic acid	C8:0	11.305	mean	1.22[a]	nd	1.01[ab]	0.84[b]	0.75[b]
			SD	0.30		0.10	0.11	0.05
Capric acid	C10:0	12.938	mean	3.04[a]	nd	2.44[ab]	2.13[b]	2.13[b]
			SD	0.52		0.07	0.42	0.30
Lauric acid	C12:0	15.272	mean	3.57[a]	nd	3.17[a]	3.04[a]	2.96[a]
			SD	0.22		0.04	0.19	0.67
Myristic acid	C14:0	18.534	mean	11.85[a]	0.05[f]	10.76[b]	9.69[c]	8.76[e]
			SD	0.27	0.00	0.49	0.62	0.67
Palmitic acid	C16:0	22.310	mean	29.70[a]	9.91[c]	29.52[a]	27.36[b]	25.99[b]
			SD	1.07	0.18	0.64	1.27	0.57
Palmitoleic acid	C16:1	23.634	mean	1.49[a]	0.18[b]	1.44[a]	1.47[a]	1.26[a]
			SD	0.45	0.00	0.04	0.06	0.30
Stearic acid	C18:0	26.242	mean	13.25[a]	4.73[c]	12.26[ab]	11.15[b]	10.99[b]
			SD	0.34	0.31	0.89	0.87	0.90
Oleic acid	C18:1	27.480	mean	23.09[c]	37.97[a]	24.82[c]	25.46[bc]	28.64[b]
			SD	0.18	1.81	0.96	0.49	3.65
	C18:1 trans	27.060	mean	3.15[a]	nd	2.60[ab]	2.26[ab]	2.11[b]
			SD	0.47		0.23	0.32	0.49
Linoleic acid	C18:2	29.432	mean	2.15[e]	32.82[a]	5.03[d]	7.92[c]	10.83[b]
			SD	0.35	0.54	0.08	0.19	2.17
	C18:2 trans	28.421	mean	0.24[a]	nd	0.17[a]	0.16[a]	0.14[a]
			SD	0.08		0.03	0.05	0.05
Linolenic acid	C18:3	30.619	mean	0.18[b]	4.75[a]	0.18[b]	0.25[b]	0.35[b]
			SD	0.06	0.55	0.01	0.02	0.06
	C18:3 trans	31.699	mean	0.22[a]	nd	0.19[a]	0.17[a]	0.14[b]
			SD	0.04		0.02	0.03	0.02
Arachidic acid	C20:0	30.861	mean	0.72[c]	1.87[a]	0.77[c]	0.85[bc]	0.99[b]
			SD	0.02	0.10	0.05	0.07	0.17
Eicosenoic acid	C20:1	31.906	mean	0.80[bc]	1.87[a]	0.73[c]	0.88[bc]	1.01[b]
			SD	0.06	0.10	0.11	0.07	0.21
Arachidonic acid	C20:4	36.069	mean	0.12[a]	nd	0.08[b]	0.07[b]	0.06[b]
			SD	0.00		0.02	0.03	0.02
	C22:0+C20:3	34.730	mean	nd	2.99[a]	0.49[c]	0.95[b]	1.23[b]
			SD		0.81	0.08	0.20	0.60
Lingoceric acid	C24:0	39.050	mean	nd	1.60[a]	0.07[c]	0.20[b]	0.34[b]
			SD		0.23	0.00	0.06	0.10
Saturated fatty acids SFA			mean	67.18[a]	18.51[e]	64.19[b]	60.37[c]	57.37[d]
			SD	0.35	1.04	0.98	0.84	0.95
Monounsaturated fatty acids MUFA			mean	30.47[b]	40.68[a]	30.57[b]	30.86[b]	31.62[b]
			SD	0.68	0.91	1.06	0.66	1.28
Polyunsaturated fatty acids PUFA			mean	2.73[e]	40.09[a]	5.64[d]	8.73[c]	10.71[b]
			SD	0.35	0.57	0.39	0.75	0.96

Figure 2: Figure displaying fatty acid composition of control processed cheese, peanuts and peanut added processed cheeses (% of total fatty acids, wt/wt).

main contributing saturated fatty acids in processed cheese were found to be palmitic (29.70%), stearic (13.25%) and myristic (11.85%) acids. Palmitic (9.91%) and stearic (4.73%) acid contents of peanuts were significantly lower ($p<0.05$) than the processed cheese while myristic (0.05%) acid was present in traces. Palmitic acid has been found to raise serum cholesterol levels. Dietary saturated fat intake has been shown to increase low-density lipoprotein (LDL) cholesterol, and it has been associated with increased risk of cardiovascular disease (CVD). This evidence, coupled with inferences from epidemiologic studies and clinical trials, has led to long standing public health recommendations for limiting saturated fat intake as a means of preventing CVD [19].

Among unsaturated fatty acids peanuts showed significantly higher ($p<0.05$) levels of oleic (37.97%), linoleic (32.82%) and linolenic (4.75%) acids than the control processed cheese (oleic acid=23.09%; linoleic=2.15%; linolenic=0.18%). The oleic acid content of peanuts was similar whereas linoleic acid was slightly lower and linolenic acid was higher than as reported by Maguire et al. [20]. However, the oleic acid content was far lower than as reported by Andersen et al. [21]. The difference in fatty acid profiles may be due to the variety or geographic location. The oleic acid, linoleic and linolenic acid contents of the processed cheese were found to be consistent with those as reported by Kim et al. [22] for different processed cheese varieties. With the addition of peanuts into the processed cheese the levels of these unsaturated fatty acids increased. Increasing the percentage of long-chain, mono- and polyunsaturated fatty acids in milk products is an interesting approach to produce healthier food because these fatty acids are linked to a reduction in the incidence of coronary heart disease in conjunction with an increase in high-density cholesterol

(HDL) [23]. Oleic acid content of control processed cheese increased significantly (p<0.05) to 28.64% at 15% peanut addition however, at 5 and 10% peanut addition no significant difference (p>0.05) was found as compared to the control. Linoleic acid content was found to be significantly higher (p<0.05) than the control processed cheese at all levels of peanut addition. Linolenic acid content also increased but no significant difference (p>0.05) found as compared to the control. Trans fatty acids decreased with the addition of peanuts into the processed cheese and at 15% peanut addition trans C18: 1 and trans C18: 3 fatty acids were significantly lower (p<0.05) as compared to the control. Trans fatty acids in the diet show a similar action to that of saturated fatty acids, making them potentially hazardous to the organism, especially with respect to coronary diseases [24].

Among other fatty acids detected, the saturated arachidic (C20:0) and unsaturated eicosenoic (C20:1) acid were present in processed cheese in lower quantity whereas in peanuts these were present relatively at a higher level of 1.87%. Addition of peanuts upto 10% had no significant influence on the arachidic (C20:0) and eicosenoic (C20:1) acid contents of the processed cheese but at 15% peanut addition these fatty acids were significantly higher (p<0.05) than the control. Arachidonic acid (C20:4n6) was not detected in peanuts however it was present in small quantity in processed cheese. C22:0+C20:3 and lingoceric acid (C24:0) were not found in processed cheese but these acids were detected in peanuts added cheeses in small quantity which might be due to the addition of peanuts into the processed cheese as these fatty acids are present at a higher level of 2.99 and 1.60 respectively in peanuts.

In general, with the addition of peanuts into the processed cheese the level of saturated fatty acids decreased. Conversely, the levels of

polyunsaturated fatty acids increased linearly with increased level of peanuts. Monounsaturated fatty acids increased but no significant difference (p>0.05) was found as compared to the control. The saturated fatty acid content of the processed cheese (SFA=67.18%) decreased significantly (p<0.05) to 64.19%, 60.37% and 57.37% and polyunsaturated fatty acids content (PUFA=2.73%) increased significantly (p<0.05) to 5.64%, 8.73% and 10.71% at 5, 10 and 15% peanut addition, respectively.

Evaluation of the texture profile

Addition of peanuts up to 10% did not show any significant influence on the textural properties of processed cheese (Figure 3). However, hardness (10.68 N) and gumminess (4.77 N) at 15% peanut addition were found to be significantly (p<0.05) lower than the hardness (20.17 N) and gumminess (0.16 N) of control processed cheese. Addition of 15% peanut also showed significantly (p<0.05) lower springiness and chewiness than the control cheese. It was found that peanut addition had no significant (p>0.05) effect on cohesiveness and adhesiveness of processed cheese. The decrease in the textural parameters can be attributed to the increase in moisture and fat content with the addition of peanuts. Alteration of ingredients affects the texture of the cheese by reducing the casein-casein and casein-fat interaction in cheese which seems to be responsible for determining its structure. Our results are in agreement with Marshall [25] and Pereira et al. [26] who reported that textural properties of processed cheese analogues decreased as the moisture content increased. Similarly, textural properties of peanut added processed cheese decreased with the increase in moisture content. This might be due to the action of water as plasticizer in the protein matrix making it less elastic and more susceptible to fracture upon compression [27]. Decrease in the textural properties might also be due to the increase in fat content with the addition of peanuts. When the fat content increased, all the above-mentioned textural properties of processed cheese samples reduced. Fat and moisture act as the filler in the casein matrix of cheese texture [28], giving it lubricity and softness. Olson and Johnson [29] also indicated that relative amounts of water, protein, and fat were the dominant factors effecting cheese hardness.

Meltability

Meltability of processed cheese decreased with increased level of peanuts but upto 10% addition no significant difference (p>0.05) was found as compared to the control (Table 2). Meltability (2.26) of 15% peanut added cheese was significantly lower than the control (4.19) even though the fat content was significantly more (p<0.05) than the control processed cheese (Table 1). The decrease in the meltability can be attributed to the addition of insoluble protein from peanuts which might have replaced the soluble caseins and interfere with the meltability of cheese. Meltability is no doubt one of the most important functional characteristics of cheese, in particular when cheese is used as a topping or is an ingredient in processed foods [30].

Sensory properties

There was no significant difference (p>0.05) in appearance among the control and peanut added cheeses (Table 3). Body and texture score did not reduce significantly (p>0.05) with the increase of peanut addition up to 10% however, 15% addition of peanuts significantly (p<0.05) reduced the score than that of the control cheese. The decrease in the body and texture scores could be attributed to the increase in the moisture content with the incorporation of peanuts. This was further confirmed from the textural studies (Figure 1) which showed that the textural properties decreased with the increase of peanut addition into the processed cheese. Flavour scores also reduced significantly

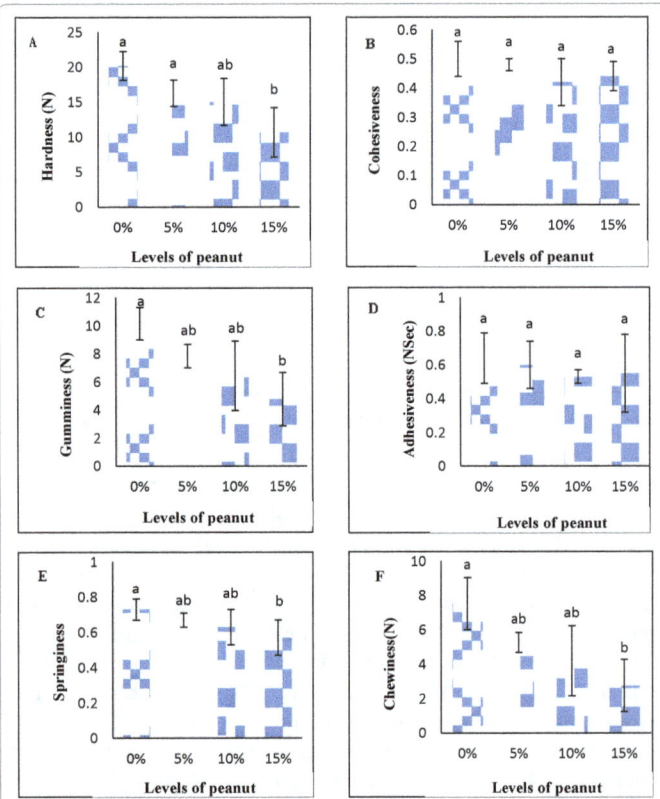

Figure 3: Effect of peanut addition on the textural properties of processed cheese A: Hardness; B: Cohesiveness; C: Gumminess; D: Adhesiveness; E: Springiness; F: Chewiness.

Samples	Meltability index
Control	4.19 ± 0.75[a]
5% Peanut	3.74 ± 0.61[a]
10% Peanut	3.06 ± 0.71[ab]
15% Peanut	2.26 ± 0.38[b]

Results are expressed as Mean ± S.D; means with different superscripts in a column differ significantly (p<0.05) (n=3).

Table 2: Effect of peanut addition on meltability of processed cheese.

Parameters	Control PC	Peanut added PC		
		5%	10%	15%
Appearance (4)	3.60 ± 0.05[a]	3.57 ± 0.59[a]	3.57 ± 0.21[a]	3.60 ± 0.20[a]
Body and texture (8)	7.10 ± 0.17[a]	6.77 ± 0.258[ab]	6.73 ± 0.17[ab]	6.70 ± 0.19[b]
Flavour (8)	7.13 ± 0.10[a]	6.52 ± 0.277[ab]	6.60 ± 0.52[ab]	6.29 ± 0.49[b]
Overall acceptability (20)	17.84 ± 0.29[a]	16.86 ± 0.56[ab]	16.90 ± 0.7[ab]	16.59 ± 0.56[b]

Results are expressed as Mean ± S.D; means with different superscripts in a row differ significantly (p<0.05) (n=3).

Table 3: Effect of peanut addition on the sensory properties of processed cheese.

(p>0.05) at 15% peanut addition. Peanut flavor was perceived in all the peanut containing cheese samples and the intensity increased with increased peanut addition. Cheese samples with 15% peanut resulted in a very strong peanut flavor which reduced its acceptability. The overall acceptability score of the peanut added cheese was lowest for 15% peanut added cheese (16.58) among the four samples which was significantly (p<0.05) lower than the control (17.84). The decreased acceptability was mainly because of the flavour characteristics while color and appearance didn't show any adverse impact on it.

Conclusion

In the current study, it was found that peanut addition up to 10% resulted in 7% to 8% reduction in the saturated fatty acid and conversely increased the unsaturated fatty acid content of the processed cheese. More reduction was found at higher levels of addition but the rheological and sensory properties showed significant adverse influence for 15% peanut added cheese. Therefore, 10% roasted peanut addition is suggested as optimum level to obtain a healthier fatty acid profile without any adverse effect on rheological and sensory properties of the processed cheese.

Acknowledgement

The first author expresses her sincere thanks to Dr. S. B. N Rao, Principal Scientist and Dr. Anjumoni Mech, Scientist, ICAR-National Institute of Animal Nutrition and Physiology, Bengaluru for their help to analyze fatty acids using Gas Chromatography.

References

1. Cerutti WG, Viegas J, Barbosa AM, Oliveira RL, Dias CA, et al. (2016) Fatty acid profiles of milk and minas frescal cheese from lactating grazed cows supplemented with peanut cake. J Dairy Res 83: 42-49.

2. Abou-zeid NA (2016) Nutraceutical ingredient poly-unsaturated fatty acids (PUFAs) fortification in milk and dairy products: A Review. Int J Adv Res Sci Eng Technol 3: 1420-1427.

3. Sanders TH (2001) Non-detectable levels of trans-fatty acids in peanut butter. J Agric Food Chem 49: 2349-2351.

4. Suchoszek-Lukaniuk K, Jaromin A, Korycińska M, Kozubek A (2011) Fatty acid content of commonly available nuts and seeds. Nuts and seeds in health and disease prevention, Elsevier, Netherlands. pp. 873-880.

5. Feldman EB (1999) Assorted monounsaturated fatty acids promote healthy hearts. America J Clini Nutr 70: 953-954.

6. Alper CM, Mattes RD (2002) Effects of chronic peanut consumption on energy balance and hedonics. J Int Assoc Stud Obes 26: 1129-1137.

7. Lopes RM, Agostini-Costa TDS, Gimenes MA, Silveira D (2011) Chemical composition and biological activities of Arachis species. J Agric Food Chem 59: 4321-4343.

8. Higgs J (2003) The beneficial role of peanuts in the diet-Part 2. Nutr Food Sci 33: 56-64.

9. Awad AB, Chan KC, Downie AC, Fink CS (2000) Peanuts as a source of β-sitosterol, a sterol with anticancer properties. Nutr Cancer 36: 238-241.

10. Isanga J, Zhang G (2009) Production and evaluation of some physicochemical parameters of peanut milk yogurt. Food Sci Technol 42: 1132-1138.

11. Razig KAA, Yousif AM (2010) Utilization of groundnut milk in manufacturing spread cheese. Pakistan J Nutr 9: 314-319.

12. Mazhuvanchery JJ (1981) IS: (SP:18) ISI handbook of food analysis. Part XI: Dairy products, Indian Standards Institution, New Delhi, India.

13. AOAC Association of Official Analytical Chemists (2005) Official methods of analysis. Washington, USA.

14. Awad S, Hassan AN, Halaweish F (2005) Application of exopolysaccharide-producing cultures in reduced-fat cheddar cheese: Composition and proteolysis. J Dairy Sci 8: 4195- 4203.

15. O'Fallon JV, Busboom JR, Nelson ML, Gaskins CT (2007) A direct method for fatty acid methyl ester synthesis: Application to wet meat tissues, oils and feedstuffs. J Anim Sci 85: 1511-1521.

16. Gunasekaran S, Mehmet Ak M (2003) Cheese texture. Cheese rheology and texture. CRC Press LLC, Boca Raton FL, USA.

17. Mleko S, Foegeding EA (2001) Incorporation of polymerized whey proteins into processed cheese analogs. Milchwissenschaft Milk Sci Int 56: 612-615.

18. Meyer (1973) Processed cheese manufacture. Food Trade Press Ltd, London, UK.

19. Siri-Tarino PW, Sun Q, Hu FB, Krauss RM (2010) Meta-analysis of prospective cohort studies evaluating the association of saturated fat with cardiovascular disease. Am J Clin Nutr 91: 535-546.

20. Maguire LS, O'Sullivan Galvin K, O' Connor TP, O'Brien NM (2004) Fatty acid profile, tocopherol, squalene and phytosterol content of walnuts, almonds, peanuts, hazelnuts and the macadamia nut. Int J Food Sci Nutr 55: 171-178.

21. Andersen PC, Hill K, Gobert DW, Brodbeck BV (1998) Fatty acid and amino acid profiles of selected peanut cultivars and breeding lines. J Food Comp Anal 11: 100-111.

22. Kim NS, Lee JH, Han KM, Kim JW, Cho S, et al. (2014) Discrimination of commercial cheeses from fatty acid profiles and phytosterol contents obtained by GC and PCA. Food Chem 143: 40-47.

23. Parodi PW (2016) Dietary guidelines for saturated fatty acids are not supported by the evidence. Int Dairy J 52: 115-123.

24. Daniel DR, Thompson LD, Shriver BJ, Chich-Kang W, Hoover LC (2005) Non-hydrogenated cottonseed oil can be used as a deep fat frying medium to reduce trans-fatty acid content in french fries. J Am Diet Assoc 105: 1927-1932.

25. Marshall RJ (1990) Composition, structure, rheological properties and sensory texture of processed cheese analogues. J Sci Food Agric 50: 237-252.

26. Pereira RB, Bennett RJ, Hemar Y, Campanella OH (2001) Rheological and microstructural characteristics of model processed cheese analogues. J Textur Studie 32: 349-373.

27. Fox PF, Guinee TP, Cogan TM, Mcsweene PLH (2000) Fundamentals of cheese Science. Aspen Publication, Gaithersburg, MO, USA.

28. Madadlou A, Khosroshahi A, Mousavi ME (2005) Rheology, microstructure and functionality of low-fat Iranian White cheese made with different concentrations of rennet. J Dairy Sci 88: 3052-3062.

29. Olson NF, Johnson ME (1990) Light cheese products: Characteristics and economics. Food Technol 44: 93-97.

30. Wang HH, Sun DW (2002) Correlation between cheese meltability determined with a computer vision method and with Arnott and Schreiber tests. J Food Sci 67: 745-749.

28

Lactic acid Bacteria Isolated from Raw Milk Cheeses: Ribotyping, Antimicrobial Activity against Selected Food Pathogens and Resistance to Common Antibiotics

Singh H[1], Kongo JM[2,3]*, Borges A[2], Ponte DJB[2,3] and Griffiths MW[1]

[1]Canadian Research Institute for Food Safety, Department of Food Science, University of Guelph, 43 Mc Gilvray Street, Guelph, ON N1G 2W1, Canada
[2]INOVA, Institute for Technological Innovation of the Azores, Road São Gonçalo, S/N 9504-540, Ponta Delgada, Portugal
[3]University of the Azores, Department of Technological Sciences and Development, Ponta Delgada, Acores, Portugal

Abstract

Fermentation with lactic acid bacteria is one of the oldest and effective methods of food preservation that meets the increasing drive of many consumers for foods perceived as minimally processed or free of unwanted chemical preservatives. The isolation and characterization of new wild LAB strains showing potential for application in food preservation may contribute to the processing of food products with good flavor and increasing safety. In the present work ninety-six LAB isolated from raw milk cheeses were subject to a preliminary characterization via identification (ribotyping) and screened for antimicrobial activity against *Listeria monocytogenes*, *Escherichia coli*, and *Salmonella* Newport, while the antibiotic resistance of the isolates against several clinic antibiotics were also studied. Most isolates of interest were identified as *Lactobacillus paracasei* ss. *paracasei*, of which, five were shown to completely inhibit *Listeria monocytogenes*, one inhibited *Escherichia coli*, five had inhibitory activity against *Salmonella* Newport, and five were active against both *Listeria monocytogenes* and *Salmonella* Newport. The *Etest* method determined that the isolates were sensitive to β-lactams, including amoxicillin and ampicillin, but were resistant to glycopeptides and aminoglycosides.

Keywords: *Lb paracasei*; Traditional cheeses; Salmonella; Bacteriocin; Food pathogens; Antibiotic resistance

Introduction

Many consumers avoid foods that have been preserved with chemicals such as sulphites and nitrites, due to their potential toxicity [1,2]. In addition, there are also issues surrounding thermal processing, as it can lead to unwanted changes of the nutritional and organoleptic properties of foods [1]. Therefore, there is an increasing interest in using biopreservation (fermentation) as a way to deliver food products perceived as minimally processed, safe, nutritious, while still having a convenient long shelf-life [2,3]. Fermentation with lactic acid bacteria (LAB), is an old method of food preservation, whereby LAB produce antimicrobial metabolites such as lactic acid, hydrogen peroxide, and eventually bacteriocins, which may be active against many undesirable food microorganisms. In particular, bacteriocins are reported as having a great potential of use in veterinary medicine, crop management and food preservation [1-4], as they exhibit such properties as being active against specific pathogens, have a narrow range of activity, may be thermo resistant, and are easily inactivated in the gastrointestinal tract [1].

However, until a bacteriocin can be fully used, a number of research steps, from identification, isolation, purification, as well as testing for their safety and technological suitability, are required. Today, nisin is one of the few bacteriocins approved by the World Health Organization for use in food [5], thus, discovery of new ones may contribute to increasing the pool of bacteriocins for food preservation. LAB optimal release of said metabolites, may be greatly affected by environmental factors such as temperature, pH and aerobic growth conditions, thus, understanding and eventually control these factors is a key part in characterizing a LAB or its metabolites, namely bacteriocins. Pathogens such as *Listeria monocytogenes*, and *Salmonella spp* are of increased concern in food safety [6], as they have been associated to many foodborne outbreaks, including some associated with cheeses [7]. Therefore, research directed to eliminate the risks associated with these pathogens has been an important activity in Food Safety [8]. The

aim of this research was to obtain a preliminary characterization of LAB isolates from local raw milk cheeses, towards undertaking depth studies for those one showing greater potential in food preservation.

Materials and Methods

Production of supernatants

Ninety six lactic acid bacteria isolates from raw-milk cheeses were aseptically streaked on de Man, Rogosa and Sharpe (MRS) agar plates (Sigma- Aldrich, St. Louis, MO, USA). Plates were then incubated at 30°C for 72 hours. A colony was then transferred to 10 mL of MRS broth, and incubated for 48 hours at 25°C under anaerobic conditions (in anaerobic jars with a 2.5 L AnaeroGen pouch (Sigma- Aldrich, St. Louis, MO, USA)), to hinder the formation of hydrogen peroxide. Following incubation, 700 μL of 1 M NaOH were added (giving a final pH of 6.5) to neutralize the pH and prevent the antimicrobial effects of acids in the broth. The suspension was centrifuged at 12000 × g at 4°C for 10 minutes, and filtered using a 0.22 μm pore size filter (EMD Millipore, Billerica, MA, USA). The supernatant was removed and stored at -80°C for later use.

Growth of indicator bacteria

Tryptic Soy Broth (TSB), or Tryptic Soy Agar (TSA) (Difco Laboratories, Detroit, MI, USA) were used to grow *Salmonella* Newport

---

*Corresonding author: Marcelino Kongo J, INOVA, Technological Innovation Institute of the Azores, Sao Goncalo Road, S/N 9504-540, Ponta Delgada, Portugal, E-mail: mkongo@uac.pt

(C1158). Brain Heart Infusion (BHI) agar or BHI broth (Difco Laboratories, Detroit, MI, USA) were used to grow *Listeria monocytogenes* (NCTC 7973), and *Escherichia coli* (ATCC 51739). All bacterial cultures were obtained from the culture collection at the Canadian Research Institute for Food Safety (Guelph, ON, Canada).

Screening for supernatants antimicrobial effect

The anti-microbial activity of the LAB supernatants were tested against *Listeria monocytogenes*, *Salmonella* Newport, and *Escherichia coli*. This was determined by measuring the optical density (OD) of these bacteria in the presence of the LAB supernatant using the Bioscreen C Microbiology Plate Reader (Labsystems, Helsinki, Finland). All 96 supernatants were tested against the three indicator micro-organisms, using a sterile 100-well honeycomb plate (Growth Curves USA, NJ, USA). The wells were inoculated with 150 µL of a diluted overnight bacterial culture (10^2 CFU/mL) and 50µL of the LAB supernatant and all samples were tested in triplicate. The overnight culture was produced by transferring one colony into 15 mL BHI or TSB, and put into a C24 incubator shaker (New Brunswick Scientific, NJ, USA) at 37°C with shaking at 200 rpm. The Bioscreen settings were: single, wide band (wb) wavelength; 37°C incubation temperature; 20 minute reheating time; kinetic measurement; 24 hour run time; medium intensity shaking for 10 seconds before reading; and reading time every 10 minutes.

Effect of temperature

LAB supernatants showing inhibitory activity were further tested to establish the effects of temperature on their activity. The LAB supernatants were heated using a digital heat block (VWR, Radnor, PA, USA) to an internal temperature of 45°C or 80°C for 30 minutes, in sterile Eppendorf tubes. In the case of 45°C, the supernatants were heated for a total of 45 minutes, as it took 15 minutes to reach the internal temperature of 45°. For 80°C, it was a total of 75 minutes, as it took 45 minutes to reach the internal temperature. The same Bioscreen procedure as described above was used to assess the antibacterial properties of the heat treated samples.

Effect of pH

The effect of pH on antimicrobial activity was determined at pH 4-5 or 8-9. Three hundred µL of the LAB supernatant were dispensed into sterile eppendorf tubes. To check the pH, pH-indicator strips (non-bleeding) pH 0-14 (colorPHast©, EMD, Germany) were used. To adjust the pH of the LAB supernatant to pH 4-5, 25 µL of 1 M HCl were added; for pH 8-9 15 µL of 1 M NaOH were added. The inhibitory activity of the supernatants against the test organisms was then determined using the Bioscreen as described above.

Proteinaceous nature of the antimicrobial substances

In order to determine if the anti-microbial activity was due to a proteinaceous substance, trypsin was added to the LAB supernatants to achieve final concentrations of 0.5 mg/mL. The resulting mixture was incubated at 37°C for 30 minutes. Following incubation, the antibacterial activity of the enzyme treated samples were determined using the Bioscreen.

Ribotyping

Considering the degree of inhibition they caused to specific indicator microorganism, selected LAB isolates were chosen and identified by automated ribotyping using the Ribo Printer (DuPont-Qualicon, Wilmington, DE, USA).

Nucleic acid was extracted from the cultures by boiling and was digested with the endonuclease EcoRI. Ribotyping was performed according to the manufacturer's instructions. The identity of the culture was determined using the Ribo Printer database.

Antibiotic resistance

The resistance of the isolates to clinically important antibiotics was tested as a way to shed light on the potential risks associated to their concentrated use in tailor-made starter cultures. The antibiotic resistance of LAB that possessed antibacterial activity and that were fully identified as indicated in 2.7, was determined using the Epsilometer (*Etest*) method [4]. Confluent growth of the LAB was produced on Lactic Acid Bacteria Susceptibility Medium agar [9]. Cells of an overnight culture of the respective LAB grown on MRS broth were harvested by centrifugation, as described above, and re-suspended in 1% saline to achieve an OD at 625 nm of 0.16-0.20 (comparable to McFarland standard 1 or 3×10^8 CFU/ml). A sterile cotton swab was immersed into the saline suspensions, and once excess fluid was removed, it was used to streak the entire surface of the agar plate in three directions. Plates were left to dry, and once dried, one E-test strip containing the antibiotics amoxicillin, ampicillin, clindamycin, erythromycin, gentamycin, kanamycin, rifampicin, streptomycin, tetracycline, or vancomycin (Biomerieux, Marcy-l'Etoile, France) was applied onto the agar, with the side displaying the minimum inhibitory concentration (MIC) facing up. Plates were then inverted (agar side facing upwards) and incubated at 37°C in an anaerobic jar containing an Anaero Gen 2.5 L pouch (Sigma- Aldrich, St. Louis, MO, USA), and readings were taken at 24 and 28 hours. An elliptical zone of inhibition is eventually produced, and the point at which the ellipse meets the strip gives a reading for the minimum inhibitory concentration (MIC) of the drug. Plates with an Etest strip were also compared to a control plate without an antibiotic strip, to see if growth occurred.

Results and Discussion

Bioscreen protocols

The Bioscreen data were analysed to determine the detection time (DT; the time taken to reach an optical density (OD) of 0.3) for each sample. The detection time was calculated as an average of triplicate readings. The average DT was then subtracted from the detection time obtained for the test bacterium grown in the absence of the LAB supernatants, which gave a detection time difference (DTD) [10]. Results were then classified as: (N): growth was similar to the control; (D): delay in growth of bacteria, DTD < 5 hours; (D+): delay in growth of bacteria, DTD >5 hours; (C): complete inhibition of bacterial growth [10]. Table 1 details the LAB supernatants exhibiting complete inhibition (C) of the respective bacterium. It can be seen that five isolates of LAB produced extracellular compounds with antibacterial activity against *Listeria monocytogenes* and *Salmonella* Newport. This is of special interest as most bacteriocins have anti-microbial activity only against gram positive bacteria, with gram negative bacteria being less susceptible [2,10]. One LAB supernatant was also found to exhibit activity against *E. coli*, which could be potentially used to control gram negative bacteria.

Effect of temperature

Table 1 describes the effect of different temperatures on the bioactivity of supernatants. The temperatures that were used are similar to those extreme values of temperatures commonly used in dairy processing. As expected, there was a decrease in inhibition activity after samples were heat treated. It is known that the antimicrobial activity

LAB isolate	Indicator microorganism	Condition[a]				
		Untreated	45°C	80°C	pH 4-5	pH 8-9
119	*Listeria monocytogenes*	C	N/A	N/A	N/A	N/A
154	*Listeria monocytogenes*	C	C	C	C	C
162	*Listeria monocytogenes*	C	C	C	C	C
171	*Listeria monocytogenes*	C	D+	C	C	N
173	*Listeria monocytogenes*	C	C	C	C	C
180	*Listeria monocytogenes*	C	D+	C	C	C
185	*Listeria monocytogenes*	C	C	C	C	C
186	*Listeria monocytogenes*	C	C	C	C	C
189	*Listeria monocytogenes*	C	C	C	C	C
220	*Listeria monocytogenes*	C	D+	C	C	C
174	*E.coli*	C	C	C	C	N
119	*Salmonella* Newport	C	N/A	N/A	N/A	N/A
130A	*Salmonella* Newport	C	D+	D+	C	N
149B	*Salmonella* Newport	C	D+	C	C	D
154	*Salmonella* Newport	C	C	C	C	D
172	*Salmonella* Newport	C	C	D	C	C
179	*Salmonella* Newport	C	D	D+	C	C
186	*Salmonella* Newport	C	C	C	C	D
189	*Salmonella* Newport	C	C	C	C	N
211	*Salmonella* Newport	C	C	C	C	N
	Salmonella Newport	C	D	C	C	D

[a]Abbreviations are: C, Complete inhibition; N, growth is similar to control; D, delay in growth of bacterium log phase by <5 hours; D+, delay in growth of bacterium log phase by≥5 hours; N/A, no activity detected.

Table 1: LAB isolates culture supernatants activity against the indicated pathogen atdifferent temperature and pH conditions.

and spectrum of activity of compounds, such as lysozyme, changes on heating due to perturbations in tertiary structure that improve access to the active moiety [11,12].

Many samples exhibited the same level of activity after treatment at 45°C as well at 80°C, while unexpectedly, a few showed a slightly increased inhibitory activity. A study by Joshi, Sharma and Rana [13] demonstrated that a partially purified bacteriocin from *Lactobacillus* sp. (CA44) retained high antimicrobial activity against *E. coli, B. cereus* and *S. aureus* after heating at 68°C for 10 and 20 minutes. Sankar et al. took a purified bacteriocin produced by *Lactobacillus plantarum* and subjected it to a variety of temperatures. For *Listeria* and *E. coli* at 40 and 50°C there was a slight decrease in the inhibition and heating at 80°C resulted in even less antibacterial activity. Sifour, Tayeb, Haddar, Namous, and Aissaoui [14] also studied a bacteriocin from *Lactobacillus plantarum* and found that following heating to 40°C and 80°C for 30 min the residual activity was almost 100%. This agrees with our results, which in general indicate that due to the properties exhibited by LAB from our study, they may be of use as bio-control agents in products that may undergo a heating process in the studied temperature ranges.

Effect of pH

The effect of the supernatant pH on its antibacterial activity is shown also in Table 1. The pH value of 4-5 was chosen as this is the pH range of many dairy products, such as cheeses and yogurts, because of the activity of starter cultures. All supernatants from the chosen LAB isolates in our study showed inhibitory activity against the indicator bacteria at pH 4-5. As a result, this LAB may be of interest to be used as adjuvants cultures to increase product safety during ripening of most cheeses. At pH 8-9, some supernatants produced a >5 hour delay in the growth of the test organism, or no inhibition was observed.

The pH range 8-9 was chosen to provide an indication of the stability of the antimicrobial compound when exposed to an alkaline environment. Sifour et al. [14] described a bacteriocin with highest residual antibacterial activity at pH 6, and between pH 4-5 the residual activity was between 80-90%. At pH 8-9 the residual activity of this bacteriocin also appeared to be approximately 90% [14]. Joshi et al. [13] found the highest antimicrobial activity of a bacteriocin against *S. aureus, B. cereus,* and *E. coli* was observed at a pH of 4 to 5. For all three bacteria, the inhibitory activity of the bacteriocin at a pH of 8-9 was noticeably decreased [13]. Similar results were reported for the effect of bacteriocin on *Listeria* spp and *E. coli* and were confirmed by this study.

Effect of the proteolytic enzyme on antimicrobial activity

Treatment with trypsin lowered the antibacterial activity of the supernatants, indicating that the supernatants' antibacterial activity was caused by a substance of proteic nature (likely a bacteriocin). This is partially in agreement with data reported by Sankar et al., who found complete inhibition of the antibacterial activity in supernatants treated with trypsin. Recall that in our study the acidity and presence of H_2O_2 were unlikely to cause the antibacterial activity exhibited by the supernatants, as they were produced under conditions designed to ensure that H_2O_2 was absent, via using anaerobic conditions for fermentation, followed by a neutralization of the supernatant pH. Sifour et al. [14] studied the effects of five enzymes (lipase, α- amylase, trypsin, pronase E, α-chymotrypsin) on a bacteriocin and found that trypsin, pronase E, and α-chymotrypsin resulted in a major reduction in activity or complete inactivation of the bacteriocin. Also a study by Ivanova, Kabadjova, Pantev, Danova, and Dousset [15] found a bacteriocin produced by *Lactococcus lactis* subsp. *lactis* b14, which activity was significantly reduced by proteinase K, pronase E, and pepsin.

Ribotyping

Ribotyping showed Lactobacillus *paracasei* ss. *paracasei* to be the dominant species among the studied isolates (Figure 1), followed by *Lactobacillus rhamnosus.* Our study agree with results reported by Kongo, Ho, Malcata, and Wiedmann [16] who also found presence of both species and dominance of *Lactobacillus paracasei* in traditional raw milk cheeses.

Antibiotic resistance

Table 2 shows the antibiotic susceptibility of the identified LAB isolates used in this study. The LAB were susceptible to amoxicillin, ampicillin, clindamycin, erythromycin, rifampicin, and tetracycline. They were found to be more resistant to streptomycin, gentamycin, kanamycin and highly resistant to vancomycin. These results are similar to other studies that determined the antimicrobial resistance of LAB [17]. Antibiotic resistance may be associated with an increasing overuse of antibiotics in the treatment of cattle and LAB are often found to be resistant to β-lactam antibiotics. The fact that our isolates were generally sensitive to β-lactams may suggest that they were isolated from products associated to an environment where these antibiotics are rarely used. However, the same isolates showed resistance to vancomycin (an antibiotic used in extreme situations in clinical applications), and this is in agreement with results from Vescovo, Morelli, and Bottazzi and de Fabrizio, Parada, and Ledford [18,19]. Considering that LAB may potentially transfer resistance to more dangerous bacteria, LAB showing multiple antibiotic resistances should be avoided as potential components of starter cultures. While

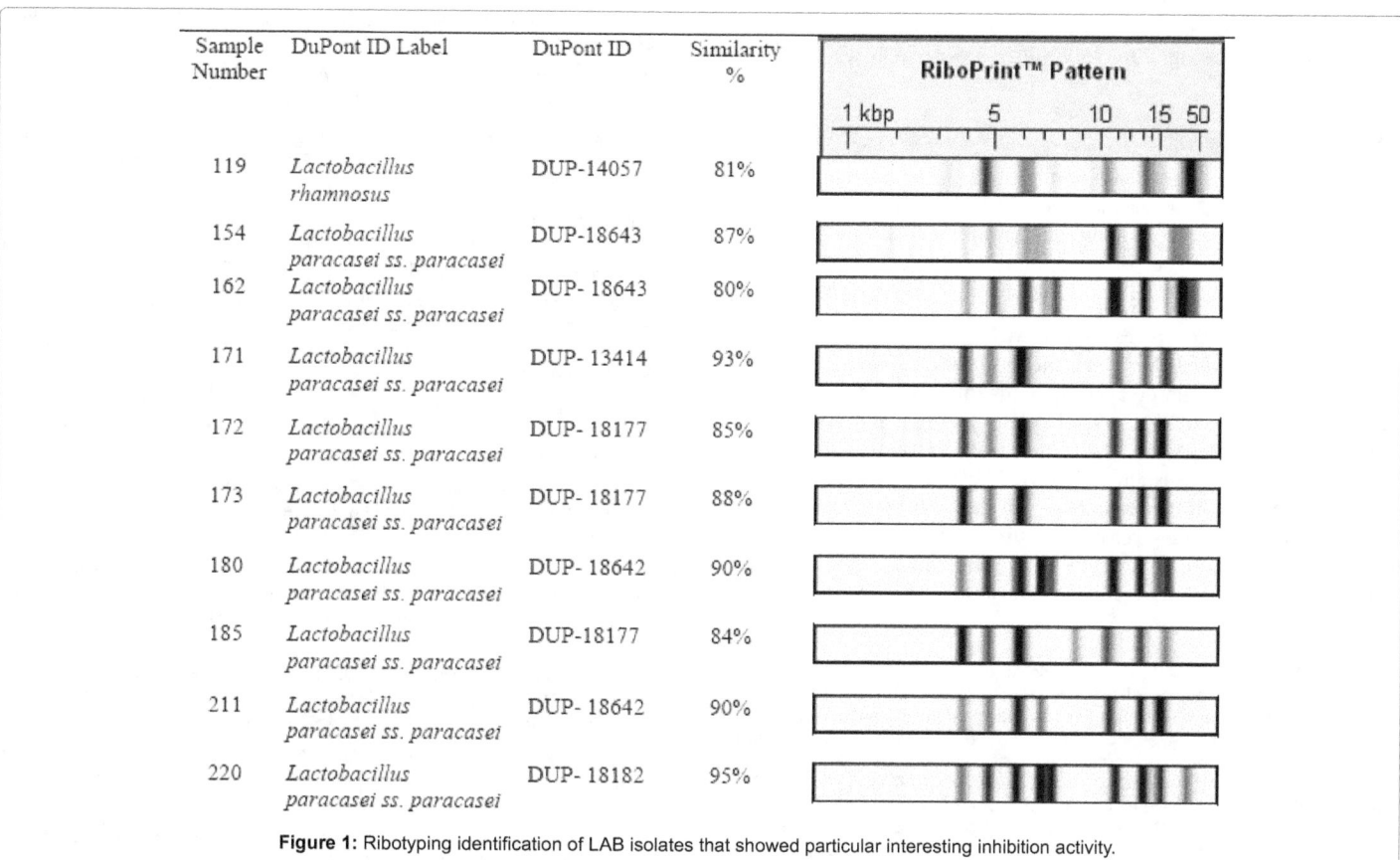

Figure 1: Ribotyping identification of LAB isolates that showed particular interesting inhibition activity.

Sample number	Reading time(hours)	MIC[a] (µg/mL)									
		Amox	Amp	Clind	Eryth	Gent	Kan	Rif	Strep	Tetra	Van
119	24	0.75	0.5	0.064	0.032	8	48	0.094	16	0.25	>256
119	48	0.5	0.5	0.19	0.094	8	96	0.064	32	0.5	>256
154	24	0.75	0.5	0.064	0.032	8	48	0.094	16	0.25	>256
154	48	0.5	0.5	0.19	0.094	8	96	0.064	32	0.5	>256
162	24	0.5	0.25	0.032	0.032	4	16	0.047	12	0.25	>256
162	48	0.38	0.25	0.094	0.064	8	32	0.047	16	0.5	>256
171	24	0.5	0.5	0.032	0.047	6	64	0.064	16	0.25	>256
171	48	0.5	0.5	0.064	0.064	8	96	0.064	24	0.75	>256
172	24	0.5	0.5	0.064	0.064	4	48	0.047	16	0.38	>256
172	48	0.5	0.5	0.064	0.094	4	64	0.047	16	1	>256
173	24	0.75	0.5	0.047	0.064	6	128	0.094	24	0.25	>256
173	48	0.75	0.5	0.047	0.094	6	>256	0.094	24	0.5	>256
174	24	0.75	0.38	0.047	0.032	6	128	0.032	12	0.25	>256
174	48	0.75	0.25	0.094	0.094	6	>256	0.047	24	0.38	>256
180	24	1	0.38	0.064	0.023	8	>256	0.094	24	0.38	>256
180	48	0.5	0.38	0.125	0.064	8	>256	0.094	24	0.38	>256
185	24	0.5	0.38	0.023	0.064	8	32	0.032	24	0.5	>256
185	48	0.5	0.38	0.047	0.125	8	128	0.032	24	0.5	>256
211	24	0.38	0.19	0.064	0.094	4	64	0.023	16	0.25	>256
211	48	0.38	0.25	0.064	0.064	6	128	0.023	16	0.38	>256
220	24	0.75	0.75	0.064	0.023	6	64	0.094	16	0.125	>256
220	48	0.5	0.5	0.125	0.094	6	96	0.094	24	0.38	>256

[a]Abbreviations are: Amox=Amoxicillin, Amp=Ampicillin, Cind=Clindamycin, Eryth=Erythromycin, Gent=Gentamycin, Kan=Kanamycin, Rif=Rifampicin, Strep=Streptomycin, Tetra=Tetracycline, Van=Vancomycin

Table 2: Antibiotic susceptibility of LAB isolates from this study.

all isolates showed similar resistance to vancomycine, there was a statistically significant difference (not shown) in MICs for most antibiotics.

Conclusion

The preliminary characterization of the 96 isolates included in the present study showed that via identification by ribotyping *Lactobacillus paracasei* ss. *paracasei*, and *Lactobacillus rhamnosus* were the dominant ones in terms of frequency of isolation. Screening of selected isolates for anti-microbial activity against *Listeria monocytogenes*,

Salmonella Newport revealed that five inhibited growth of *Listeria*, five inhibited *Salmonella* Newport, and five inhibited both *Listeria* and *S.* Newport. The antimicrobial compounds they released were thermo stable at both 45°C and 80°C and all isolates inhibited bacterial growth at pH 4-5, while at pH 8-9 the antimicrobial activity of many was decreased. Antibiotic resistance tested via *Etest* found most LAB isolates were sensitive to β-lactam antibiotics, yet they were more resistant to glycopeptides and aminoglycosides. Future studies will be undertaken to fully characterize and identify the antimicrobial substances that showed inhibitory activity against *Listeria* and *Salmonella*.

Acknowledgments

The authors would like to thank Mazin Matloob for his technical expertise in helping with the ribotyping of the isolates and Hany Anany for helping with Bioscreen C Microbiology Plate Reader use. Additionally, the authors would like to acknowledge the financial support of Dairy Farmers of Ontario and the Natural Sciences and Engineering Research Council of Canada. This research is part of project SEPROQUAL INOVAÇÃO funded by PROCONVERGENCIA, FEDER (Fundo Regional para o Desenvolvimento Regional) from the European Union, and the Azorean Regional Government via the Secretariats for Agriculture and Forestry, Fisheries and Economy, and the Secretariat of Science and Technology who supported author J. Marcelino Kongo.

References

1. Ananou S, Maqueda M, Martinez-Bueno M, Valdivia E (2007) Biopreservation, an ecological approach to improve the safety and shelf-life of foods. Communicating Current Research and Educational Topics and Trends in Applied Microbiology 1: 475-487.

2. Stiles ME (1996) Biopreservation by lactic acid bacteria. Antonie Van Leeuwenhoek 70: 331-345.

3. De Martinis E, Publio M, Santarosa PR (2001) Antilisterial activity of lactic acid bacteria isolated from vacuum-packaged brazilian meat and meat products. Brazilian Journal of Microbiology 32: 32-37.

4. Danielsen M, Wind A (2003) Susceptibility of lactobacillus spp. to antimicrobial agents. International Journal of Food Microbiology 82: 1-11.

5. Leroy F, Vuyst LD (2010) Bacteriocins of lactic acid bacteria to combat undesirable bacteria in dairy products. Australian Journal of Dairy Technology 65: 143-149.

6. Roberts TA, Pitt JI, Farkas J, Grau FH (Eds) (1998) Micro-organisms in Food : Microbial Ecology of Food Commodities, 6. London, UK.

7. De Buyser ML, Dufour B, Maire M, Lafarge V (2001) Implication of milk and milk products in food-borne diseases in France and in different industrialized countries. International Journal of Food Microbiology 67: 1-17.

8. Wu Y, Griffiths M, McKellar R (2000) A comparison of the bioscreen method and microscopy for the determination of lag times of individual cells of *Listeria monocytogenes*. Letters in Applied Microbiology 30: 468-472.

9. Klare I, Konstabel C, Mueller-Bertling S, Reissbrodt R, Huys G, et al. (2005) Evaluation of new broth media for micro dilution antibiotic susceptibility testing of *lactobacilli, pediococci, lactococci,* and *bifidobacteria*. Applied and Environmental Microbiology 71: 8982-8986.

10. Anany H, Lingohr EJ, Villegas A, Ackermann HW, She YM, et al. (2011) A *shigellaboydii* bacteriophage which resembles *Salmonella* phage ViI. Virology Journal 8: 242-252.

11. Ibrahim HR, Higashiguchi S, Jeneja LR, Kim M, Yamamoto T (1996) A structural phase of heat-denatured lysozyme with novel antimicrobial action. J Agric Food Chem 44: 1416-1423.

12. Masschalck B, Houdt R, Haver EGR, Michiels CW (2001) Inactivation of gram-negative bacteria by lysozyme, denatured lysozyme, and lysozyme-derived peptides under high hydrostatic pressure. Applied and Environmental Microbiology 67: 339-344.

13. Joshi VK, Sharma S, Rana NS (2006) Production, purification, stability and efficacy of bacteriocin from isolates of natural lactic acid fermentation of vegetables. Food Technology and Biotechnology 44: 435-439.

14. Sifour M, Tayeb I, Haddar HO, Namous H, Aissaoui S (2012) Production and characterization of bacteriocin of *Lactobacillus plantarum* F12 with inhibitory activity against *Listeria monocytogenes*. TOJSAT 2: 55-61.

15. Ivanova I, Kabadjova P, Pantev A, Danova S, Dousset X (2000) Detection, purification and partial characterization of a novel bacteriocin substance produced by *Lactoccous lactis* subsp. *Lactis* b14 isolated from boza-bulgarian traditional cereal beverage. Biocatalysis: Fundamentals and Application 41: 47-53.

16. Kongo JM, Ho AJ, Malcata FX, Wiedmann M (2007) Characterization of dominant lactic acid bacteria isolated from Sao Jorge cheese, using biochemical and ribotyping methods. Journal of Applied Microbiology 103: 1838-1844.

17. Ammor MS, Belen Florez A, Mayo B (2007) Antibiotic resistance in nonenterococcal lactic acid bacteria and bifidobacteria. Food Microbiology 24: 559-570.

18. Vescovo M, Morelli L, Bottazzi V (1982) Drug resistance plasmids in *Lactobacillus acidophilus* and *Lactobacillus reuteri*. Applied and Environmental Microbiology 43: 50-56.

19. De Fabrizio SV, Parada JL, Ledford RA (1994) Antibiotic resistance of *Lactococcus lactis* -An approach of genetic determinants location through a model system. Microbiology, Foods and Nutrition 12: 307-315.

29

Efficacy of Different Washing Treatments on Quality of Button Mushrooms (*A. bisporus*)

Gupta P[1]* and Bhat A[2]

[1]*Department of Food Science and Nutrition, Lovely Professional University, Punjab, India*

[2]*Division of Post-harvest Technology, Sher-e-Kashmir University of Agricultural Sciences and Technology of Jammu, Jammu, India*

Abstract

To enhance the quality and to obtain a better colour, fresh button mushrooms were given washing treatments with different chemical solutions of different concentration for 10 minutes and stored under refrigerated conditions to check their shelf life. The chemicals used were citric acid (0.5%, 1.5%, 2.5%), H_2O_2 (1.5%, 2.5%, 3.5%) and EDTA (2%, 4%, 6%). Of all the treatments used, 2.5% citric acid was found to be the most effective in controlling weight loss, maturity index and microbial growth for up to a period of 12 days and was found to be highly acceptable according to the scores after sensory evaluation. With the advancement in storage period, significant difference in the color values (L*, a* and b*) and browning index was observed and an increase in weight loss, maturity index and microbial growth of mushrooms was observed.

Keywords: Physiological weight loss; Maturity index; Microbial growth; Citric acid; Hydrogen peroxide; EDTA

Introduction

White button mushrooms also called as *Agaricus bisporus* are an important part of human diet since antiquity because of its attractive colour, flavor and aroma. Also the nutritional composition of white button mushrooms is high. *Agaricus* on fresh weight basis contains 2.9 percent crude protein 5 percent carbohydrate, 0.9 percent crude fibre, 0.8 percent ash and 0.3 percent fat Rai [1] K and P are the dominant elements among the minerals. K to Na ratio found in mushrooms is desirable for hypertension patients [2]. The white button mushrooms are low in calorie, where the carbohydrates are stored as glycogen, chitin and hemicellulose instead of starch. Mushrooms are deliciously palatable non-conventional source of protein, which can bridge the protein gapping in Indian diet [3]. But white button mushroom are highly perishable because of high respiration, which leads to color change due to enzymatic oxidation and decreases their marketability. Loss of whiteness during storage is particularly deleterious for the mushroom industry [4]. There are several indicators that determine the quality of mushrooms, such as visual appearance, size, colour, maturity stage, development stage, microbial growth and weight loss. Discoloration reactions of mushrooms are complex and depend on raw material condition, washing, cutting, handling and packaging practices and bacterial growth during storage. Of these, colour is the most important parameter because it is first perceived by consumers and discoloration decreases the commercial value [5]. Color, fresh and clean appearance and uniform closed buttons have high importance for mushroom quality and consumer preferences [6,7]. The maturity index of mushrooms is a sign of maturity which increases after harvest and can be observed on the basis of cap opening. A good quality mushroom should be free from open veils, disease, spots, insect injury and decay. Guthrie (1984) gave description about different stages of sporophore development in mushrooms which include 1) veil intact (tight), 2) veil intact (stretched), 3) veil partially broken(< half), 4) veil partially broken (>half), 5) veil completely broken, 6) cap open, gills well exposed, 7) cap open, gill surface flat. Treatment with antibrowning agents is an effective and frequently employed method for controlling the enzymatic browning in several fresh-cut fruits and vegetables [8]. In order to improve the color and enhance the shelf life washing of mushrooms with various anti browning inhibitors is recommended. Citric acid is widely used as an additive and as an antimicrobial by

virtue of their low pH in food industry. Brennan et al. [9] proposed washing with 40 g/L citric acid in mushrooms to be sliced. Like citric acid EDTA is also has the potential to inhibit microbial growth [10]. The immersion in hydrogen peroxide solution followed by dipping in a solution of enzymatic browning inhibitor was found to be beneficial for extending the shelf life of whole mushrooms. The aim of the present study is to check the efficacy of different washing treatments for minimizing the color change, enhancing the shelf life and giving them a fresh appearance under refrigerated storage.

Material and Methods

Button mushrooms were procured from the department of Plant Pathology and the research was conducted in the department of Food Science and Technology, SKUAST-Jammu. Freshly harvested mushrooms of uniform size and intact veil were chosen and washed with tap water to remove surface contamination. After this, whole mushrooms were dipped in different browning inhibitor solutions viz., T_1 (H_2O_2, 1.5%), T_2 (H_2O_2, 2.5%), T_3 (H_2O_2, 3.5%), T_4 (citric acid, 0.5%), T_5 (citric acid, 1.5%), T_6 (citric acid, 2.5%), T_7 (EDTA, 2%), T_8 (EDTA, 4%) and T_9 (EDTA, 6%) for 10 minutes. The mushrooms were then surface dried, packed in PP with five pin holes on each packet and stored under refrigerated condition and evaluated for change in colour, veil opening, physiological weight loss, total plate count and sensorial changes after 3 days interval for up to a period of 12 days. Approximately, 100 g of mushroom per pack were taken for each treatment with four replicates.

Maturity index

The maturity index was measured on the basis of scale ranging

*Corresponding author: Gupta P, Department of Food Science and Nutrition, Lovely Professional University, Jalandhar-Delhi GT Road, Punjab-144411, India
E-mail: prerna.gskuastj@gmail.com

from 1 (veil intact) to 7 (cap open, gills flat) as suggested by Guthire [11]. Veil opening rate can be used to characterize the aging of white mushrooms. It is defined as the ratio of the number of mushrooms with cap opening out to the total number of mushrooms. Randomly, six mushrooms were taken from a packet and were evaluated manually by visual observation.

Physiological weight loss

Weight loss was determined by weighing the contents of the package before and after storage and was expressed as the percent loss of weight with respect to the initial weight.

$$\text{Weight loss } (\%) = \frac{W_i - W_f}{W_i}$$

Where,

W_i is the initial weight of the mushrooms before treatment

W_f is the final weight of mushrooms after storage

Microbial analysis

Microbial analysis was done according to total plate count method using nutrient agar. Nutrient agar was prepared and sterilized in an autoclave at 121°C for 20 minutes. Serial dilution of mushrooms (up to six dilutions) was then carried out and 0.1 ml of the aliquot was added on the petri plate containing nutrient agar as medium (using micropipette). The plates were then covered and sealed with the help of porcelain wax and incubated at 37°C for 48 hours. The total microbial count was recorded as cfug^{-1} [12].

Color

The surface colour of mushrooms was measured with the help Hunter Lab Mini Scan XE Colorimeter with an 8-mm-diameter diaphragm calibrated with a white tile (X=81.1, Y=86.0 and Z=91.8) [6]. Where, L* indicates (whiteness/darkness), a* (greenness/redness) and b* (yellowness/blueness). Colour was measured from all the three sides of mushrooms and mean value was taken.

Browning index

The browning index was calculated using the following expression [13]:

$$BI = 100 \times \left[\frac{X - 0.31}{0.71} \right]$$

Where,

$$X = \frac{(a* + 1.75L*)}{(5.645L* + a* - 3.012b*)}$$

Sensory analysis

The sensory analysis of the treated mushrooms was done after three days interval. Mushrooms were slightly sautéed in little oil and a pinch of salt was added and they were evaluated on the basis of appearance, flavor, texture, taste and overall acceptability using 9-point hedonic scale from 9=like extremely to 1=dislike extremely. The data obtained was analyzed statistically.

Statistical analysis

The results were analyzed using Factorial Completely Randomized Design (FCRD).

Results and Discussion

The quality of fresh mushrooms is influenced by stage of development and pre-harvesting factors. In order to prevent the deterioration of mushrooms and to increase the shelf life, mushrooms were subjected to various postharvest treatments and changes in physiology and biochemical components were studied. The results obtained from present study are summarized here under: White button mushrooms of uniform size, intact veil and free from mechanical damage were selected. The most successful strategies aimed to prevent browning occurring on fresh-cut fruit and vegetables are based on treatments with reducing agents, acidifying agents, chelating substances and calcium solutions Martin-Diana et al. [14] Various chemical treatments helped in delaying cap opening and keeping the veil intact (Table 1). All the treatments were effective in keeping the veil intact up to 6th day of storage. Among the various treatments used citric acid (2.5%) was the most effective in delaying cap opening. However the mushrooms treated with 2.5% citric acid (T$_9$) recorded the lowest (1.19) veil opening (i.e. veil intact) whereas the highest maturity index of 2.11 was observed in T$_6$ (6% EDTA) on 3rd day of storage. Maturity index increased with the increase in storage period with a value of 3.70 in T$_9$ to 4.83 in T$_6$ on 12th day of storage. Weight loss in mushrooms is a common phenomenon which occurs mainly due to moisture loss and loss of carbon reserves due to respiration [15,16]. Treatment of mushrooms with various chemicals had a significant effect in controlling weight loss and it might be because these treatments helped in reducing the rate of respiration and transpiration (Table 2). Use of citric acid and hydrogen peroxide was found to be effective in reducing weight loss as compared to EDTA. Bayoumi [17] also reported that H$_2$O$_2$ treatment significantly reduced fresh weight loss of pepper fruits during storage under room temperature and fridge conditions. The antimicrobial effect of citric acid, EDTA and H$_2$O$_2$ helped in controlling the bacterial growth. Among the various treatments citric acid was found to be most effective in inhibiting microbial growth (Table 3). On 3rd day of storage it was 4.02 cfu/g in treatment T$_9$ and on 12th day of storage the total plate count was found to be 6.42 cfu/g in T$_9$ whereas the count was highest in T$_6$ (6% EDTA) ranging from 5.29 cfu/g on 3rd day to 8.48 cfu/g on 12th day of storage. The significant effect of citric acid in limiting the development of food spoilage and pathogenic microorganisms is widely accepted [18,19]. The total microbial

Treatment	Storage period (days)			
	3	6	9	12
T$_1$ (1.5% H$_2$O$_2$)	1.42	2.60	3.75	4.25
T$_2$ (2.5% H$_2$O$_2$)	1.33	2.48	3.61	4.13
T$_3$ (3.5% H$_2$O$_2$)	1.24	2.32	3.45	3.75
T$_4$ (2% EDTA)	1.36	2.48	3.80	4.28
T$_5$ (4% EDTA)	1.62	2.67	4.05	4.46
T$_6$ (6% EDTA)	2.11	3.46	4.29	4.83
T$_7$ (0.5% C.A)	1.36	2.52	3.35	4.07
T$_8$ (1.5% C.A)	1.28	2.38	3.26	3.82
T$_9$ (2.5% C.A)	1.19	2.41	3.31	3.70

All values are mean significant values, C.D (p=0.05), Treatment=0.10, Storage=0.05, T × S=0.16

Table 1: Change in the maturity index of mushrooms under refrigerated condition.

Treatment	Storage period (days)			
	3	6	9	12
T₁ (1.5% H₂O₂)	0.42	1.72	3.47	4.19
T₂ (2.5% H₂O₂)	0.49	1.24	3.12	3.27
T₃ (3.5% H₂O₂)	0.34	0.68	1.33	2.76
T₄ (2% EDTA)	0.97	1.89	3.52	5.8
T₅ (4% EDTA)	1.36	3.52	5.54	7.89
T₆ (6% EDTA)	1.84	3.85	6.23	8.22
T₇ (0.5% C.A)	0.40	1.65	2.99	3.65
T₈ (1.5% C.A)	0.24	0.59	1.84	3.19
T₉ (2.5% C.A)	0.23	0.38	1.28	2.83

All values are mean significant values, C.D (p=0.05), Treatment=0.10, Storage=0.05, T × S=0.16

Table 2: Change in the physiological weight loss (%) of mushrooms stored under refrigerated condition.

Treatment	Storage period (days)		
	3	9	12
T₁ (1.5% H₂O₂)	5.23	6.19	7.65
T₂ (2.5% H₂O₂)	4.85	5.17	7.42
T₃ (3.5% H₂O₂)	4.25	6.25	7.13
T₄ (2% EDTA)	4.82	6.59	7.54
T₅ (4% EDTA)	5.08	7.10	8.35
T₆ (6% EDTA)	5.29	7.32	8.48
T₇ (0.5% C.A)	4.33	6.48	7.27
T₈ (1.5% C.A)	4.16	5.2	6.81
T₉ (2.5% C.A)	4.06	4.58	6.42

All values are mean significant values, C.D (p=0.05), Treatment=0.10, Storage=0.08, T × S=0.16

Table 3: Microbial count in white button mushrooms (cfug-1) stored under refrigerated condition.

count though increased with storage, but was lesser than the count of ISI specification (IS: 7463-2004) [20]. All the treatments helped in maintaining the whiteness of mushrooms. Though with the increase in storage period the whiteness decreased and as storage progressed, the L* value (lightness/darkness) decreased and a* (redness/greenness) and b*value (yellowness/blueness) increased. The rate of browning during storage might be influenced by the length of immersion of samples in the treatment solution. The initial value as observed for colour of fresh mushrooms was of L* value equal to 86.32, a* value equal to 0.15 and b* value equal to 2.12. Among the various treatments used, 3.5% H₂O₂ was the most effective in retaining the whiteness of mushrooms because of its bleaching effect (Table 4). Treatment with high concentration of citric acid led to slight change in color of mushrooms to yellow, though browning was significantly controlled and BI was effectively reduced during storage (Table 5). The browning index of mushrooms is related to change in color of mushrooms. As the L* value decreased

and a* and b* value increased, the browning index increased. Brennan et al. (2000) also observed a yellowing of the surface cap when whole mushrooms were washed with 40 g/L citric acid for 10 min. However, these authors proposed this treatment to reduce microbial counts of mushrooms before slicing, having no deleterious effect on the sensory property of sliced mushrooms after cooking. Erbay et al. [21] also documented that treatment with citric acid showed the least reduction in L* value since the beginning of storage and decreased when storage progressed. Appearance of mushrooms is one of the important quality

Treatment		Storage period (days)			
		3	6	9	12
T₁ (1.5% H₂O₂)	L*	83.21	81.64	80.47	77.63
	a*	0.70	1.13	2.23	4.53
	b*	6.16	13.22	19.25	27.35
T₂ (2.5% H₂O₂)	L*	84.12	81.67	80.19	78.21
	a*	0.81	1.47	2.19	2.63
	b*	4.18	9.26	13.76	28.42
T₃ (3.5% H₂O₂)	L*	84.64	83.54	81.49	79.75
	a*	0.75	1.26	2.13	2.33
	b*	3.39	9.13	13.67	23.67
T₄ (2% EDTA)	L*	84.32	82.68	80.29	78.39
	a*	0.51	1.22	2.60	4.59
	b*	6.68	11.79	20.21	33.85
T₅ (4% EDTA)	L*	83.41	81.10	78.41	77.62
	a*	0.75	2.19	2.30	3.28
	b*	8.37	15.71	25.80	34.56
T₆ (6% EDTA)	L*	83.29	80.61	79.29	77.23
	a*	1.06	2.59	4.29	6.07
	b*	8.64	16.63	27.39	35.97
T₇ (0.5% C.A)	L*	84.29	82.61	81.54	79.79
	a*	0.73	1.21	1.52	2.42
	b*	3.38	7.60	12.73	21.81
T₈ (1.5% C.A)	L*	83.51	82.76	79.16	77.63
	a*	1.41	1.46	1.51	2.46
	b*	4.20	8.18	13.39	19.13
T₉ (2.5% C.A)	L*	83.24	82.18	81.63	78.73
	a*	0.81	1.18	2.34	2.77
	b*	5.57	9.61	15.58	25.30

All values are mean significant values, L* (Whiteness/darkness), a* (redness/greenness), b* (yellowness/blueness) C.D (p=0.05)

Table 4: Color changes in mushrooms stored under refrigerated conditions.

Treatment	Storage period (days)			
	3	6	9	12
T₁ (1.5% H₂O₂)	1.94	5.14	6.89	11.20
T₂ (2.5% H₂O₂)	1.35	3.08	4.89	11.13
T₃ (3.5% H₂O₂)	1.10	2.97	4.76	8.75
T₄ (2% EDTA)	2.03	3.86	7.37	9.47
T₅ (4% EDTA)	2.63	5.52	9.85	14.41
T₆ (6% EDTA)	2.78	5.98	10.78	15.94
T₇ (0.5% C.A)	1.10	2.51	4.30	8.01
T₈ (1.5% C.A)	1.49	2.94	6.08	7.12
T₉ (2.5% C.A)	1.78	3.15	5.47	9.69

All values are mean significant values, C.D (p=0.05), Treatment=0.10, Storage=0.08, T × S=0.16

Table 5: Change in browning index value of mushrooms stored under refrigerated condition.

Treatment	Appearance	Flavor	Texture	Taste	Overall acceptability
T_1 (1.5% H_2O_2)	6.91	6.93	6.80	6.80	6.85
T_2 (2.5% H_2O_2)	7.06	7.03	7.17	7.00	7.06
T_3 (3.5% H_2O_2)	7.40	7.37	7.35	7.28	7.21
T_4 (2% EDTA)	6.94	6.96	7.04	6.75	6.92
T_5 (4% EDTA)	6.82	6.61	6.70	6.10	6.37
T_6 (6% EDTA)	6.63	6.40	6.33	6.15	6.41
T_7 (0.5% C.A)	6.93	7.27	6.91	6.88	6.99
T_8 (1.5% C.A)	7.16	7.52	7.05	7.12	7.21
T_9 (2.5% C.A)	7.36	7.64	7.38	7.38	7.45
C.D. (p=0.05)	Treatment=0.08 Storage=0.04 T x S=0.20	Treatment=0.11 Storage=0.07 T x S=0.25	Treatment=N.S Storage=0.90 T x S=N.S	Treatment=0.19 Storage=0.10 T x S=0.38	Treatment=0.15 Storage=0.08 T x S=0.30
All values are mean significant values taken from mean of three replicates					

Table 6: Sensory evaluation of mushrooms stored under refrigerated condition.

attribute. At the beginning, on the basis of hedonic rating, the scores of sensory evaluation were: (8.85) for appearance, (8.57) for flavor, (8.44) for texture and (8.72) for taste and (8.64) for overall acceptability. Treatment with various chemicals helped in preventing off-odor in mushrooms during storage. Among the different treatments, citric acid helped in prolonging its shelf life, maintaining, freshness and eating quality during storage at refrigerated temperatures as compared to control. These results are in agreement with the findings of Rosario [22] who observed that increasing storage time cause progressive degradation, which leads to decrease in overall acceptability. The scores decreased with storage as the mushrooms were not liked by panelists as evident by change in colour of mushrooms and increased maturity index (Table 6). Loss of firmness was observed in all the treatments. This loss of firmness is a gradual process of catabolism catalyzed by different enzymes.

Conclusion

To maintain the quality and availability of mushrooms and to extend its acceptability, they were treated with aqueous solution of different chemicals and stored under refrigerated condition. Treatment of mushrooms with various chemicals viz: hydrogen peroxide, ethylene diaminetretaacetic acid and citric acid were effective in retaining the quality parameters of mushrooms. From the observations it was observed that citric acid is the most effective in controlling weight loss, maturity index and microbial growth. Citric acid showed slightly antibacterial potential but induced a slight yellowness on mushroom surface. It was also observed that higher concentration of citric acid led to more yellowness on the surface of mushrooms as recorded by hunter color lab but browning was significantly reduced. Hydrogen peroxide is more effective in maintaining the whiteness of mushrooms and was found to be at par with citric acid. From different concentrations of EDTA used it was recorded that the lesser percentage of EDTA is more effective. Jafri et al. [23] in 2013 had showed the senescence inhibition of oyster mushrooms by the combined effect of chemical treatment and MAP, resulting in maintenance of tissue firmness and sensory quality, inhibition of lipid peroxidation and better retention of phenolics and antioxidant ability as compared to the control samples [24-28]. These results suggested that a combination of chemical treatment

and MAP had good promise in maintaining oyster mushroom quality and extending their postharvest life up to 25 days when stored at 4°C. Another important observation made was that the use of chemical treatment alone was more effective than the use of MAP alone. High weight loss with visible evidence of deterioration and senescence were the main effects observed in the chemically untreated samples.

Acknowledgment

This study was funded by Sher-e-Kashmir University of Agricultural Sciences and Technology of Jammu, Chatha, and Jammu.

References

1. Rai RD (1995) Nutritional and medicinal value of mushrooms. Advances in horticulture Mushroom 13: 537-551.

2. Chang ST, Miles PG (1989) Biology and cultivation of edible mushrooms. Academic Press London.

3. Desayi D (2012) Development and sensory evaluation of mushroom fortified noodles. Int J Food Agri Vet Sci 2: 187-189.

4. Jolivet S, Arpin NL, Wichers HJ, Pellon G (1998) Agaricus bisporus browning A review. Mycol Res 102: 1459-1483.

5. Weijn A, Tomassen MMM, Bastiaan-Net S, Wigham MLI, Boer EPJ, et al. (2011) A new method to apply and quantify bruising sensitivity of button mushrooms. Food Sci Technol 47: 308-314.

6. Gonzalez- Fandos E, Gimenez M, Olarte C, Sanz S (2000) Effect of packaging condition on the growth of microorganisms and the quality characteristics of fresh mushrooms (A. bisporus) stored at inadequate temperature. J Applied Microbiology 889: 624-632.

7. Vizhanyo T, Jozsef F (2000) Enhancing color differences in images of diseased mushrooms. Comp Elec Agri 26: 187-198.

8. Suttirak W, Manurakchinakor S, Walailak J (2010) Potential application of ascorbic acid citric acid and oxalic acid for browning inhibition in fresh-cut fruits and vegetables. Sci Technol 7: 5-14.

9. Brennan MH, Le Port G, Gormley TR (2000) Post harvest treatment with citric acid or hydrogen peroxide to extend the shelf life of fresh sliced mushrooms. Lebensmittel Wissenschaft Technol 33: 285-289.

10. Breenan HM, Gormley RT (1998) Extending the shelf life of fresh sliced mushrooms. The National Food Centre Dunsinea Castleknock Dublin.

11. Guthire BD (1984) Studies on the control of bacterial deterioration of fresh washed mushrooms (Agaricus bisporus/ brunescens). The Pennsylvania State University.

12. Nwachukwu E, Ezeigbo CG (2013) Changes in the microbial population of pasteurized sourop juice treated with benzoate and lime during storage. African J Microbiology Research **7**: 3992-3995.

13. Bozkurt H, Bayram M (2006) Color and textural attributes of sucuk during ripening. J Meat Science73: 344-350.

14. Martin-Diana AB, Rico D, Frias JM, Barat JM, Henehan GT M et al. (2007) Calcium for extending the shelf life of fresh whole and minimally processed fruits and vegetables: a review. Trend Food Sci Tech 18: 210-218.

15. Jauathunge L, Illeperuma C (2001) Extension of post harvest life of oyster mushroom under ambient conditions by modified atmosphere packaging. Journal Tropical Agricultural Research 13:78-89.

16. Du J, Fu M, Li M, Xia W (2007) Effect of chlorine dioxide gas on post harvest physiology and storage quality of green bell pepper (*Capsicum frutescens L var Longrum*). Agricultural Sciences in China 6: 214-219.

17. Bayoumi AY (2008) Improvement of post harvest keeping quality of white pepper fruits (*Capsicum annuum* L) by hydrogen peroxide treatment under storage conditions. Acta Biologica Szegdiensis 52: 7-15.

18. Banwart GJ (1989) Basic food microbiology. The University of Michigan Avi Publishing Co Westport Conn.

19. Beuchat LR, Golden DA (1989) Antimicrobials occurring naturally in food. Food Technology 43: 134-142.

20. Singh M, Vijay B, Kamal S, Wakcaure GC (2011) Mushrooms-cultivation, marketing and consumption. Directorate of Mushroom Research, Solan .

21. Erbay B, Kucukoner E, Orhan H (2011) Colour and some physical properties of frozen mushroom (*Agaricus bisporus*) which dipped in different antioxidant solutions. International J Health and Nutrition.

22. Rosario MJG (1996) Formulation of ready to drink blends from fruits and vegetable juices. J Philippines Agriculture Research 9: 201-209.

23. Jafri M, Jha A, Bunkar DS, Ram RC (2013) Quality retention of oyster mushrooms (*Pleurotus florida*) by a combination of chemical treatments and modified atmosphere packaging. Post Harv Biol Technol 76: 112-118.

24. Langnika C, Zhang M, Wang S (2011) Effect of high argon pressure and modified atmosphere packaging on the white mushroom (*Agaricus bisporus*) physico-chemical and microbiological properties. J Food and Nutrition Research 50: 167-176.

25. Mattila P, Konko K, Eurola M (2001) Contents of vitamins mineral elements and some phenolic compounds in cultivated mushrooms. J Agricultural and Food Chemistry 49: 2343-2348.

26. Ranganna S (2005) Sensory evaluation handbook of analysis and quality control for fruit and vegetables products. Tata Mc Graw Hill Education Private Ltd New York USA.

27. Miller CF, Cooke HP, Choi WS (1994) Enzymatic browning control in minimally processed mushrooms. J Food Science 59: 1042-1047.

28. Simon A, Gonzalez-Fandos E (2009) Effect of washing with citric acid or sodium hypochlorite on the visual and microbiological quality of mushrooms (*Agaricus bisporus*). J Food Quality 33: 273-285.

30

Effects of Fermentation on the Nutritional Quality of QPM and Soybean Blends for the Production of Weaning Food

Emire SA[1]* and Buta MB[2]

[1]Food, Beverage and Pharmaceutical Industry Development Institute, Ministry of Industry, Addis, Ababa, Ethiopia
[2]Food Process Engineering Department, Addis Ababa Science and Technology University, Ethiopia

Abstract

The purpose of this work was to study the effects of fermentation on quality protein maize (QPM) and soybean blends with respect to the nutritional quality including physico-chemical and functional properties; microbiological and sensory analyses, minerals and antinutrients composition. Quality protein maize-soybean blend flours were fermented for 24 and 48 hrs by natural and controlled fermentations. In contrary concentration of tannins and phytate were reduced significantly due to the fermentation process. Micronutrients increment in (mg/100 g) for P, Fe and Zn was 32.57 to 61.9; 3.98 to 7.20 and 2.61 to 4.21; respectively were revealed. Fermentation significantly ($p < 0.05$) decreased the antinutrients which resulted a significant increase in micronutrients. Microbiological result revealed significant reduction of undesirable coliform count and increment of LAB with increase in fermentation time. Sensory quality result showed that gruel prepared from the fermented blended flours at 24 hrs of fermentation time and <250 μm particle size was found acceptable. In line with the result of this study, natural and controlled fermentation uniformly reduced antinutrients composition and improved the nutritional quality of the weaning blends via increased energy and nutrient densities. Fermentation of cereal and legume blends is low-cost and safe technique to save life of children suffering from protein-energy malnutrition.

Keywords: Antinutrients; Fermentation; Micronutrient; Nutritional quality; Blends; Weaning food

Introduction

Fermentation is widely applied in the processing of cereals for the preparation of a wide variety of dishes in Africa and it contributes to the development of acceptable texture, flavour and improves the nutritional quality, digestibility and safety of foods [1]. Fermentation has also been identified to significantly improve the nutritional value (protein quality) of maize-based foods and as well reduce their antinutrients [2].

The high dependence on maize as a staple food in tropical Africa, coupled with the low nutritive value of the commodity has led to the investigation of simple traditional methods in the improvement of the chemical and functional qualities of maize-based foods [3]. The use of legumes such as soybean has been successfully used to increase the nutritional value of cereal foods. Soybean has high protein content and constitutes the natural protein supplements to staple diets. Protein quality is synergistically improved in cereal-legume blends because of the lysine, tryptophan and methionen contributed by the quality protein maize [4]. Soybean products are frequently incorporated into products used for the treatment or prevention of malnutrition. Enriching weaning foods with soy is a convenient, inexpensive, and highly effective way to upgrade the quality of traditional weaning foods and to provide the nutrition a growing child needs. Soy works together with grain proteins to achieve an overall increase in the value of the protein. Adding even small quantities of soy can greatly increase protein content and quality of weaning foods [5]. However, raw ingredients such as cereals and legumes that are used to prepare fermented foods contain significant amounts of antinutreints. These components may decrease the nutritional value of foods by interfering with mineral bioavailability and the digestibility of proteins and carbohydrates [6].

There is urgent need for provision of weaning foods rich in protein, low-cost and suitable for provision of infants' nutritional needs in order to reduce child malnutrition which is a major global health problem [7]. Blends with a cereal-legume ratio of 70: 30 have been introduced in many communities for use in the preparation of complementary foods with augmented protein quality. These foods should meet World Health Organization estimated energy and nutrient needs from complementary foods [8]. The rationale of this study was to investigate the influence of fermentation on antinutrients composition, functional properties, physico-chemical and sensory characteristics of QPM-based soybean blended weaning foods in order to increase energy, nutrient densities and mineral bioavailability in a weaning gruel.

Materials and Methods

Research materials collection and sample preparation

Bako Hybrid Quality Protein Maize (BHQPY-545) and soybean (Afgat) varieties were collected from Bako and Hawassa Agricultural Research Centers; respectively. All experiments were performed at Addis Ababa Institute of Technology, Food Engineering laboratory, Ethiopian Health and Nutrition Research Institute and Kality Food Processing Share Company. Quality protein maize and soybean samples were prepared according to the method described [9].

Blend formulation

Formulations of high-protein-energy weaning blends were based on the material balance method, targeting 18% protein, 59% carbohydrates [10] and minimum energy value of 380 kcal per 100 g dry matter, according to WFP requirement specifications for particularly the age group of 6 to 18 months. Therefore, the blend ratio of 82:18; QPM: soybean was formulated.

*Corresonding author: Emire SA, Food, Beverage and Pharmaceutical Industry Development Institute, Ministry of Industry, Addis, Ababa, Ethiopia
E-mail: shimelisemire@yahoo.com

Natural and controlled fermentation process

Fermentations (NF, CF) were performed using the methods described [11]. After fermentation the samples were dried and grounded and finally placed into plastic containers and stored at 4°C.

Proximate composition and determination of energy value

After blend formulation both NF and CF fermentation process for 0, 24 and 48 hours with particle size distribution of <500 μm and <250 μm. The dried samples were analyzed for the parameters such as protein, crude fat and crude fiber using the method of AOAC and total carbohydrates were calculated by difference [12].

Determination of nutrient and anti-nutrient concentration

The micronutrients: iron, zinc and phosphorus were analyzed using atomic absorption spectoscopy. The antinutrients such as tannins and phytate were determined by the modified Vanillin assay [10].

Determination of functional properties of blends

Bulk density was determined by the method of Narayana [13]; dispersibility in water was determined by the method of Kulkarni [14]; water and oil absorption capacities were determined according to method described by Nwosu, Coffman [15,16]; respectively.

Microbiological analysis

Analyses of mold and yeast; aerobic plate count (APC), coliform count, fecal coliform count and *E.coli* were done using standard methods of ISO: 7954 [17].

Sensory quality assessment

The prepared weaning foods as gruel from famix provided from FAFA, QPM-Soybean blend flour and blend fermented for 24 and 48 hrs were evaluated by semi-trained panelists. The gruels were prepared by mixing about 10 g of blend flour dissolved in 200 ml distilled water and cooked at 92°C for 15 min. The panelists were asked to rank the gruel on the basis of appearance (color), odor, and taste using a nine point hedonic scale (where 1 = dislike extremely and 9 = like extremely). Overall acceptability of the samples was also rated on same scale with 9 = extremely acceptable and 1 = extremely unacceptable [18].

Results and Discussion

Effect of fermentation on proximate composition and energy value of QPM-soybean blends

Fermentation time significantly (p<0.05) affects protein content of blend flour (Table 1). As fermentation time increased from 0 to 24 h and 48 hrs, protein content of blends were 14.71%, 17.43% and 17.52%; respectively. The protein content of blends before and after fermentation (0,12 and 24 hrs) was higher than the minimum protein requirement (14%) of WFP specification for corn-soya blend and within the range (16.00%-19.97%) reported by Lalude [19].

As fermentation time increases the fat content of fermented blends increases from 8.42% to10.9% with respect to the fermentation time of 0, 24 and 48 hrs (Table 1). This is due to the removal of soluble carbohydrates during fermentation. According to the findings of Amankwah the fat content of formulation of weaning food from fermented maize, rice, soybean and fishmeal was 9.38% and 8.75% [20]. The experimental value is within the range with this value. The value of crude fat content of the blend before and after fermentation (8.42%, 10.2%, and 10.9% to 8.86%) is higher than the value of WFP specification for the minimum requirement of 6% fat of corn-soya

blend. And it is comparable with the value of famix infant food (≥7%). The fat content of current study is also within the range with the value (9.0%) of Nutrend- commercially sold Nigerian weaning food.

The value of crude fiber after fermentation (4.49%, 5.32% for particle size distribution of <250 μm and 5.19%, 4.67% for particle size distribution of <500 μm for natural fermentation) and (5.91%, 5.63%) for particle size distribution of <250 μm and (6.88%, 5.96%) for particle size distribution of through <500 μm for controlled fermentation) at 24 and 48 h fermentation time respectively (Table 1). Fermentation time is significantly (p<0.05) decreased the crude fiber contents of blend. According to high fiber contents of weaning foods may inhibit mineral absorption and reduce the digestibility of proteins in foods [21]. According to WHO specification the maximum requirement is (5%). Therefore, the experimental values of the blends after fermentation are close to this value. Fermentation time had significantly (p<0.05) decreasing effect on the total carbohydrates content of blend (66.63%, 64.16%, 63.9%) and (66.63%, 61.91%, 60.49%) for particle size distribution of <250 μm and <500 μm at 0, 24 and 48 h fermentation time respectively as can be seen from the This is possibly due to the degradation of carbohydrates by microorganisms and the decreasing effect of fermentation upon the amount of Non-Digestible Carbohydrates (NDC) which are fibers (Table 1). All the experimental values of blend before and after fermentation are comparable with values (67.21% and 63.32%) research findings by Mbata for fermented maize flour and Bambara groundnut-maize fortified flour [22]. It was reported that the total carbohydrate content of maize-soybean blend for the production of Ogi is (61.76%); that is lower than the value of the current study [23].

In the case of controlled fermentation, the calorific values of the blends are decreased, even though there are some fluctuations in natural fermentation. This may be due to the value difference of the nutrients (protein, fat and carbohydrate) by the effect of fermentation process. The calorific value of the blend shown in the Table 1, are in agreement with the researcher reported by Griffith [24]. The calorific value for weaning food revealed in the range of 395 to 509 kcal which is in agreement with the value (398.9 kcal) for "Nutrend" (Nestle, Nigeria-weaning diet) obtained commercially and the research finding reported by (441 kcal) [19]. The value is also within the range provided by WHO specification (380 kcal) for the weaning food from corn-soya blend.

Influence of fermentation process on antinutrients reduction of QPM-soybean blends

During the preparation of many fermented foods, tannins are reduced before the fermentation step because of their presence in the seed coats of raw ingredients. In several fermented foods, the seed coat or testa is removed from the substrate before fermentation. Consequently, the antinutrients potential caused by the presence of tannins is of little concern [11]. Fermentation process further reduced tannins content of the blend (Table 2). Therefore, fermentation time significantly (p<0.05) affected tannins content of the blend. The phytate concentration present in raw materials and foods of plant origin are suggested to be a major factor responsible for lowering the availability of minerals and some proteins as reported [11]. Fermentation duration, type and particle size distribution greatly affects the composition of antinutients.

Effect of reduction of antinutrients on micronutrients composition of blends

The P, Fe, Zn and Ca concentrations of the blends before fermentation were (32.57 mg/100 g, 3.98 mg/100 g, 2.61 mg/100 g,

Fermentation type and particle size distribution	Flour samples	Crude protein (%)	Crude fat (%)	Crude fiber (%)
Before fermentation	Blend	14.72^g± 0.03	8.42^{ef}± 0.01	7.33^{ac}± 0.02
(<250 µm), NF	Blend24	17.43^e ± 0.02	10.20^b ± 0.14	4.49^h ± 0.08
	Blend48	17.52^{de} ± 0.01	10.90^g ± 0.28	5.32^{fg}± 0.01
(<500 µm), NF	Blend24	17.57^{de} ± 0.06	10.20^b±0.07	5.19^g ± 0.06
	Blend48	17.85^b ± 0.04	10.80^{de}± 0.21	4.67^h ± 0.06
(<250 µm), CF	Blend24	17.30^f ± 0.21	9.40^c ± 0.14	5.91^e ± 0.02
	Blend48	17.67^{cd} ± 0.04	8.50^{ef} ± 0.35	5.63^{ef}± 0.05
(<500 µm), CF	Blend24	17.72^{bc} ± 0.01	9.36^c ± 0.06	6.88^b ± 0.04
	Blend48	19.44^k ± 0.05	8.86^d ± 0.02	5.96^e ± 0.04
Fermentation type and particle size distribution	**Flour samples**	**Carbohydrates (%)**	**Calorific value kcal/100g**	
Before fermentation	Blend	66.63^d ± 0.02	400.81^b	
(<250 µm), NF	Blend24	66.90^d ± 0. 28	412.67^f	
	Blend48	65.01^e ± 0. 08	412.25^f	
(<500 µm), NF	Blend24	66.74^d ± 0.03	412.63^f	
	Blend48	67.97^d ± 0.05	404.01^h	
(<250 µm), CF	Blend24	64.16^f ± 0. 11	410.75^d	
	Blend48	63.90^g ± 0.42	387.06^g	
(<500 µm), CF	Blend24	61.91^h ± 0.36	387.54^g	
	Blend48	60.49^i ± 0. 06	384.60^j	

All values are means ± SD, Values in the same column with different superscripts for each type of analysis are significantly different (P < 0.05).

Where CF-Controlled fermentation and NF-Natural fermentation

QPMf$_{24}$- QPM flour fermented for 24 hours, QPMf$_{48}$- QPM flour fermented for 48 hours, Blend$_{24}$- Blend fermented for 24 hours and Blend$_{48}$- Blend fermented for 48 hours.

Table 1: Proximate chemical composition of blends

Fermentation type and particle size distribution	Flour Samples	Tannins (mg/100g)	Phytate (mg/100g)
Before fermentation	Blend	21.95^c±0.67	249.2^a ± 0.14
<250µm, NF	Blend24	BDL	155.75^d ±0.53
	Blend48	BDL	133.06^e ±0.04
<500µm, NF	Blend24	3.10^c ± 0.21	155.51^d ±0.36
	Blend48	BDL	147.50^f ±0.35
<250µm, CF	Blend24	6.93^e ± 0.02	143.20^g ±0.14
	Blend48	4.98^d ± 0.06	139.22^i ±0.16
<500µm, CF	Blend24	8.94^f ± 0.04	146.64^h ±0.45
	Blend48	5.05^d ± 0.11	138.65^i ±0.46

Values in the same column with different superscripts for each type of analysis are significantly different (P < 0.05).

Where-BDL: Below detection limits

Table 2: Content of tannins and phytate of blends.

and 34.08 mg/100 g) respectively. In the case of NF, for particle size distribution of <250 µm and <500 µm P of the blend is (32.57 mg/100 g, 61.90 mg/100 g, 61.20 g/100 g) and (32.57 mg/100 g, 59.60 mg/100 g, 55.30 mg/100 g) during 0, 24, and 48 h fermentation time respectively. As can be seen from the Table 3, P as fermentation increases from 0 to 24 h and 0 to48 h increases too. The same is true for minerals: Fe; Zn and Ca that shows similarly increasing in values as fermentation time increased. This is due to the minerals of the grain that are not readily available for microorganisms as they are complexed with phytate, at pH values of <5.5 the endogenous grain phytase hydrolyses phytate and minerals are released from the complex [25].

Largely, fermentation time significantly (p<0.05) affect the mineral composition of the blend. The Ca, P, and Fe content are higher than the values (22 mg/100 g, 26 mg/100 g, 1.0 mg/100 g) reported by for Nutrend [19]. Values of iron and zinc are within the range of WHO standards (3.25 mg/100 g, 5 mg/100 g) for the manufacture of corn soya blend for infants. The micronutrients values (17 to 25 mg/100 g for Ca; 7.19 to 10.98 mg/100 g for Fe and 1.78 to 2.01 mg/100 sg for Zn) are in agreement with the values for sorghum based weaning food [11].

Effect of fermentation viscosity of QPM-soybean blends

As indicated in Table 4, the viscosity of the blend is (5.31×10^{-3}Pa.s) and the viscosity of blend gruel during fermentation of 24 h and 48 h is (4.38×10^{-3}Pa.s, 4.21×10^{-3}Pa.s); respectively. From the results, fermentation time had significantly (p<0.05) decreasing effect on the viscosity of the blend. A prolonged time of starch gelatinization was observed while preparing gruels of fermented blend. This possibly is due to the degradation of starch granules during fermentation and this might cause for the reduction of starch swelling while cooking.

Effect of fermentation on viscosity of QPM-soybean blends

As indicated in Table 5, the viscosity of the blend is (5.31×10^{-3}Pa.s) and the viscosity of blend gruel during fermentation of 24 h and 48 h is (4.38×10^{-3} Pa.s, 4.21×10^{-3} Pa.s) respectively. From the results, fermentation time had significantly (p<0.05) decreasing effect on the viscosity of the blend. A prolonged time of starch gelatinization was observed while preparing gruels of fermented blend. This possibly is due to the degradation of starch granules during fermentation so that

Fermentation type and particle size distribution	Flour Samples	P (mg/100g)	Fe (mg/100g)	Zn (mg/100g)	Ca (mg/100g)
Before fermentation	Blend	32.57[c] ± 0.40	3.98[ef] ± 0.06	2.61[f] ± 0.08	34.08[c] ± 0.17
<250 µm, NF	Blend24	61.90[a] ± 0.64	7.20[a] ± 0.14	4.21[g] ± 0.15	24.91[d] ± 0.64
	Blend48	61.20[ah] ± 0.14	4.74[c] ± 0.03	3.81[cd] ± 0.08	21.34[e] ± 0.24
<500 µm, NF	Blend24	59.60[d] ± 0.42	6.22[d] ± 0.16	3.98[gc] ± 0.06	24.92[d] ± 0.65
	Blend48	55.30[e] ± 0.21	3.42[b] ± 0.03	3.65[d] ± 0.06	22.53[f] ± 0.37
<250 µm, CF	Blend24	58.50[f] ± 0.35	4.31[e] ± 0.22	2.89[e] ± 0.06	65.02[g] ± 0.01
	Blend48	56.30[g] ± 0.21	4.21[ef] ± 0.15	2.81[e] ± 0.04	64.97[g] ± 0.02
<500 µm, CF	Blend24	60.60[h] ± 0.42	3.99[ef] ± 0.06	2.84[e] ± 0.03	65.42[g] ± 0.30
	Blend48	54.30[i] ± 0.21	3.88[f] ± 0.07	2.71[ef] ± 0.04	64.77[g] ± 0.54

Values in the same column with different superscripts for each type of analysis are significantly different (P < 0.05).

Table 3: Micronutrient composition of blends.

Blend with respective fermentation time	Viscosity (Pa.s)
Blend	5.31×10^{-3b} ± 0.22
Blend24	4.38×10^{-3e} ± 0.26
Blend48	4.21×10^{-3e} ± 0.15

Values in the same column with different superscripts for each type of analysis are significantly different (P < 0.05).

Table 4: Viscosity of gruel from blends at a temperature of 50°C.

cause for the reduction of starch swelling while cooking. According to Plahar in terms of starch stability, fortification with soy flour is a cause for strengthening of the starch granules [26].

Impact of fermentation on functional properties of QPM-soybean blends

Bulk density and dispercebility: As fermentation time increase from 0 to 24 and 48 h, the bulk density significantly decreases for QPM-Soybean blends. The bulk density of QPMf before fermentation is (0.84 and 0.92) and after fermentation of 24 and 48 h is (0.62, 0.59 and 0.63, 0.6) for particle size distribution between <250 and <500 µm; respectively. The bulk density of QPMf before fermentation is significantly (p<0.05) higher than that of after fermentation. The values (0.66 and 0.61) of the current research finding is in agreement with that of reported Lalude [19], of a weaning food prepared from sorghum based and Nutrend. Similarly, the values are comparable with value (0.68) reported by Mesfin [27]. As fermentation time increased the value of dispercebility significantly (p<0.05) increased from 0.65%, 69% to 66%, 67% and 65%, 69% for fermentation time of 24 and 48 h and particle size distribution of <250 and <500 µm; respectively. The values of both bulk density and dispersability of QPMf and blend are higher than the values (0.55 gm/ml and 32.93%) reported by Ahima respectively [28].

Water and oil absorption: As indicated in Table 5, water absorption of both QPMf and blend is significantly (p<0.05) decreased when fermentation time increased from 0 to 24 h and from 0 to 48 h. Ahima reported that the water absorption of flour from commercially sold floury maize and Maize-soy flour blend is (194.65% and 172.98% respectively) [28]. This result indicated that the water absorption of QPMf and blend is lower than that of the researches findings. In the current study the unfermented and fermented blends found to contain comparable amount of water absorption with (134%) that is reported by Emmanuel, cowpea-fortified nixtamalized food. Oil absorption capacity of QPMf (1.4 ml, 2.3 ml) is higher than of blend (1.2 ml, 1.7 ml) with particle size distribution of <250 µm and <500 µm. Similarly, QPMf fermented for 24 hrs and 48 hrs (4.2 ml, 5.2 ml and 4.8 ml, 5.0 ml) is significantly higher than that of fermented blend (2.2 ml, 3.0 ml and 1.9 ml, 2.8 ml) for particle size of <250 µm and

<500 µm respectively. These results indicated that blending QPMf with soybean flour significantly (p<0.05) decreased the oil absorption of the blend. Therefore, blend ratio and fermentation time has significantly (p<0.05) a decreasing effect on the oil absorption. The values obtained from current study are higher than that of (1.22 ml-2.23 ml) reported by Mensah of extrusion cooking of full-fat soy flour [29]. Similarly, oil absorption of both fermented QPMf and blend are higher than the value (1.82 ml, 1.44 ml) for different varieties reported by Assefa of improved varieties of soybean in Ethiopia [30].

Sensory quality attributes of value added products

The average results of taste evaluation by the panelist for the control, fermented for 24 h, unfermented blend and fermented for 48 h were 8.9 (like extremely); 7.8 (like very much); 7.2 (like moderately) and 5.8 (like slightly) respectively. The overall acceptability of gruel fermented for 24 h with the average value of 8.5 is not significantly different (p>0.05) compared with that of the control. The result of the others ranged 5.9 to 7.1. This might be due to all the products are evaluated without flavoring agents. Similar results on weaning and commentary food products were report by Abbey, Beruk [31,32] (Table 6).

Microbiological quality of blends

In Table 7 mould count, yeast count, Aerobic Bacteria plate Count (APC), coliform count and others are shown. The result indicated that as fermentation time increased, the undesirable microorganism, mould count, decreased significantly for both type of fermentation. The molds isolated in the current study are commonly present as contaminants and do not appear to play any significant important role in the fermentation process. This shows clearly that the importance of fermentation in the aspect of food preservation.

At 0 hr fermentation time, the yeast count was found to be $<1 \times 10^1$ cfu/g, which is considered to be no yeast colonies in the count, but during 24 and 48 h fermentation, the values were increased upon both Natural and Controlled Fermentation. This shows that fermentation time significantly (p<0.05) affect the yeast count. The coliform count at the start of fermentation was found to be 4.3×10^2 cfu/g and upon increasing fermentation time, 24 and 48 h the count was decreased to 3.1×10^2 cfu/g and 3.2×10^2 cfu/g in the case of NF and almost eliminated ($<1 \times 10^1$ cfu/g) in the case of CF. The expected decrease or elimination of coliform is in agreement with the value (2.85 cfu/g, 0 cfu/g, 0 cfu/g) for 0, 24 and 48 hrs fermentation time reported by Mbata of fermented maize flour fortified with bambara groundnut [22]. The aerobic bacteria plate count, (APC) as shown in the Table 7 is 1.8×10^2 cfu/g at 0 h fermentation time and (2.6×10^3 cfu/g, 2.4×10^3 cfu/g) in the case of NF and (6.7×10^3 cfu/g, 5.7×10^3 cfu/g) for CF during 24 and 48 hrs fermentation time; respectively. Aerobic plate counts taken at 24 h

Types of flour	Bulk density (g/ml)	Dispersibility (%)	Water absorption (%)	Oil absorption (ml/g of sample)
QPMF$_{<250}$	0.84[a] ± 0.01	65[a] ± 0.71	139.09[a] ± 0.06	1.4[af] ±0.07
QPMF$_{<500}$	0.92[b] ± 0.02	69[c] ± 0.35	136.27[b] ± 0.19	2.3[b] ±0.11
QPMF$_{<250, 24}$	0.62[efg] ± 0.01	66[a] ± 1.41	129.23[c] ± 0.16	4.7[c] ±0.21
QPMF$_{<250, 48}$	0.59[fg] ± 0.02	67[ab] ± 2.12	137.17[d] ± 0.12	5.2[d] ±0.12
QPMF$_{<500, 24}$	0.63[efg] ± 0.01	65[ab] ± 0.71	139.73[a] ± 0.52	4.8[ce] ±0.14
QPMF$_{<500, 48}$	0.60[g] ± 0.02	69[c] ± 1.41	144.80[e] ± 0.57	5.0[de] ±0.07
Blend$_{<250}$	0.77[c] ± 0.01	57[e] ± 1.06	142.00[f] ± 0.35	1.2[f] ±0.11
Blend$_{<500}$	0.79[c] ± 0.03	64[a] ± 0.35	136.33[b] ± 0.23	1.7[ag] ±0.12
Blend$_{<250, 24}$	0.74[d] ± 0.01	67[bc] ± 1.41	140.37[g] ± 0.26	2.2[bh] ±0.14
Blend$_{<250, 48}$	0.67[e] ± 0.02	60[d] ± 0.71	146.00[h] ± 0.21	3.0[i] ±0.13
Blend$_{<500, 24}$	0.75[cd] ± 0.01	66[ab] ± 0.02	126.13[i] ± 0.09	1.9[gh] ±0.07
Blend$_{<500, 48}$	0.65[ef] ± 0.03	65[a] ± 2.12	132.83[j] ± 0.59	2.8[i] ±0.21

Values in the same column with different superscripts for each type of analysis are significantly different (P < 0.05).

Table 5: Bulk density, dispensability, water and oil absorption of blends.

Sample code	Appearance	Odor	Taste	Overall acceptability
Control (Famix)	8.30[a] ± 0.65	8.80[a] ± 0.24	8.90[a] ± 0.23	8.90[a] ± 0.34
Blend$_{< 500}$	6.00[b] ±1.73	6.67[b] ± 1.15	7.30[b] ± 0.58	7.00[b] ± 0.00
Blend$_{<250}$	6.40[b] ± 1.20	6.69[b] ± 0.84	7.10[b] ± 0.51	7.20[b] ± 0.57
Blend$_{24<500}$	7.70[cd] ± 0.00	7.33[c] ± 0.58	7.70[c] ± 0.56	8.30[c] ± 0.43
Blend$_{48<500}$	7.00[e] ± 0.58	5.70[d] ± 1.13	5.70[d] ± 0.60	6.00[d] ± 0.26
Blend$_{24<250}$	7.80[ac] ± 0.44	7.54[e] ± 0.78	8.10[e] ± 0.47	8.50[ac] ± 0.62
Blend$_{48<250}$	7.50[d] ± 0.39	5.90[d] ± 0.95	6.00[d] ± 0.86	5.80[d] ± 0.18

All values in the same column with different superscripts for each type of analysis are significantly different (P < 0.05).

Table 6: Sensory quality evaluation of weaning food.

Isolated microorganisms	Microbial count load (cfu/g) of blend samples				
	0	24, NF	48, NF	24, CF	48, CF
Mold count at 25 °C /5-7 days	2.1×10^4	4×10^2	3.2×10^2	5×10^2	4×10^2
Yeast count at 22°C /5-7 days	Nil	2.9×10^2	3.2×10^2	2.0×10^2	3.5×10^2
APC at 30°C/72 h	1.8×10^2	2.6×10^3	2.4×10^3	6.7×10^3	5.7×10^3
Coliform count	4.3×10^2	3.1×10^2	3.2×10^2	Nil	Nil
Fecal coliform count	Nil	Nil	Nil	Nil	Nil
E.coli count	Nil	Nil	Nil	Nil	Nil
S.coccus spp	Nil	Nil	Nil	Nil	Nil

Values in the same column with different superscripts for each type of analysis are significantly different at P < 0.05.

Table 7: Microbiological quality of blends at different fermentation time for NF and CF.

intervals of fermentation indicated that the increased growth of yeasts and lactic acid bacteria gradually throughout fermentation while the decrease in numbers of molds and coliforms.

Conclusions

In this research, it was observed that fermentation process significantly changed the nutritional value of the weaning blends by reducing antinutrients. The maximum reduction of phytate due to fermentation effect can lead to the increment of the bioavailability of micronutrients. Subsequently, 47.4% (P), 47.9% (Ca), 44.7% (Fe) and 38% (Zn) increment was observed. The fermentation process affected the energy value and nutrient densities of the weaning blends. Gruels prepared from fermented blend flours were less viscous and the dietary bulkiness nature was improved. The microbial load of blend samples at different fermentation time for NF and CF found acceptable; and the sensory quality evaluation of the prepared gruel was acceptable

by panelists. An infant food of higher energy and nutrient density formulated and prepared from a formulation of 82% quality protein maize and 18% soybean had the strongest impact on nutritional quality; and should be viewed as an option in the development of infant weaning foods in order to reduce child malnutrition problem in the Ethiopian context.

Accordingly, the use of cereal based legume blends with augmented protein quality could help rural communities predominantly to make better use of their available resources; and the same time can improve nutrient density and fulfill the minimum daily requirements for energy, protein and micronutrients. Promotional efforts should be made to transfer technique of this high-protein-energy weaning blends formulation and processing technology to food manufacturing industries and household women in order to diminish child mortality in Sub-Saharan African Countries.

References

1. Campbell PG (1994) Fermented foods - a world perspective. Food Res Int 27: 253-268.

2. Reddy NR, Pierson MD, Salunkhe DK (1986) Legume-based Fermented Foods. In: Reddy NR, Pierson MD, Salunkhe DK (Eds) CRC Press, Boca Raton, Florida.

3. Bressani R, Benavides V, Acevedo E, Ortiz MA (1990) Changes in selected nutrient content and in protein quality of common and quality protein maize during tortilla preparation. Cereal Chem 67: 515-518.

4. Emmanuel OA, Sefa-Dedeh S, Agnes SB, Sakyi-Dawson E, Justice A (2007) Influence of spontaneous fermentation on some quality characteristics of maize-based cowpea-fortified nixtamalized foods 7.

5. Louis St (2006) Weaning Foods: Characteristics, Guidelines, and the Role of Soy foods. WISHH World Initiative for Soy in Human Health.

6. Reddy NR, Pierson MD (1994) Reduction in antinutritional and toxic components in plant foods by fermentation. Food Res Inter 27: 281-290.

7. Kim F, Camilla H, Nanna R, Pernille K, Maria S, et al. (2009) Choice of foods and ingredients for moderately malnourished children 6 months to 5 years of age. Food Nutr Bull 30: 343-403.

8. Ejigui J, Savoie L, Marin J, Desrosiers T (2007) Improvement of the nutritional quality of a traditional complementary porridge made of fermented yellow maize (Zea mays): Effect of maize-legume combinations and traditional processing methods. Food Nutr Bull 28: 23-34.

9. Bryce JB, Shibuya K, Black RE (2005) WHO six reports on the World Health Situation, Part I. Global analysis, Geneva. 365: 1147- 1152.

10. Butler LG, Price ML, Brotherton JE (1982) Vanillin assay for Proanthocyanidins (Condensed Tannins): Modification of the Solvent for Estimation of the Degree of Polymerization. J Agric Food Chem 30: 1087-1089.

11. Emire SA, Rakshit SK (2005) Proximate composition and physico-chemical properties of improved dry bean (Phaseolus vulgaris L.) varieties grown in Ethiopia. LWT 38: 331-338.

12. AOAC (2000) Association of Official Analytical Chemists. (17thedn) AOAC International. Washington, DC, USA.

13. Narayana K, Narasinga Rao MS (1984) Effect of partial proteolysis on the functional properties of winged pea flour. J Food Sci 49: 944-947.

14. Kulkarni KD, Kulkarni DN, Ingle UM (1991) Sorghum malt based weaning food formulations, preparations, functional properties and nutritive values. Food Nut Bull 13: 322-329.

15. Nwosu JN, Onuegbu NC, Kabuo NO, Okeke MO (2010) The effects of steeping with chemicals (Aluma and Tona) on the proximate and functional properties of pigeon pea (Cajanus cajan) flour. Pak J Nutr 9: 762-768.

16. Coffman CW, Gracia VV (1977) Functional properties of amino acid content of protein isolate from mung bean flour. J Food Technol 12: 473-484.

17. ISO: 7954 (1987) Microbiology: general guidance for enumeration of yeasts and moulds - Colony count technique at 25°C.

18. Inyang CU, Idoko CA (2006) Assessment of the quality of ogi made from malted millet. Afri J Biotech 5: 2334-2337.

19. Lalude LO, Fashakin JB (2006) Development and nutritional assessment of a weaning food from sorghum and oil-seeds. Pak J Nutr 5: 257-260.

20. Amankwah EA, Barimah J, Acheampong R, Addai LO, Nnaji CO (2009) Effect of fermentation and malting on the viscosity of Maize-Soyabean weaning blends. Pak J Nutr 8: 1671-1675.

21. Amuna P, Zotor F, Sumar S, Chinyanga YT (2000) The role of traditional cereal/legume/fruit-based multi-mixes in weaning in developing countries. J Nutr Food Sci 30: 116-122.

22. Mbata TI, Ikenebomeh MJ, Alaneme JC (2009) Studies on the microbiological, nutrient composition and antinutritional contents of fermented maize flour fortified with bambara groundnut (*Vigna subterranean* L). Afri J Food Sci 3: 165-171.

23. Bolaji OA, Olubunmi OA, Samuel AG (2010) Quality assessment of selected cereal-soybean mixtures in ogi production. New York Science Journal 3: 17-26.

24. Griffith LD, Castell-Perez ME, Griffith ME (1998) Effects of blend and processing method on the nutritional quality of weaning foods made from select cereals and legumes. American Association of Cereal Chem Inc 75: 105-112.

25. Hammes PW, Brandt JM, Francis LF, Rosenheim SH, Vogelmann AS (2005) Microbial ecology of cereal fermentations trends. Food Sci Technol 16: 4-11.

26. Plahar WA, Nti CA, Annan NT (1997) Effect of soy fortification method on the fermentation characteristics and nutritional quality of fermented maize meal. Plant Foods Hum Nutr 51: 365-380.

27. Mesfin W, Shimelis AE (2013) Effect of soybean/cassava flour blend on the proximate composition of Ethiopian traditional bread prepared from quality protein maize. Afri J Food, Agric Nutri Develop 13: 7985-8003.

28. Ahima KJ (2005) Formulation of weaning food using composite of maize groundnut and soybean and assessing its nutritional effect using animal model.

29. Mensah P, Drasar BS, Harrison TJ, Tomkins AM (1991) Fermented Cereal Gruels: Towards A Solution of The Weanling's Dilemma. Food Nutr Bull 13: 50-57.

30. Yimer A, Admassu S (2008) Effect of processing on some antinutritional factors of improved soya bean (*Glycine max*) varieties grown in Ethiopia. Addis Ababa University, Addis Ababa.

31. Abbey BW, Nkanga UE (1988) Production of high quality weaning products from Maize-Cowpea-Crayfish mixtures. Nutr Rept Int 37:952-957.

32. Beruk BD, Kebede A, Esayas K (2015) Effect of blending ratio and processing technique on physicochemical composition, functional properties and sensory acceptability of quality protein maize (QPM) based complementary food. Inter J Food Sci Nutri Eng 5: 121-129.

31

Evaluation of Physical and Compositional Properties of Horse-chestnut (*Aesculus indica*) Seed

Syed IR*, Sukhcharn S and Saxena DC

Department of Food Engineering and Technology, Sant Longowal Institute of Engineering and Technology, Longowal, Punjab, India

Abstract

An investigation was carried out on the physical and compositional properties of Horse Chestnut seeds found in Kashmir valley, India. Physical properties such as seed shape and size, geometric and arithmetic mean diameter, sphericity, aspect ratio, bulk and true density, density ratio, porosity, angle of repose and static friction coefficient were determined. The average seed length, width and thickness were 4.7, 4.0 and 3.2 cm, respectively. Geometric and arithmetic mean diameter was 3.92 and 3.97 cm. Average sphericity and aspect ratio was 83.4 and 85.11%, respectively. The average true density, bulk density, density ratio and porosity of the variety was 1072 g/cm^3, 518 g/cm^3, 48.32% and 51.68%, respectively. The mechanical property viz., angle of repose values obtained on plywood, mild steel, stainless steel and galvanized sheet were 22.3°, 20.3°, 18.3° and 22.79° for the seed variety. The coefficient of static friction obtained on plywood, mild steel, stainless steel and galvanized sheet were 0.54, 0.52, 0.51 and 0.52, respectively. The force required to break the seed was 328 N. Color and compositional analysis of the seed powder was conducted. The *L*, a and b value were found to be 92.07, 3.47 and 13.70 with a whiteness value of 83.78. The powder was having moisture (12.71%), protein (6.78%), fat (3.27%), ash (3.16%), fibre (6.34%) and carbohydrate (67.74%) with energy value of 327.51 cal/100 g.

Keywords: Horse chestnut; Sphericity; Porosity; Bulk density; Force

Introduction

Horse Chestnut (*Aesculus indica* Caleb.) known as *handun* is a fast growing tree species mainly found in temperate regions of Asia particularly in India, Nepal, Pakistan and Afghanistan. In India, the tree occupies moist and shady ravines of Jammu and Kashmir, Himachal Pradesh and Uttar Pradesh [1,2], with an average height of about 22.5 m, having an upright straight cylindrical bole with spreading crown, and produces huge quantity of seeds every year. There are two varieties of Horse Chestnut viz, *budh handun* and *lakut handun* found in Kashmir valley. Seed is covered with a capsule and single seed being present in each capsule [3]. The seeds are about 3.5 cm in diameter, round in shape, with a hard shiny black rind from outside and lime white cotyledons inside [4].

Flour from the seeds is used for making halwa and also mixed with wheat flour to prepare chapattis [5]. It has also been used as food during the times of famine by various tribes of North and North-Eastern India. Besides playing a significant role in food and non-food applications, it is well known for its medicinal value [6-13]. Seeds are used to extract oil which is among prominent nine oil yielding tree species used to cure Rheumatic pains in Kashmir [14]. Crushed seeds when fed to cattle are reported to improve the quality and quantity of milk. Oils and fats from the seeds are essential and indispensable ingredients of human and animal diet. The physical properties of agricultural produce are important in designing and constructing equipment and structures for handling, transportation, processing and storage and also for assessing the product quality [15-17]. Physical and mechanical properties of fruit, nut, seed and kernel are important to design equipment for dehulling, nut shelling, drying, oil extraction, and other processes like transportation and storage [18]. Physical properties such as size, shape, sphericity, aspect ratio, true density, bulk density and porosity, and mechanical properties such as coefficient of friction, angle of repose as well as fracture resistance are very important in the design of processing machines for major agricultural crops [19]. Many studies have been reported on the chemical and physical-mechanical properties of fruits and kernels, such as apricot kernel, berries, cherry laurel, cornelian cherry, fresh okra fruit, orange, rose fruit, wild plum etc.

Till date no information is available on physical and mechanical properties of Horse Chestnut seed for design and development of equipment needed for processing of Horse Chestnut seeds into desirable products. The objective of this study was to determine the engineering properties like physical and mechanical properties of Horse Chestnut seeds so that knowledge gained will be used in optimizing machine design parameters for handling, transport, processing and storage of the chestnut seeds. Compositional analysis of the seed powder was also conducted.

Materials and Methods

Material collection

Fully matured Horse Chestnut seeds were harvested from the trees located in Anantnag area of J & K, India and any bruised or diseased seeds were discarded. The seeds were then washed under running potable water to remove surface dirt and air dried and used as basic raw material for the study.

---

*Corresponding author: Syed IR, Research Scholar, Department of Food Engineering and Technology, Sant Longowal Institute of Engineering and Technology, Longowal-148 106, Punjab, India, E-mail: syedinsha12@gmail.com

Determination of physical and mechanical properties

Weight: Randomly selected ten seeds were taken for weight measurement on digital weighing balance (M/s. Sartorius Mechatronics Pvt Lt., Bangalore, India) to an accuracy of 0.01 g.

Determination of kernel ratio and shell ratio: The same seeds selected above for weight measurement were taken and deshelled and the kernels obtained thereafter were weighed on digital weighing balance (M/s. Sartorius Mechatronics Pvt Lt., Bangalore, India) [20]. The weight of the single kernel was calculated by taking average of the ten kernels. Kernel ratio was determined according to the Turkish Standard Institute [21] by the formula:

$$\text{Kernel ratio (\%)} = \frac{\text{Kernel wt. (g)}}{\text{Seed wt. (g)}} \times 100$$

However, the shell ratio was also calculated as follows:

$$\text{Shell ratio (\%)} = \frac{\text{Seed wt. (g)} - \text{Kernel wt. (g)}}{\text{Seed wt. (g)}} \times 100$$

Determination of size: For determining the physical characteristics 10 seeds were selected randomly and for each seed three linear dimensions were measured, that is (a) length, (b) width, and (c) thickness, using a Vernier calliper (Kanon Instrument, Japan) reading to 0.01 cm.

Determination of shape: The shape was expressed in terms of its sphericity index and aspect ratio. The higher the sphericity value the closer is the shape to a sphere. For the sphericity index (Φ), the dimensions obtained for the 10 selected seeds above for size measurement were used to compute the sphericity index based on the equation [22] as:

$$\Phi \, (\%) = \frac{(X.Y.Z)^{\frac{1}{3}}}{X}$$

Where,

X = length (L) in cm

Y = width (cm)

Z = thickness

For the aspect ratio same seeds were selected for conducting the experiment. The aspect ratio (R_a) was calculated as recommended by Mohsenin [23]:

$$R_a \, (\%) = \frac{Y}{X} \times 100$$

Where,

Y = Width (cm) and

X = Length (cm)

Determination of bulk and true density: The bulk density is the ratio of the mass of the sample to its container volume occupied. For bulk density measurement, an empty cylindrical container was filled with seeds to a known volume. Tapping during the filling was done to obtain uniform packaging and to minimize the wall effect, if any. The filled sample was weighed and the bulk density was calculated using the below equation [24];

$$\text{Bulk density} \left(\frac{g}{cm^3} \right) = \frac{M}{V}$$

Where,

M = mass of the sample (g)

V = Volume of the filled sample (cm³)

The true density is defined as the ratio of mass of the sample to its true volume. It was determined by the toluene displacement method in order to avoid absorption of water during experiment [25]. Five randomly selected seeds were weighed and immersed into a 1000 ml measuring cylinder containing 500 ml of toluene [26,27]. It was ensured that the seeds were submerged during immersion. The net volumetric displacement was recorded from the graduated scale of the cylinder. The true density was then calculated using equation below:

$$\text{True Density (g/cm}^3) = \frac{\text{Weight of seed (g)}}{\text{Rise in tolune level (cm}^3)}$$

Determination of density ratio and porosity of seed: The density ratio (D_r) is the ratio of true density to bulk density expressed as percentage as follows:

$$D_r = \frac{\tilde{n}_t}{\tilde{n}_b} \times 10$$

Porosity, ε (%) indicates the amount of pores in the bulk material and was calculated as per Demir et al. [24]. The porosity of the seed was calculated from the average values of bulk density and true density using the relationship.

$$\text{å} \, (\%) = \left[1 - \frac{\tilde{n}_b}{\tilde{n}_t} \right] \times 100$$

Where,

ρ_b = Bulk density (g/cm³)

ρ_t = True density (g/cm³)

Arithmetic and geometric mean diameter: The arithmetic mean diameter (D_a) and geometric mean diameter (D_g) of seeds were calculated from the geometrical dimensions by the formula given by Dutta et al. [28]. Both arithmetic and geometric mean are expressed in cm and calculated by the following equations:

$$D_a = \frac{X + Y + Z}{3}$$

$$D_g = (X.Y.Z)^{\frac{1}{3}}$$

Where,

X = Length (L) in cm

Y = Width (W) in cm

Z = Thickness (T) in cm

Surface area: Surface area (S) was estimated by the formula corresponding to the geometrical shape. The surface area of seed was determined by using the following equation [28]:

$$S \, (cm^2) = \pi D_g^2$$

Where,

D_g = Geometrical mean diameter (cm)

Angle of repose: The dynamic angle of repose is the angle with the horizontal at which the material will stand when piled. It was determined on four surfaces viz., plywood, stainless steel, galvanised steel sheet and mild steel sheet. A regular cylindrical container opened at both ends was placed on each flat surface and filled up to the top with seeds. The container was then lifted up gradually from the surface. The gradual lifting continued until a conical heap was formed [20]. The angle of repose (Φ) was calculated from the height and base radius of the heap formed by following relation [23].

$$\ddot{O} = \frac{h}{R} \text{ Or } \quad \ddot{O} = \tan^{-1}\frac{h}{R}$$

Where, Φ = angle of repose (°);

h = height of the pile (cm);

D = diameter of the pile (cm)

Coefficient of static friction: The coefficient of static friction (μ) was determined against surfaces of plywood, stainless steel, galvanised steel sheet and mild steel sheet using the inclined plane apparatus [29]. The table was gently raised with the seed on the plane and the angle of inclination to the horizontal at which the seed started sliding was read off with the help of protractor attached to the apparatus. The tangent of the angle was reported as the coefficient of friction [29]. The measurement was done 10 times and calculated as:

$$\mu = \tan\theta$$

Breakage test: The Breakage test for the seeds was performed to know the magnitude of the force required to break the seed. The force required for breaking individual Horse Chestnut seeds was done by using a Stable Micro System Texture Analyzer (Model TA-XT2i, Texture Technologies Corp., Scarsdale, New York, U.S.A.). The individual seeds were placed centrally beneath the probe (p/75 cylinder probe) and breaking force was determined. The compression test was selected in texture analysis using a 50 kg load cell and sample was compressed to breakage. The strain required for breakage was recorded using the following conditions: pretest speed: 1.0 mm/s, test speed: 2.0 mm/s, post-test speed: 10 mm/s, compression distance: 15 mm. The values reported were the average of ten readings.

Processing of seeds: The seeds were taken and dehulled manually and the kernels obtained were sliced and dried in tray dryer (M/s. Balaji Enterprises, Saharanpur, India) at 50°C. The dried chips were then grinded into fine powder in Laboratory grinding mill (M/s. Philips India Limited, Kolkata, India). All chemicals were used of analytical grade.

Color characteristics: Color measurement of powder was carried out in triplicates, using Color Flex Spectrocolorimeter (Hunter Lab Colorimeter D-25, Hunter Associates Laboratory, Ruston, USA). The results were expressed in terms of standard L*, a* and b* values. The functions ΔE and whiteness index were also calculated.

Compositional properties: Moisture content, crude protein, fat, and ash content of flour was determined according to AOAC methods [29]. Total carbohydrate content was determined according to the difference method by Mathew et al. [30] that is, by subtracting the sum of the percentages of crude protein, lipid, crude fiber and ash content from 100 [31].

Results and Discussion

Physical properties

The results of the determined physical parameters of the Horse Chestnut seeds are shown in Table 1. The weight of the variety was found to be 40.8 g while kernel and shell ratio was found to be 85.71 and 14.29%, respectively. The mean length, width and thickness were found to be 4.7, 4.0, and 3.2 cm, respectively.

The geometric mean diameter and arithmetic mean diameter were expressed in terms of length, width and height and the values calculated

Property	Horse Chestnut variety
Weight of Seed, g	40.80 ± 1.3
Kernel ratio (%)	85.71 ± 1.2
Shell ratio (%)	14.29 ± 1.1
Length, cm	4.7 ± 0.29
Width, cm	4.0 ± 0.28
Thickness, cm	3.2 ± 0.10
Geometrical mean diameter (cm)	3.92 ± 0.02
Arithmetic mean diameter(cm)	3.97 ± 0.03
Sphericity, %	83.4 ± 0.01
Aspect ratio, %	85.11 ± 0.12
True density, g/cm³	1072 ± 2.4
Bulk density, g/cm³	518 ± 2.8
Density ratio, %	48.32 ± 1.2
Porosity, %	51.68 ± 0.5
Seed mass, g	33.75 ± 2.10
Surface area (cm²)	48.25 ± 0.17

Results are expressed as Mean ± Standard deviation of ten observations

Table 1: Physical characteristics of Horse Chestnut seed.

are 3.92 and 3.97 cm, respectively. These parameters depend upon the size of the seed, with bigger size seeds having higher values [20]. The parameters are of great value as the dimensions are important in determining aperture size in material separation [32]. These parameters of Horse Chestnut seed may be useful in determining the size of the components of machine for deshelling.

Sphericity and aspect ratio were found to be 83.4 and 85.11%, respectively. Higher sphericity values indicate that seed shape is closed to a sphere, while high aspect ratio indicates that seeds will roll than slide on flat surfaces. This is very important in design of hoppers for machines [31]. However, if aspect ratio value is being close to the sphericity values the seed will undergo a combination of rolling and sliding action on flat surfaces.

The true density, bulk density, density ratio and porosity were 1072 g/cm³, 518 g/cm³, 48.32% and 51.68%, respectively. Both the true and bulk density characteristics are useful in the estimation of load and hence in the design of load shafts for processing machine. The porosity in the seed is related to sphericity and aspect ratio, which ensure a more compact arrangement of the seeds. These properties may be useful for the bulk storage and transportation of the seed. The average seed mass of the variety was 33.75 g, depending on the size or dimensions of seed. The surface area was 48.25 cm², respectively and depends on the size of the seed, with higher value for larger sized seed.

Mechanical properties

The results for static coefficient of friction and angle of repose of Horse Chestnut seeds obtained on plywood, mild steel, stainless steel and galvanized sheet are presented in Table 2. The coefficient of static friction obtained for the variety was 0.54 on plywood, 0.52 on mild steel, 0.51 on stainless steel and 0.52 on galvanized sheet. The static coefficient of friction was higher on plywood followed by mild steel and galvanized steel and lower on stainless steel for both varieties. There was a small difference in the coefficient of friction obtained on mild steel, stainless steel and galvanized steel. The seeds move easily on surfaces due to higher hardness and slippery surface.

The dynamic angle of repose was higher on galvanized steel followed by plywood and mild steel and lower on stainless steel for both varieties. The angle of repose was 22.3° on plywood, 20.3° on mild steel, 18.3° on stainless steel and 22.79° on galvanized sheet. The values of angle of repose are higher due to the smooth seed surface. The average breaking force required for breaking the seeds are 328 N as shown in Table 2. The average breakage force required to break the seeds are higher due to the thick seed shell.

Color characteristics

Color is an important quality factor that typically relates to the acceptability, marketability and wholesomeness of foods. Color characteristics of seed powder are shown in Table 3. The L, a, b values of HCN seed powder were found to be 91.23, 3.10 and 15.07 with a whiteness value of 82.28.

Compositional analysis

The data for compositional analysis of seed powder is presented in Table 3. Chemical composition is an important criterion to have an idea about the overall composition and nutritional status of any ingredient. Moisture content of flour is an important quality parameter which determines its shelf life. The results showed that HCN seed powder contain 12.71% moisture content (Table 3). Usually, low moisture content helps in enhancing the storage stability at ambient temperature as moisture is an important parameter in the storage of flour/powders, very high levels greater than 12% allow for microbial growth and thus low levels are favourable and give relatively longer shelf life. The seed powder was found to contain about 3.16% ash. Ash content reflects the quantity of mineral matter present in a substance and higher values of ash content indicates that higher amount of minerals are present. Fat content was found to be 3.27%. Fibre is an important component of many complex carbohydrates. It is always found only in plants particularly vegetables, fruits, nuts and legumes. As shown in Table 3, crude fibre content of the seed was found to be 6.34%. Total protein of HCN seed was quantified to be 6.78%. Carbohydrate content was found to be 67.74% highest from other components.

Conclusion

The physical properties including seed shape, size, geometric mean diameter, sphericity, bulk density, porosity, surface area, static coefficient of friction and angle of repose were investigated. The study reveals that physical properties of Horse Chestnut seed may be useful in designing equipment for postharvest handling and processing operations like deshelling of seed, pulverisation of seed kernel into flour etc. These properties are necessary for the design of equipments for harvesting, separating, processing, packing and transportation. Compositional analysis of the seed powder conducted revealed the seeds are rich in carbohydrates 67.74%, protein 6.78% and fibre 6.34%. Therefore, it may be used as potentially attractive source of carbohydrates, protein and crude fibre.

Acknowledgments

The first author is grateful to university grants commission, New Delhi, India for providing financial assistance in the form of Maulana Azad National Fellowship (MANF).

Property	Horse Chestnut variety
Coefficient of static friction	
Plywood	0.54 ± 0.04
Mild steel	0.52 ± 0.07
Stainless steel	0.51 ± 0.05
Galvanised steel	0.52 ± 0.06
Dynamic angle of repose (degrees)	
Plywood	22.3 ± 1.4
Mild steel	20.3 ± 1.6
Stainless steel	18.3 ± 2.5
Galvanised steel	22.79 ± 1.2
Breaking force (N)	328 ± 5.12

Results are expressed as Mean ± Standard deviation of ten observations

Table 2: Mechanical characteristics of Horse Chestnut seed.

Parameters	Seed powder
Color characteristics	
L*	92.07 ± 0.49
a*	-3.47 ± 0.21
b*	13.70 ± 0.26
ΔE	NA
Whiteness index	83.78 ± 0.06
Compositional properties	
Moisture (%)	12.71 ± 1.23
Protein (%)	6.78 ± 1.19
Crude Fat (%)	3.27 ± 0.39
Ash (%)	3.16 ± 0.05
Crude fiber (%)	6.34 ± 0.22
Carbohydrate (%)	67.74
Energy value (Cal/100 g)	327.51

Results are expressed as Mean ± Standard deviation of triplicate observations

Table 3: Color and compositional characteristics of Horse Chestnut seed powder.

References

1. Singh B (2006) Simple process for obtaining beta Aescin from Indian horse chestnut. United States patent application pub. no. US 2006/0030697.

2. Zhang Z, Li S, Lian XY (2010) An overview of genus Aesculus L.: ethnobotany, phytochemistry, and pharmacological activities. Pharmaceutical Crops 1: 24-51.

3. Rafiq SI, Jan K, Singh S, Saxena DC (2015) Physicochemical, pasting, rheological, thermal and morphological properties of horse chestnut starch. J Food Sci Technol 52: 5651-5660.

4. Parmar C, Kaushal MK (1982) Aesculus indica. Wild fruits, Kalyani Publishers, New Delhi, India.

5. Rajasekaran A, Singh J (2009) Ethnobotany of Indian horse chestnut (Aesculus indica) in Mandi district, Himachal Pradesh. Indian Journal of Traditional Knowledge 8: 285-286.

6. Kaul MK (1997) Medicinal plants of Kashmir and Ladhak. Indus Publishing Company, New Delhi.

7. Kaur L, Joseph L, George M (2011) Phytochemical analysis of leaf extract of Aesculus indica. International Journal of Pharmacy and Pharmaceutical Sciences 3: 232-234.

8. Chakraborthy GS (2009) Evaluation of Immuno modulatory action/activity of Aesculus indica. International Journal of Pharma Technological Research 1: 132-134.

9. Chakraborthy GS (2009) Free radical scavenging activity of Aesculus Indica leaves. International Journal of Pharma Technological Research 1: 524-526.

10. Singh B, Katoch M, Raja R, Zaidi A (2007) Inst of Himalayan Bioresource Tech A, Palampur. A new original agent from Indian Horse Chestnut Aesculus Indica, European Patant EP1489910.

11. Ikram M, Gilani SN (1986) Anti-inflammatory activity of Aesculus indica fruit oil. Fitoterapia 57: 455-456.

12. Qayum A, Ahmed N, Ahmad KD, Gilani NU (1988) Pharmacological screening of indigenous medicinal plants. Pak J Pharm Sci 1: 37-39.

13. Bhatt JP (1992) Neurodepressive action of a piscicidal glycoside of plant, Aesculus indica (Colebr.) in fish. Indian J Exp Biol 30: 437-439.

14. Sharma PK (1991) Herbal remedies for treating rheumatic pains in Jammu and Kashmir. Indian Journal of Forestry 14: 206-210.

15. Sirisomboon P, Pornchaloeampong P, Romphophak T (2007) Physical properties of green soybean: Criteria for sorting. Journal of Food Engineering 79: 18-22.

16. Kashaninejad M, Mortazavi A, Safekordi A, Tabil LG (2006) Some physical properties of Pistachio (Pistacia vera L). nut and its kernel. Journal of Food Engineering 72: 30-38.

17. Plange BA, Baryeh EA (2003) The physical properties of Category B cocoa beans. Journal of Food Engineering 60: 219-227.

18. Sirisomboon P, Kitchaiyab P, Pholphoa T, Mahuttanyavanitcha W (2007) Physical and mechanical properties of Jatropha curcas L. fruits, nuts and kernels. Biosystems Engineering 97: 201 - 207.

19. Owolarafe OK, Olabige MT, Faborode MO (2007) Physical and mechanical properties of two varieties of fresh oil palm fruit. Journal of Food Engineering 78: 1228-1232.

20. Calisir S, Aydin C (2004) Some physical-mechanical properties of cherry laurel (Prunus lauracerasus L.) fruits. Journal of Food Engineering 65: 145-150.

21. TSI (1990) Methods TS 1275 and TS 1276. Ankara, Turkey.

22. Karababa E (2006) Physical properties of popcorn kernels. Journal of Food Engineering, 72: 100-107.

23. Mohsenin NN (1980) Physical Properties of Plant and Animal Materials. Gordon and Breach Science Publishers, New York.

24. Demir F, Dogan H, Ozcan M, Haciseferogullari H (2002) Nutritional and physical properties of hackberry (Celtis australis L.). Journal of Food Engineering 54: 241-247.

25. Mwithiga G, Sifuna MM (2006) Effect of moisture content on the physical properties of three varieties of sorghum seeds. Journal of Food Engineering 75: 480-486.

26. Ovelade OJ, Odugbenro PO, Abiove AO, Raji NL (2005) Some physical properties of African star apple (Chrysophyllum alibidum) seeds. Journal of Food Engineering 67: 435-440.

27. Goyal RK, Kingsly ARP, Kumar P, Walia H (2007) Physical and mechanical properties of aonla fruits. Journal of Food Engineering 82: 595-599.

28. Dutta SK, Nema VK, Bhardwaj RK (1988) Physical properties of gram. Journal of Agricultural Engineering Research 39: 259-268.

29. AOAC (2006) Official Methods of Analysis. Association of Official Analytical Chemists, AOAC Press, Gaithersburg.

30. Mathew T, Al-Bader M, Bou-Resli MN, Dashti HM, Al-Zaid NS, et al. (2006) Alteration of brain zinc level in rat pups of zinc supplemented mothers. Trace Elementary Electrolytes 23: 231-236.

31. Moshsenin NN (1978) Physical Properties of animal and plant material. Gordon and Breach Science Publishers, USA.

32. Omobuwajo TO, Sanni LA, Balami YA (2000) Physical properties of Sorrel Seed. Journal of Food Engineering 45: 37-41.

Effects of Green Tea Extracts on the Caffeine, Tannin, Total Polyphenolic Contents and Organoleptic Properties of Milk Chocolate

Aroyeun SO* and Jayeola CO

Cocoa Research Institute of Nigeria, Ibadan, Nigeria

Abstract

Chocolate and green tea contains antioxidants that may be used as health promoting foods. A new product incorporating green tea in Chocolate and termed green tea chocolate was developed in this project. Green tea was produced using the Chinese method viz: Plucking, Fixing, Rolling and Drying. The green tea produced was milled into powder and mixed at different ratios: 10:90, (GTCE) 20:80 (GTCD), 30:70 (GTCC), 40:60 (GTCB) and 50:50 (GTCA) (w/w) of the green tea powder to the Chocolate. Conventional milk chocolate (GTCF) without green tea served as the control for the production of chocolate followed by the standard method. Proximate Analysis, Total Phenol and antioxidant properties were determined using standard methods; Sensory Analysis was done by panel of tasters who measured the Taste, Odour, flavour and general acceptability.

The average protein of the control chocolate was 8.05 which varied between 7, 24% (10% supplementation) and 8.39% (50% supplementation). The % crude fibre for the control chocolate was 1.17% and for other supplemented levels, it varied from 0.93% for 10% green tea addition to 1,23%. All the crude fibre increased significantly ($p < 0.05$) on increasing the green tea in the chocolate. It was found that at 10% inclusion of green tea, there were increases of 9.7%, 9.6%, 4.4% and 3.2% in crude fibre as compared to the control. This may be due to higher contents of dietary fibre in the green tea. The percentage total ash of the control was 2.43% and increased significantly and consistently from 2.28 at 10% level of inclusion to 2.55 at the 50% green tea supplementation.

The chocolate containing the smallest quantity of green tea powder seemed to have the lowest amount of crude fat and rose steadily with increase in the green tea supplementation. Although at these levels the effect of the crude fat on the chocolate was not significant while the significance was observed only at 50% green tea inclusion. The control chocolate had 162.39 mg/100 g gallic acid equivalent and it increased significantly ($p < 0.05$) with rise in green tea powder in the recipe. This study also established that at the organoleptic threshold of 10%, green tea supplementation, there was no significant difference in the polyphenol content of the chocolate. It became significant at higher level of inclusion, i.e. 40-50% at which the taste of the chocolate and the overall acceptability and colour and sweetness became impaired. The caffeine content increased with increase in green tea powder in the chocolate. Green tea inclusion at 20%, 30%, 40% and 50% were showing significantly better chelating properties than the control samples and the 10% samples. The iron chelating ability increased with high contents of green tea and in the order 50% > 40% > 30% > 20% > control > 10%. The L* value reduced significantly from 10% to the 50% green tea powder.

In conclusion, replacement of cocoa nibs with green tea powders up to 20-50% impaired the taste of the chocolate.

Keywords: Chocolate; Tea powders; Caffeine

Introduction

Chocolate and green tea contains antioxidants that may be used as health promoting agents. Tea and tea products mainly contain tea polyphenols, which are natural antioxidants and have been demonstrated to show antioxidative, anti-carcinogenic and anti-microbial properties by many researchers [1,2]. These health benefits of teas, in particular green tea, are gaining increased attention in recent years. Green tea contains the most abundant tea polyphenols, namely tea catechins. The major nutraceutical Compounds in green teas are tea catechins, which are flavanols. Flavanols are a class of flavonoids which are polyphenols. Green tea is rich in flavanols (300-400 mg/g) which are of interest to human health [3]. Tea catechins have the most effective antioxidant activity compared to other Tea polyphenols. The major green tea catechins are (−)-epigallocatechin gallate (EGCG), (−)-epicatechin gallate (ECG), (−)-epigallocatechin (EGC) and (−)-epicatechin (EC). These epicatechins can change to their epimers that are non epicatechins, i.e. (−)-gallocatechin gallate (GCG), (−)-catechin gallate (CG), (−)-gallocatechin (GC) and (±)-catechin C) Figure 1. EGCG is the most abundant and active catechin and it is usually used as a quality indicator [4-6]. In addition, green tea contains other polyphenols such as gallic acid, quercetin, kaempferol, myricetin and their glycosides, but at lower concentration than EGCG [3-7].

Tea catechins are an efficient free radical scavenger due to their one electron reduction potential. A lower reduction potential has a tendency to lose electron or hydrogen [8]. The rate of reaction with free radicals and the stability of the resulting antioxidant radicals contribute to the reactivity of antioxidant. Guo [9] reported the scavenging ability of tea catechins on superoxide anions (O^{2-}), singlet oxygen, the free radicals generated from 2,2P-azobis (2-amidinopropane) hydrochloride (AAPH) and 1,1-diphenyl-2-picrylhydrazyl (DPPH) radicals. They suggested that the scavenging ability of EGCG and GCG was higher than that of EGC, GC, EC and C due to their gallate group.

They are nowadays utilized in a wide range of applications, such as food, beverage, cosmetics toiletries etc. [10]. The consumption of green tea in the form of hot beverages is common in China and Japan

*Corresponding author: Aroyeun SO, Cocoa Research Institute of Nigeria, PMB, 5244, Ibadan, Nigeria, E-mail: aroyeun2000@yahoo.co.uk

and the incorporation of it either in the form of powder or in the form of extracts in foods are available in the literature. Somboonvecharkarn [11] reported the effect of green tea extract in soy bread and observed that green tea fortified soy bread was similar to the regular soy bread in some physical properties. In the food industry, both chocolate and green tea is popular choices for health benefits and also their reasonable price compared to other dietary supplements such as Ginseng and Ginko leaves. Chocolate and green tea has been associated with antioxidant properties which are also linked to prevention of cancer, cardiovascular disease and as antiobesic effects [12]. Catechin and polyphenol compounds in green tea are excellent antioxidants that are enhanced by the presence of metals [13]. Studies have shown that 2 grams of green tea have equivalent antioxidant activity of 109-147 mg epigallocatechin gallate (EGCG), 14 muM for Catechin and 22 muM for vitamin C [14]. Although the popularity of green tea and chocolate is growing in the United States, there are no records in the literature where the two were combined. The objectives of this work, however, are to create a novel food product incorporating the combination of chocolate and green tea and determine the chemical and organoleptic qualities of the newly developed product.

Materials and Methods

Fresh tea leaves were harvested from the Mambilla Highland; Taraba State, Nigeria Green Tea was processed using the Chinese method viz: Plucking, withering, Fixing, Rolling and Drying. Ingredients of Chocolate were mixed together during Conching and the weight was taken. The incorporation of green tea was based on different combination such as 10:90, 20:80, 30:70, 40:60, 50:50 (w/w) of the green tea powder and chocolate, the flow chart of chocolate processing was shown below in Figure 1 and the Recipe for the production of chocolate was shown in Table 1.

Proximate Analysis

Moisture content

The moisture content of chocolate was determined by drying the samples in an oven at 105°C until a constant weight was obtained (AOAC).

Crude protein

Crude protein content was calculated by converting the nitrogen content, determined by Kjedahl's method ($6.25 \times N$)

Crude fat

This was determined by the method described by the AOAC, using the Soxhlet methods

Total ash

Content was determined by dry ashing in a furnace at 525°C for 24 h.

Caffeine determination

This was done in accordance with Yao [15]. Tannin Analysis was carried out using the method of Somboonvecharkarn [11].

Total Phenol Determination

Extraction procedure

Known weight of the green tea chocolate was ground and transferred into a test tube and mixed with 10 ml of 80% methanol. Suspension was vortexed and centrifuged for 10minutes. The mixture was sonicated for 5 minutes, then shaken at 120 rpm at 70°C for 2 hours. It was then centrifuged for 10minutes. Supernatants were collected and filtered.

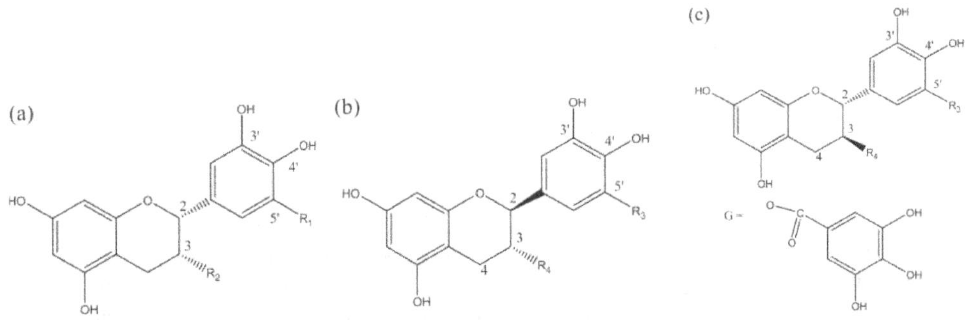

Figure 1: General chemical structures of green tea catechins: (a) epi-catechins; (b) non-epicatechins and (c) (+)-catechin.

Samples %	Tea	Nibs	Milk	Sugar	Cocoa Butter	Lecithin
10%	2.8	25.96	20.77	47.3	2.5	0.51
	11.25g	101.25	81.00	184.5	10	2.0
20%	5.77%	23.08	20.77	47.30	2.5	0.51
	22.5g	90	81.0	184.5	10	2.0
30%	8.65%	20.19	20.77	47.3	2.5	0.51
	33.75g	78.75	81.0	184.5	10	2.0
40%	11.56%	17.31	20.77	47.30	2.5	0.51
	45.10g	67.50	81.0	184.5	10	2.0
50%	14.42%	14.42	20.77	47.3	2.5	0.51
	56.25g	56.25	81.0	184.5	10	2.0

Table 1: Recipe used in green tea production in this study.

Folin-Dennis Ciocalteau procedure

Total phenolic content was estimated by the Folin-Ciocalteau colorimetric method based on the procedure of Singleton and Rossi, 1965 using Gallic acid as a standard phenolic compound. 100 µl of the filtered extracts were mixed with 400 µl of 80% methanol and 2.5 ml of 0.2N Folin Ciocalteau phenol reagent. After 5 minutes, 2 ml of 7.5% sodium carbonate was added. The absorbance of the resulting blue-coloured solution from yellow solution was measured at 765 nm spectrophotometrically after 30 minutes in the dark at room temperature. Quantitative measurements were performed based on a standard curve of Gallic acid [16]. The total phenolic content was expressed as Gallic acid equivalents (GAE) in mg/g dry material

Chelating effect on ferrous ions

The ferrous ion chelating activity of the green tea chocolates was assessed as described by Tawaha K [17]

Colour analysis

For each sample, colour was determined with the portable Minolta Chromameter CR (Minolta, Osaka, Japan). The Lab values follow the Hunter Lab color scale

Sensory analysis

The sensory Analysis was carried out in accordance with standard methods (Figure 2).

Statistical analysis

Samples were analyzed in triplicates. Means were separated using ANOVA.

Results and Discussion

From the univariate statistical analyses, all chemical properties measured varied significantly (p < 0.05) among the chocolate samples.

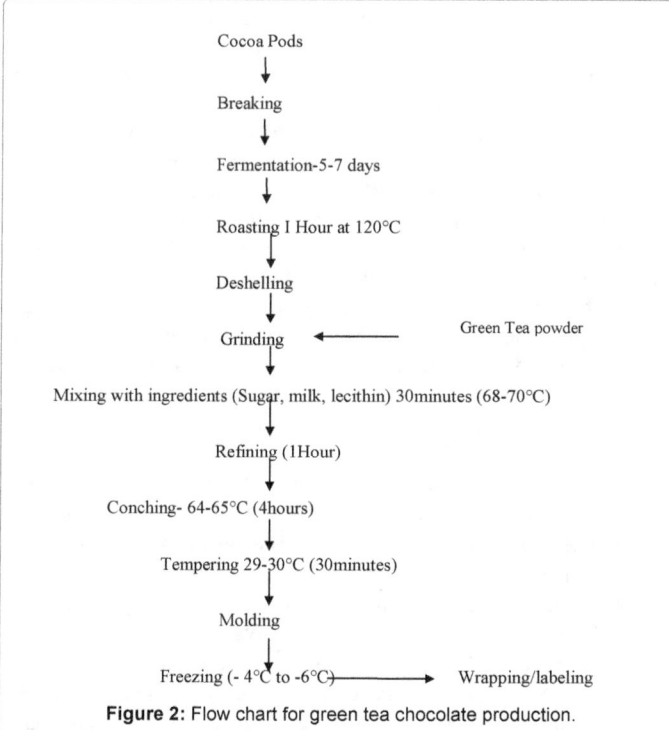

Cocoa Pods

↓

Breaking

↓

Fermentation-5-7 days

↓

Roasting I Hour at 120°C

↓

Deshelling

↓

Grinding   ←   Green Tea powder

↓

Mixing with ingredients (Sugar, milk, lecithin) 30minutes (68-70°C)

↓

Refining (1Hour)

↓

Conching- 64-65°C (4hours)

↓

Tempering 29-30°C (30minutes)

↓

Molding

↓

Freezing (- 4°C to -6°C)   →   Wrapping/labeling

Figure 2: Flow chart for green tea chocolate production.

Some of the results were shown in Tables 1–4 while the summary has been displayed in the boxplots (Figure 1). This directly reflected in the smaller thickness of the boxes and deviation from the means (depicted by the protruding bars) of the former compared to the latter. According to Table 1, the proximate chemical composition of the chocolates such as the crude protein, crude fibre, total ash, moisture content, sucrose and crude fat differed significantly (p < 0.0-5) with increase in the levels of the green tea added. The average protein of the control chocolate was 8.05 whereas that of chocolate supplemented with green tea varied accordingly. The protein values varied from 7.24 for 10%, 7.65 for 20%, 7.93% for 30%, 8.27% for 40% and 8.39% for 50% green tea supplementation respectively.

The % crude fiber for the control chocolate was 1.17% and for other supplemented levels, it varied from 0.93% for 10% green tea addition, 1.03 for 20% addition, 1.14 (30%) 1.18 (40%) and 1.23 (50:50). The percentage total ash of the control was 2.43% and increased significantly and consistently from 2.28 at 10% level of inclusion to 2.55 at the 50% green tea supplementation. The sucrose level of the control was 48.3% and the sucrose decreased significantly and steadily from 10% green tea inclusion to the 50% level. This is an expected trend; however, as green tea is bitter and the reduction in sucrose level is an indication of the significant impact of the bitterness on the reduction in sweetness of the chocolate. As for the % crude fat, the control chocolate had the highest % amount while the samples in which green tea has been supplemented increased steadily and significantly with increased amount of green tea powder.

The chocolate containing the least inclusion seemed to have the lowest amount of crude fat and rose steadily with increase in the green tea supplementation. Although at these levels the effect on the crude fat on the chocolate was not significant while the significance was observed only at 50% green tea inclusion. This might be due to the fat content in green tea which reflected in the crude fat of the chocolate. All the crude fibre increased significantly (p < 0.05) on increasing the green tea in the chocolate. It was found that at 10% inclusion of green tea, there were increases of 9.7%, 9.6%, 4.4% and 3.2% in crude fibre as compared to the control. This may be due to higher contents of dietary fibre in the green tea. This result is in agreement with the work of Guo [9]

Polyphenols and tannin

The control chocolate had 162.39 mg/100 g gallic acid equivalent and it increased significantly (p < 0.05) with rise in green tea powder in the recipe Table 2. The maximum increase of 28.4% in polyphenol over the control chocolate was found in 50% green tea inclusion while an increase of 12.45% of the polyphenol was found in chocolate with green tea supplemented at 40%. There was no significant effect of green tea addition on the polyphenol profile of the chocolate at 10%, 20% and 30% respectively. This in effect confirmed that chocolate itself contains polyphenols which made the addition of green tea powder

Samples	% Crude Protein	% Crude Fibre	% Total Ash	% Moisture Content	% Sucrose	% Crude Fat
10%	7.24	0.93	2.28	2.15	49.18	31.28
20%	7.65	1.03	2.41	2.06	48.25	33.42
30%	7.93	1.14	2.37	2.02	46.13	35.84
40%	8.27	1.19	2.49	1.24	45.74	37.52
50%	8.39	1.23	2.55	1.15	43.25	39.34
Control	8.05	1.17	2.43	1.41	48.36	38.12

Table 2: Proximate analysis of chocolate formulated with different levels of green tea powder.

to be insignificant on the polyphenol of the chocolate. However, no report is found in the literature confirming which of the two samples i.e. green tea or chocolate has the higher polyphenols content. It was only established that both of them are polyphenolic products.

This study also established that at the organoleptic threshold of 10%, green tea supplement, there was no significant difference in the polyphenol content of the chocolate. It became significant at higher level of inclusion, i.e. 40-50% at which the taste of the chocolate and the overall acceptability and colour and sweetness became impaired. The polyphenol contents of chocolate at these two levels had earlier been reported by Mexis [18] who reported that the beneficial effect of tea derived primarily from ingredients such as antioxidant substances (polyphenols).This means that the possibility of 40% and 50% Green tea powder addition in chocolate recipe can cause an increase in the beneficial effect of chocolate consumption, because, there is a synergy of polyphenol between green tea and chocolate [19-21]. The tannin increased as green tea powder increased but it was not significant until at 40-50% when it became significant. In our study, it was also observed that % tannin of all the examined green tea chocolates and the control was lower (Tables 3-5).

Caffeine

The caffeine content in Table 2 increased with increase in green tea powder in the chocolate. Up to 30% inclusion, the green tea did not have any significant effect on the chocolate ($p < 0.05$). Caffeine is a major component of tea, coffee, kola and cocoa [14] although it

Samples	Iron Chelating agent mg/100 g	Tannin (%)	Caffeine (%)	Total Polyphenol mgGAE/100 g
10%	1.84	0.0021	0.053	118.46
20%	1.76	0.0029	0.062	124.84
30%	1.69	0.0035	0.069	151.65
40%	1.73	0.0043	0.074	185.49
50%	1.65	0.0051	0.081	226.68
Control	1.79	0.0039	0.067	162.39

Table 3: Polyphenols profiles of green tea chocolate.

Samples	L*	A	b
10%	57.77	2.68	20.22
20%	52.4	2.76	18.64
30%	51	2.84	18.22
40%	48.90	2.88	17.4
50%	42.32	3.42	16.35
Control	58.54	2.66	22.54

Table 4: Colour analysis of green tea chocolate.

Chocolates	Colour	Taste	Smoothness	Sweetness	Overall Acceptability
GTC_a	5.21^c	4.10^c	4.23^d	4.40^d	4.20^d
GTC_b	5.23^c	5.46^b	4.62^d	4.62^d	4.81^d
GTC_c	5.34^c	5.41^b	5.11^c	5.12^c	5.63^c
GTC_d	6.21^b	5.42^b	5.44^b	6.32^b	6.11^b
GTC_e	7.62^{ab}	7.45^a	7.62^a	7.45^{ab}	7.52^{ab}
GTC_f	8.42^a	7.68^a	8.26^a	8.22^a	8.58^a

GTC_a- 50%; Green tea: Cocoa (w/w); GTC_b- 40%; GTC_c- 30%; GTC_d- 20%; GTC_e- 10%; GTC_f- 0%

a, b, c and d : Means along the same vertical columns with different alphabets are significantly different at $p < 0.05$ than those reported in the literature [13]. Decreased tannin concentrations are significantly useful for iron absorption and improved digestion [11].

Table 5: Sensory evaluations of green tea chocolate.

is a stimulant; excess of it can cause impairment of the mechanical properties of growing bone in early life. According to Table 2, the 20.8% increase in caffeine levels of the control chocolate over the 10% supplementation is desirable for low caffeine consumers. Since the sensory threshold of the chocolate remains at 10%, the reduction of the caffeine level is a welcome development. This has also been supported by Chand [19], who reported that it is practically impossible to avoid caffeine as it is present in various foods and beverages and over the counter medications [14]. On balance, it is better to reduce the daily intake of caffeine.

Iron chelators

The ability to chelate transition metals can be considered as an important antioxidant mode of action. In fact, the chelation and deactivation of transition metals prevent these species from participating in hydroperoxidation and decomposition reaction [15]. Green tea inclusion at 20%, 30%, 40% and 50% were showing significantly better chelating properties than the control samples and the 10% samples at 1.76, 1.69, 1.67, 1.65 mg/100mg respectively (Table 2). Since ferrous ions were the most effective prooxidants and are commonly found in vegetables, the high ferrous ions chelating abilities of the 20%, 30%, 40% and 50% of green tea in chocolate would be beneficial. The iron chelating ability as found in this study increased with high contents of green tea and in the order 50% > 40% > 30% > 20% > control > 10%. In fact, the chelating effects of green tea on the chocolate in the listed order were stronger when compared to vitamin C, BHT, and BHA [15]. These data revealed that our chocolate products with green tea demonstrate an interesting capacity for iron binding. In fact, numerous other studies indicated that plant extracts enriched in phenolic compounds are capable of complexing with and stabilizing Transition metal ions rendering them unable to participate in metal-catalyzed initiation and hydroperoxide decomposition reactions [16]

Colour

According to Table 3, Changes in colour parameters L* of chocolate with green tea is a function of aesthetic value. The L* value reduced significantly from 10% to the 50% green tea powder inclusion in the order Control > 10% > 20% > 30% > 40% > 50% representing 58.54, 57.77%, 52.4%, 51.0% 48.9% and 42.32% in values. Different trend was observed in a* value and the b* values without any significant difference. Lower L* value indicated increasing darkness and lower b* value suggested decreasing yellowness. This result was in agreement with McKay [1] who reported that the addition of green tea extract to soy bread increased the darkness of the crumbs and the bread faded from bright yellow to medium brown. This result is not in agreement with Wang [6] who reported an increase in L* values of commercially packaged dark chocolate. The difference in our study and Wang was directly related to the formation of large white spots on the surface of the chocolate known as fat bloom which is not present in our samples.

Sensory of evaluations

The effect of green tea supplementation on the sensory characteristics of chocolate is presented in Table 4. With increase in the levels of green tea in the formulation, the sensory scores for colour, taste, sweetness, flavor and overall acceptability of the chocolate decreased sharply. Replacement of cocoa nibs with green tea powders up to 20-50% impaired the taste of the chocolate. Control samples had the highest scores of 7.68 in taste, which decreased significantly from 7.45 to 4.10 due to the bitter taste of the green tea. The colour of the control samples was scored highest when compared to other chocolate samples where green tea powder was incorporated. The dullness of the chocolate increased with addition of green tea powder in the chocolate mix.

The texture of the chocolate in terms of smoothness decreased according to the green tea percentage inclusion meaning that the higher the green tea powder the lower is the smoothness or the higher is the roughness. The control samples had maximum overall acceptability whereas chocolate containing 40 to 50% were found to be unacceptable to the panelists. The overall acceptability score for control was 8.58 on a 9 point hedonic scale. Chocolates made from blends containing 10% level of green tea powder did not differ significantly from the control sample in Taste, odour, flavor and overall acceptability ($p < 0.05$) (Table 5).

Similar observation with supplementation of soy flour [20], bajol grain flour and wheat flour have also been reported. McKay and Somboonvecharkarn [1,11] also reported similar work on the effect of green tea extract in soy bread physical properties and total phenolic content. For the overall acceptability, ratings, it was concluded that green tea powder could be incorporated into chocolate to 10% without necessarily affecting their sensory quality

Conclusion

In conclusion, replacement of cocoa nibs with green tea powders up to 20-50% impaired the taste and the colour of the chocolate.

Acknowledgement

Authors wish to acknowledge the executive Director/Chief Executive of the Cocoa Research Institute of Nigeria professor, M.O, Akoroda for permission to publish this paper. Mr. Obatoye and Mrs. Ogunlusi of the chocolate laboratory are also appreciated for their efforts to get the chocolate produced despite all odds.

References

1. McKay DL, Blumberg JB (2002) The role of tea in human health update. J Am Coll Nutrit 21: 1-13.

2. Rietveld A, Wiseman S (2003) Antioxidant effects of tea: evidence from human clinical trials. The J Nutrit 133: 3285-3292.

3. Dubick MA, Omaye ST (2007) Grape wine and tea polyphenols in the modulation of atherosclerosis and heart disease. Boca Raton, CRC Press.

4. Lakenbrink C, Lapczynski S, Maiwald B, Engelhardt UH (2000) Flavonoids and other polyphenols in consumers brews of tea and other caffeinated beverages. J Agric Food Chem 48: 2848-2852.

5. Wang H, Helliwell K (2000) Epimerization of catechins in green tea infusions. Food Chem 70: 337-344.

6. Wang R, Zhou W, Jiang X (2008) Reaction kinetics of degradation and epimerization of epigallocatechin gallate (EGCG) in aqueous system over a wide temperature range. J Agric Food Chem 56: 2694-2701.

7. Sakakibara H, Honda Y, Nakagawa S, Ashida H, Kanazawa K, et al. (2003) Simultaneous determination of all polyphenols in vegetables fruits and teas. J Agric Food Chem 51: 571-581.

8. Higdon JV, Frei B (2003) Tea catechin and polyphenols: Health effects metabolism and antioxidant functions. Crit Rev Food Sci Nutrit 43: 89-143.

9. Guo Q, Zhao B, Shen S, Hou J, Hu J, et al. (1999) ESR study on the structure antioxidant activity relationship of tea catechins and their epimers. Biochimica et Biophysica Acta 1427: 13-23.

10. Wang H, Helliwell K, You X (2000) Isocratic elution system for the determination of catechins caffeine and gallic acid in green tea using HPLC. Food Chem 68: 115-121.

11. Somboonvecharkarn C (2011) The effects of green tea extract on soy bread physical properties and total phenolic content. The Ohio State University Department of Food Science and Technology.

12. Doss MX (2005) Trapping of growth factors by catechins a possible therapeutical target for prevention of proliferative diseases. J Nutr Biochem 16: 259-266.

13. Hai NY, Jun JY, Sheng RS (2006) Effects of EGC-3 cell cytoplasimic membrane in the presence of Cu2+. J Food Chem 95 :108-115.

14. Oh JH, Kim EH, Kim JL, Kang JS, Jung SK (2004) Study on antioxidant potency of green tea by DPPH method. Food and Agricultural Organization, Journal of Korean Society of Food Sci Nutr 33: 1079-1084.

15. Li GY, McCulloch RD, Fenton AL, Cheung M, Meng L, et al. (2006) Structure and identification of ADP-ribose recognition motifs of APLF and role in the DNA damage response. PNAS 107: 9129-9134.

16. Bourgou S, Ksouri E, Belitia A, Skaudrant A, Fallein M, et al. (2008) Phenolic composition and biological activities of Tunisia Nigella sativa L shoots and roots. Compte Rendu deBiologies 331: 46-55

17. Tawaha K, Alali F, Gharabibeh M, Mohammad M, El-Elimat T, et al. (2007) Antioxidant activity and total phenolic content of selected Jordanian plant species. Food Chem 104: 1372-1378.

18. Mexis SF, Badeka AV, Riganakos KA, Kontominas MG (2010) Effect of active and modified atmospheres packaging on Quality Retention of dark chocolate with hazelnuts. Innovative Food Sci Emerg Tech 11: 177-186.

19. Chand P, Gopal R (2005) Nutritional and medicinal improvement of black tea by yeast fermentation. Food Chem 89: 449-453.

20. Zhao H, Dong J, Lu J, Chen J, Y Shan Lin, et al. (2006) Effect of extraction solvent mixtures on antioxidant activity evaluation and their extraction capacity and selectivity for free phenolic compounds in Barely (Hordeum vulgare L.) J Agric Food Chem 54: 7277-7286.

21. Shalini H, Sudesh J (2005) Organoleptic and nutritional evolution of wheat biscuits supplemented with untreated and treated fenugreek flour. Food Chem 90: 427-435.

33

Microbial and Physicochemical Qualities of Pasteurized Milk

Woldemariam HW[1]*and Asres AM[2]

[1]Department of Food Process Engineering, College of Biological and Chemical Engineering, Addis Ababa Science and Technology University, Addis Ababa, Ethiopia
[2]Department of Food Technology and Food Process Engineering, Faculty of Chemical and Food Engineering, Bahir Dar Institute of Technology, Bahir Dar, Ethiopia

Abstract

A study was conducted to investigate the microbial and physicochemical qualities of pasteurized milk. The result of the microbial investigation indicated that Total Plate Counts were not significant ($p < 0.05$) between the samples but Coliform Counts (3.1×10^6 cfu/ml) of sample S were significant. Physicochemical quality analysis revealed total solids and protein contents were not significant ($p < 0.05$) for all pasteurized milk samples. On the contrary, fat contents of samples M (4.9%) and S (4.75%) were significant ($p < 0.05$), total ash (0.6%) of sample S was significant ($p < 0.05$) with the control (0.8%) and lactose contents of samples H (2.07%) and S (1.14%) were also significant ($p < 0.05$) with the control (4.7%). Results of this study suggest that pasteurized milk samples tend to increased in microbial population is suggestive of unsanitary practices during processing and poor pasteurization efficiency. Besides, the variation in physicochemical compositions could be due to failure of regular standardization throughout the production period.

Keywords: Microbial quality; Pasteurized milk; Physico-chemical quality

Introduction

Milk is considered as nature's single most complete food [1]. Milk is a complex mixture of fat, protein, carbohydrates, minerals, vitamins, and other miscellaneous constituents dispersed in water, make it a complete diet [2]. The nature of milk and its chemical composition renders it one of the ideal culture media for microbial growth and multiplication [3]. The safety of dairy products with respect to food-borne diseases is a great concern around the world. This is especially true in developing countries where production of milk and various dairy products take place under rather unsanitary conditions and poor production practices [4]. Different heat and treatments are given to raw milk in order to remove pathogenic organisms, to increase the shelf life, to help subsequent processing, e.g. for warming before separation and homogenization or as an essential treatment before cheese making, yoghurt manufacture and production of evaporated and dried milk products [5].

Pasteurization, sterilization (in bottle) and UHT (ultra-high-temperature) treatment integrated with aseptic packing are two such treatments. Sterilization is the term applied to a heat treatment process which has a bactericidal effect greater than pasteurization. Although it does not result in sterility, it gives the processed milk a longer shelf life. Because of the long holding time at this elevated temperature, the product has a cooked flavour and a pronounced brown colour. Unlike sterilization, pasteurization is not intended to kill all pathogenic microorganisms in the food or liquid. Instead, pasteurization aims to reduce the number of viable pathogens so they are unlikely to cause disease. Physiochemical analysis is important tool to monitor the quality of milk and dairy products. Milk is an important source of all basic nutrients for mammals. Milk from various mammals are used for producing different dairy products including milk cream, butter, yogurt, ghee, sour milk, etc. [5,6]. Consumers always demands nutritionally enriched milk and dairy products [7,8]. According to World Health Organization (WHO) standards and other scientific works, quality milk contains 2.6% fat, 3.5% protein, 0.17% Titratable Acidity (TA), 7.71% Solids-Not-Fat (SNF) and SG 1.030, total bacterial count 1.3×10^6 cfu/ml. The pH 6.6 ensures the milk freshness at boiling point 100°C to 117°C [9-13].

The other consumers concern is that there may be adulteration of milk with different substances especially with water. When adulteration is made there is question of safety. The milk may contain hazardous chemicals that may be added which cause health risk. The most common adulterant is water [9]. When the water is added, we are paying extra money that we should not pay in addition to this the added water quality may lead to health risk. The current processes for milk collection from many subsistence farmers are time-consuming, costly and prone to adulteration. Adulteration of milk can cause the deterioration of dairy products and to ensure milk quality requires the necessity and greater emphasis on regulatory aspects with advanced methods of analysis and monitoring milk production and processing, and the new product ideas such as genetically modified foods and the nutraceuticals have set new goals for quality assurance and food safety [14]. Fresh milk considered as a complete diet because it contains the essential nutrients as lactose, fat, protein, mineral and vitamins in balanced ratio rather than the other foods [9]. Recently, consumer's health concerns are developed to the milk properties i.e., SNF, TS, acidity, and bacterial count along with protein and fat content.

Pasteurized milk is not durable for long period due to poor time-temperature management, inappropriate storage conditions, adulteration and contamination. The other main problem of the producers is to examine routinely the efficiency of pasteurization and miss labeling. Therefore, the key objective of this study was to assess different pasteurized milks in terms of microbial and physicochemical quality due to the reason that there is supply of low quality milk due to high degree of milk adulteration and poor hygiene practice during milking, transporting, distributing, processing and serving.

*Corresponding author: Woldemariam HW, Department of Food Process Engineering, College of Biological and Chemical Engineering, Addis Ababa Science and Technology University, Addis Ababa, PO Box 16417, Ethiopia
E-mail: henymarbdr@gmail.com

Materials and Methods

Sample collection

Different brands of pasteurized milk samples were collected from Bahir Dar and Addis Ababa and stored in to a refrigerator at 4°C until analysis. The study was carried out from September 2014 to June 2015.

Microbial analysis of pasteurized milk samples

Microbial analyses were carried out to investigate the quality of pasteurized milks using standard methods [15-17]. One millilitre of pasteurized milk from each sample was homogenized in 9 mL of standard Maximum Recovery Diluent before undertaking the microbial analysis (Figure 1). Maximum Recovery Diluent and media prepared for each test (except Violet Red Bile Agar (VRBA) was autoclaved for 15 min at 121°C [17]. Media used were prepared according to the directions given by the manufacturers on the packaging materials.

The number (N) of cfu/mL of a sample is calculated according to the method described by Roberts and Greenwood [18]:

$$N = \frac{C}{v(n_1 + 0.1n_2)d}$$

Where,

C is the sum of colonies on all plates counted.

v is the volume applied to each plate.

n_1 is the number of plates counted at the first dilution.

n_2 is the number of plates counted at the second dilution.

d is the dilution from which the first count was obtained.

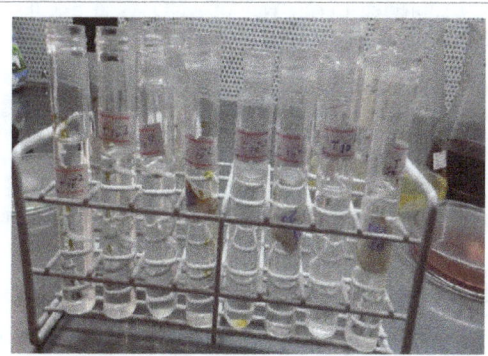

Figure 1: Preparation of dilutions.

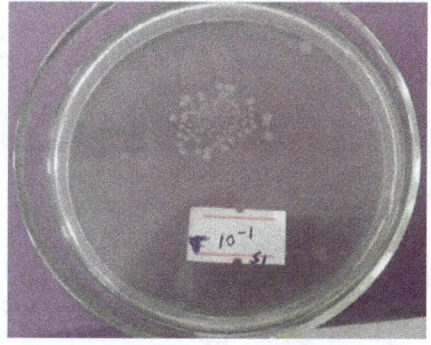

Figure 2: Total plate counts of pasteurized milk samples.

Figure 3: Coliform counts of pasteurized milk samples.

Microbial counts

The pasteurized milk samples were assessed for Total Plate Counts (TPC) and Coliform Counts (CC). Dilutions were selected so that total number of colonies on a plate was between 30 and 300 for TPC, while for CFC; dilutions were selected for plate counts between 15 and 150 [16,17].

Total plate count (TPC)

Homogenized pasteurized milk sample was serially diluted by adding 1mL into 9 mL of Maximum Recovery Diluent, until a solution is obtained that is expected to give a plate count between 30-300. One millilitre of the sample from a chosen dilution was placed on the petri dish with pour plated molten plate count agar (10-15 ml) allowed to solidify for 15 min and incubated for 48 hours at 37°C (Figure 2). Finally, the counts were made using digital colony counter. The plate counts were calculated by multiplying the count on the dish by 10n, in which n stands for the number of consecutive dilutions of the original sample [15-17].

Coliform count (CC)

One ml of milk sample serially diluted as 1:10 was transferred into sterile plates. Molten violet red bile agar (15 ml) having temperature of 45°C was added to the milk sample, mixed thoroughly, and allowed to solidify for 5-10 minutes. The mixture was then overlaid with plating agar to inhibit surface colony formation and incubated at 37°C for 24 hours (Figure 3). Counts were made using colony counter [15-17].

Physico-Chemical Analysis

Milk fat

Gerber method was used to determine the milk fat content. Milk samples were kept at 37°C for 30 minutes in a water bath. Ten millilitres of concentrated sulphuric acid was pipetted into a butyrometer. Then, 11 ml of milk was added using milk pipette into a butyrometer having the sulphuric acid, and then one millilitres of amyl alcohol was added. The butyrometer stopper was put on and the sample was shaken and inverted several times until all the milk was digested by the acid. Then, the butyrometer was placed in a water bath at 65°C for five minutes. The sample was placed in a Gerber centrifuge (Model: NOVA/3670-2631/, Germany) for four minutes at 1100 rpm. Finally, the sample was placed in to water bath for 5 minutes at 65°C and fat percentage was read from the butyrometer (Figure 4). The average of triplicate readings was computed and recorded [15,16].

Total solids

To determine the total solids, five grams of milk sample was placed

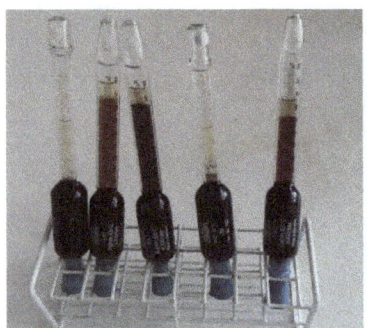

Figure 4: Fat readings of pasteurized milk samples.

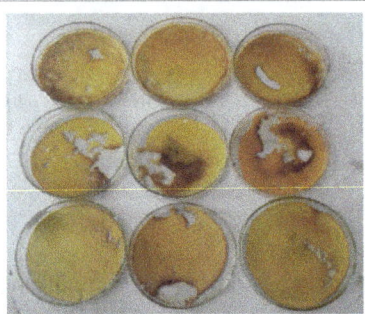

Figure 5: Total solids of pasteurized milk samples.

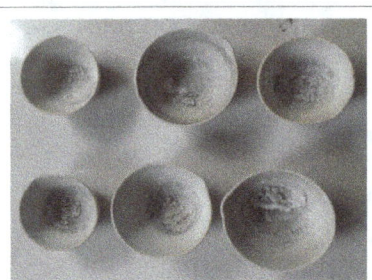

Figure 6: Ashe of pasteurized milks.

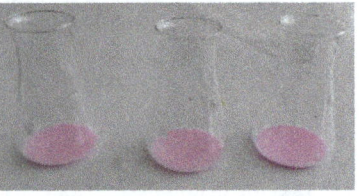

Figure 7: Total protein determination of pasteurized milks by titration method.

in a pre-weighed and dried triplicate of crucibles. The samples were kept at 102°C in a hot air oven (Model: DHG-9140, China) overnight. Then, the dried samples were taken out of the oven and placed in a desiccators (Figure 5). Then the dry sample was weighed and total solids were calculated using the following formula [19]:

$$Total\ Solids = (\frac{Crucible\ weight + oven\ dry\ weight - Crucible\ weight}{Sample\ weight}) \times 100$$

Total ash

Total ash was determined gravimetrically by igniting the dried milk samples in a muffle furnace (Model: FSL 340-0100, UK) in which the

temperature was slowly raised to 550°C. The samples were ignited until carbon (black color) disappears or until the ash residue becomes white (Figure 6) and total ash was calculated using the following formula [16]:

$$Total\ ash = (\frac{Weight\ of\ residue}{Weight\ of\ sample}) \times 100$$

Total protein

Formaldehyde titration method was used to determine the total protein content. Ten ml of milk was added into a beaker. Then, 0.5 ml of 0.5 percent phenolphthalein indicator and 0.4 ml of 0.4 percent Potassium Oxalate was added into the milk. Then, the sample was titrated with 0.1N Sodium Hydroxide solution. The titration was continued until pink color becomes intense (Figure 7). Finally, the burette reading was recorded. The reading was multiplied by a factor 1.74 and total protein was calculated using the formula below [20].

$$Total\ protein = Burette\ reading \times 1.74$$

Solids- not –fat (SNF)

The solids not fat (SNF %) was determined with the equation below by subtracting the percent fat from total solids [20].

$$SNF = (TS - fat) \times 100$$

Lactose

Percent lactose was determined by subtracting the fat, protein and total ash percentages from the percentage of the total solids [1].

$$Percent\ lactose = Percent\ total\ solids - (\%fat + \%protein + \%total\ ash)$$

Data analysis

Triplicate data collected for microbial and physicochemical qualities were subjected to one way of ANOVA at $p<0.05$ using SAS version 9 [21].

Results and Discussion

The study was aimed to assess the microbial and physicochemical qualities of different brands of pasteurized milk samples. Accordingly, the samples were tested for Total Plate Counts (TPC), Coliform Counts (CC), fat, total solids, total ash, total protein, solids-not-fat (SNF) and lactose.

Microbial analysis of milk samples

Different brands of pasteurized milk samples were analyzed for TPC and CC and results are displayed in Table 1.

Results revealed that Total Plate Counts were not significant between the samples. However, one of the samples (S) is significant at $p<0.05$ in terms of Coliform Counts (Table 1). Total plate count is the most accurate method for counting live microorganisms in raw milk and heat-treated milk [17]. Milk produced under hygienic conditions from healthy cows should not contain more than 5×10^4 bacteria per millilitre [22]. Therefore, the total bacterial counts of the collected samples were higher than acceptable standard, which could be associated with poor efficiency of pasteurization.

Coliform bacteria could contaminate milk from dung, bedding materials, polluted water used for cleaning, soil and inadequately cleaned milking utensils [17,23]. This could possibly expose the milk to high risk of contamination, which in turn increases the microbial count. Coliform count is especially associated with the level of hygiene

Samples	Microbial quality parameters	
	TPC (cfu/ml)	CC (cfu/ml)
M	2.6×10^6	<30
H	3.6×10^6	<30
S	2.1×10^6	$3.1 \times 10^{6***}$
R	<30	<30

Comparisons significant at p<0.05 are indicated by ***M, H, S and R are codes for different pasteurized milk samples.

Table 1: Microbial analysis result of milk samples.

Samples	Physico-chemical quality parameters					
	Fat (%)	Total Solids (%)	Total Ash (%)	Total Protein (%)	SNF (%)	Lactose (%)
M	2.9***	9.8	0.63	3.42	2.27	2.85
H	4.4	11.93	0.69	4.79	3.72	2.07***
S	4.75***	10.04	0.6***	4.0	4.16	1.14***
R	3.25	11.05	0.66	4.35	2.59	2.79
Ctrl	2.7	11.7	0.8***	3.5	1.9	4.7***

Comparisons significant at p<0.05 are indicated by ***. M, H, S, R and Ctrl are codes for different pasteurized milk samples.

Table 2: Physicochemical analysis result of milk samples.

during production and subsequent handling since they are mainly of faecal origin [17,24]. Coliforms do not survive pasteurization and their presence in the pasteurized milks indicates recontamination after pasteurization. If coliform count of any milk is higher than a certain level, say over ten coliform organisms per millilitre of pasteurized milk, it means the milk was produced under improper procedures [25]. The existence of coliform bacteria in high proportion is suggestive of unsanitary condition or practices during processing [16].

Physicochemical analysis of milk samples

The results of milk fat, total solids, total ash, total protein, solids-not-fat (SNF) and lactose of pasteurized milk samples are presented in Table 2. The total solids and protein contents were found to be not significant at p<0.05 for all types of pasteurized milk samples (Table 2). On the other hand, the fat content of pasteurized milk samples (M) and (S) were significant at p<0.05 with other samples and the control (label on the package by the manufacturer). Besides, the ash content of sample (S) is significant at p<0.05 only with the control (Table 2).

Lactose contents of samples (H) and (S) are significantly different at p<0.05 with the control. In addition, samples (M) and (S) are significant at p<0.05 in terms of their lactose and solid-not-fat contents with each other (Table 2).

Conclusion

The milk samples under consideration failed to maintain the standard quality of milk both microbiologically and chemically. The presence of bacterial population in processed milk indicates defect in processing plants. The actual microbial counts indicated poor microbial quality of milk samples which might be due to poor pasteurization efficiency or recontamination. The chemical compositions of the milk analyzed were not similar with that of the control except total solids content. From the study, it is possible to conclude that the producers were not properly standardizing the milk components based on their standards and routinely examining the qualities of pasteurized milks immediately before the products are released in to the market. Thus, frequent inspection of pasteurized milks by responsible bodies is vital to check whether the products meet the minimum legal standards and should monitor the overall hygienic condition surrounding the production and handling of milk. Realistic standards for pasteurized milk need to be devised and appropriate training should be given to producers in hygienic handling of milk. It is recommended therefore, that adequate sanitary measures be observed at all stages of processing to consumption of the pasteurized milk to protect the milk from spoilage which eventually be posing a serious health risk to the consumers.

Acknowledgement

The authors gratefully acknowledge the School of Research and Graduate Studies of Bahir Dar Institute of Technology for financial support and the School of Chemical and Food Engineering for supporting with laboratory facilities for the study.

References

1. O'Mahoney F (1988) Rural dairy technology-experiences in Ethiopia. ILCA Manual No 4, Dairy Technology Unit ILCA, Addis Ababa, Ethiopia.

2. Haug A, Hostmark AT, Harstad OM (2007) Bovine milk in human nutrition: A review. Lipids in Health and Disease 6: 25.

3. Soomro AH, Arain MA, Khaskheli M, Bhuto B (2002) Isolation of E. coli from raw milk and milk products in relation to public health sold under market conditions at Tandonjam. Pakistan J Nutri 1: 150-152.

4. Zelalem Y, Faye B (2006) Handling and microbial quality of raw and pasteurized cow's milk and Irgo-fermented milk collected from different shops and producers in Central Highlands of Ethiopia. Eth J Anim Prod 6: 67-82.

5. Singh HS (1993) Heat induced interactions of proteins in milk. Protein and fat globule modifications. IDF seminar Special Issue 93:191.

6. Bylund G (1995) Dairy processing handbook. Tetrapak Processing Systems, AB S 22186 Lund, Sweden.

7. Akhtar S, Zahoor T, Hashmi M (2003) Physico-chemical changes in UHT treated and whole milk powder stored at ambient temperature. Pakistan J Research Science 14: 68-82.

8. Manji B, Khakuda K (1988)The role of protein denaturation, extent of proteolysis and storage temperature on the mechanism of age-gelation in a model. J Dairy Science 71: 194-201.

9. Clare DA, Bang WS, Cartwright G, Drake MA, Coronel P, et al. (2005) Comparison of sensory microbiological, and biochemical parameters of microwave verses indirect UHT fluid skim milk during storage. J Dairy Science 88:4172- 4182.

10. Davies FL (1975) Heat resistance of bacillus species. Int J Dairy Technology 28: 69-72.

11. Wilson HK, Herreid EO, Whitney RM (1960) Ultracentrifugation Studies of milk heated to terilization temperatures. J Dairy Science 43: 165-174.

12. Westhoff DC, Dougherty SL (1981) Characterization of bacillus species isolated from spoiled ultrahigh temperature processed milk. J Dairy Science 64: 572- 577.

13. Ramsey JA, Swartzel KR (1984) Effect of UHT processing and storage conditions on rates of sedimentation and fat separation of aseptically packaged milk. J Food Science 49: 257-262.

14. Datta N, Elliot AJ, Perkin ML, Deeth HC (2002) UHT treatment of milk: comparison of direct and indirect methods of heating. Australian J Dairy Technology 57: 211-257.

15. Marth EH (1978) Standard methods for the examination of dairy products (14th edn). American Public Health Association Washington DC, USA.

16. Richardson GH (1985) Standard methods for the examination of dairy products (15th edn). American Public Health Association, Washington DC, USA.

17. Van Den Berg JCT (1988) Dairy technology in the tropic and subtropics. Center for Agricultural Publishing and Documentation (Pudoc). Wagneningen, The Netherlands.

18. Roberts D, Greenwood M (2003) Practical food microbiology (3rdedn).

19. O'Connor CB (1994) Rural dairy technology: ILCA Training manual. International Livestock Research Institute, Addis Ababa, Ethiopia 1:119.

20. Foley YJ, Buckley J, Murphy MF (1974) Commercial testing and product control in the dairy industry. University College Cork.

21. Gomez KA, Gomez AA (1984) Statistical procedures for agricultural research. John Wiley and Sons Inc, New York.

22. O'Connor CB (1993) Traditional cheese making manual. ILCA (International livestock center for Africa) Addis Ababa, Ethiopia.

23. Kalogridou-Vassiliaduo D (1991) Mastitis-related pathogen in goat milk. Small Rum Res 4: 203-212.

24. Omore A, Arimi S, Kaugethe E, McDermoh J, Staal S (2001) Assessing and managing milk-born health risk for the benfit of consumers in Kenya. Smallholder Dairy (R and D) project (SDP), Nairobi, Kenya.

25. Walstra P, Geurts TJ, Omen A, Jellema D, Van Boekel MAJS (1999) Dairy technology principles of milk properties and processes. Marcel Dekker Inc, New York.

34

The Effects of Tannic Acid on Some Properties of Cow Gelatin's Film

Dana E[1], Ardestani SS[2] and Khodabandehlo H[1]*

[1]Department of Chemical Engineering, Shahrood Branch, Islamic Azad University, Shahrood, Iran
[2]Department of Chemical Engineering, Food industry, Islamic Azad university of Science and Research unit, Tehran, Iran

Abstract

This paper mainly aimed on the effect of tannic acid's effect on chemo-physical and mechanical characteristic properties of cow's gelatin film. Through this research the cow gelatin film along with tannic acid with different viscosity for; 0, 250, 500, 1000 ppm via costing solvent methods examined and experimented mechanically and physico-chemical with National American Standard methods. The increasing mechanical pulling resistance test shows the reduction percentages of pulling properties due to the increasing viscosity of the tannic acid. The physicochemical properties of solvent such as: water absorbing, water solution capability, water steam influence tendency, oxygen attraction will surprisingly decrease when tannic acid concentration increases ($P<0.05$). In general, according to our research; in food industries, "the eatable films are used as an active packing".

Keywords: Tannic acid; Cow gelatin; Costing method; Mechanical properties

Introduction

Gelatin films, with its many potential benefits, are widely used in the pharmaceutical/medical and cosmetic products, as well as in food industries to improve the shelf life of the content by controlling water absorption, oxygen permeation and aroma loss [1-4]. Despite this unique property, gelatin film does not have ideal mechanical properties and water vapor barrier, which could limit its applications [5-7]. Interactions between gelatin and phenolic molecules, such as ferulic acid and tannin acid, have been reported to modify optical transparency [8]. Enhanced mechanical properties could also be obtained by treating gelatin with glyoxal and genipin, but high cost of material and potential toxicity limits its possible application [9-11]. In recent years, biodegradable edible films from natural biological materials are gaining importance. Among the natural materials, protein based edible films are the most appealing for their nutritional value and enhancement of the quality and safety of food products [12]. Environmental concerns and a more reasonable use of fossil resources have amounted to a growing interest for renewable and biodegradable packaging materials [13]. However, one of the main challenges of using cellulose-based packaging materials (such as cups, plates, trays and food containers) is the poor formability of cellulose [14].

Materials and Methods

Materials

Cow gelatin purchased from Canadian Sigma co. Liang Traco (the eatable glycerol) from Penang-Malaysia magnesium nitrate and P_2O_5 to control the humidity off of Sigma Aldrich (Kuala Lumpur–Malaysia). Other necessary materials provided from local lab. Cow gelatin and Fluka tannic acid purchased from Sigma Co of Germany. Eatable glycerol sorbitol, magnesium nitrate and calcium chloride the isotherm salts including sulfuric acid, lithium chloride, magnesium chloride, potassium carbonate, magnesium nitrate, potassium iodide, sodium chloride, potassium bromide, and potassium chloride purchased from Germen Marquee co.

Methods

Preparations and examinations: Picking up different weights of 0.1, 0.05, 0.025 grams of Tonic acid and reach to volume in balloon of 100. 20 gram of NaOH along with 50 cc distilled water reach to the volume. We placed the 0.1, 0.05, 0.025 grams of Tannic acid has reached to the volume, adjacent to a magnet inside the shrilled containers together we put them inside bigger dish contained a thermometer on a hot plate. By adding few drops of NaOH we reach the solution to pH 10. Once the temperature to 60 Celsius, we enter the Oxygen to the solution. Then we add 9.2 grams of Cow gelatin to the compound, after 30 minutes we add the plasticizer. Then adjust the frame and the plate properly. Now we pour 50 grams of solution of ben-miry in center part of plate to spread out in plate. After 24 hours we separate when the dried film and place in desiccator so the film gets damped evenly. The thickness of film measured with hand held microsites with the sensitivity of 0.01 millimeter. And randomly average gained from 5 points of plate according to WVP calculations in Japan –Mitutoyo-Tokyo. An examination consists of: mechanical specialty, physical properties (the capability of solution n absorbing water with different Tannic acid viscosity) evaluations.

Statistics analysis: Computerized Enova, Tokay or donken tests are used for physical or mechanical specialties tests of different kinds of films at 5%. Statistics analysis has done by SPSS17.

Results Analysis

Visual analyzing

In Recent inspection, the gelatin films turning from yellow to brown as the tannic acid concentration increases.

The effects of acid tannic on the cow's gelatin film thicknesses analysis

Experiments results show there no significant effect on different film thicknesses. However the general thickness is between 13-15 mm (Table 1).

*Corresonding author: Khodabandehlo H, Department of Chemical Engineering, Shahrood Branch, Islamic Azad University, Shahrood, Iran
E-mail: hessamkhodabandehlo@yahoo.com

Sample	thickness
reference	$(13 \pm 0.1)^a$
250PPM acid Tannic	$(14 \pm 0.1)^a$
500 PPM acid Tannic	$(13 \pm 0.1)^a$
1000 PPM acid Tannic	$(15 \pm 0.1)^a$

Letter in table (a) presenting the tolerances of 0.05%.

Table 1: Reviews cow gelatin film thickness as compared with the control sample.

Tannic acid (ppm)	Tensile Strength (MP)	Elongation at break (%)	Young's modulus (GPa)
0	33.63 ± 0.74^d	27.1 ± 4.2^a	0.395 ± 0.042^c
250	34.96 ± 0.33^c	22.1 ± 2.5^b	0.402 ± 0.061^b
500	35.33 ± 0.43^b	18.4 ± 4.1^c	0.413 ± 0.024^b
1000	37.22 ± 0.95^a	11.3 ± 3.8	0.491 ± 0.054^a

Letters in table (a,b,c,d) presenting the tolerances of 0.05%.

Table 2: Effect of different concentration of tannic acid of mechanical properties of film.

Tannic acid (ppm)	WVP × 1011 [g m⁻¹ s⁻¹ Pa⁻¹]	O. P [cm³m/(m⁻²day)]
0	8.9 ± 0.31^a	226.18 ± 7.92^a
250	7.10 ± 0.33^b	200.11 ± 874^b
500	3.98 ± 0.450^c	143.57 ± 2.25^c
1000	161 ± 0.13^d	97.23 ± 4.40^d

Letters in table (a,b,c,d) presenting the tolerances of 0.05%.

Table 3: Effect of tannic acid to permeability oxygen and water vapor on gelatin films.

Analysis and comparison mechanical properties of control film with different concentration of tannic acid

As it is shown in Table 2, the mechanical properties of cow film will increases when the concentration of Tannic Acid increases between zeros to 1000 PPM the stretch resistance of film also increases (between 33 mega Pascal to about 37 mega Pascal (p<0.05). In this research it also shows the film stretches reduce from 27 to 11 percent depending on tannic acid concentration increasing value (p<0.5). Data presenting the average standard ± tolerances Letters in table (a,b,c,d) presenting the tolerances of 0.05%.

The water steam absorbent potential (WVP)

The Water steam absorbent potential (it better to write water resistance ability) has negative effect on film life. Table 3 shows these effects. Gelatin Films with different concentration of tannic acid against water effects will improve with increasing the concentration of tannic acid. The later chemical bods created by tannic acid in compare of matrix bio composite are the cause of resistance. Therefore these bonds create the way out for water molecules.

Oxygen influence resistance on film

Influence ability in respect to gas and other volatile materials in poly saccharide are under variety of effects as- region characterizations in compare amorphous region, the level of polymeric chain sensitivity and inside chaining concentration tendency of polymers. In this research as it shows in Table 4 in corrected gelatin films, as the tannic acid concentration increases, the film tendency towards absorbing oxygen decreases. Data presenting the average standard ± tolerances letters in table (a, b, c, d) presenting the tolerances of 0.05%.

Effect of tannic acid on physiochemical property of the gelatin films

Moisture content, solubility in water and water adsorption capacity: Table 4 indicates the moisture content, solubility in water and water absorption capacity for gelatin films and samples treated with tannic acid. Data with ± shows the average tolerance. English letters (a,b,c,d) is the 5% meaningful probable average variation.

As it is specified in the table with adding tannic acid to the gelatin film, film water absorbing capacity (WAC) greatly reduces as acid increases (p<0.05). The reason is because hydroxyl group exist in gelatin film are water solvable and the presence of tannic acid will block such relation between water and the gel-film. Instead the tannic acid repels the water on bio-polymer matrix and stops to absorb it by gel. The Bio-polymer composites water Solvability can be also reduced by lipid, protein based, and Nano particles compound adding to gelatin film (Figures 1 and 2).

Evaluation of the transmission and absorption of light in the visible and ultraviolet zone from the gelatin films: In following Figures 3 and 4, the absorption and trans-passing the light value with 200 to 1100 Nanometer wave length of bio-composite film contain tannic acid with different concentration is showing. As it is shown in conclusion by adding tannic acid to the film visible light has passed through and ultra violet signal gradually blocked as the acid concentration has increased as at 1000 ppm ultra-wave length has completely blocked through.

Finding bond with FTIR method: FTIR spectrum of matrix Gelatin films in following Figure 5 and corrected matrix gelatin film with tannic acid in Figure 6, clearly shows active groups with using acid is changed and some recently appeared. This shows the transaction is entirely chemical with lateral chemical bonds and has been the element of actual cause of gelatin film's improvement.

Tannic Acid (ppm)	Damp (58%)	%Solubility	Water Absorbent (a)
0	14.88 ± 0.34^a	30.10 ± 1.23^z	9.93 ± 0.33^a
250	14.88 ± 0.34^a	28.16 ± 1.41^a	$8.78 \pm .18^b$
500	13.88 ± 0.08^b	26.85 ± 1.29^b	8.74 ± 0.11^b
1000	13.61 ± 0.23^b	22.40 ± 2.01^c	$7.32 \pm .25^c$

Letters in table (a,b,c,d) presenting the tolerances of 0.05%. (a) - (gr wet material/ gr dry material).

Table 4: Moisture content, Solubility in water and water adsorption capacity.

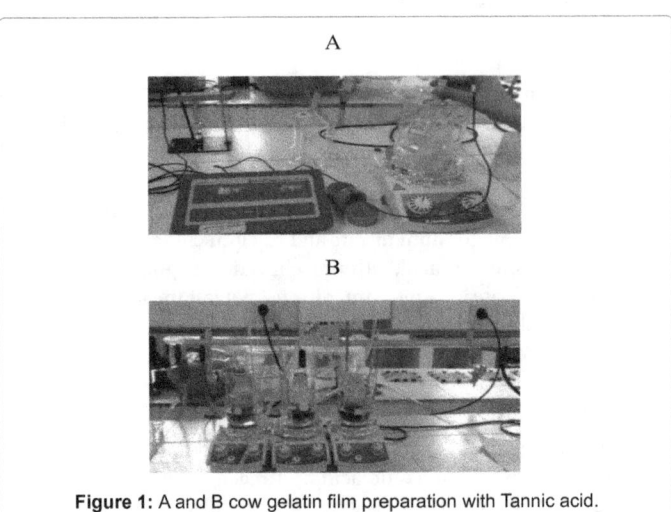

Figure 1: A and B cow gelatin film preparation with Tannic acid.

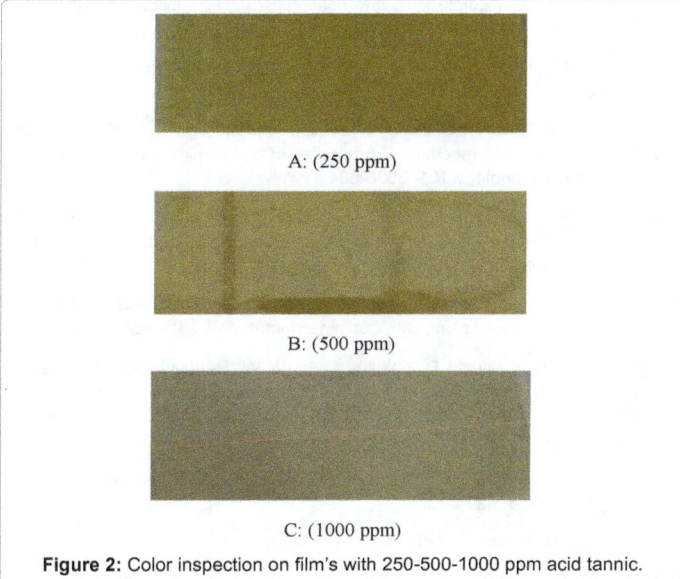

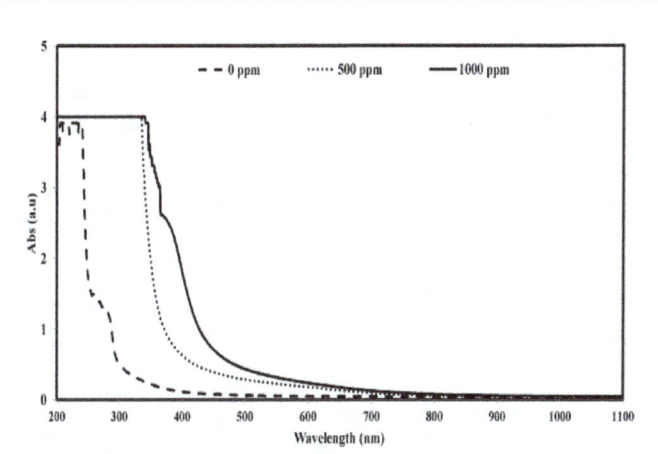

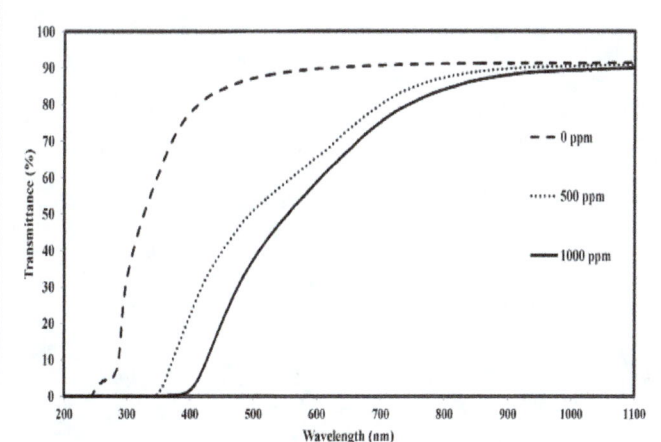

Figure 2: Color inspection on film's with 250-500-1000 ppm acid tannic.

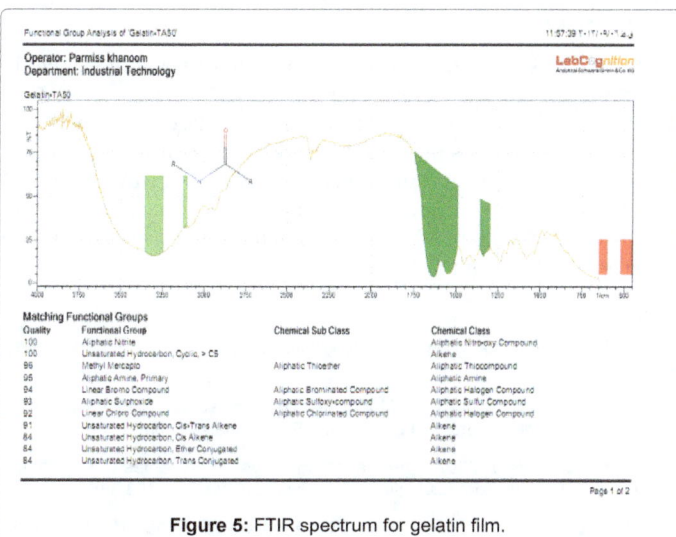

Figure 5: FTIR spectrum for gelatin film.

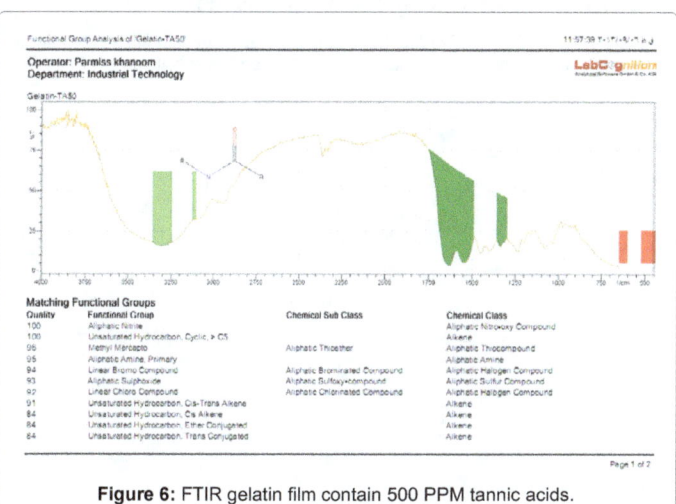

Figure 6: FTIR gelatin film contain 500 PPM tannic acids.

Figure 3: Amount of light absorption by bio-composite film that treatment with tannic acid in wavelengths of 200-1100 nm.

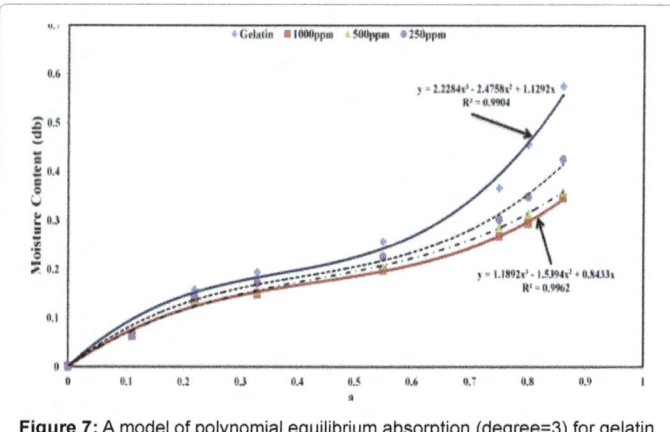

Figure 7: A model of polynomial equilibrium absorption (degree=3) for gelatin films in comparison with gelatin bio-composite contain acid tannic.

Study of polynomials models absorbing equilibrium

Figure 7 shows a modeled curve of polynomial for Gelatin film and films contain tannic acid equilibrium absorption. As it is illustrated in this model with lateral bonds and with tannic acid's help could affect to absorbing equality and pulling it down instead.

Figure 4: The Bio-composite gelatin films contained tannic acid light absorption capacity through wave lengths of 200-1100 nm.

Conclusion

The results show with increase of acid tannic concentration, Tensile strength have significant grow up from 33 to 37 Mpa ($p<0.05$). Also, found that the elongation percentage significant decrease from 27% to 11% ($p<0.05$) as well as the results showed with increase the particle, Some parameters similar, water solubility, permeability to oxygen and the rate of water absorption significant decreased ($p<0.05$). However with increase of tannic acid concentration the amount of permeability to water vapor and light passed decreased ($p<0.05$).

References

1. Achet D, He XW (1995) Determination of the renaturation level in gelatin films. Polymer 36: 787- 791.

2. Bigi A, Panzavolta S, Rubini K (2004) Relationship between triple-helix content and mechanical properties of gelatin film. Biomaterials 25: 5675-5680.

3. Cao N, Fu Y, He J (2007) Preparation and physical properties of soy protein isolate and gelatin composite film. Food Hydrocolloids 21: 1153-1162.

4. Irwandi J, Faridayanti S, Mohamed ESM, Hamzah MS, Torla HH, et al. (2009) Extraction and characterization of gelatin from different marine fish species in Malaysia. International Food Research Journal 16: 381-389.

5. Cao N, Fu Y, He J (2007) Mechanical properties of gelatin films cross-linked, respectively, by ferulic acid and tannin acid. Food Hydrocolloids 21: 575-584.

6. Bourtoom T (2008) Edible films and coatings: Characteristics and properties. International Food Research Journal 15: 237-248.

7. Bourtoom T (2009) Edible protein films: Properties enhancement. International Food Research Journal 16: 1-9.

8. Peña C, De la Caba K, Eceiza A, Ruseckaite R, Mondragon I (2010) Enhancing water repellence and mechanical properties of gelatin films by tannin addition. Bioresource Technology 101: 6836-6842.

9. Bigi A, Cojazzi G, Panzavolta S, Rubini K, Roveri N (2001) Mechanical and thermal properties of gelatin films at different degrees of glutaraldehyde cross linking. Biomaterials 22: 763-768.

10. Bigi A, Cojazzi G, Panzavolta S, Roveri N, Rubini K(2002) Stabilization of gelatin films by cross linking with genipin. Biomaterials 23: 4827-4832.

11. Spanneberg R, Osswald F, Kolesov I, Anton W, Radusch HJ, et al. (2010) Model studies on chemical and textural modifications in gelatin films by reaction with glyoxal and glycolaldehyde. Journal of Agricultural and Food Chemistry 58: 3580-3585.

12. Gennadios A, Weller CL, Hanna MA (1997) Soy protein/fatty acid films and coatings. Inform 8: 622-623.

13. Rhim J (2010) Food Sci Biotechnol 19: 243-247.

14. Vishtal E, Retulainen E (2012) Deep-drawing of paper and paperboard: The role of material properties bioresources 7: 4424-4450.

35

Studies on the Composition and *In-Vitro* Antioxidant Activities of Concentrates from Coconut Testa and Tender Coconut Water

Geetha V, Bhavana KP, Chetana R, Gopala Krishna AG and Suresh Kumar G*

Department of Traditional Foods and Sensory Science, CSIR-Central Food Technological Research Institute, Mysore, India

Abstract

Tender coconut water is known to have health beneficial effects. Antioxidants are the bioactives which helps in the amelioration of wide range of diseases like diabetes, obesity, cancer etc. In the current study two concentrates were prepared from coconut testa, as phenolic concentrate (PHE) and tender coconut water concentrate (TCW).These concentrates were evaluated for proximate composition, phenolic acids and antioxidant activities. Protein (3.7% and 5.2%), carbohydrates (56.6% and 53.5%), phenolics (3.4% and 2.6%) and flavonoids (1.9% and 1.4%) were observed in PHE and TCW respectively. Phenolic acids composition was estimated by HPLC and major phenolic acids were found to be gallic/tannic, protocateuchic and ferulic acid. Both concentrates had good reducing power ability with IC50 values of 68.4 µg (PHE) and 73.5 µg (TCW). Further, DNA protection assay evidenced the dose dependence protection for free radical induced oxidation by PHE and TCW. Hence, concentrates of PHE and TCW are useful in preventing stress induced ailments. Since the concentrates are stable it can be used in different food preparations.

Keywords: Coconut; Tender coconut water; Phenolic concentrate; Free radicals; DNA protection

Introduction

Coconut palm (*Cocos nucifera* L.) is one of the most useful tropical trees having religious and also traditional values in Asian countries. Jean et al. [1] described tender coconut water as a natural sterile liquid found in the young immature coconuts also known as liquid endosperm and is slightly translucent, it contains good amount of minerals, sugars, amino acids, proteins, vitamins and antioxidants, it is used to treat dehydration especially in diarrhea; vitamin C and riboflavin present in tender coconut water plays a major role. It reduces blood pressure; hepatoprotective in nature [2], it eliminates mineral poisoning [3]. Based on the compositional and functional properties it is also considered as sport beverage [4] therefore has drawn the attention of manufacturers and industries. However, perishability of coconut water is very high when exposed to air; it develops sour, off flavor and taste. Its natural freshness is lost within 24 h to 36 h even under cold, unless treated scientifically. Therefore an attempt was made to convert the TCW into stable concentrate.

Thin brown outer skin of coconut kernel known as coconut testa, coconut testa is removed during the processing of virgin coconut oil, preparation of desiccated coconut powder [5], therefore it is byproduct of virgin coconut oil processing industry, it contains plenty of bioactives such as phenolic acids, flavonoids that are known to have health beneficial effects and are under-utilized [5]. We made an attempt to prepare phenolic rich concentrate (PHE) from testa material.

Free radicals are reactive molecules and are short lived species having unpaired electrons. Bansal and Bilaspuri [6] described free radicals induce cell damage by attacking them resulting in oxidation of cell components and molecules. Oxygen is the element of life; however in pathological condition it generates reactive oxygen species (ROS) through various processes within the biological system that ultimately induces cell death *via* necrosis. Oxidative stress causes loss of structure and functionality of healthy cells. Pathogenesis of about more than 50 diseases has been implicated by free radicals [7]. Interventions of antioxidants limit the damage caused to DNA, proteins, and other macromolecules. Oxidation by free radicals to the tissues results in wide variety of diseases, most notably cancer and heart disease [8]. The most effective way is to prevent the damaging activity of free radicals

in the system is by consuming diets rich in polyphenols, polyphenols are found mostly in plant materials [9] these antioxidants offer some protection against development of cancers, cardiovascular diseases, obesity and diabetes [10,11]. In the present study, tender coconut water concentrate (TCW) and phenolic concentrate from coconut testa (PHE) were prepared at CFTRI, Mysore. They were analysed for proximate composition, phenolic acids and antioxidant assays. The main objective of the present work is the analysis of the developed concentrated and their nutraceutical value, which can be further used for the preparation of health beneficial foods.

Materials and Methods

Materials

Coconut testa was obtained from the local industry in Mysore, tender coconut water from local market. Standards like bradford reagent, 1,1-Diphenyl-2-picryl hydrazyl (DPPH), calf thymus DNA, gallic acid, p-coumaric, ferulic, caffeic, gentisic, protocatechuic, vanillic acids, syringic etc., from Sigma chemical co. (St. Louis Missouri, USA). HPLC column was Shimadzu C_{18} column (250 mm × 4.6 mm, 5µm). The other chemicals such as ferric chloride, hydrogen peroxide, trichloro acetic acid and solvents used were of the analytical grade purchased locally.

Preparation of concentrates

The concentrates; phenolic concentrate from mature coconut testa (PHE) and tender coconut water concentrate(TCW) from tender coconut water, based on process developed from CSIR-CFTRI, Mysore.

*Corresponding author: Suresh kumar G, Department of Traditional Foods and Sensory Science, CSIR- Central Food Technological Research Institute, Mysore 570 020, India, E-mail: sureshg@cftri.res.in

Estimation of moisture, reducing sugar, total carbohydrates and protein contents

Moisture was estimated by oven methods, a method that dries samples at 135°C for 2 h [12], reducing sugar [13] total carbohydrate [14] and proteins [15]

Estimation of total polyphenols

Total phenolics in the sample were measured calorimetrically according to Singleton and Rossi [16] an aliquot of appropriately diluted samples and standard gallic acid (0-100 µg) in methanol were taken in clean dry test tubes in triplicates. Standard solution was prepared using gallic acid (0-100 µg/mL). 2.5ml of (1:10) diluted Folin- Ciocalteu's reagent was mixed, after 5 minutes 2.0 ml of Na_2CO_3 (7.5%) was added and mixed, the samples were incubated in the dark at room temperature for 2 hours. The absorbance was taken at 765 nm. The concentration of polyphenols in the sample expressed as g/100 g of gallic acid equivalent (GAE).

Estimation of total flavonoids

Total flavonoids in the sample were measured calorimetrically according to Zhishen et al. [17], an aliquot of suitably diluted sample were taken in clean dry test tubes in triplicates. Distilled water (4 ml) was added, at different interval of time viz., at 0th min. 0.3 ml $NaNO_2$ (5%), after 5 min. 0.3 ml $AlCl_3$ (10%), at 6th min. 2 ml 1M NaOH were added to the mixture. Immediately, the reaction flask was diluted to volume of 10 ml using distilled water and mixed thoroughly. Absorbance read at 510nm using water as blank. Total flavonoids were determined using standard graph of catechin (0-100 µg/mL) and expressed as g/100g of Catechin equivalents (CE).

Antioxidant activities

Free radical scavenging effect was studied according to Lai et al. [18] using 1,1-diphenyl-2picrylhydrazyl (DPPH), the samples (5-80 µg , GAE) were mixed with 750 µM DPPH in methanol and the final volume made to 500 µl. The mixture was incubated for 15 min. at room temperature and the absorbance was read at 517 nm against sample blank. Gallic acid was used as standard. The scavenging activity was calculated using the following equation:

$$\text{Scavenging activity (\%)} = \frac{\text{Absorbance of control} - \text{absorbance of sample} \times 100}{\text{Absorbance of control}}$$

Reducing power assay

Different concentrations of the samples were mixed with 2.5 ml of phosphate buffer (0.2M) and 2.5 ml of potassium ferricyanide (1% w/v). This mixture was heated at 50°C in water bath for 20 minutes. On cooling 2.5 ml of trichloro acetic acid (10%) was added and centrifuged for 10 minutes at 6000 rpm. The upper layer (2.5 ml) of solution was mixed with equal volume of distilled water, mixed well and 0.5 ml of freshly prepared ferric chloride solution was added, absorbance was measured at 700 nm and graph prepared and compared with standard GA.

DNA damage study

Protection of calf thymus DNA from oxidation by free radicals was analyzed according to the method by Rodriguez and Akman [19] DNA from calf thymus (sigma) was oxidized using Fenton's reagent (30 µM H_2O_2, 50 µM ascorbic acid and 80 µM $FeCl_3$) in presence and absence of test sample and relative difference in the migration of oxidized DNA was observed on 1% agarose gel by electrophoresis after staining

with ethidium bromide. Gels were documented and protection were analysed based on native and oxidized means.

Analysis of phenolic acids using HPLC

The phenolic acid compositions were determined by HPLC (model LC-10 AVP Shimadzu Corp., Tokyo, Japan) coupled with UV detector connected to spherisorb C-18 reverse-phase column (250 mm × 4.6 mm, 5 µm). Following conditions were used λ 280nm, mobile phase; water: methanol: acetic acid (80:18:2), isocratic with flow rate of 1 mL/min. Standards of various phenolic acids; gallic acid (GA), protocateuchic acid (PC), caffeic acid (CA), ferulic acid (FA), syringic acid(SA), vanillic acid (VA), p-coumaric acid(pCA) and cinnamic acid (CiA) were used for the identification of phenolic acids in the samples and expressed in relative percentage.

Identification of peaks by LC- Mass spectrometry (LC-MS)

The unknown peak fraction was collected from the HPLC were subjected to LC-MS for phenolic acid identifications. The mass of the peak was identified using the instrument Q-Tof Ultima, Waters corp. UK, Alliance HPLC system was equipped with PDA detector, Waters 2996. The source used was ESI –ve, Capillary (kV): 3.5, Cone: 100 V, Source temperature was 120°C, Desolventation temperature: 350°C, Cone gas: 50 L/Hr, Desolventation gas: 500 L/Hr

Statistical analysis

All the analyses were carried out in triplicate and the average values were expressed mean ± SD. The significance of difference was calculated by Student's t-test and $p<0.05$ were considered to be significant. The correlations between antioxidant activity and phenolic contents were calculated using trial version graph pad prism software.

Results and Discussion

Proximate composition of phenolic and tender coconut water concentrates of coconut

Prepared concentrates of phenolic (PHE) and tender coconut water (TCW) subjected to proximate composition, and the results indicated as in (Table.1) showed that the moisture (22.7% and 36.4%), protein (3.7% and 5.2%), total sugar (56.6% and 53.5%), reducing sugar (31.0% and 34.3%), total polyphenols (3.4% and 2.6%) and total flavonoids (1.9% and 1.4%), PHE and TCW respectively The moisture and protein content of TCW was slightly higher than PHE. However total sugar and total phenolics were slightly less in TCW when compared with PHE concentrate.

Proximate composition of TCW showed higher protein content due to the process of concentrating,which increased the soluble protein. Phenolics/flavonoids and other free sugars are generally extracted using 70% ethanol [20] and hence phenolic extracts contain more free sugars and phenolics (small molecule antioxidants). Presence of high amount of sugar in the concentrate acts as preservative, with effective amount of phenolics, it can be used to treat stress induced ailments

Sl.no	Parameters	PHE	TCW
1	Protein (%)	03.7 ± 0.5	05.2 ± 0.8
2	Reducing sugars (%)	31.0 ± 1.2	34.3 ± 1.4
3	Total sugars (%)	56.6 ± 2.1	53.5 ± 2.1
4	Moisture (%)	22.7 ± 1.1	36.4 ± 1.8
5	Total Polyphenols (%)	03.4 ± 0.4	02.6 ± 0.2
6	Total Flavonoids (%)	01.9 ± 0.1	01.4 ± 0.1

Table 1: Proximate composition of PHE and TCW concentrates.

[21]. Thereby the concentrates may be useful in preparation of foods for the stress induced diseases.

In-vitro antioxidant assays of phenolic (PHE) and tender coconut water (TCW) concentrate

Free radical scavenging assay was performed using a colored free radicals DPPH and will be decolorized by neutralizing it. IC_{50} value is a concentration at which 50% of the free radicals will be scavenged by the test sample. The IC_{50} values was found in the order GA < PHE < TCW (Figure 1A) indicates PHE has the scavenging ability at 68.4 µg/mL which is lesser than that of TCW 73.5 µg/mL and hence PHE concentrate has a better antioxidant activity.

Reducing power assay shows the reduction of free radicals by antioxidants, in this method, antioxidant compound forms a coloured complex with potassium ferricyanide, trichloro acetic acid and ferric chloride, which is measured at 700 nm. Increase in absorbance of the reaction mixture indicates the reducing power of the samples [22]. PHE concentrate showed dose dependent reducing ability which is better than the TCW concentrate (Figure 1B). Gallic acid (GA) is well known phenolics having good antioxidants ability and has highest reducing power ability when compared to the test samples.

DNA protection assay wherein DNA was oxidized using fenton's reagent containing different source of free radicals along with and without test samples as presented in the gel images of PHE (Figure 2a) and TCW (Figure 2b).

Free radicals are continuously produced in our body and play a major role in the pathogenesis and tissue damage in many clinical disorders [23-25] they are generated by various processes in biological system which are also the by-products of normal metabolism. Free radicals like Reactive oxygen species (ROS), Superoxide(O_2^-), hydrogen peroxide (H_2O_2), and hydroxyl radicals (OH-) are highly reactive and attack the essential biomolecules, destabilizes them, which further results in the collapse of the cell [26]. Ability of TCW and PHE concentrate in bringing down the free radicals to some extent is very important which was shown by our prepared concentrates, there are no studies showing the radical scavenging activity of coconut concentrates. Mantena et al. [27]; Leong and Shui [28] studied antioxidant activities in tender coconut water. There are many methods to test antioxidant properties, DPPH is the most commonly used assay for the determination of IC_{50} values (50% of free radicals scavenging concentration) recently, DNA protection assay is gaining importance. The use of more than one

method is recommended for quantifying the antioxidant activity of any samples [1] such *in-vitro* studies would be cost effective and can be screened quickly for the potency of the samples.

Reducing properties of the active components present in the samples is important, even at lower concentration the extracts should be effective. Though the biological system has the mechanisms to eliminate free radicals by either natural enzymatic and non-enzymatic antioxidant defense system, external antioxidants are useful in minimizing the stress induced abnormalities. Vishakh et al. [29] have shown the beneficial effects of phytochemicals present in tender coconut water and also studied their antioxidants activities *invitro*. Tender coconut water has been tested in CCL_4 induced liver damage in rats, which showed that antioxidant enzymes activities were ameliorated [2]. It also reverses the blood pressure, improves insulin sensitivity by ameliorating antioxidant activity [30,31]. TCW reduces the nicotine and its metabolite cotinine cytotoxic effect on spermatozoa DNA damage [32]. Anurag and Rajamohan [33] reported the beneficial effects of TCW with several factors viz. potassium, calcium, magnesium, L-arginine.

Phenolic compositions of concentrates of PHE and TCW by HPLC and LC-MS

HPLC analysis for the PHE and TCW were carried out and the chromatogram presented as in Figure 3. Standard phenolic acids mixture was resolved as shown in the figure followed by the samples PHE and TCW. The phenolics acids such as GA, PC, CA, and FA, SA, VA, pCA and CiN were identified in the fractions. The PC, GA, SA and FA are the major in PHE whereas GA, FA and PC in TCW (Table 2). The peak unidentified (U) in PHE was fractionated and subjected to Mass spectroscopy as shown in Figure 3 fragmentation was at 198.32 whose mass pattern was similar to standard synergic acid (197.4).

Prepared TCW and PHE concentrate from coconut have GA/ TA, pCA and FA. Phenolics and flavonoids are present in the kernel part of the nuts [34]. Since our concentrates were kernel origin therefore showed good antioxidant activity by scavenging free radicals. FA

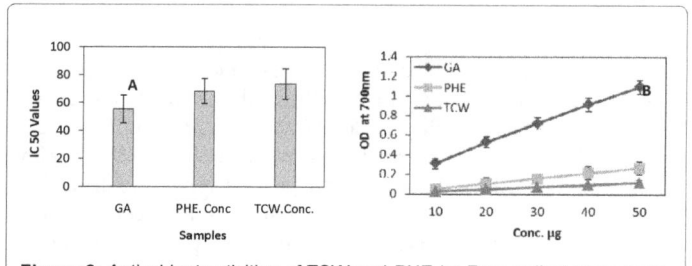

Figure 2: Antioxidant activities of TCW and PHE by Free radical scavenging activity. (A) DPPH (B) Reducing power assay.

Peak number	Phenolic acids	PHE Conc. (µg/mg)	TCW Conc. (µg/mg)
1	Gallic acid	33.05 ± 0.13	38.94 ± 1.34
2	Protocateuchic acid	41.43 ± 1.23	15.59 ± 0.45
3	Caffeic acid	06.34 ± 0.03	04.20 ± 0.05
4	Ferulic acid	19.83 ± 0.06	24.35 ± 1.57
6	Vanillic acid	nd	02.65 ± 0.03
7	p-coumaric acid	18.80 ± 0.48	05.80 ± 0.06
8	Syringic acid	28.25 ± 0.53(U)	nd
10	Cinnamic acid	10.80 ± 0.21	nd
*nd: not defined			

Table 2: Phenolic acid composition of phenolic concentrate and tender water concentrate by HPLC.

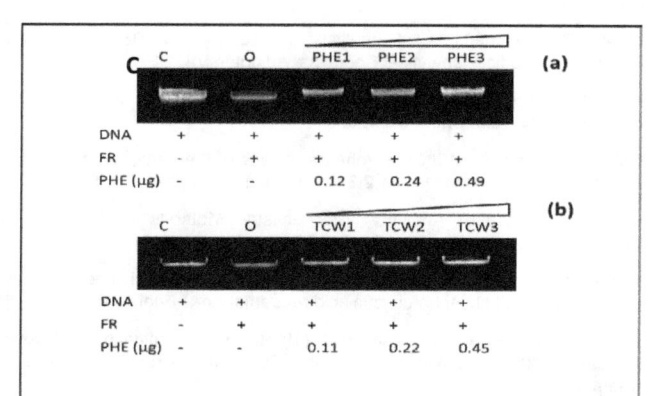

Figure 1: HPLC chromatogram of phenolic concentrate and tender water concentrate a) Standard phenolics, b) Testa phenolic concentrate, c) Tender coconut water concentrate.

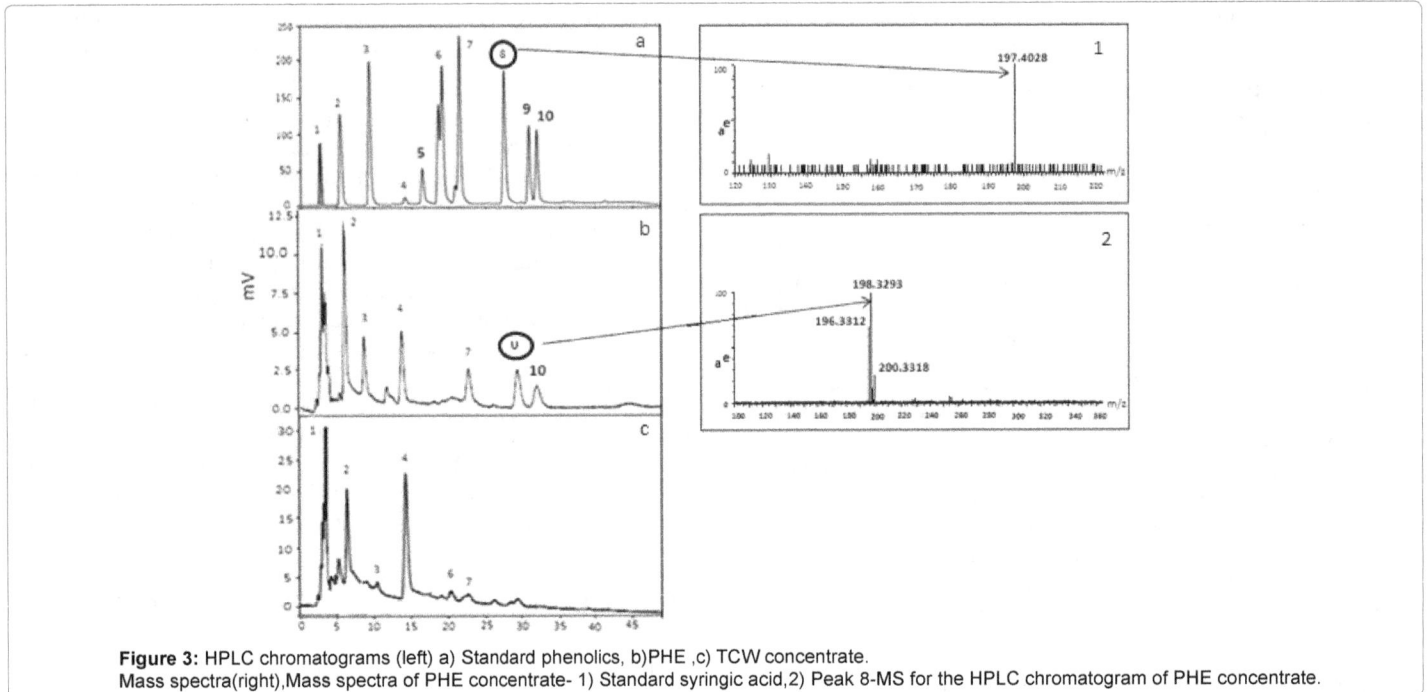

Figure 3: HPLC chromatograms (left) a) Standard phenolics, b)PHE ,c) TCW concentrate.
Mass spectra(right),Mass spectra of PHE concentrate- 1) Standard syringic acid,2) Peak 8-MS for the HPLC chromatogram of PHE concentrate.

(Ferulic acid), a well-known compound, reduces various metabolic disorders including obesity and diabetes and is available in both free and bound forms [35,36]. The availability of FA in TCW along with L-arginine amino acid is important [37], which has health beneficial effects apart from other bioactives. Busarakorn, et al. [38] have shown the presence and antioxidant activity in TCW *in- vitro*. The present study of the prepared concentrates (PHE and TCW) for the phenolic content and its potentiality as antioxidants is useful, which confirmed the scientific background for development of food product from the concentrates.

Conclusion

Prepared tender coconut water (TCW) and phenolic concentrates (PHE) retained most of bioactives, had free radical scavenging and DNA protecting properties, which confirms the antioxidant property. Phenolic acids such as gallic acid, ferulic acid, Protocatechuic acid etc., were identified in the concentrates and they are responsible for the antioxidant properties. Hence TCW can be used as ready to drink food product having natural health beneficial nutrients. PHE concentrate is rich in phenolics, can be blended with any food products for health beneficial effect.

Acknowledgement

Thanks to the Director, CSIR-Central Food Technological Research Institute, Mysore. The authors also thank the Coconut Developing Board, Kochi, Kerala for the finance assistance.

References

1. Jean WHY, Liya G, Yan FN, Swee NT (2009) The chemical composition and biological properties of coconut (*Cocos nucifera* L.) water molecules 14: 5144-5164.

2. Loki AL, Rajamohan T (2003) Hepatoprotective and antioxidant effect of tender coconut water on carbon tetra chloride induced liver injury in rats. Indian J Biochemistry and Biophysics 40: 354-357.

3. Osim EE, Dikko H (1990) The effect of coconut water (*Cocos nucifera*) on phenobarbitone induced hypnosis in the rat. Indian Coconut Journal 21: 2-4.

4. Martina C, Giovanna F, Isabella E, Eugenio A, Emanuela B, et al. (2015) High Pressure carbon dioxide pasteurization of coconut water: A sport drink with high nutritional and sensory quality. J Food Engineering 145: 73-81.

5. Prakruthi A, Sunil L, Prasanth KPK, Gopala Krishna AG (2014) Composition of coconut testa, coconut kernel and its oil. J American Oil Chemists' Society 91: 917-924.

6. Bansal AK, Bilaspuri GS (2010) Impacts of oxidative stress and antioxidants on semen functions. Vet Med Int 7.

7. Tran B, Oliver S, Rosa J, Galassetti P (2012) Aspects of inflammation and oxidative stress in pediatric obesity and type-1 diabetes: an overview of ten years of studies. Exp Diabetes Res 7.

8.   Halliwell B (1994) Free radicals antioxidants and human disease: curiosity cause or consequence? Lancet 344: 721-724.

9. Jin, D, Russell JM (2010) Plant phenolics: extraction analysis and their antioxidant and anticancer Properties. Molecules 15: 7313-7352.

10. Robert BR (2003) Flavonoids in health and disease. Rice-Evans CA, Packer L. (eds.) American society for Clinical Nutrition 458.

11. Shahidi F, Wanasundara PK (1992) Phenolic antioxidants. Crit Rev Food Sci Nutr 32: 67.

12. (2005) Official methods of analysis. AOAC method 930.15 (18thedn) Association of official analytical Chemists. Arlington VA, USA.

13. Miller GL (1959) Use of dinitrosalicylic acid reagent for determination of reducing sugar. Anal Chem 31: 426.

14. Mckelvy JF, Lee YC (1969) Microheterogeneity of the carbohydrate group of *Aspergillus oryzae* A-amylase 1,2,3. Arch Biochem Biophys 132: 99-110.

15. Bradford MM (1976) Analytical Biochemistry: Methods in the Biological Sciences today! Anal Biochem 72: 248-254.

16. Singleton VL, Rossi JA Jr (1965) Colorimetry of total phenolics with phosphomolybdic phosphotungstic acid reagents. Am J Enol Vitic 16: 144-158.

17. Zhishen J, Mengcheng T, Jianming W (1999) The determination of flavonoid contents in mulberry and their scavenging effects on superoxide radicals. Food Chem 64: 555-559.

18. Lai LS, Chous ST, Chao WW (2001) Studies on the antioxidant activities of Hsian-tsao (*Mesona procumbens Hemsl*) leaf gum. J Agri Food Chem 49: 963-968.

19. Rodriguez H, Akman S (1998) Mapping oxidative DNA damage at nucleotide

level. Free Radical Research 29: 499-510.

20. Ayumi H, Masatsune M, Seiichi H (1999) Analysis of free and bound phenolics in rice. Food Sci Technol Res 5: 74-79.

21. Pandey KB, Rizvi SI (2009) Plant polyphenols as dietary antioxidants in human health and disease. Oxid Med Cell Longev 2: 270-278.

22. Suresh KG, Harish N, Shylaja MD, Salimath PV (2006) Free and bound phenolic antioxidants in amla (*Emblica officinalis*) and turmeric (*Curcuma longa*). J Food Composition and Analysis 19: 446-452.

23. Halliwell B, Gutteridge GMC, Cross CE (1992) Free radicals antioxidants and human disease: Where are we now? J Laboratory and Clinical Investigation 119:589-620.

24. Levy Y, Bartha P, Ben AA, Brook JG, Dankner G, et al. (1998) Plasma antioxidants and lipid peroxidation in acute myocardial infarction and thrombolysis. J Am Coll Nutr 17: 337-41.

25. Slater TF, Cheesman KH, Davies MJ, Proudfoot K, Xin W (1987) Free radical mechanisms in relation to tissue injury. Proceedings of Nutritional Society 46: 1-12.

26. Satish BN, Dilipkumar P (2015) Free radicals natural antioxidants and their reaction mechanisms. 5: 27986-28006.

27. Mantena SK, Jagadis Badduri SR, Siripurapu KB, Unnikrishnan MK (2003) Invitro evaluation of antioxidant properties of *Cocos nucifera* Linn. water. Nahrung 47: 126-131.

28. Leong LP, Shui G (2002) An investigation of antioxidant capacity of fruits in Singapore markets. Food Chemistry 76: 69-75.

29. Vishakh R, Shailaja MS, Suchetha KN (2014) Evaluation of antioxidant property of West Coast tall variety tender coconut water and synthetic Trans Zeatin - An in-vitro study. J Pharmacognosy and Phytochemistry 3: 155-159.

30. Preetha PP, Devi VG, Rajamohan T (2012) Hypoglycemic and antioxidant potential of coconut water in experimental diabetes. Food Funct 3: 753-757.

31. Bhagya D, Prema L, Rajamohan T (2012) Therapeutic effects of tender coconut water on oxidative stress in fructose fed Insulin resistant hypertensive rats. Asian Pacific J Trop Med 5: 270-276.

32. Seema P, Swathy SS, Indira M (2007) Protective effect of selenium on nicotine-induced testicular toxicity in rats. Biological Trace Element Research 120: 212-218.

33. Anurag P, Rajamohan T (2003) Beneficial effects of tender coconut water against isoproterenol induced toxicity on heart mitochondrial activities in rats. Indian J Biochemistry and Biophysics 40: 278-280.

34. Nevin KG, Rajamohan T (2006) Virgin coconut oil supplemented diet increases the antioxidant status in rats. Food Chemistry 99: 260-266.

35. Choi ER, Kim BH, Naowaboot J, Lee MY, Hyun MR, et al. (2011) Effects of ferulic acid on diabetic nephropathy in a rat model of type-2 diabetes. Exp Mol Med 43: 676-83.

36. Hüseyin B (2015) Ferulic acid in cereals - A review. Czech J Food Sci 33: 1-7.

37. Sandhya VGN, Rajamohan T (2014) The role of coconut water on nicotine-induced reproductive dysfunction in experimental male rat model. Food Nutri Sci 5: 1121-1130.

38. Busarakorn M, Intira K, Pramote K, Prasong S, Marcus N, et al. (2015) Phenolic compounds antioxidant activity and medium chain fatty acids profiles of coconut water and meat at different maturity stages. Int J Food Prop 14: 1532-2386.

36

Study of HACCP Implementation in Milk Processing Plant at Khyber Agro Pvt. Ltd in Jammu & Kashmir

Tabeen Jan*, Yadav KC and Sujit Borude

Department of Food Process Engineering, Sam Higginbottom Institute of Science and Technology, Allahabad, UP, India

Abstract

The goal of the study was to set up a HACCP plan for milk processing industry in Pulwama Jammu & Kashmir to abolish and diminish the hazards for safe and sound milk and cheese production and to appraise the degree of conformity to food safety and to investigate the actual intricacy that occurs during the HACCP implementation process. Hazard Analysis and Critical Control Point System (HACCP) has been indicated as an effective and rational means of assuring food safety from principal production to final consumption, it is appreciated as a worldwide systematic and defensive tactic to address biological, chemical and physical hazards through deterrence and anticipation instead of end-product testing and inspection. The study was based on actual conditions in the milk processing plant, the seven principles of HACCP and several existing standard models of HACCP were practically applied using qualitative approach to eliminate the hazards and to guarantee safe dairy products as HACCP can enhance the responsibility and degree of control for hazards for ensuring food security at food industry level. CCPs were identified in the milk and cheese production using the decision tree the most important identified CCPs were pasteurization temperature, working of UV light, cold storage temperature, and metal detector. The prerequisite program was provided to deal with hazards so as to reduce the number of CCPs before the production to simplify the HACCP plan.

Keywords: CCPs; Food safety; HACCP; Milk processing

Introduction

The food processing industry in India is one of the leading industrial sectors in terms of production, consumption, export and growth prospects. Significant sub-sectors in the food processing industries are fruit and vegetable processing, fish processing industries, milk, meat and poultry manufacturing, packaged /convenient food, alcoholic beverages and aerated drink and grain milling and processing industrial sectors. As per national dairy development board of India the annual milk production status in Jammu & Kashmir state in the year 2014-2015 was 1951000 tones. Milk and milk products are highly nutritious but are mostly prone to health risks due to microbial contamination. Availability of standard Hygienic Milk is a subject of matter in many parts of the country counting the Kashmir valley. Unhealthy practices in dairy farm units, at milk reception centers, processing lines and during post processing handling are allied with a potential health risk to consumers due to the presence of pathogens in the milk and due to environmental contamination. The microbial contamination of milk and Milk products should not surpass the quantity that could badly have effect on shelf life and, if it does, it makes the milk insalubrious and hence unhealthy for human consumption [1] Food safety in dairy industry is a technical discipline depicting milk acquirement, processing, handling, storage and marketing of milk and milk products in manner that prevent food borne illness. This is mainly due to unhygienic conditions and inappropriate processing, handling and storage conditions, the human resources in the industry are poorly educated, unlicensed, and inexpert in food hygiene and they work under unsophisticated and unsanitary conditions with little or no knowledge about the reasons and roots of food borne diseases If HACCP is applied to milk marketing it should consider the advancement to milk safety at all phases of the chain. Hazard analysis and critical control point system (HACCP) has been internationally acknowledged and accepted as the system for the effective food safety management [2]. The main hazards in dairy industry are Microbiological as studied by Tranter [3]. There is always an ever increasing consumer demand for safe and high quality foods of protracted life. It is important to develop a food safety policy and plan for the implementation of HACCP because most of dairy foods are sensitive have less shelf life and are prone to foods borne diseases due to destitute handling and manufacturing practices. The purpose of the study was to design a HACCP plan to control hazards and to ensure safe and secure production of milk and milk products the implementation of HACCP system can provide safe food to the consumer and can improve the quality, safety and customer confidence. As the milk and milk products are consumed by all age groups, infants, adolescents, old aged and even by the immune suppressed ones so a necessity of implementing Food safety programs like HACCP should be made mandatory for safety of the one's consuming the milk and milk products. There is a need to endorse and enhance implementation of hazard analysis critical control point system and consumer food safety education efforts at all stages of milk and cheese production and marketing chains (Figure 1).

Methods

The research work on 'Study of HACCP implementation in milk processing plant" was conducted in Khyber Agro farms Private ltd. Lethpora, Pulwama, Jammu and Kashmir during December 2015 to May 2016, the methodology used for this research is mentioned.

Implementation of HACCP in milk and cheese processing plant

The purpose of this study was to design a HACCP system in milk processing plant. This study is based on qualitative approach rather than quantitative approach and based on HACCP checklist and CCP decision tree given by FAO. The HACCP system was implemented on the twelve steps given by codex alimentarious commission mentioned as the following.

*Corresponding author: Tabeen Jan, Department of Food Process Engineering, Sam Higginbottom Institute of Science and Technology, Allahabad, U.P., India
E-mail: tabeen.tw@gmail.com

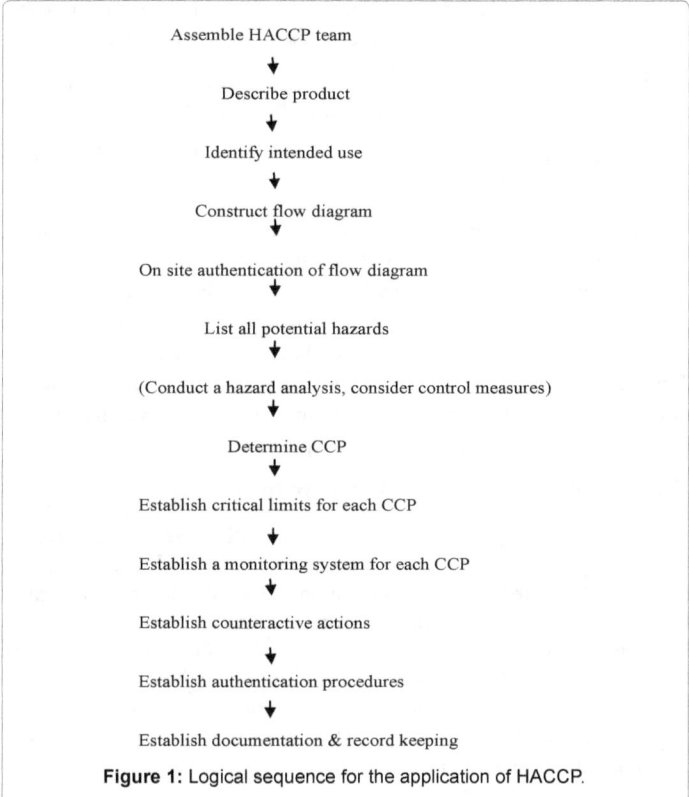

Assemble HACCP team

Describe product

Identify intended use

Construct flow diagram

On site authentication of flow diagram

List all potential hazards

(Conduct a hazard analysis, consider control measures)

Determine CCP

Establish critical limits for each CCP

Establish a monitoring system for each CCP

Establish counteractive actions

Establish authentication procedures

Establish documentation & record keeping

Figure 1: Logical sequence for the application of HACCP.

Assemble the HACCP team: The first task in the application of HACCP in the milk processing plant was to create a team having the knowledge and expertise to develop a HACCP plan. The team formed was multidisciplinary and included plant personals from production/sanitation, quality assurance, microbiologist, engineering and inspection all these experts were internal. Highly effective HACCP teams had well defined role clarity and ensured proper depiction of the team.

Describe the product: The HACCP team made a complete description of the product on the basis of ingredients/processing method /packaging materials/etc. used in the product Preparation to help out in the identification of all probable hazards associated with the product this is shown in Table 1.

Identification of intended use: The intended use of the product refers to its normal use by end-users or consumers, it was identified on the basis of normal use of the product by the consumers, including infants, elderly people immune suppressed ones the products i.e., Processed milk was used by infants as baby foods, health drinks, beverages like tea, coffee, etc., it was obligatory to be placed in ice-cold conditions or boiled, and cheese (coagulated product) used in pizzas, instantly ready to consume foods, customary and convectional foods with shelf life of 1 week and could be consumed with no risk.

Construct a flow diagram: It was easier to identify paths of potential contamination, to suggest method of control so a flow diagram of whole process was made. The function of a flow diagram was to provide a clear, simple outline of the paths involved in the process. The range of the flow diagram covered all the steps in the process which were openly under the charge of the company. The flow diagram of the products was developed for the specific products i.e., milk and cheese on the possible parts of sequence and their processing steps.

On-site confirmation of flow diagram: The HACCP team leader after the formation of a flow diagram had to verify the flow diagram on-site for accuracy and completeness. The HACCP team leader along with the other HACCP team members scrutinized the flow diagram constructed to confirm with the authentic operations it represents on site. The various changes, proceedings, or activities that would entail an on-site confirmation included, shifting of raw material and ingredients through processing pathways and apparatus, redeployment of equipment. New ingredient used or product developed, movement of product from one line to next line or equipment, packaging conditions storage etc.

List all the potential hazards conduct a hazard analysis: Hazard analysis is the most important aspect of HACCP plan to ensure the safety of product during and after processing and to improve the product shelf life and make it safe to consume. Hazard analysis was conducted by the HACCP team on the basis of HACCP checklist (as per FAO) and all the feasible hazards correlated with unprocessed material, ingredients, process operations, post process operation were identified, and marked as Biological (B), Chemical (C), Physical (P). Hazard identification is beneficial to identify potential biological, chemical and physical hazards that may arise during each step of processing.

Determine CCPs: The identification of CCPs is the most important aspect of the HACCP plan the CCPs were detected on the basis of decision tree given by Codex mentioned in the appendices, it requires a logical reasoning approach. The application of decision tree was flexible according to the type of operation i.e., production, processing, storage, and distribution or other.

Establish critical limit for each CCP: For each CCP identified a critical limit was established and specified. A critical limit represented the boundaries that were used to judge whether an operation was producing safe products. Critical limits were set for factors such as temperature, time, product measurements, water activity (ah), humidity level, etc. if these parameters are maintained within boundaries then the safety of the product will be confirmed.

Establish monitoring procedures: Monitoring is the process that the producer depends upon to show that the HACCP plan is being followed. It provides the manufacturer with accurate reports enabling the producer to prove that the conditions of production are in compliance with the HACCP plan Mostly time-temperature treatments (thermograph), pH, moisture level, equipment's and proper processing techniques were monitored by the HACCP team these activities were monitored on weekly and monthly basis. Monitoring procedures performed during operation were recorded in documents for future information and allowed taking action in the event of loss of control or for a process adjustment to be made if there is a tendency towards a loss of control.

Product name	Processed milk	Cheese
Product characteristics	Weight of packet 500 ml, 1litre	500gm
	Water -86.6 %	42.50%
	Fat- 4.5 %	40%
	SNF-8.5%	-
	Protein- 3.4 %	21-23%
	Ash -0.7%	3%
Packaging	LDPE poly packs	Food grade AL. coated PE
Shelf life	7 days	14 days
Labelling instruction	Store in refrigerated condition <4°C	<4°C
Distribution conditions	<5°C	<5°C

Table 1: Description of the products.

Establish corrective action: For each CCP identified in the process a specified critical limit was set. When monitoring activities identified a deviation associated to a CCP, corrective actions were completed to bring the process back into control.

Establish verification procedure: Verification refers to the diligence of methods, procedures, checks, and other appraisal, in addition to scrutinizing to determine compliance with the HACCP plan. Authentication ensures that adequate contingency procedure plans are in place when critical limits are exceeded, the verification was done by quality analyst and supervisor of food safety team on daily weekly and monthly basis.

Establish documentation and record keeping: Records are essential for assessment of the adequacy of the HACCP plan and the adherence of the HACCP system to HACCP plan. Records were maintained for the whole HACCP plan and included processing charts, written records, computerized records, records generated by the HACCP system microbial and analytical testing records verification and validation records. These records were well maintained in record books of the industry.

Results and Discussion

In this study the HACCP system for milk and cheese processing was developed step by step on the basis of twelve steps mentioned in the materials and methods as per Code Alimentarious commission (CAC) at Khyber Agro farms. The pre-requisite program was provided to minimize hazards to simplify the HACCP plan. Based on industrial standards and Govt. regulations the hazard identification, critical limits monitoring and validation corrective actions were performed The CCPs were identified by the HACCP team by logically answering questions given in the decision trees. The decision tree technique was put into practice to decide the CCPs as this method is visual easy to understand substitute to numerical charts and statistical probabilities used in other decisions For the hazards identified control measures were recommended and for critical control points identified proper monitoring procedures and corrective actions were put forward.

Pre-requisites

This first step was fetching in all existing pre-requisite programs under the cover of HACCP program and giving them a widespread route of achieving zero defects within the end product so as to guarantee that are no health concerns within the final product. Some precondition programs created the basis of the HACCP model for ensuring a strong system of checks hostile to possible failures of critical control points. The pre-requisite programs used in the milk processing are the sanitation programs to maintain sanitary condition in the building, premise equipment's, to maintain a clean and hygienic environment, essential for the production of highest quality and safe food products, Good manufacturing practices in the industry to control hazards The Pest Control Program was designed to allow no pests in the plant. This included rodents, insects and birds The Product Recall Program was developed to protect customers from the probable events of product safety malfunction by eradicating all wary products from the distribution channels in the minimum time, once a product recall or withdrawal is defensible and commenced for the product The aim of the Chemical Control Management Program was to reduce the possibility of chemical contamination of ingredients contact surfaces and finished products, as well as shielding the work area and the workforce from revelation to hazardous chemicals.

Possibility of hazard occurrence in milk and cheese processing and suggestive measures to control hazards as per HACCP

Hazards identified in the milk and cheese processing steps and corrective measures suggested are shown in Tables 2 and 3.

To determine the critical control points (CCPs), their critical limits and monitoring frequency

The critical control points in milk and cheese processing steps were determined on the basis of CCP decision tree. The CCPs identified are mentioned in Tables 4 and 5. For each CCP critical limits were identified and monitoring procedures and frequency was decided in case of any failure at CCP corrective actions can be taken which include Temperature control checks, maintaining proper pH, aw control, microbial testing of end product, equipment calibration when failure of CCP occurs owing to improper working of equipment.

Constraints in adopting HACCP by the industry

There are many constraints in implementing HACCP system by an industry which include the Need of awareness and responsiveness of HACCP, No apparent reimbursement, lack of industrial personnel training, lack of management commitment, unevenness of production lines and individuality of each product, lack of Government support in implementation of food safety management programs like HACCP, lack of Technical expertise and inadequate personnel and broad-scale improvement and advancement of the plant required before HACCP could be set in the industry.

Processing steps	Hazard			Control measure
	Microbial	**Physical**	**Chemical**	
Receiving of milk	Unhygienic contacts, Salmonella, Staphylococcus	Extraneous matter	Starch	Implementation of GMP OPRP, Effective filtering
Cooling (OPRP)	Unhygienic contacts	Extraneous matter	Not usually	Effective cleaning of cooling tanks, implementation of GMP
R.O treatment	Faecal, coli forms	Heavy metals, calcium Hardness of water	Not usually	Filter changing and effective cleaning membrane filtration,
Standardization	Unhygienic contacts	Extraneous matter	Not usually	Effective cleaning
Pasteurization CCP B1	Unhygienic contacts, Salmonella, Staphylococcus, Staphylococcus	Extraneous matter	Not usually	Implementation of GMP, proper pasteurization
Poly packing (UV)	NON working of UV light	Extraneous matter	Not usually	Monitoring of UV light
Cold storage	Unhygienic contacts	Extraneous matter	Not usually	Effective cleaning, pest control
Crate washing (OPRP)	Unwashed can lead to microbial hazards	Extraneous matter	Not usually	Effective crate washing with tested water
Dispatch	Not usually	Extraneous materials	Not usually	Effective cleaning and maintenance of hygienic conditions

Table 2: Hazard analysis of milk processing steps.

Processing steps	Hazard			Control measure
	Microbial	Physical	Chemical	
Receiving of milk	Unhygienic contacts	Extraneous matter	None	Implementation of GMP
Standardization	Unhygienic contacts	Extraneous matter	None	Implementation of GMP
Pasteurization	Unhygienic contacts improper pasteurization	Not usually	None	Proper pasteurization
Cooling	Unhygienic contacts	None	None	Implementation of GMP
Coagulation by 1% citric acid	Unhygienic contacts	None	Adulterants	Proper verification of citric acid quality from a certified buyer
Draining of whey	Unhygienic contacts	Extraneous matter	Not usually	Proper cleaning and GMP implementation
Milk solids filled in blocks and pressed by weight	Unhygienic contacts	Extraneous matter	Not usually	Implementation of GMP
Cutting into pieces	Not usually	Metal dust	Not usually	Implementation of GMP and use of metal detector
Dipped in chilled water 4°C	Unhygienic contacts	Extraneous matter	Not usually	Implementation of GMP
Draining of water	Unhygienic contacts	Not usually	Not usually	Implementation of GMP
Packed into desired weight and stored at 4°C(OPRP)	Unhygienic contacts	Pests	Not usually	Implementation of GMP, and packing under UV light
Dispatch in chilled conditions (OPRP)	Unhygienic contacts	Extraneous materials	Not usually	Strict Control of sanitary conditions during dispatching to retailers

Table 3: Hazard analysis of cheese processing steps.

Product	Hazard	Ccp	Critical limit	Monitoring	Corrective action	Verification
Milk	Microbial	CCP B1 Pasteurization temperature	76-80°C for 15 sec	Pasteurized milk& pasteurization Temperature	Sent for re-pasteurization, effective monitoring, study thermographs	Proper temperature control and working of pasteurizer
	Microbial	CCP B2 Working of UV light	Non-Working of UV light	Working of UV light	Re-processing of whole lot and repair of UV light(preventive maintenance)	Microbial load & working of UV tubes during packing
	Microbial,	CCP B3	>4°C	Cold storage temperature and hygiene	Effective temperature control, Cold storage structure to be modified to maintain proper temperature	Cold storage temperature

Table 4: HACCP plan for pasteurized milk.

Product	Hazard	Ccp	Critical limit	Monitoring	Corrective action	Verification
Cheese	Microbial	CCP B1 pasteurization of milk for cheese	76°C -80°C for 15 sec	Pasteurized milk & Pasteurization temperature	Sent for re-pasteurization, effective monitoring, study thermographs	Proper temperature of pasteurization by Lab testing & studying thermographs
	Physical (metal pieces)	CCP P1 Cutting of cheese (metal detector)	Fe material: 0.4mm Non Fe material: 0.5mm SS material: 0.7mm	Metal pieces by Metal detector x-ray scanning, Each time the product is cut into pieces	Check the sliced Cheese for metal contamination & use of certified cheese cutting machines	Proper working of Metal detector Each time the product is cut into pieces by production manager
	Microbial	CCP B2	Cold storage temperature for storing cheese	Cold storage temperature & hygiene Hourly by production manager	Effective temperature control, Cold storage structure to be modified to maintain proper temperature	Cold storage temperature Lab testing Every 4 hrs after corrective action by Production manager

Table 5: HACCP plan for cheese processing.

Conclusion and Recommendations

Conclusion

From the research study it can be concluded that that application of HACCP system can improve the quality of pasteurized Milk and cheese by control of critical points The pre-requisites programs (GMPs) and the operational pre-conditional programs that form the sturdy pillars of a stout and sturdy HACCP plan must be made mandatory to ensure good quality and hygiene in the plant as well as product the direct application of HACCP is difficult in industries that are not producing food products. But for industries that are associated with the food production industry the implementation of HACCP provides familiar value. The lessening of identified CCP number is necessary since it will ensure the safety of food products for consumption and safety of the consumer, and will decrease the overall cost for monitoring hence increase in the net outcome of the company Further studies are needed to verify and validate the HACCP system in milk processing plant.

Recommendations

Proper education and training of the management bodies, workers, employees and handlers allied with milk and cheese manufacturing in the plant is the basic necessity required. Application of HACCP system throughout the chain of pasteurized milk and Cheese process should be encouraged to ensure better control and safe food. Enhancement of storage and marketing conditions of dairy products are required. To shun the spoilage of sensitive products like cheese and milk, the manufacturer should apply the CIP (clean in place) and COP(clean out place) after every production process, guidance curriculum of GMP and HACCP should be designed and deliberately conducted by training centers and institutes for entrepreneurs of dairy industry to smoothen the process of adoption of these food safety programs and formulate special program for building up the food safety awareness to consumers and to make it mandatory for dairy producers to adopt food quality and food safety management programs.

References

1. CAC (2003) Guidelines for the Application of the Hazard Analysis Critical Control Point (HACCP) system. Codex Alimentarious Commission, FAO, Rome.

2. International Livestock Research Institute (ILRI) (2005) Bridging the regulatory space for small-scale milk traders. ILRI Briefing Paper.

3. Tranter HS (1990) Food borne staphylococcal illness. Lancet 336: 1044-1046.

37

Whey Protein Concentrate as a Substitute for Non-fat Dry Milk in Yogurt

Berber M[1], González-Quijano GK[2] and Alvarez VB[1]*

[1]*Department of Food Science, The Ohio State University, Columbus, OH 43210, USA*

[2]*Departamento de Graduados en Alimentos, Escuela Nacional de Ciencias Biológicas, Instituto Politécnico Nacional, Distrito Federal, México*

Abstract

Whey is a liquid by-product produced during cheese manufacturing. Whey was once considered a waste product but it is converted to food ingredients through processing steps that include membrane filtration, heat or enzyme modification, and fractionation. These processes have made it possible to improve sweet whey utilization. Whey proteins have good nutritional properties and enhance the textural properties of food when they are used as ingredients. The objective of this study was to evaluate non-fat, low-fat and full-fat stirred style strawberry flavored yogurt formulated with whey protein concentrate 80 (WPC80) replacing non-fat dry milk (NFDM). Levels of total solid were adjusted to 14.8%, 15.7% and 17.3% for non-fat, low-fat and full-fat yogurts, respectively. Yogurts formulated with non-fat dry milk were used as controls for all fat levels. Batches of 17 pounds of yogurt at 0% fat, 1% and 3.25% fat were made by mixing milk with powdered ingredients in a liquefier and homogenized at 2,300 psi for first and 500 psi for second stage, respectively. Double stage homogenized yogurt milk was pasteurized at 92°C for 30 seconds. Following cooling, yogurts were fermented to a final pH of 4.5. Yogurts were analyzed for their chemical and physical properties following the standard methods of analysis. Sensory evaluation was done by descriptive method with hedonic and monadic scales by trained judges. Whey protein yogurts at all fat levels showed better water holding capacities (ca. 10%) than controls with increased hardness (ca. 20%) and viscosity (ca. 40%). Sensory results revealed that whey protein yogurts had higher flavor and overall liking scores than controls, while controls had better scores for the yogurts' texture. Results showed that WPC80 is a good alternative to replace NFDM completely in yogurt. Whey yogurts had equal or greater quality than yogurt products made with NFDM.

Keywords: Whey protein; Yogurt; Non-fat dry milk; WPC80

Introduction

Yogurt is a fermented milk product with a good reputation due to its probiotic cultures and its reported beneficial effects on health [1]. In theory, only milk and starter culture activity are needed to make a yogurt product. However, in practice, the total solids content of yogurt needs to be increased to prevent syneresis [2,3]. A preheating step at high temperatures is also needed to enhance gel formation by the whey proteins [4]. Yogurt can be fortified with milk protein powders to increase the total solids content to desired levels [5]. However, a study reported the importance of controlling the texture of cultured products when formulating the mix with lower milk solids and added stabilizers. The authors suggested that excessive use of stabilizer can negatively influence the sensory properties of the yogurt by producing an over stabilized texture and mouth-feel [6]. The common practice for yogurt manufacturing is to formulate yogurt with nonfat dry milk or whey protein concentrates, combined with heating to denature the whey proteins [4,5]. Caseins denature at 160°C, however whey proteins start to denature above 70°C [7]. Since whey protein denaturation enhances the gel structure of yogurt, it will decrease the need for the addition of nonfat dry milk. Therefore, fortifying yogurt milk with whey protein powders may result in a better textured yogurt product [8].

Whey proteins are byproducts of the cheese making process and were once considered a waste product, but now are considered a valuable byproduct, due the properties and uses as mentioned in a review by Smithers [9]. Novel production techniques such as drying and membrane filtration allow a decrease in the lactose content while increasing protein content [10,11]. Whey protein structure is rich in branched chained amino acid (BCAA) such as leucine, valine, and isoleucine [12,13]. Whey proteins are available as acid whey, sweet whey, whey protein concentrate (protein content range 34% to 80%) and whey protein isolate (protein content>90%) [13]. Whey proteins are also used as functional ingredients and as milk replacers in dairy products such as ice cream and as optional ingredient in yogurt

products [14]. Since WPC is a dairy ingredient, which comes directly from milk, it can be used in dairy applications and labeled as a natural ingredient. Previous studies show that WPC addition to yogurt may improve texture and water holding [15]. Fortified yogurt with WPC to replace skim milk powder in the formulation resulted in a firmer coagulum, higher viscosity and less syneresis [16]. However, the U.S Code of Federal Regulations allows WPC to be used only as a secondary ingredient in yogurt applications [17]. Regulations state that the solids should come from either evaporated milk or non-fat dry milk (NFDM) to increase the non-fat milk solids level to the required 8.25%. A recent study reported that variations in fat content and changing the casein-to-whey protein ration had a decreasing effect on flavor and increasing effect on graininess [18] Thus, the use of WPC as a major ingredient is limited [15]. The objective of this study is to investigate the replacement of NFDM with WPC in yogurt formulations at three fat levels. Products were analyzed for their compositional, textural and sensory properties.

Whey protein yogurt

Whey proteins were used to substitute non-fat dry milk in yogurt formulations at three fat levels; non-fat, low-fat and full-fat. Yogurt products were made at pilot plant scale following the commercial

*****Corresonding author:** Alvarez VB, Department of Food Science, The Ohio State University, Columbus, USA, E-mail: alvarez.23@osu.edu

processing procedures. The finished products were tested for chemical, textural and sensory properties. The results shown that whey proteins were suitable to replace non-fat dry milk and produced high protein yogurts that were acceptable for the consumers.

Materials and Methods

Formulation

Control non-fat, low-fat and full fat products were produced with ingredient formulations similar to commercial products. Whey protein products were developed by replacing NFDM completely in control products with 80% whey protein concentrate. Control and whey protein product formulations and ingredients are presented in Table 1. Both control and whey protein products were formulated at the same total solids content for comparison purposes. The ingredients for yogurt formulation were the following: whey protein concentrate (WPC80, Agrimark, Lawrence, MA), yogurt stabilizer (Crest 41-1444, Crest Foods Inc., Ashton, IL), direct-vat-set yogurt culture (*Streptococcus thermophilus, Lactobacillus bulgaricus, Lactobacillus acidophilus,* and *Bifidobacterium* spp.) (Yo-Mix 205, Danisco, Madison, WI), strawberry base (FRD-12-25794, Fruitcrown Products Co., Farmingdale, NY), non-fat dry milk solids (NFDM) (Kroger, Cincinnati, OH) and sucrose (Kroger, Cincinnati, OH).

Yogurt processing and pilot plant scale-up

Raw milk (4% milkfat), obtained from a dairy farm (The Ohio State University, Columbus, OH) was used for yogurt processing. Milk was separated into skim milk (<0.5% milkfat) and cream (20% milkfat) using an Alfa-Laval 29AI separator (Stockholm, Sweden). Milk was kept at 4 ± 2°C at all times. Yogurt mixes were standardized at 0%, 1% and 3.25% milkfat using the ingredients listed in Table 1. Three 17 pound batches of yogurt were made for each fat level formulation. Each yogurt batch was homogenized at a pressure of 2,300 psi first stage and 500 psi second stage using a Lab 100 M-G two-stage valve homogenizer (Lubeck-Schlutut, Germany), and then pasteurized at 92°C for 30 sec in an AVP Junior HTST system (Tonawanda, NY). Pasteurized yogurt mix was placed in a 4°C walk in refrigerated cooler for overnight storage. The next morning, each yogurt mix was placed into six separate 2.5 gallon 304 Stainless steel containers (Hamby Dairy supply, Maysville, MO) and warmed up to 42 ± 1°C in a heated water bath (Fisher Scientific, Hampton, NH). Each yogurt batch was inoculated following the manufacturer's (Danisco Co. Inc., Madison, WI) recommended inoculation rate (0.02% w/w). Yo-Mix 205 starter culture (frozen pellets) was poured directly into the pasteurized mix. The mixture was agitated for 5 minutes to distribute the culture evenly. Temperature was maintained at 42 ± 1°C during the fermentation process. The fermentation was stopped when the pH of the mixture reached 4.5 in about 4 to 4.5 hours. Strawberry base flavor (Fruitcrown, Farmingdale, NY) was added to the final yogurt at a rate of 15% (w/w).

The finished product was placed into 8 ounce sanitized (200 ppm sodium hypochlorite) plastic containers, labeled and stored in the cooler at 4 ± 2°C for 1 week.

Chemical, textural and sensory analyses

Chemical composition analyses: Yogurt samples were analyzed for moisture content using a CEM Lab Wave 9000 moisture/solids analyzer (Matthews, NC). Fat content was determined by the Babcock method (AOAC method number 989.04) [19]. Protein content was measured using a micro Kjeldahl total nitrogen (TN) analyzer (AOAC method number 991.20; 33.2.11) [19]. All analyses were conducted in triplicate.

pH: pH was measured using a pH meter WTW-pH 330 (Weilheim, Germany) with a glass electrode standardized at 25°C in the range 4.0 to 7.0 with commercial buffers (Fisher Scientific).

Hardness: An Instron 5542 series single column testing system (Norwood, MA) was used to measure yogurt hardness [20]. Samples were removed from refrigeration (4 ± 1°C) just before analysis. Gels were penetrated using a 35 mm diameter flat probe at a crosshead speed of 0.83 mm/s. Hardness (N) was defined as the maximum mean force necessary to penetrate up to 50% compression of the gel's anvil height. Anvil height of yogurt samples was set at (6.0 cm), and anvil diameter was set at (6.8 cm).

Water holding capacity: A 20 g yogurt sample (Y) was centrifuged for 10 min at 1792 g at 4°C. The whey expelled (W) was removed and weighed, and water holding capacity was calculated by the method of Sodini [21].

Viscosity: Apparent viscosity was measured using a viscometer (Brookfield DV-II+, Middleboro, MA) with an LV spindle number 3 rotated at 1.5 rpm. Samples were kept in a water bath at 23°C [22].

Sensory analysis: The sensory panel consisted of 5 expert members who had previous experience in sensory evaluation of yogurts. Products were presented in 3-digit coded, white plastic isothermal cups stored at 4°C. The samples were at approximately 10°C when they were tested according to the procedure of Tribby [22]. Panelists were provided with mineral water for palate cleansing between samples. The sessions were carried out in a temperature controlled room at 20°C under white lighting in individual booths. Data acquisition was assisted by CompuSense Five software (CompuSense Inc., 2010). Both monadic and hedonic scales were used to rate the flavor and the texture attributes of products. The attributes were evaluated in the following order: visual texture with a spoon, odor, aroma, taste, and texture-in-mouth.

Experimental design: The experimental design was a randomized complete block, performed in triplicate on separate days (runs), with run as the blocking variable. Variables were the fat content and the day effect for homogenous results. All analyses were performed in

Ingredients(%)/Products	Control Non-fat	Control Low-fat	Control Full-fat	Whey Non-fat	Whey Low-fat	Whey Full-fat
Cream	-	4.8	15	-	4.3	14.5
Skim milk	79.25	74.45	64.25	79.25	74.95	64.75
WPC80	-	-	-	2	2	2
NFDM	2	2	2	-	-	-
Sugar	3	3	3	3	3	3
Stabilizer	0.75	0.75	0.75	0.75	0.75	0.75
Flavor (strawberry)	15	15	15	15	15	15
Total (%)	100	100	100	100	100	100

Table 1: Formulations for the developed control and whey protein products.

triplicate and were analyzed by one way ANOVA with sample formula for yogurt and run as the main effects, using PASW (formerly known as SPSS) statistical software (Release 18.0.2). Means comparisons were made when the effect was significant (P < 0.05) using Tukey's HSD procedure. All results were reported as the combined means of 3 repeated measures from each of the 3 runs made.

Results and Discussion

Composition

Total solids include fat, protein, carbohydrate and minerals for the yogurt product. Total solids content was similar in control and whey yogurts with the same levels of fat but it was significantly different (*P<0.05*) among products with different levels of fat (Table 2). Non-fat control and non-fat whey protein yogurts had the same total solids content of 14.8%. Low-fat and full-fat yogurts showed the same levels of total solids content, in both control and 80% WPC containing samples, 15.7% and 17.3%, respectively. The effect of total solids on the properties of yogurt will be discussed later in the corresponding sections. The code of federal regulations [17] require that yogurt products should have levels of 0% to 0.5%, 0.5% to 2% and more than 3.25% in order to be named as non-fat, low-fat and full-fat, respectively. Low-fat control and whey protein yogurts had around 1.1% fat content (Table 2) while full-fat had 3.25% fat content. There were no differences (*P<0.05*) in fat content between control and whey samples of both low-fat and full-fat yogurts. However, non-fat whey yogurt had 0.1% fat content while control non-fat yogurt had no detectable fat content. The reason for this difference in the fat content may be attributed to the WPC used in the formulation, which contains around 5-8% of fat. The presence of fat is important to both texture and flavor of yogurt. Non-fat yogurts had lower texture and sensory results than low-fat and full fat yogurts. Other studies such as Sandoval-Castilla, et al. [20] reported similar observations related to texture and flavor of yogurt, which were enhanced by homogenization of fat globules. In addition to the presence of fat, the interactions of fat globules with protein molecules are important for the textural properties of the finished product [2,23]. Higher sensory and texture scores of control low-fat yogurt (1.1% fat content) can be explained by this interaction [2]. Because control low-fat yogurt was fortified with casein and double stage homogenized at 2300 psi and 500 psi, this would result in more reformed casein micelles. Protein is the one of main solid ingredients in the yogurt products. The protein content was significantly different (*P<0.05*) among all control and whey yogurts (Table 2). Control non-fat yogurt had 3.68% protein content and non-fat whey yogurt had the highest protein content of 4.61%. This was due to the high protein content of WPC (80%). Likewise, the reason for the low protein content for the control yogurts was the protein source, which is NFDM (34%). Similar results were observed for low-fat and full-fat control and whey containing

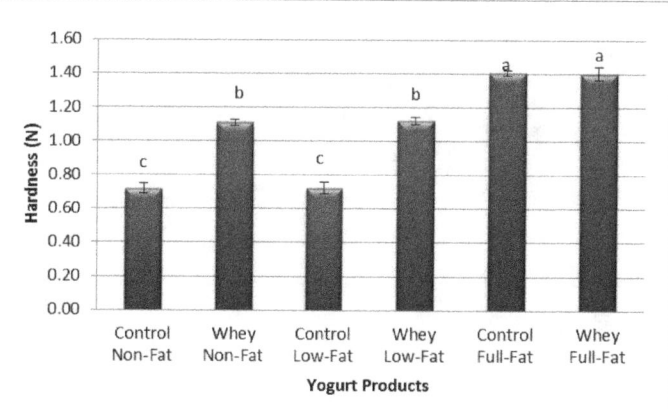

Figure 1: Hardness values of yogurt products.
[a-d]Different letter among the bars indicates significant difference (*P<0.05*) (Tukey's test).
*Results are mean of triplicate analyses of three different batches.

yogurt products. This suggests that high protein concentration in the added 80% WPC was the reason for whey protein yogurts to have higher protein content than control yogurts. The presence of protein is essential for good body and texture of yogurt products Sodini, et al. [21]. Whey low-fat yogurts had 4.55% protein content while control low-fat yogurts had 3.59%. The main reason for the addition of non-fat milk solids in yogurt formulations is to add milk protein. Increased levels of protein resulted in higher values in hardness, WHC and viscosity scores for whey protein yogurts versus controls. Henriques, et al. [3], as well as Bong and Moraru [5] also reported that increased protein content improved the textural properties of yogurt products; these findings may explain the higher texture results for whey protein yogurts (Figure 1) due to the high protein content in the formulation.

Heat induced protein denaturation is a treatment that affects texture, flavor and chemical properties of yogurt products [24]. Improved texture of 80% WPC containing yogurt could have been the result of the denaturation of whey proteins. However, increased protein content combined with protein denaturation in whey protein yogurts might be the reason for their lower texture sensory evaluation scores compared to the controls. Full-fat whey yogurt had 4.39% protein content while full-fat control yogurt had 3.45%. The variation was significantly different (*P<0.05*), and full-fat yogurts had the lowest protein content when compared to non-fat and low-fat yogurts, due to the fact that they contained the highest fat content of about 3.25%.

When comparing the pH results among yogurt products at the same fat level, the variations for pH were significantly different (P < 0.05) for all non-fat, low-fat and full-fat control whey protein yogurts (Figure 2). The differences in flavor perception are shown in (Figure 3). An important factor for these differences could be the pH, which results from acid formation during fermentation that is necessary to coagulate proteins to form the typical coagulum of yogurt. The acid level present gives the acid taste of yogurt. Soukoulis, et al. [25] reported that pH affected the flavor of yogurt when WPC were added in the formulation. Furthermore, other authors found that higher whey protein content in yogurts increased the buffering capacity in the product and also influenced the flavor perception [24,26,27].

Control full-fat yogurt had a pH of 4.39 while whey full-fat yogurt had a pH of 4.37. As in the case of the low-fat yogurt results, the pH of full fat products decreased as fat content increased. Low pH is reported to cause textural defects in yogurt products since it decreases the gel

	Total solids (%)	Fat content (%)	Protein content (%)
Control non-fat	14.8c	ND*	3.68d
Whey non-fat	14.8c	0.1c	4.61a
Control low fat	15.7b	1.1b	3.59e
Whey low-fat	15.7b	1.1b	4.55b
Control full-fat	17.3a	3.3a	3.45f
Whey full-fat	17.3a	3.3a	4.39c

*ND: not detectable

[a-f] Different letter among the bars indicates significant difference (*P<0.05*) (Tukey's test).

* Results are mean of triplicate analyses of three different batches.

Table 2: Total solids, fat and protein content of control and whey yogurts.

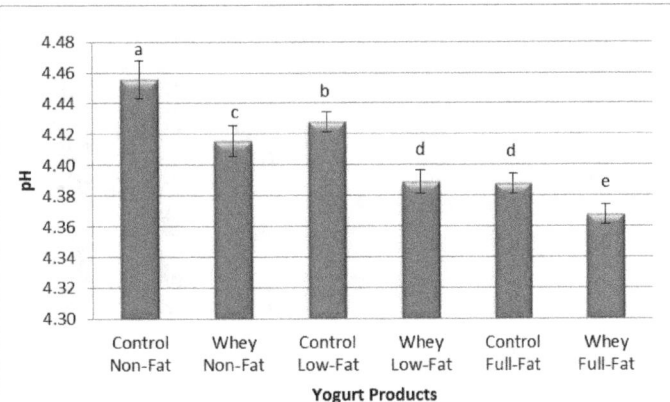

Figure 2: pH of yogurt products.
a-eDifferent letter among the bars indicates significant difference (*P<0.05*) (Tukey's test).
* Results are mean of triplicate analyses of three different batches.

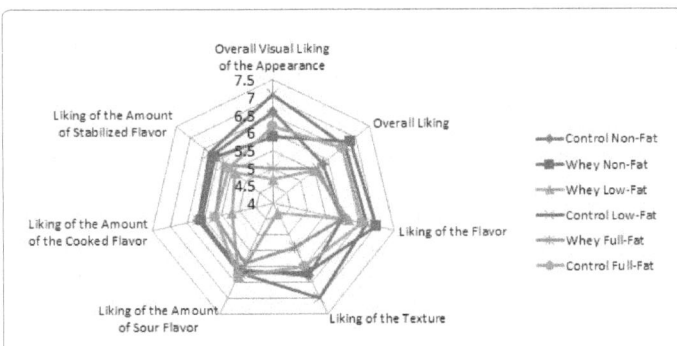

Figure 3: Hedonic scale sensory scores of yogurt products.
* Results are mean of triplicate analyses of three different batches

strength due to excessive charge repulsion [27].

Texture

Hardness was used to measure the textural properties of yogurt products, in this case, hardness values of non-fat, low-fat and full-fat whey protein yogurt samples at non-fat, low-fat and full-fat were significantly different (*P<0.05*) than the controls (Figure 1). Control non-fat and low-fat yogurts had the same lowest hardness values of 0.72 N each while whey non-fat and whey low-fat yogurts had values of 1.11 N and 1.12 N, respectively. The increased difference in hardness between control and whey protein yogurts may been associated with the denaturation of whey proteins.

These observations are supported by the results of a study reporting that yogurt containing whey proteins had firmer texture than casein fortified yogurt [28]. A similar effect was observed in control full-fat yogurt that had a hardness value of 1.40 N, which was lower than 1.47 N for whey full-fat yogurt. Increased levels of fat content produced greater increase in the hardness value of both control (1.40 N) and whey protein yogurt (1.47 N). Other studies reported similar results and concluded that the increase in yogurt hardness was directly related to fat and total solids content [6]. As indicated earlier, total solids content also influenced the hardness of both control and whey protein yogurts. Non-fat and low-fat yogurts had 14.8% and 15.7% total solids content and their respective hardness values were 0.72 N and 1.12 N. These total solids values are lower than those for full-fat product

that had 17.3% total solids and 1.47 N. The optimum total solids level was in the range of 14 to 16% to produce acceptable products; some studies reported results where total solids content affects the textural properties of yogurt products [6]. WHC is an indicator of how much water can be held in a yogurt product. Control non-fat yogurt had the lowest water holding capacity at 582 gm/kg while whey full-fat had the highest water holding capacity of 768 gm/kg (Figure 4). There are many factors associated with yogurt's water holding capacity. Total solids content, protein and fat content, source of protein, selection of starter culture (either ropy or non-ropy) and processing conditions are important factors that decrease or increase the water holding capacity in yogurts [24]. Non-fat whey yogurt had higher a WHC score of 682 gm/kg than control non-fat, which had a value of 582 gm/kg Since both products have the same fat and total solids content, the higher WHC can be attributed to addition of whey proteins. Sodini, et al. [24] reported that protein denaturation was responsible for the increase of water holding capacity in yogurt products. Similar studies also stated that whey protein denaturation enhanced the gelling properties and hence the WHC with adequate heat treatment [29].

Whey low-fat yogurt had a higher water holding capacity of 730 gm/kg than control low-fat, which had 622 gm/kg Similarly, the water holding capacity score of 768 gm/kg for whey full-fat yogurt was higher than the control full-fat yogurt, which had a WHC score of 725 gm/kg. Increase in total solids and fat content resulted in an increase of water holding capacity for all yogurt products. Some studies also concluded that the origin and total solids content affect the water holding capacity of yogurt products as well the viscosity [30]. An additional factor that influences the WHC of yogurt is the homogenization of the mix during yogurt making. Homogenization increases the interaction of protein molecules with fat globules and free water in the yogurt mix prior to coagulation [8]. The viscosity values of all yogurts increased with increased fat content (Figure 5). Control non-fat yogurt had the lowest viscosity score while whey full-fat yogurt had the highest viscosity score. Whey proteins also increased viscosity of yogurts at all fat levels. Control non-fat yogurt had a viscosity score of 19,355 cPs that was lower than the 31,847 cPs value for non-fat whey yogurts. Similar results were observed in low-fat yogurt products. Control low-fat yogurt had 21,340 cPs while whey low-fat yogurt had 33,011 cPs.

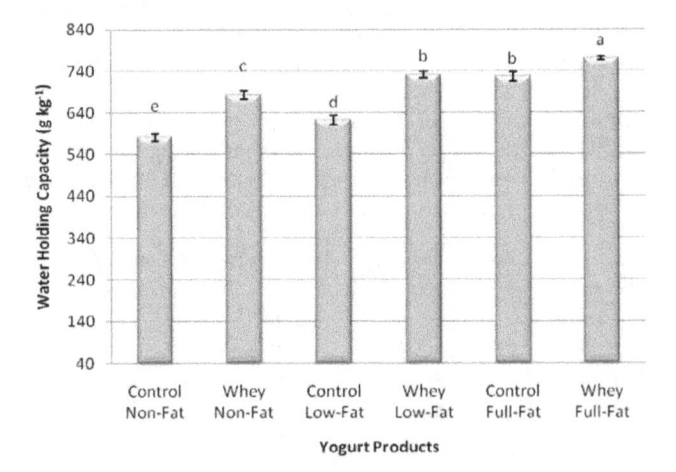

Figure 4: Water holding capacity of yogurt products.
a-eDifferent letter among the bars indicates significant difference (*P<0.05*) (Tukey's test).
* Results are mean of triplicate analyses of three different batches.

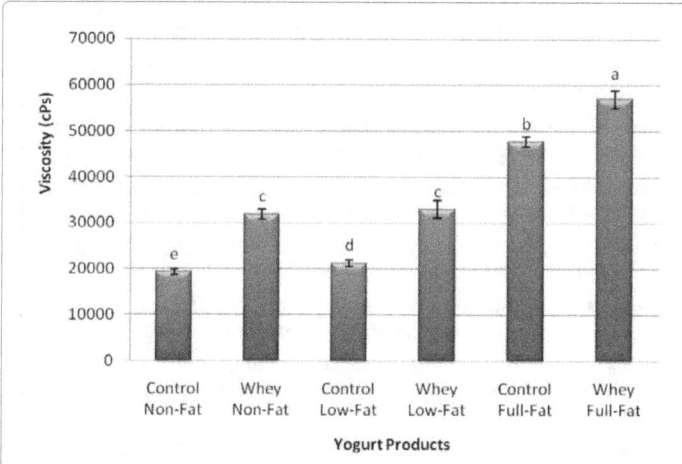

Figure 5: Viscosity values of yogurt products.
a-eDifferent letter among the bars indicates significant difference (*P<0.05*) (Tukey's test)
* Results are mean of triplicate analyses of three different batches.

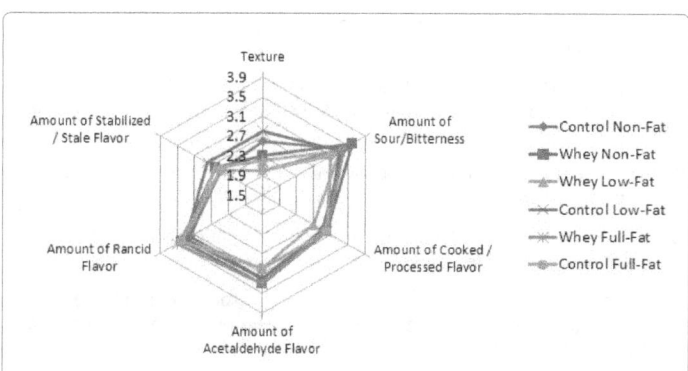

Figure 6: Monadic scale sensory scores of yogurt products.
* Results are mean of triplicate analyses of three different batches.

These viscosity results may be due to the high total solids content and whey protein denaturation of the yogurt mix. Isleten and Karagul-Yuceer [23] reported the addition of whey protein isolate to increase total solids resulted in the increase in viscosity of yogurts. Whey full-fat yogurt had the highest viscosity score of 56,977 cPs when compared with the control full-fat score of 47,722 cPs. The increase in viscosity was significantly different (*P<0.05*) from low-fat and non-fat yogurts. Brauss et al. [31], Soukoulis, et al. [25], and others, noticed that the heat treatment also affects the viscosity, as well as the fat and protein content.

Sensory

Sensory results on a monadic scale (1: low amount to 5: excess amount) showed differences between products (Figure 6). Another name for the monadic scale is the metric scale and it is used for non-comparative measurements. Texture scores had the most variations among whey yogurts. Control low-fat yogurt had the most desirable texture score of 2.8 compared to other yogurts.

Whey low-fat and control full-fat products had the best scores for acetaldehyde flavor with a score of 3.0. Acetaldehyde is the major flavor compound of the fermentation process in yogurt applications and gives yogurt its famous tarty flavor [32]. González-Martínez, et al. [15]

reported more production of acetaldehyde in yogurt with whey protein powder used to fortify the yoghurt mix at levels ranging between 0.6 and 4%. In our case whey protein is the main added solid, so the change in formation of acetaldehyde during the fermentation process may be due to the interactions between the nutrients and the starter cultures [33]. Also the whey proteins β-lactoglobulin (β-LG) and α-lactalbumin (α-LB) interacted differently with fat content as observed in previous work by Corredig and Dalgleish [34]. They found that both α-LB and β-LG were bound to the surfaces of fat globules when milk was heated in the temperature range 65-85 °C. For sour/bitter scores, control and whey protein products scored between 3.1 and 3.6. The results showed that the acidity of yogurts was well developed. The pH of yogurts was also in agreement with the sour results since they ranged between pH 4.36 and 4.46.

The most widely used measuring scale for food acceptability is the 9-point hedonic scale. This scale starts with dislike extremely (scale 1) and ends with like extremely (scale 9). Whey protein non-fat product had the highest overall likeability score of 6.8 compared to other yogurts followed by control low-fat (6.6) and control non-fat yogurt (6.5) (Figure 3). These scores suggest that 80% WPC may work well as an added ingredient in low fat yogurt, especially when today's low fat and high protein diets are taken into account. For overall visual likeability appearance, whey protein products had lower scores than the control yogurts. Control low-fat yogurt had the highest score of 7.1 in this category which was in agreement with the monadic texture score; it may also be associated with the texture liking. Control low-fat had the highest texture liking score at (7.0) followed by control non-fat (6.3) and whey non-fat (6.2) respectively. Whey low-fat and full-fat yogurts had the lowest scores of 4.3 and 6.0, respectively. This result indicates that textural issues might be related to whey protein denaturation which caused more firm gel formation of yogurt. Non-fat whey yogurt had highest score of 7.0 in the flavor liking category followed by control low-fat (6.7) and control full-fat (6.6). This result may be related to the volatile high acetaldehyde composition of non-fat whey yogurts. Liking scores of the stabilized, cooked and sour flavor were not different (*P<0.05*). Sensory scores revealed that control low-fat yogurt had better textural properties than others, while non-fat whey yogurt had higher scores than others for the flavor.

Conclusion

The results of this study have shown that it is possible to completely replace NFDM in yogurt formulations with 80% WPC while enhancing textural properties such as water holding, hardness and viscosity. Also, fat was shown to increase water holding, hardness and viscosity. Additionally, whey protein addition improved the protein content and reduced syneresis. Yogurt can be made with 80% whey protein concentrate as added solids without compromising the overall acceptability. Readers should be aware that the effect of WPC on yogurt properties depends on the type and hence manufacturing conditions of the WPC. In this study only one type WPC was used and the results may not be representative of all commercial WPCs.

Acknowledgments

The author Murat Berber gratefully acknowledges the support given by the Ministry of National Education (Turkey) for scholarship. Additionally, the authors Murat Berber and Valente B. Alvarez want to thank the support of The Ohio State University Wilbur A. Gould Food Industries Center. The author Génesis K. González-Quijano wants to acknowledge the support given by CONACyT and the Escuela Nacional de Ciencias Biológicas, IPN-México.

References

1. Wang H, Livingston KA, Fox CS, Meigs JB, Jacques PF, et al. (2013) Yogurt

consumption is associated with better diet quality and metabolic profile in American men and women. Nutr Res 33: 18-26.

2. Aziznia S, Khosrowshahi A, Madadlou A, Rahimi J, Abbasi H, et al. (2009) Texture of nonfat yoghurt as influenced by whey protein concentrate and Gum Tragacanth as fat replacers. Int J Dairy Technol 62: 405-410.

3. Henriques MF, Gomes DGS, Pereira CD, Gil MM (2013) Effects of liquid whey protein concentrate on functional and sensorial properties of set yogurts and fresh cheese. Food Bioprocess Tech 6: 952-963.

4. Xu ZM, Emmanouelidou DG, Raphaelides SN, Antoniou KD (2008) Effects of heating temperature and fat content on the structure development of set yogurt. J Food Eng 85: 590-597.

5. Bong DD, Moraru CI (2014) Use of micellar casein concentrate for Greek-style yogurt manufacturing: Effects on processing and product properties. J Dairy Science 97: 1259-1269.

6. Lucey JA (2004) Cultured dairy products: An overview of their gelation and texture properties. Int J Dairy Technol 57: 77-84.

7. Dissanayake M, Ramchandran L, Piyadasa C, Vasiljevic T (2013) Influence of heat and pH on structure and conformation of whey proteins. Int Dairy Journal 28: 56-61.

8. Loveday SM, Sarkar A, Singh H (2013) Innovative yoghurts: Novel processing technologies for improving acid milk gel texture. Trends Food Sci Tech 33: 5-20.

9. Smithers GW (2008) Whey and whey proteins-From 'gutter-to-gold'. Int Dairy Journal 18: 695-704.

10. Hausmann A, Sanciolo P, Vasiljevic T, Kulozik U, Duke M, et al. (2014) Performance assessment of membrane distillation for skim milk and whey processing. J Dairy Science 97: 56-71.

11. Espina V, Jaffrin MY, Frappart M, Ding LH (2010) Separation of casein from whey proteins by dynamic filtration. Desalination 250: 1109-1112.

12. Wit JN (1998) Nutritional and functional characteristics of whey proteins in food products. J Dairy Sci 81: 597-608.

13. Ha E, Zemel MB (2003) Functional properties of whey, whey components, and essential amino acids: mechanisms underlying health benefits for active people (review). J Nutr Biochem 14: 251-258.

14. Gauche C, Tomazi T, Barreto PLM, Ogliari PJ, Bordignon-Luiz MT, et al. (2009) Physical properties of yoghurt manufactured with milk whey and transglutaminase. LWT- Food Sci Technol 42: 239-243.

15. González-Martínez C, Becerra M, Cháfer M, Albors A, Carot JM, et al. (2002) Influence of substituting milk powder for whey powder on yoghurt quality. Trends Food Sci Tech 13: 334-340.

16. Guirguis N, Broome MC, Hickey M (1984) The effect of partial replacement of skim milk powder with whey-protein concentrate on the viscosity and syneresis of yogurt. Aust J Dairy Technol 39: 33-35.

17. Code of Federal Regulations (CFR) (2013) Tittle 21. Food and Drugs.

18. Tomaschunas M, Hinrichs J, Kohn E, Busch-Stockfisch M (2012) Effect of casei-to-whey protein ratio, fat content and protein content on sensory properties of stirred yoghurt. Int. Dairy Journal 26: 31-35.

19. Association of Official Analytical Chemists (AOAC) (2002) Official methods of analysis. Arlington, VA, USA.

20. Sandoval CO, Lobato CC, Aguirre ME, Vernon CEJ (2004) Microstructure and texture of yogurt as influenced by fat replacers. Int. Dairy Journal 14: 151-159.

21. Sodini I, Montella J, Tong PS (2005) Physical properties of yogurt fortified with various commercial whey protein concentrates. J. Sci. Food Agric 85: 853-859.

22. Tribby D (2009) The Sensory Analysis of Dairy Products: Yogurt. (2nd edn), Springer Science, New York.

23. Isleten M, Karagul-Yuceer Y (2006) Effects of dried dairy ingredients on physical and sensory properties of nonfat yogurt. J Dairy Sci 89: 2865-2872.

24. Sodini I, Mattas J, Tong PS (2006) Influence of pH and heat treatment of whey on the functional properties of whey protein concentrates in yoghurt. Int Dairy Journal 16: 1464-1469.

25. Soukoulis C, Panagiotidis P, Koureli R, Tzia C (2007) Industrial yogurt manufacture: Monitoring of fermentation process and improvement of final product quality. J Dairy Sci 90: 2641-2654.

26. Kelly M, Vardhanabhuti B, Luck P, Drake MA, Osborne J, et al. (2010) Role of protein concentration and protein–saliva interactions in the astringency of whey proteins at low pH. J Dairy Sci 93: 1900-1909.

27. Vardhanabhuti B, Kelly MA, Luck PJ, Drake MA, Foegeding EA, et al. (2010) Roles of charge interactions on astringency of whey proteins at low pH. J Dairy Sci 93: 1890-1899.

28. Guggisberg D, Eberhard P, Albrecht B (2007) Rheological characterization of set yoghurt produced with additives of native whey proteins. Int. Dairy Journal 17: 1353-1359.

29. Mao R, Tang J, Swanson BG (2001) Water holding capacity and microstructure of gellan gels. Carbohyd. Polym 46: 365-371.

30. And JL, Guo M (2006) Effects of polymerized whey proteins on consistency and water-holding properties of goat's milk yogurt. J. Food Sci 71: 34-38.

31. Brauss MS, Linforth RST, Cayeux I, Harvey B, Taylor AJ, et al. (1999) Altering the fat content affects flavor release in a model yogurt system. J. Agric. Food Chem 47: 2055-2059.

32. Antunes AEC, Cazetto TF, Bolini HMA (2005) Viability of probiotic micro-organisms during storage, postacidification and sensory analysis of fat-free yogurts with added whey protein concentrate. Int J Dairy Technol 58: 169-173.

33. Zisu B, Shah NP (2003) Effects of pH, Temperature, supplementation with whey protein concentrate, and sdjunct cultures on the production of exopolysaccharides by *Streptococcus thermophilus* 1275. J Dairy Sci 86: 3405-3415.

34. Corredig M, Dalgleish DG (1996) Effect of different heat treatments on the strong binding interactions between whey proteins and milk fat globules in whole milk. J Dairy Res 63: 441-449.

38

Occurrence of Lactose Intolerance among Ethiopians

Habtamu LD[1]*, Ashenafi M[2], Taddese K[3], Birhanu K[1] and Getaw T[4]

[1]*Wolaita Sodo University, PO Box 128, Wolaita Sodo, Ethiopia*
[2]*Addis Abba University, College of Veterinary Medicine and Agriculture, Ethiopia*
[3]*Ethiopian Development Research Institute, Ethiopia*
[4]*International Food Policy Research Institute, Ethiopia*

Abstract

Milk has been known as nature's most complete Food. Among the components of milk is lactose. Some people who lack the enzyme needed to digest lactose suffer from lactose intolerance and therefore cannot digest milk. Worldwide, nearly 70 percent of the adult population is thought to be lactose intolerant, and the condition is very common among American Indians and those of Asian, African, Hispanic, and Mediterranean descent. Lactose intolerance is one of the factors affecting milk demand and consumption, which subsequently influencing milk market/ productivity. But no information is available on lactose intolerance in Ethiopia. Therefore, this paper is the first attempt to study the incidence of lactose intolerance in the country. The aim of the paper, therefore, was to generate information on lactose intolerance through case study approach including questionnaire survey and document analysis. In the case study area of Ada district of Eastern Shoa of Ethiopia, out of 188 households/individuals surveyed, 7.45% of the respondents reported that they don't consume milk but consume fermented milk ('Ergo'). The majority reported symptoms of lactose intolerance (71.4% get vomiting upon consuming milk, 28.6% feel abdominal pain). The percentage of occurrence of lactose intolerance found in the survey is believed to affect the milk consumption in the country. Dairy processors need to make yoghurt or remove lactose, for consumers with special dietary requirements/lactose intolerance/to improve milk consumption. Awareness about lactose intolerance also needs to be created in the society by health extension workers.

Keywords: Milk consumption; Lactose intolerance; Lactose-free milk products; Ethiopians

Introduction

Milk has been known as nature's most complete Food. More than 100 different components have been identified in cow's milk. Intake of cow's milk and milk products contributes to health throughout life. Experimental studies indicate that cow's milk protein may help to increase bone strength, enhance immune function, reduce blood pressure and risk of some cancers, and protect against dental caries. Milk fat is also a source of energy, essential fatty acids (linoleic and linolenic), fat-soluble vitamins (A, D, E, and K), and several health-promoting components such as conjugated linoleic acid, sphingomyelin, and butyric acid. For example, emerging scientific findings reveal that CLA may protect against certain cancers and cardiovascular disease, enhance immune function, and reduce body fatness/increase lean body tissue. Milk and other dairy foods are an important source of many vitamins (riboflavin, vitamin B12 and vitamin A) and minerals (calcium, phosphorus, potassium, zinc, magnesium). A sufficient intake of calcium helps to reduce the risk of osteoporosis, hypertension, some cancers, and some types of kidney stones, and may have a beneficial role in weight management [1-3]. Milk lipids also contribute to the palatability of the diet [4].

Billions of people around the world consume milk and dairy products every day. Not only are milk and dairy products a vital source of nutrition for these people, they also present livelihoods opportunities for farmers, processors, shopkeepers and other stakeholders in the dairy value chain. But to achieve this, consumers, industry and governments need up-to-date information on how milk and dairy products can contribute to human nutrition and how dairying and dairy-industry development can best contribute to increasing food security and alleviating poverty [5].

Despite the slow growth seen in developing countries, the ILRI, along with other agencies, projected that the demand for dairy and dairy products will more than double in developing countries by the year 2025 with an estimated annual growth in consumption of 3.3% per year. This period of growth is referred to as the Livestock Revolution and is based on the projected increase in demand through population growth, urbanization and increased income generation. With this growth there will be changes in eating habits. Urban populations are also more likely to include milk and meat products in their diets based on their preference for increased variety and convenience [6,7].

In order to keep up with this increase in demand, there has been a push to focus on a more productive and market oriented dairy system throughout developing countries. A variety of international organizations such as Land O'Lakes, Send a Cow, and Heifer Project International have been working to promote milk production in SSA with the objectives of improving nutrition through increased milk consumption and increasing income generation for smallholder dairy farmers. Among the constraints and challenges of dairy development in developing countries is promotion of marketing and consumption of milk and dairy products [8].

Dairy production is one of livestock production system prevalent in Ethiopia. The urban and peri-urban milk production systems are major milk suppliers to urban areas. Considering the important prospective for smallholder income generation and employment opportunities from the high value dairy products, the development of the dairy sector can contribute immensely to poverty alleviation and improved nutrition in the country [9]. The per capita consumption of milk estimated at 19 kg, which is lower than African averages of 27 kg/

*Corresponding author: Habtamu Lemma Didanna, Wolaita Sodo University, PO Box 128, Wolaita Sodo, Ethiopia, E-mail: abebeh09@gmail.com

year [10]. In addition, dairy production in Ethiopia is anticipated to increase rapidly in response to the fast growing demand for livestock products resulting from increasing human population, especially in urban areas, and rising consumer income, which is an opportunity for dairy producers in sourcing their livelihoods. However, among the potential factors affecting milk demand are lactose intolerance and fasting season, which subsequently influencing milk consumption/ nutrition security and livelihoods.

Worldwide, nearly 70 percent of the adult population is thought to be lactose intolerant, and the condition is very common among American Indians and those of Asian, African, Hispanic, and Mediterranean descent [11]. But no information is available on lactose intolerance in Ethiopia and there exist gap in the research and literature. Therefore, this paper is the first attempt to know the incidence of lactose intolerance in the country.

The aim of this paper, therefore, was to address information on lactose intolerance through case study approach including interviews and document analysis. Brief questionnaire survey of 188 households/ individuals was made for the case study (Table 1) in Ada district of Eastern Shoa of Ethiopia, which is one of the largest market-oriented dairy producing areas.

Milk Composition/Lactose

The nutritional value of milk is mainly determined by water, protein, fat, sugar (lactose), vitamins, and micronutrients [12]. The disaccharide lactose comprises 4.8-5.2 wt% of milk. In addition to its nutritional relevancy, the determination of lactose is important for people with lactose intolerance [13]. In Ethiopia, there are various figures on lactose content of cattle milk including 4.53% [14]; 4.18% for Boran-Friesian cross and 4% for Borana cows [15]; 5.17% for crossbred and 5.29% local cows [16].

Cow milk contains approximately 5 g of lactose/100 g. As well as providing energy, lactose (along with milk oligosaccharides) supports growth, aids in softening of stools and enhances water, sodium and calcium absorption [17].

Lactose must be broken down by lactase (an enzyme/β-galactosidase found in the intestine) before the body can use it. If there is not enough lactase, undigested milk sugar remains in the intestine. Bacteria in the colon then ferment this sugar. Gas, cramping, diarrhea, nausea, and vomiting can follow. Most of us begin to lose intestinal lactase as we age. Lactose intolerance occurs around 3-5 years of age and primarily

SNo.	Age	Sex	Symptoms	Residence
1	1 yr and 5 months	F	Vomiting	Town
2	19 yr	M	Abdominal pain	"
3	21 yr	M	"	"
4	60 yr	F	Vomiting	"
5	15 yr	M	"	"
6	NA	M	"	"
7	25 yr	F	"	"
8	18 yr	M	Vomiting	Rural
9	6 yr	M	Abdominal pain	"
10	NA	NA	Vomiting	"
11	38 yr	F	"	"
12	NA	NA	Vomiting and Abdominal pain	"
13	12 yr	F	Vomiting	"
14	51 yr	M	Diarrhea	"

Table 1: Information on respondents found to have lactose intolerance (out of 188 people participated in the case study).

affects adults. Cows'. This loss of intestinal lactase activity is not a disease, but rather a normal pattern in human physiology and is transmitted by a recessive gene [18].

Fermented milk products may be better tolerated by people with lactose intolerance, primarily because they contain less lactose [19]. In particular, yoghurt containing live bacteria may be better tolerated by lactose malabsorbers because of the β-galactosidase in the yoghurt or the presence of bacteria in the yoghurt that produce β-galactosidase in the small intestine. Furthermore, yoghurt takes longer to pass through the digestive system than does milk, thus allowing more effective breakdown of lactose [20]. In other words, the bacteria in the yoghurt partially digest the lactose into glucose and galactose (and the glucose to lactic acid); in addition, yoghurt's semisolid state slows gastric emptying and gastrointestinal transit, resulting in fewer symptoms of lactose intolerance [21,22]. Aged cheeses tend to have lower lactose content than other cheeses and, thus, may also be better tolerated. Predigested milk or dairy products with lactase are available in some countries and will often permit a lactose-intolerant individual to be able to take some or all milk products freely [21].

A promising recent development is the introduction of a new lactose-free dairy drink produced by a special filtration process that removes half of the milk lactose. A lactase enzyme is then added to the milk to break down the remaining milk sugars into simpler forms that the body can absorb. The ability to remove lactose from milk and milk products could capture the non-trivial market share of people who are lactose intolerant. Lactose-free milk could be an important factor, driving higher milk consumption [23]. This includes Lactose-free milk such as GalactominTM and lactose-hydrolyzed milk are prepared (from fungal lactase) for people suffering from lactose intolerance [24].

Lactose Intolerance

Lactose intolerance is the inability to digest lactose into its constituents, glucose and galactose, because of low levels of lactase enzyme in the brush border of the duodenum [25]. Lactose intolerance is defined as a metabolic disorder when people are unable to digest significant amounts of lactose due to the genetically insufficient production of the lactase enzyme (β-galactosidase).

In adults with lactase deficiency (also called lactase nonpersistance, LNP), lactose is not digested in the upper bowel and reaches the lower bowel, where it is fermented by gut micro-organisms, which produces hydrogen, carbon dioxide and methane gas. Undigested lactose also draws water into the intestinal lumen through its osmotic effect, which increases motility and can cause diarrhoea. Symptoms include abdominal pain, bloating and flatulence. Thus, low lactase levels cause lactose malabsorption (or lactose maldigestion). When lactose malabsorption gives rise to symptoms, this is called "lactose intolerance", i.e. lactose malabsorption is the physiologic problem that manifests as lactose intolerance [21].

Lactose intolerance and its prevalence in other countries

Lactase deficiency in adults is a normal developmental phenomenon characterized by the down-regulation of lactase activity, which occurs soon after weaning in most ethnic groups [26]. Lactose maldigestion increases with age during adulthood [6]. The lactase persistence trait is more common in populations that practice cattle herding and dairying [27], and is related to genetic selection of individuals with the ability to digest lactose [21].

Children of some ethnic groups commonly lose lactase at one to two years of age (e.g. Thai children) while in others lactase persists until later in life (10-20 years of age) (e.g. Finnish children) [28,29]. According to some estimates, approximately 70 percent of the world's population has primary lactase deficiency [21].

Problems of lactose intolerance are especially severe in African and Asian populations. In Asia, more than half of the population is believed to have some form of lactose intolerance [22].

The frequency and intensity of lactose intolerance/malabsorption varies widely among populations from _100% in south-east Asia to 5% in north-west Europe [30,31]. Lactose deficiency in Europe has been reported to vary between 4% (Denmark and Ireland) and 56% (Italy) [32]. In India, the frequency of lactase persistence is higher in the north than the south, and in the rest of the world, lactase persistence frequency is generally low. In Africa, the distribution is patchy, with some pastoralist nomadic tribes having high frequencies of lactase persistence compared with the neighbouring groups inhabiting the same country [33], with a similar pattern observed between Bedouin and neighbouring populations in the Middle East [34,35].

Symptoms of lactose intolerance

People who have either primary or secondary lactase deficiency are lactose intolerant, as judged by a lactose tolerance test, and may exhibit symptoms of lactose intolerance [36].

The symptoms of lactose intolerance due to lactose malabsorption and caused when milk is consumed by a lactase non-persistent person, vary greatly from person to person, but if they are evident, they usually manifest themselves within 1-2 h of ingestion. Ingestion of dairy products resulting in symptoms of lactose intolerance generally leads to transient symptoms without causing harm to the gastrointestinal tract [21].

Diagnosis of lactose intolerance

Several indirect methods have been developed for the purpose of diagnosis, all of which utilise lactose digestion to inform on an individual's lactose tolerance status, and by implication lactase persistence status. The general practice is to give a lactose load after an overnight fast. The two most widely used methods are described below [32].

Diagnosis of lactase non-persistence/persistence

The blood glucose test: A baseline measurement of blood glucose is taken before ingestion of a lactose load, and then at various time intervals (usually every 30 min) for the following 2 h. An increase in blood glucose indicates lactose digestion (lactase cleaves the lactose molecules into glucose and galactose, allowing absorption into the bloodstream and subsequent detection), and no increase, or a 'flat line', is indicative of a lactose non-digester/maldigester or intolerant phenotype [32].

The breath hydrogen test: Undigested lactose remains in the intestine and is fermented by colonic bacteria, producing hydrogen gas, carbon dioxide, and methane in some individuals. A portion of the hydrogen produced in the colon diffuses into the blood and is excreted via the lungs. The breath test measures the excretion of this hydrogen. Typically, a subject is given an oral dose of lactose following an overnight (12 h) fast. Breath samples are collected at regular intervals for a period of 5-8 h and analyzed by gas chromatography. The historical test used 50 g of lactose as a challenge dose, and an increase of 20 parts per million (p.p.m.) or greater above the fasting level as an indicator

of lactose maldigestion. More recently, it has been shown that using a sum of hydrogen from hours 5, 6, and 7 and a _15 p.p.m. above-fasting criterion for maldigestion resulted in 100% sensitivity and specificity for carbohydrate maldigestion [36].

Concluding Remarks

Lactose intolerance is one of the factors affecting milk market demand and subsequently milk productivity. To bridge the gap of information on incidence of lactose intolerance, a survey was conducted in Ada'a district of Ethiopia, which is among the largest market-oriented dairy producing areas. 7.45% of the respondents reported that they don't consume milk but consume fermented milk ('Ergo'). The majority reported that they get vomiting upon consuming milk, and feel abdominal pain. The percentage of occurrence of lactose intolerance found in the survey is believed to affect the milk consumption in the country [37,38].

With the above concluding remarks, the following recommendations were drawn:

I. A large scale national study that assess wider population with more elaborate clinical diagnosis need to be done in future studies to know the exact prevalence/ distribution of lactose intolerance in the country. Lactase activity is usually assessed indirectly from measurements of lactose absorption via breath tests. Measurement of the hydrogen in the breath after taking lactose is a useful test because large amounts of hydrogen indicate that lactose is not being fully digested [36].

II. Dairy processors need to make yoghurt or remove lactose/lactose-free milk for consumers with special dietary requirements/lactose intolerance to improve milk consumption.

III. Awareness about lactose intolerance needs to be created among the society by health extension workers.

IV. Taking lactase tablets or drops, such as Lactaid or Dairy Ease is an option, which contain the enzyme that breaks down lactose, reducing the amount that your body must digest on its own are among the dietary strategies effectively manage lactose intolerance [36].

References

1. Mc Carron DA, Heaney RP (2004) Estimated healthcare savings associated with adequate dairy food intake. Am J Hypertens 17: 88-97.

2. Huth PJ, DiRienzo DB, Miller GD (2006) Major scientific advances with dairy foods in nutrition and health. J Dairy Sci 89: 1207-1221.

3. Jarvis JK, Lois D, Mc Bean GDM (2007) Handbook of Dairy Foods and Nutrition. 3rd eds. National Dairy council. CRC Press, USA.

4. Taylor MW, MacGibbon AH (2002) Lipids. Elsevier Sc. Ltd, New Zealand.

5. FAO (2013) Milk and dairy products in human nutrition, Rome, Italy.

6. Delgado C, Rosegrant M, Steinfeld H, Ehui S, Courbois C, et al. (1999) Livestock to 2020: The Next Food Revolution. Food Agriculture and the Environment Discussion Paper 28. International Food Policy Research Institute, Washington, D.C, USA.

7. Thorpe W, Muriuki HG, Omore A, Owango MO, Staal S, et al. (2000) Development of smallholder dairying in Eastern Africa with particular reference to Kenya.

8. Ndambi O, Hemme T, Latacz-Lohmann U (2007) Dairying in Africa-Status and recent developments. Livestock Research for Rural Development.

9. Ahmed MM, Ehui S, Assefa Y (2004) Dairy development in Ethiopia. IPFRI-

EPTD, Discussion paper 123, Environment and production technology division. Washington DC, USA.

10. FAOSTAT (2009) FAO statistical yearbook. Rome: Food and Agriculture Organization of the United Nations.

11. (2002) Encyclopedia of Foods: A guide to Healthy Nutrition. Dole Food Company, Inc. Published by Elsevier 3.

12. Walstra P, Wouters JTM, Geurts TJ (2006) Dairy Science and Technology. CRC Taylor & Francis Group, New York.

13. Rasooly A, Herold KE (2011) Biosensors. Elsevier Ltd. Encyclopedia of dairy sciences, USA.

14. Nebiyu R (2008) Traditional and improved milk and milk products handling practices and composition and microbial quality of raw milk and butter in Delbo watershed area of Wolaita zone. Hawassa University, Ethiopia.

15. Mesfin R, Getachew A (2007) Evaluation of grazing regimes on milk composition of Borana and Boran-Friesian crossbred dairy cattle at Holetta research center, Ethiopia. Livestock Research for Rural Development.

16. Alemu S, Tamiru F, Almaw G, Tsega A (2013) Study on bovine mastitis and its effect on chemical composition of milk in and around Gondar Town, Ethiopia. Journal of Veterinary Medicine and Animal health 5: 215-221.

17. Hernandez-Ledesma B, Ramos M, Gomez-Ruiz JA (2011) Bioactive components of ovine and caprine cheese whey. Small Ruminant Res 101: 196-204.

18. Campbell GM (2003) Aerated foods. Encyclopedia of food science and nutrition. Academic Press, USA.

19. Panesar PS (2011) Fermented dairy products: starter cultures and potential nutritional benefits. Food Nutr Sci 2: 47-51.

20. Buttriss J (1997) Nutritional properties of fermented milk products. Int J Dairy Technol 50: 21-27.

21. Heyman MB (2006) Lactose intolerance in infants, children, and adolescents. Pediatrics 118: 1279-1286.

22. Encyclopedia of Dairy Sciences (2011) (2ndedn), Elsevier Ltd. USA.

23. OECD/FAO (2006) OECD-FAO Agricultural Outlook. Rome, Italy.

24. Bender DA (2006) Benders' Dictionary of nutrition and food technology. (8thedn), Woodhead publishing limited. Cambridge, England.

25. Rusynyk RA, Still CD (2001) Lactose intolerance. J Am Osteopath Assoc 101: S10-20.

26. European Food Safety Authority/ EFSA (2010) Scientific opinion on lactose thresholds in lactose intolerance and galactosaemia. EFSA Journal.

27. Swallow DM (2003) Genetics of lactase persistence and lactose intolerance. Annu Rev Genet 37: 197-219.

28. Sahi T (1994) Genetics and epidemiology of adult-type hypolactasia.Scand J Gastroenterol Suppl 202: 7-20.

29. Wang Y, Harvey CB, Hollox EJ, Phillips AD, Poulter M, et al. (1998) The genetically programmed down-regulation of lactase in children. Gastroenterology 114: 1230-1236.

30. Paige DM, Davis LR (1985) Nutritional significance of lactose. I. Aspects of lactose digestion. In: Developments in Dairy Chemistry, Elsevier Applied Science, London 3: 111-132.

31. Mustapha A, Hertzler SR, Savaino DA (1997) Lactose: nutritional significance. In: Advanced Dairy Chemistry, Lactose, Water, Salts and Vitamins, (2ndedn). Chapman and Hall, London: 127-154.

32. Ingram CJE, Swallow DM (2009) Lactose Malabsorption. In: Advanced Dairy Chemistry Volume 3: Lactose, Water, Salts and Minor Constituents (3rdedn.). Springer ScienceþBusiness Media, LLC, Ireland.

33. Bayoumi RA, Saha N, Salih AS, Bakkar AE, Flatz G, et al. (1981) Distribution of the lactase phenotypes in the population of the Democratic Republic of the Sudan.Hum Genet 57: 279-281.

34. Hijazi SS, Abulaban A, Ammarin Z, Flatz G (1983) Distribution of adult lactase phenotypes in Bedouins and in urban and agricultural populations of Jordan. Trop Geogr Med 35: 157-161.

35. Dissanyake AS, El-Munshid HA, Al-Qurain A (1990) Prevalence of primary adult lactose malabsorption in the eastern province of Saudi Arabia. Ann Saudi Med 10: 598-601.

36. Suarez F, Shannon C, Hertzler S, Savaiano D (2003) Lactose Intolerance. In: Encyclopedia of food sciences and nutrition. Elsevier Science Ltd. USA.

37. Lomer MC, Parkes GC, Sanderson JD (2008) Review article: lactose intolerance in clinical practice--myths and realities.Aliment Pharmacol Ther 27: 93-103.

38. FAO (1990) The technology of traditional milk products in developing countries. Rome, Italy.

39

The Influence of Heat Treatment in Liquid Whey at Various pH on Immunoglobulin G and Lactoferrin from Yak and Cows' Colostrum/Milk

Shimo Peter Shimo[1,3], Wu Xiaoyun[1], Ding xuezhi[1], Xiong Lin[2] and Yan Ping[1]

[1]Lanzhou Institute of Husbandry and Pharmaceutical Sciences, Chinese Academy of Agricultural Sciences, Lanzhou P.R. China
[2]Laboratory of Quality and Safety Risk Assessment for Livestock Product (Lanzhou), Ministry of Agriculture, Lanzhou, China
[3]Government Chemist Laboratory Agency, Dar es Salaam, Tanzania

Abstract

Yak milk is gaining popularity yearly, as a source of nutritious, immune, less likely to cause allergies than cows' milk and a means of generating income, though ranked behind bovine milk in China. The main objective of the research was to compare the Immunoglobulin G (IgG) and Lactoferrin (LF) concentrations of liquid whey from Yak and dairy Cows' colostrum or milk. Thereafter, the thermal stability of liquid whey was evaluated by measuring the concentration change in IgG and LF using an ELISA technique, as influenced by temperature and pH after centrifugation.

IgG and LF concentrations in liquid whey from colostrums both in yak and cow were significantly higher ($p < 0.05$) compared to milk. IgG did not differ significantly ($p > 0.05$) between yak and cow's milk, as well as between yak and cow's colostrum. Also, LF content was not significant different between yak and cow milk. However, LF concentration was significantly higher in cows' colostrum than in yak. With reference to IgG content, yak breed observed to produce high quality colostrum for all samples tested compared to cows.

The extent of IgG and LF denaturation confirmed to increase with increased protein concentration, temperature and pH change near to pI. Colostral IgG and LF were more denatured compared to milk in both yak and dairy cows. There was significant ($p < 0.05$) influence of pH changes resulted in either partial or complete denaturation or increased tendency to aggregate, which was removed during centrifugation. Liquid whey less affected at lower pH and at mild heat temperatures.

Keywords: ELISA; Yak colostrum/milk; Heat stability; Liquid whey; Immunoglobulin G; Lactoferrin

Introduction

Whey, a soluble fraction of milk and by-product liquid from cheese or casein manufacturers is widely recognized to contain many valuable constituents such Lactoglobulin (β-Lg), Lactalbumin (α-Lac) and BSA. IgG and LF are natural occurring antimicrobial proteins available in colostrum and milk as the major antimicrobial agent for a wide range of pathogens or spoilage micro-organisms [1-6]. Colostrums contain natural defensive antibodies from mother to neonate. It is highly concentrated with immune, growth and tissue repair factors, with a distinct level of protein concentration particularly in Immunoglobulins (IgG, IgA and IgM) and nutritive components. IgG being the major Immunoglobulin present in ruminant milk, and IgA being the major Immunoglobulin present in human milk [7,8].

LF, a glycoprotein with molecular weight of about 80 kDa (~700 amino acids) belongs to the transferrin family, commonly known as lactotransferrin [9]. It is an iron (Fe^{+3} ions) binding glycoprotein which is considered as an essential element of innate immunity with antigen-nonspecific defense mechanism after exposure to an antigen [10,11]. Antibacterial activities of LF is effective against gram positive and gram negative bacteria, viruses (non-capsular and capsular) and several parasites and fungi [12].

Processing conditions applied during production involving heating, adjustment of pH, minerals, sugars, protein composition and processing time determine retentions of nutritional value and physiological importance of whey proteins [6,13,14]. Whey proteins are comparatively thermal labile, therefore; heat treatment of milk contribute to denaturation and aggregation of the denatured proteins with caseins [15].

The high level of intramolecular disulphide bridges plus other whey components including fats, lactose, carbohydrates, salts, and other proteins assist to stabilize antibodies during heat treatment [16]. Immunoglobulins thermal resistance varies between classes, i.e. high stability in a decreasing order from IgG, IgA and IgM as well as varies among species [17-19]. The conformation changes in the IgG molecule which occurs during heat exposure reduces antigen-binding activity of this protein Calmettes et al. [20] and antigen-binding site of IgG is more sensitive to heat compared to other part of the molecule [21,22]. On the other hand, LF are denatured by heat treatment too, therefore pasteurizing milk might affect LF unless mild heat treatment is applied [23,24].

Yaks (*Bos grunniens*) are distributed at high altitudes ranging from 2000 m to 4500 m. Though the milk yield is low but with good quality due to high level of fat, proteins, sugar content as well as carotene and vitamin E compared to cows' milk. Yak milk are easily digested and absorbed by infants due to low α-caseins (about 40%) and higher β-casein (more than 45%) with small increase in κ-casein (15%) compared to cows' milk which contains higher α-caseins (50%) and lower β-casein (35%) [25-27].

*Corresonding author: Shimo Peter Shimo, Lanzhou Institute of Husbandry and Pharmaceutical Sciences, Chinese Academy of Agricultural Sciences, Lanzhou P.R. China, E-mail: shimope_2000@yahoo.com

In China, Yaks and their products are mostly dominated in northwestern China [13,14,28]. Due to the increase demand in new functional food development and therapeutic value of LF and IgG (such as infant formulas fortification from different milk sources), more studies are needed from different animal species. Therefore, we aimed to compare the IgG and LF concentrations from Yak and dairy Cows' milk or colostrum. Furthermore, we evaluated the effect of heat at various pH on antibacterial proteins (IgG and LF) in liquid whey from yak colostrum and milk in comparison to cows' as applied during dairy processing temperatures.

Apart from IgG and LF in liquid whey, other major whey proteins namely β-Lg, α-Lac, BSA and total whey proteins (TWP) also were analyzed as might contribute to denaturation of IgG and LF. But, our discussion focused only on antimicrobial proteins (IgG and LF). Turbidity and free sulfhydril (-SH) on liquid whey supernatants were also evaluated to monitor the effect of heat, pH. Therefore, optimization of environmental processing conditions during extraction of liquid whey expected to retain reasonable amount of antimicrobial proteins which is beneficial in new functional food development.

Materials and Methods

Materials

Collection of samples: Random sampling of colostrum and milk samples from individual yak were collected from Gannan and Qinghai Plateau (Gansu and Qinghai Provinces, China); while Holstein Friesian cow milk and colostrum were collected from Lanzhou dairy Farmer. A total of 80 samples (20 samples each group) were collected for this study. Approximately 80 mL sample for each animal was collected immediately after milking into sterile plastic test tubes and kept on ice before sent to the laboratory. All samples were stored at -80°C (Ultra Low Temperature Freezer, Haier-Biomedical, China) before extraction and analysis.

Standard ELISA kits for IgG and LF were purchased from Biotechnology Co. Ltd, Shanghai-China. A Coomassie (Bradford) Protein Assay Kit for total protein analysis and all other analytical grade chemicals were purchased from Sigma Aldrich (St Louis, MO, China).

Methods

Liquid whey preparation: Liquid whey samples were prepared according to Chen et al. [29] and Cozma et al. [30] with some modifications. Frozen colostrum/milk samples were thawed in running tap water; then, 40 mL of colostrum/milk samples were skimmed by centrifuge (Biofuge stratus centrifuge, Thermo, Germany) at 5000 rpm for 15 min at 4°C. Skimmed samples were adjusted to pH 4.6 using 1 M HCL or 1 M NaOH, warmed at 40°C in water bath (Julaba F12, Germany) for 30 min. and centrifuged at 5000 rpm for 15 min at 4°C. Furthermore, liquid whey supernatants obtained (1.5 mL) were centrifuged at 17,000 rpm for 30 min at 4°C to obtain the clear supernatants which were stored in fridge at 4°C (Ronshen, BCD-209YMB, China) for short-term or frozen at -18°C (Sanyo Medical Freezer, Japan) for long term analysis, respectively.

Total whey proteins assay: The Total Whey Protein (TWP) concentration was assayed by using the Modified Bradford Assay protocol as applied by Cozma et al. [30] using Coomassie (Bradford) Protein Assay Kit Manufacturer's instructions. The liquid whey samples extracted from milk and colostrum were diluted 50 times and 1000 times with PBS, respectively. For each sample, 20 μL aliquot was transferred into microplate wells followed by addition of 200 μL Bradford reactive and incubated for 10 min at room temperature.

Thereafter, with 15 minutes, absorbance was measured at 595 nm (λ max) using microplate reader (Multiskan FC, version 1.00.96, Finland). The BSA standard was serially diluted with PBS according to Manufacturers' instructions at the range of 10-150 μgmL⁻¹ to prepare a standard curve for quantifying TWP for each sample.

ELISA assays: IgG and LF concentrations in liquid whey samples were quantified by Enzyme-Linked Immunosorbent Assay (ELISA) technique using Bovine ELISA quantification kits according to the Manufacturer's protocols with minor modifications pertaining sample dilutions. 50 μL of each diluted specific Bovine standard and liquid whey sample were transferred into a 48-microplate wells coated with primary antigen, incubated in oven (THZ-C-1, China) at 37°C for 30 min., followed by washing five times using 20-fold washing buffer solution. Then, 50 μL HRP conjugate reagent was added to each microplate well except blank and incubated at 37°C for 30 min. After 30 min of incubation, microplate wells were washed five times using 20-fold washing buffer solution followed by addition of 50 μL Chromogenic agent A and B, respectively, to each well and incubated at 37°C for 10 min. in dark place. The reaction was stopped by addition of 50 μL stop solution reagent after incubation, whereby colour changed from blue to yellow. Within 15 min. samples were assayed at 450 nm using microplate reader (Multiskan FC, version 1.00.96, Finland). Using specific Bovine standards, the same technique was applied to quantify β-Lg, α-Lac and BSA in liquid whey samples from milk and colostrum.

Bovine standards for IgG (range; 5-80 μgmL⁻¹) and LF (range; 12.5-800 μgL⁻¹) were diluted as per Manufacturer's instructions to prepared calibration curves for quantification in each sample and all samples were analyzed in duplicate.

Heat treatment of liquid whey: From each five liquid whey samples either colostrum or milk, 2 mL were selected randomly and mixed together to obtain one homogenous sample (approximate 10 mL). Heat stability of liquid whey from yak and cow's milk/colostrum were evaluated according to Chevalier et al. [31] and Laleye et al. [32] with some modifications. Liquid whey samples were adjusted to pH 3.5, 4.6, and 5.5 (pH meter 55, Martini instruments, Mauritius) by 1 M HCL or 1 M NaOH. After pH adjustment, 5 mL samples in 15 mL capacity closed plastic test tube was heated in digital temperature-controlled water bath (Julaba F12, Germany) at 60°C/30 min, 63°C/30 min, 72°C/15 sec and 90°C/5 min. Different heating up times were allowed to equilibrate to reach the water bath temperature before measuring experimental time. One extra sample was used to monitor the temperature by inserting a thermometer. Immediately, heat treated samples were cooled in an ice water and at each pH studied; a control sample was maintained at room temperature (22°C). An aliquot of heat treated sample (1.5 mL) was centrifuged at 10,000 rpm for 15 min at 4°C and each thermal treatment was done in duplicate.

Turbidity: Optical density (absorbance at 20°C) of the supernatant was measured the 280 nm wavelength by UV/VIS 2550 Spectrophotometer (Shimadzu, Japan) to evaluate the denatured proteins with comparison to untreated sample as applied by Laleye et al. [32].

Free sulfhydryl content: Free sulfhydryl (-SH) content was determined according to Monahan, et al. [33] and Hoffmann and Mil [34]. 0.02 mL of 10 mM DTNB was added to 2 mL of the liquid whey supernatant and the absorbance (at 20°C) measured by UV/VIS 2550

Spectrophotometer (Shimadzu, Japan) at 412 nm. Free sulfhydryl content was calculated according to Manufacturer's protocol using Molar absorptivity (E=14,150 M^{-1}cm^{-1}).

Statistical analysis

All samples were analyzed in duplicate, and the results were reported as the means ± SD. Results were evaluated statistically using the Statistical Package for Social Sciences, v16.0 (SPSS, Chicago, IL, USA). The data for IgG and LF concentration from different sources of liquid whey were analyzed statistically by one-way analysis of variance (ANOVA). A three factor analysis of variance with interaction was used to evaluate the different sources of liquid whey, effect of heat treatment and various pH on denaturation of proteins in liquid whey. Thereafter, multiple comparisons of means were analyzed using Least Significance Difference (LSD) and Duncan comparison at a α-level of 5%.

Results and Discussion

IgG concentration in liquid whey

Results for liquid whey from yak and cows'colostrum as well as milk respectively are presented in Figure 1. Colostral IgG concentration were 59.40 ± 4.44 mgmL^{-1} and 65.99 ± 25.90 mgmL^{-1} from yak and cows' liquid whey samples, respectively. IgG did not differ significantly (p>0.05) between yak and cows' colostrum. Yak and cows' milk IgG concentration were 0.58 ± 0.16 mgmL^{-1} and 0.49 ± 0.08 mgmL^{-1}, respectively. IgG content both yak and cows' milk were not significantly differently (p>0.05). However, results indicated that colostral IgG concentration were significantly higher (p<0.05) compared to milk both in yak and cow.

Analyses of colostrum and milk from individual animal species (yak and cow) indicated considerable variations in their IgG content. The colostral IgG concentration ranged from 54.44-72.72 mgmL^{-1} (70.98% of TWP) and 18.95-102.21 mgmL^{-1} (83.88% of TWP) for yak and cows' colostrum, respectively. The milk IgG concentration of liquid whey ranged from 0.35-0.87 mgmL^{-1} (8.72% of TWP) and 0.39-0.65 mgmL^{-1} (9.50% of TWP) for yak and cow, respectively (Figure 1).

Our results are in agreement with other studies, bovine colostrum contain high IgG content compared to normal milk which has been reported to vary from 40-200 mgmL^{-1} IgG and 0.7-1.0 mgmL^{-1} in colostrum and normal milk, respectively [3,8,35-37].

Several factors may be contributed to concentration variations of IgG in colostrum and milk, such as species, breed, health status, pathogen exposure, feeding practices and environmental conditions (season), prepartum diet/nutritional management and the stage of lactation/parity [38-43].

Other factors which might be attributed to variation of IgG content within yak breed or between yak and cows includes number of lactation and length of the non-lactating [42,44,45]. Heat stress also reported to affect the animal that attribute to reduction of colostral levels of IgG and IgA, fat, lactose, protein, and energy [39].

According to Hansen [46] and Godden [47], high quality colostrum should contain more than 50 gL^{-1} of IgG. From our results showed that 100% of yak produced high quality colostrum compared to cow Holstein Friesian cow which was 73.33%.

LF concentration in liquid whey

Experimental results are presented in Figure 1, LF content measured by ELISA technique were 0.58 ± 0.16 mgmL^{-1} and 0.49 ± 0.08 mgmL^{-1} from yak and cows' milk, respectively. Both yak and cows' milk were not significant different (p>0.05). Colostral LF concentration were 1.52 ± 0.56 mgmL^{-1} and 2.48 ± 0.63 mgmL^{-1} from yak and cows' liquid whey supernatants, respectively. Cows' colostrum contained significantly higher LF content (p<0.05) compared to yak colostrum. In both yak and cow, colostrum contained significant higher (p<0.05) LF concentration compared to milk.

The mean LF concentration from yak colostrum ranged 0.94-2.47 mgmL^{-1} (1.84% of TWP) and 1.04-3.11 mgmL^{-1} (3.15% of TWP) for cows. While, mean LF concentration ranged from 0.24-0.43 mgmL^{-1} (4.69% of TWP) and 0.15-0.35 mgmL^{-1} (4.32% of TWP) for yak and cows' milk respectively.

Our results obtained were in agreement with other studies reported from different bovine, whereby LF concentration is significantly higher in colostrum than in milk. According to Haenlein, [48] the LF concentration was 1.5 mgmL^{-1} and 0.02-0.5 mgmL^{-1} in cow colostrum and milk respectively. The same results of LF content in cows' milk were reported by Zimecki and Kruzel [36]. Different factors may contribute to variation of LF concentration such as different animal health status, level of BSA, lactation stage as well as daily milk production, some studies reported a big variation of LF content in milk among individual cows, for example from 0.06-1.0 mgmL^{-1} LF [28,49].

TWP composition

The TWP were 6.70 mgmL^{-1} and 5.20 mgmL^{-1} for yak and cows'

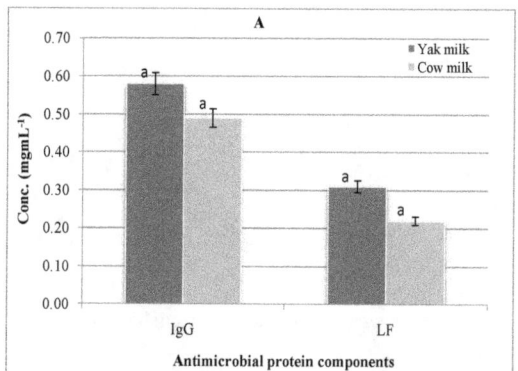
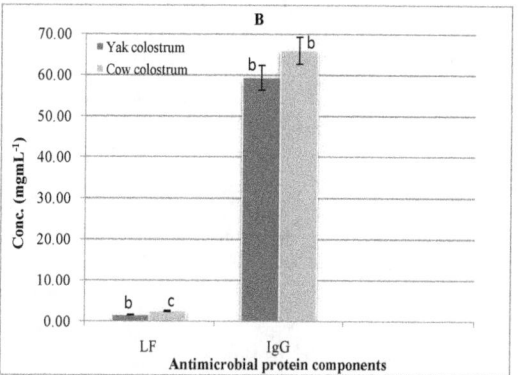

Figure 1: (a) Yak and cows' milk (b) Yak and cows' colostrum: Concentration of IgG and LF in liquid whey supernatant from yak and cows' milk as well as colostrum respectively after centrifugation. ^{a,b,c}Different superscripts within one protein type in both graphs indicate significant difference between different source (P < 0.05). Each value is the average of 20 samples determinations, error bars indicates standard errors.

milk, respectively, whereas 83.68 mgmL⁻¹ and 78.67 mgmL⁻¹ for yak and cows' colostrum respectively. High content of TWP observed may be attributed to high content of IgG in colostrum than in milk. The study revealed that, the TWP was significantly higher (p<0.05) in colostrum compared to milk in both yak and cows. By comparison of means, TWP between yak and cows' colostrum did not differ significantly (p>0.05), the same applies to yak and cows' milk. These results obtained from cows' milk are in agreement with literatures [50].

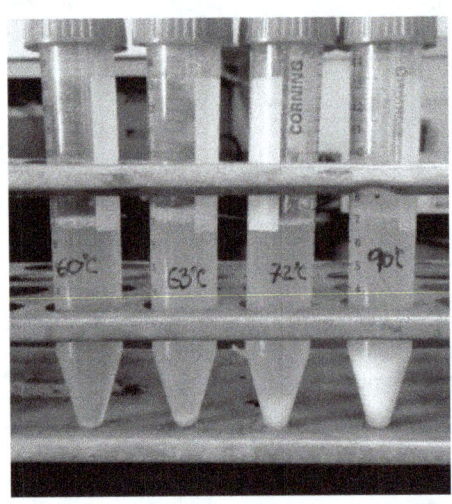

Figure 2: Appearance of liquid whey from yak colostrum after heat treatment at 60°C/30 minutes, 63°C/30 minutes, 72°C/15 seconds and 90°C/5 minutes heat treatment at pH 5.5 before centrifugation.

Effect of heat on liquid whey turbidity at various pH

Appearance and results of turbidity as the apparent optical density at 280 nm of liquid whey are presented in Figures 2 and 3, respectively. Slightly turbid and transparent (whitish in colour-viscous) and some small soluble aggregates increased with temperature increase particularly at 72°C/15 sec and 90°C/5 min., indicating whey proteins were denatured with temperature increase. Statistics analysis on heating liquid whey at each pH treatment followed by centrifugation of the supernatants showed that 60°C/30 min had no significant effect on turbidity. Also, comparison between 63°C/30 min v/s 60°C/30 min treatment showed no significant different on turbidity (p>0.05) too. However, more effect noted in both samples at 72°C/15 sec and 90°C/5 min, though were not significantly different between the two treatments (p>0.05), suggesting that denaturation temperature for β-Lg probably was reached.

Except cows' colostrum was more affected, probably due to high content of IgG and β-Lg that was noted on raw samples, turbidity results showed that comparison between different sources of liquid whey processed did not differ significantly (p>0.05) among other sources.

The pH change on the liquid whey influenced turbidity decrease significantly (p<0.05) when supernatants centrifuged after heat treatment at various pH, as the sign of denaturation and aggregation [51]. Turbidity (%) decreased significantly from pH 3.5 to pH 5.5 in both milk and colostrum supernatants. The liquid whey supernatant from yak/cows' milk or colostrum treated at 60-63°C at pH 3.5-5.5 after centrifuge, the turbidity reduced by 2.89-26.51% but high effect was noted at pH 5.5. Serious effect was noted 72°C/30 min and 90°C/5 min reduced the turbidity from 20.74-67.40% when supernatants were

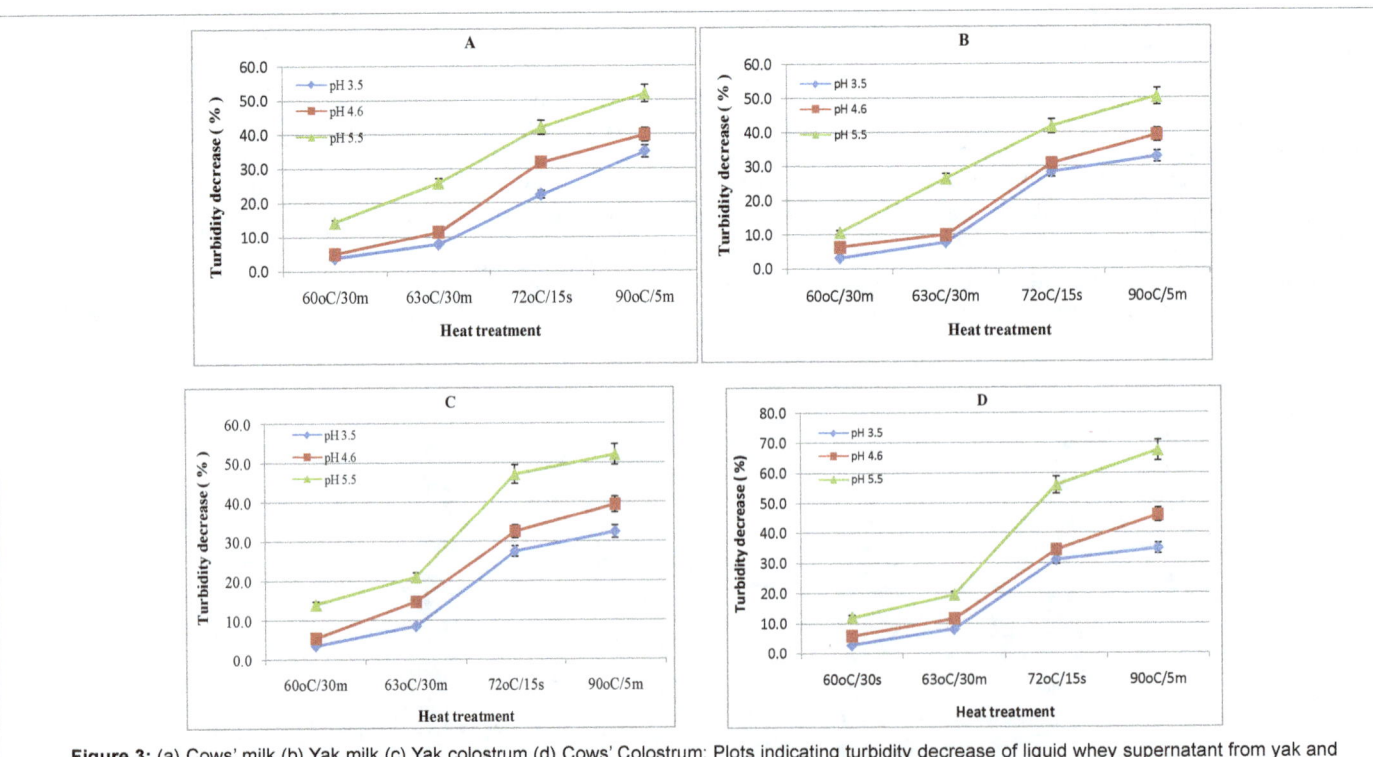

Figure 3: (a) Cows' milk (b) Yak milk (c) Yak colostrum (d) Cows' Colostrum: Plots indicating turbidity decrease of liquid whey supernatant from yak and cow after heat treatment (60°C/30 minutes, 63°C/30 minutes, 72°C/15 seconds and 90°C/5 minutes) at various pH (pH 3.5, 4.6 and 5.5) followed by centrifugation. Each value is the average of two determinations, error bars indicates standard errors.

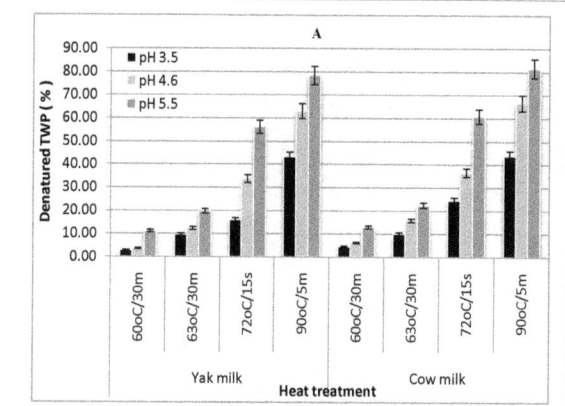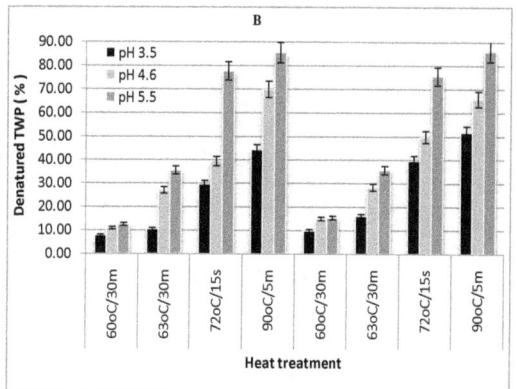

Figure 4: (A) Milk (B) Colostrum: Graphs indicating percentages reduction of Total Whey Proteins (TWP) in liquid whey supernatant from yak and cow after heat treatment (60°C/30 minutes, 63°C/30 minutes, 72°C/15 seconds and 90°C/5 minutes) at various pH (pH 3.5, 4.6 and 5.5) followed by centrifugation. Each value is the average of two determinations, error bars indicates standard errors.

centrifuged. Changes on liquid whey supernatant turbidity correlated with changes on TWP after heat treatment. TWP concentration in centrifuged supernatants was significantly reduced with increase pH and temperatures (Figure 4). The TWP concentration of Liquid whey from yak/cow milk or colostrum treated at 60-63°C/30 min at pH 3.5-5.5 reduced by 2.80-35.52%; while when the temperature raised to 72°C/15 sec and 90°C/5 min TWP reduced by 15.94-85.81%. High effect was noted at pH 5.5 and at high temperatures (72°C/15 sec and 90°C/5). An increase in pH influenced formation of soluble protein aggregates with large particle sizes which was removed out after centrifugation. The same explanation evidence reported recently by Chandrapala et al. [52].

The turbidity decrease of heated and centrifuged supernatants might be due to the processing condition near to pI of β-Lg. The temperature applied probably caused the native monomer β-Lg to reversibly interchange into a non-native monomer and non-native disulfide-bonded dimers of β-Lg and then interacted with other whey proteins such as α-LA, BSA [32,53-55] . Thereafter the denatured proteins were removed during centrifugation that attributed to reduce the liquid whey turbidity. The same phenomena was revealed when Bernal and Jelen [56] studied thermal stability of whey by a calorimetric at pH below pH 3.7, protein precipitation was prevented during heating at 95°C for 5 min; although protein denaturation occurred in various whey protein fractions. The protein concentration in the supernatant as a function of pH and temperature has been also reported by Chandrapala et al. [52], whereby soluble aggregates with large particles increased with pH and temperature increase as well as a decrease in proteins content by 7 to 20% noted from pH 3-10.5 at 15-90°C, respectively.

The effect of temperature and pH on free SH

It was observed that, denaturation and aggregation of liquid whey process as function of temperature and pH. Free -SH significantly decreased (p<0.05) with pH increase (pH 3.5 to 5.5) and temperature increase to indicate that more free -SH involved during denaturation process at pH 5.5 i.e. no more active sites after being denatured at pH 5.5 compared to low pH during heat treatment [57,58].

From our results it was observed that, liquid whey had low free -SH concentration at pH 5.5 compared to pH 3.5 and 4.6 (results not shown) indicating that more denaturation and aggregation occurred that involved SH and S-S bonds at pH 5.5.

Free -SH content decreased as proteins denaturation increased

after heat treatment at various pH followed by centrifugation. Liquid whey studied comprised all major proteins including β-Lg, α-La, BSA, IgG and LF together, in that way heat treatment may easily induce interactions that result to sized aggregates by covalent (S-S) and non covalent bonds [59]. Presence of BSA on the liquid whey up on heating at pH 5.5 might be facilitated conformation of different monomers which expose -SH group to the outer structures with another reactive BSA to form a dimers and trimers through -SH/S-S interchange reactions [59].

Effect of heat on liquid whey IgG content at various pH

Results showed that IgG in milk was significantly (p<0.05) less denatured compared to colostrum in both yak and cows. Effect of heat treatment at various pH on liquid whey followed by centrifugation did not differ significantly (p>0.05) on heat stability between yak and cows' liquid whey (milk IgG concentration). However, the liquid whey from yak colostrum was less denatured significantly (p<0.05) compared to cow's colostrum. Probably, due to high IgG and β-Lg which were observed in cows colostrum though did not differ significantly. Generally, as reported previous, the extent of IgG denatured increased with increasing protein concentration, temperature and pH near to pI of IgG and β-Lg.

The pH sensitivity resulted in either partial or complete denaturation and increased the tendency to aggregate, which observed and were removed during centrifugation. The pH change in liquid whey significantly influenced denaturation (p<0.05) as approaching to pI. More serious aggregation observed and reflected with the calculated denatured IgG percent at 72°C/15 sec and 90°C/5 min (Figure 5). For mean IgG concentration in all liquid whey sources assayed, at 60°C/30 min treatment in all pH tested (pH 3.5-5.5) had no significant effect on IgG content reduction. Moreover, the effect of pH change from pH 3.5 to 5.5 during heat treatment of liquid whey reduced significantly IgG concentration as the temperature and pH increased. The mean reduction in IgG concentration as depicted in Figure 5 was less than 10% in milk for all pH tested at 60°C/30 min, except at pH 5.5 where reduction was higher in colostrum that rose to 12.95% and 13.48% in yak and cows' colostrum, respectively. When liquid whey from yak/cow milk or colostrum heated at 60-63°C/30 min. at pH 3.5-5.5, the IgG concentration reduced by 2.67-21.64% whereas at 72°C/15 sec and 90°C/5 min reduced by 14.45-55.67%.

High percent reduction in IgG concentration observed at pH 5.5

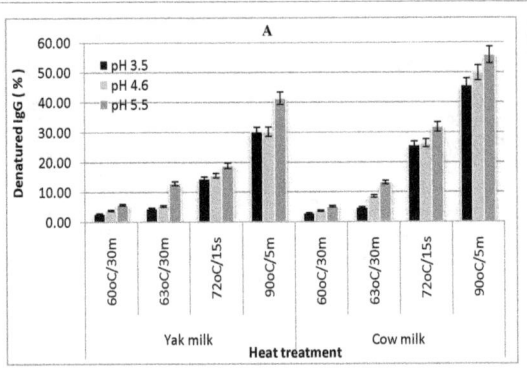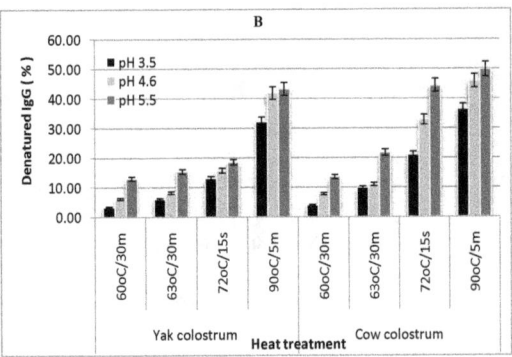

Figure 5: (A) Milk (B) Colostrum: Graphs indicating percentages of denatured IgG of liquid whey supernatant from yak and cow after heat treatment (60°C/30 minutes, 63°C/30 minutes, 72°C/15 seconds and 90°C/5 minutes) at various pH (pH 3.5, 4.6 and 5.5) followed by centrifugation. Each value is the average of two determinations, error bars indicates standard errors.

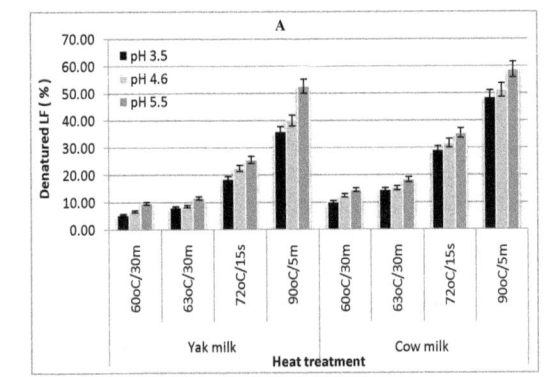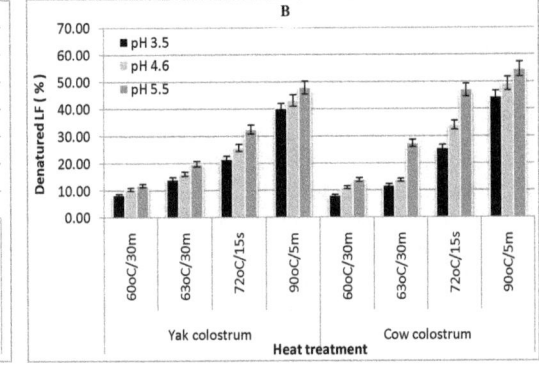

Figure 6: (A) Milk (B) Colostrum: Graphs indicating percentages of denatured LF of liquid whey supernatant from yak and cow after heat treatment (60°C/30 minutes, 63°C/30 minutes, 72°C/15 seconds and 90°C/5 minutes) at various pH (pH 3.5, 4.6 and 5.5) followed by centrifugation. Each value is the average of two determinations, error bars indicates standard errors.

at all temperatures used, suggesting that the effect was due to proteins concentration and processing condition near to pI of IgG and β-Lg which contributed to initiate coagulation and denaturation process [55,58].

High denaturation effect observed in colostrum liquid whey may be due to high IgG content and TWP compared to milk. Our results are in line with those reported by Levieux et al. [60], colostrum was more sensitive to heat treatment compared to milk due to high IgG content (12·6 mgmL⁻¹ v/s 0·5 mgmL⁻¹ for the mature milk) that influenced coagulation even at low temperature.

The decrease in IgG content, TWP assayed by ELISA technique and turbidity decrease (after centrifugation) may be due to aggregation and denaturation/unfolding of IgG molecules effected by heat treatment and pH increase (close to pI) that influences in loss of antigenicity [16,61]. The conformation change in the IgG molecule occurred during heat exposure reduces antigen-binding activity of this protein Calmettes et al. [62], whereas, antigen-binding site of IgG is more sensitive to heat compared to other part of the molecule [21,22]. Trujillo et al. [19] reported that Immunoglobulins, are species specific on heat sensitivity particularly IgG such as on bovine, caprine and ovine. It may be concluded that less effect IgG denaturation observed in acidic medium.

The presence/absent of casein or fat and different processing technique applied to prepare liquid whey might also influence the reduction of IgG and other whey proteins [62]. Walstra et al. [63]

demonstrated the effect of removal of fat from colostrum. Heat treatment of colostrum at 60°C/30 min followed by centrifugation reduced 28% IgG due to flocculation of the fat globules by the immunoglobulins. According to Elfstrand et al. [17], reduction of IgG and other growth factors influenced by interaction of fat globules and high IgG concentration particularly in colostrum entrapped in the caseins matrix or cleaved by proteolytic enzymes.

On the other hand, Elfstrand et al. [17] showed that, heat treatment (60°C, 45 min) of liquid whey from Colostrum without the filtration process there was no change on concentration of IgG1 and TGF-b2. The same observation was noted even when whole colostrum was pasteurized (60°C/30 min), that no reduction on the concentration of the Immunoglobulins. Immunoglobulins in whole colostrum and colostrum whey were more resistant to heat processing temperatures compared to colostrum concentrate without fat, casein, salts and lactose, means that such components may protect the Immunoglobulins during heat processing treatment and filtration [16].

Effect of heat on liquid whey LF content at various pH

Heat treatment of liquid whey at various pH from different sources showed that LF in milk was less denatured ($p < 0.05$) compared to liquid whey from colostrum both in yak and cows species. In comparison, the effect of heat treatment at various pH on liquid whey showed no significance difference ($p > 0.05$) in LF denatured between yak and cows' milk liquid whey. However, LF in yak colostrum liquid whey was less denatured compared to cow's colostrum liquid whey. The extent of LF denaturation increased with increased protein concentration, temperature and pH near to pI of LF and β- Lg; thus why colostrums were more denatured compared to milk in both yak and cow.

LF concentration decreased significantly as temperature increased ($p<0.05$), although there was no significant difference between 60°C/30 min and 63°C/30 min. The pH change of the liquid whey decreased significantly ($p<0.05$) from pH 3.5 to 5.5 or near to pI of LF and β-Lg, serious coagulation observed at such conditions which reflected with the calculated denatured LF percent (Figure 6). In general, the mean LF content showed significantly a gradual reduction of concentration from pH 3.5-5.5 for all liquid whey sources analyzed.

The mean reduction in LF concentration was about 10% in both milk and colostrum for all pH tested at 60°C/30 min, except at pH 5.5 where reduction was more noticed in colostrum that rose to 13.98% and 11.82% in cow and yaks' colostrum, respectively. When liquid whey from yak/cow milk or colostrum treated at temperature 60-63°C/30 min at pH 3.5-5.5, LF concentration reduced by 5.45-27.34%; Moreover, at 72°C/15 sec and 90°C/5 min reduced by 18.59-58.69%. High percent reduction in LF concentration noted at pH 5.5 at all temperatures used, suggesting that the effect was due to processing near to pI of LF and β-Lg which initiated coagulation and denaturation process.

Heat stability of LF is very crucial when it comes to bioactive components of foods. Several studies have been researched the influence of pH during pasteurization of LF in solutions. Saito et al. [64] studied the influence of pH (pH 2-11) observed that 5% bovine LF solution in distilled water at 80-120°C for 5 min gelled at neutral and alkaline pH while in acidic condition remained soluble and clear. Wakabayashi et al. [24] reviewed that bovine LF are comparatively stable to heating (90-100°C) at pH 4 pertaining iron binding capacity and antigenic activity.

Ueno et al. [65] studied the effects of temperatures (50-80°C for 10 min) on the iron-solubilizing capacity of LF in the presence of sodium bicarbonate observed that the temperatures above 70°C caused precipitation in the presence of Fe(III). The precipitation observed was related to the degree of thermal denaturation of LF, formed high-molecular-weight aggregates as disulphide bonds. Naturally, LF is a glycosylated protein, whereby the number and location glycosylation sites differ from one species to another. While the proportionality of glycan of oligomannosidic and of N-acetyllactosamine type also varies with lactation stage. Therefore, the primary structure of the specific glycans bound to LF determines the thermal stability [66].

However, discrepancies in our results obtained on effect of heat treatment at various pH of the liquid whey with other studies might be due to different methods to measure its denaturation and thermal treatment. Additionally, the effect of IgG and LF content in liquid whey may be influenced by sample volume and type of the container used to heat samples. Based on the processing condition, our study was conducted on unpurified proteins, whereas in the most mentioned and reported researches used pure, desalted proteins and in buffer solutions. Therefore, probably lactose, mineral, impurities and other proteins influenced our results.

To our knowledge this is the first time yak milk/colostrum liquid whey supernatants was heat processed at various pH to study the effect on antimicrobial proteins retained in comparison to cow.

Conclusion

According to the experimental results, liquid whey pH and thermal

processing temperatures have significant effect on physical properties as well as antimicrobial proteins composition in manufacturing and formulation of functional foods. We confirmed that at low pH and mild heat treatment retained reasonable amount of IgG and LF. High level of aggregation and denaturation of the liquid whey proteins observed to be at pH 5.5 and at temperatures above 63°C. These results indicate that IgG and LF at pH 3.5 and 4.6 at mild heat temperatures condition is suitable as a practical method for pasteurization. Further studies are highly encouraged to verify the effects of heat treatment at various pH on purified IgG and LF from yak colostrum/milk liquid whey.

Acknowledgement

Research supported by The Agricultural Science and Technology Innovation Program of China (CAAS-ASTIP-2014-LIHPS-01) and the Key Technologies R&D Program of China during the 12th Five -Year Plan period (2012BAD13B05).

References

1. Meisel H (1998) Overview on Milk Protein-derived Peptides. International Dairy Journal 8: 363-373.

2. De Wit JN (1998) Marschall Rhône-Poulenc Award Lecture. Nutritional and functional characteristics of whey proteins in food products. J Dairy Sci 81: 597-608.

3. Korhonen H, Marnila P, Gill HS (2000) Milk immunoglobulins and complement factors. Br J Nutr 84 Suppl 1: S75-80.

4. Florisa R, Recio I, Berkhout B, Visser S (2003) Antibacterial and antiviral effects of milk proteins and derivatives thereof. Curr Pharm Des 9: 1257-1275.

5. Ha E, Zemel MB (2003) Functional properties of whey, whey components, and essential amino acids: mechanisms underlying health benefits for active people (review). J Nutr Biochem 14: 251-258.

6. Michaelidou AM (2008) Factors influencing nutritional and health profile of milk and milk products. Journal of Small Ruminant Research 79: 42-50.

7. Farrell HM Jr, Jimenez-Flores R, Bleck GT, Brown EM, Butler JE, et al. (2004) Nomenclature of the proteins of cows' milk--sixth revision. J Dairy Sci 87: 1641-1674.

8. Stelwagen K, Carpenter E, Haigh B, Hodgkinson A, Wheeler TT, et al. (2009) Immune components of bovine colostrum and milk. J Anim Sci 87: 3-9.

9. García MIA, Cendón TS, Arévalo-Gallegos S, Rascón-Cruz Q (2012) Lactoferrin a multiple bioactive protein: an overview. Biochim Biophys Acta 1820: 226-236.

10. Legrand D, Pierce A, Elass E, Carpentier M, Mariller C, et al. (2008) Lactoferrin structure and functions. Adv Exp Med Biol 606: 163-194.

11. Legrand D, Mazurier J (2010) A critical review of the roles of host lactoferrin in immunity. Biometals 23: 365-376.

12. Szwajkowska M, Wolanciuk A, Barlowska J, Król J, Litwinczuk Z, et al. (2011) Bovine milk proteins as the source of bioactive peptides influencing the consumers ' immune system-a review. Animal Science Papers and Reports 29: 269-280.

13. Liu HN, Zhang C, Zhang H, Guo HY, Wang PJ, et al. (2013) pH treatment as an effective tool to select the functional and structural properties of yak milk caseins. J Dairy Sci 96: 5494-5500.

14. Li Q, Ma Y, He S, Elfalleh W, Xu W, et al. (2014) Effect of pH on heat stability of yak milk protein. International Dairy Journal 35: 102-105.

15. Boye JI, Alli I (2000) Thermal denaturation of mixtures of alpha-lactalbumin and beta-lactoglobulin?: a differential scanning calorimetric study. Journal of Food Research International 33: 673-682.

16. Indyk HE, Williams JW, Patel HA (2008) Analysis of denaturation of bovine IgG by heat and high pressure using an optical biosensor. International Dairy Journal 18: 359-366.

17. Elfstrand L, Lindmark-Mansson H, Paulsson M, Nyberg L, Akesson B, et al. (2002) Immunoglobulins , growth factors and growth hormone in bovine colostrum and the effects of processing. International Dairy Journal 12: 879-887.

18. McMartin S, Godden S, Metzger L, Feirtag J, Bey R, et al. (2006) Heat treatment of bovine colostrum. I: effects of temperature on viscosity and immunoglobulin

G level. Journal of Dairy Science 89: 2110-2118.

19. Trujillo AJ, Castro N, Quevedo JM, Argüello A, Capote J, et al. (2007) Effect of heat and high-pressure treatments on microbiological quality and immunoglobulin G stability of caprine colostrum. J Dairy Sci 90: 833-839.

20. Calmettes P, Cser L, Rajnavölgyi E (1991) Temperature and pH dependence of immunoglobulin G conformation. Arch Biochem Biophys 291: 277-283.

21. Mainer G, Sanchez L, Ena J, Calvo M (1997) Kinetic and Thermodynamic Parameters for Heat Denaturation of Bovine Milk IgG , IgA and IgM. Journal of Food Science 62: 1034-1038.

22. Domínguez E, Pérez MD, Puyol P, Sanchez L, Calvo M, et al. (2001) Effect of pH on antigen-binding activity of IgG from bovine colostrum upon heating. J Dairy Res 68: 511-518.

23. Tomita M, Wakabayashi H, Yamauchi K, Teraguchi S, Hayasawa H, et al. (2002) Bovine lactoferrin and lactoferricin derived from milk: production and applications. Biochem Cell Biol 80: 109-112.

24. Wakabayashi H, Yamauchi K, Takase M (2006) Lactoferrin research, technology and applications. Journal of International Dairy 16: 1241-1251.

25. Li H, Ma Y, Dong A, Jiaqi W, Li Q, et al. (2010) Protein composition of yak milk. Journal of Dairy Science and Technology 90: 111-117.

26. Li H, Ma Y, Li Q, Wang J, Cheng J, et al. (2011) The chemical composition and nitrogen distribution of Chinese yak (Maiwa) milk. Int J Mol Sci 12: 4885-4895.

27. Rang Z, Xiuying W, Xiaohua L (2014) Nutrition and safety of yak milk as infant formula source. In Y. Ping (Ed.), Proceedings of the 5th International Yak Conference, Agricultural Sciences and Technology. Lanzhou China: China, 209-213.

28. Cheng JB, Wang JQ, Bu DP, Liu GL, Zhang CG, et al. (2008) Factors affecting the lactoferrin concentration in bovine milk. J Dairy Sci 91: 970-976.

29. Chen CC, Tu YY, Chang HM (2000) Thermal Stability of Bovine Milk Immunoglobulin G (IgG) and the Effect of Added Thermal Protectants on the Stability. Journal of Food Chemistry and Toxicology 65: 188-193.

30. Cozma A, Andrei S, Miere D, Filip L, Loghin F, et al. (2011) Proteins Profile in Milk from Three Species of Ruminants. Journal of Notulae Scientia Biologicae 3: 26-29.

31. Chevalier F, Chobert JM, Popineau Y, Nicolas MG, Haertlé T, et al. (2001) Improvement of functional properties of ß-lactoglobulin glycated through the Maillard reaction is related to the nature of the sugar. International Dairy Journal 11: 145-152.

32. Laleye LC, Jobe B, Wasesa AA (2008) Comparative study on heat stability and functionality of camel and bovine milk whey proteins. J Dairy Sci 91: 4527-4534.

33. Monahan FJ, German JB, Kinsellat JE (1995) Effect of pH and Temperature on Protein Unfolding and Thiol/Disulfide Interchange Reactions during Heat-Induced Gelation of Whey Proteins. Journal of Agric. and Food Chem 43: 46-52.

34. Hoffmann MAM, Mil PJJMV (1997) Heat-Induced Aggregation of -Lactoglobulin?: Role of the Free Thiol Group and Disulfide Bonds. Journal of Agric and Food Chem 45: 2942-2948.

35. Madureira AR, Pereira CI, Gomes AMP, Pintado ME, Malcata XF, et al. (2007) Bovine whey proteins-Overview on their main biological properties. Journal of Food Research International 40: 1197-1211.

36. Zimecki M, Kruzel ML (2007) Milk-derived proteins and peptides of potential therapeutic and nutritive value. J Exp Ther Oncol 6: 89-106.

37. Senda A, Fukuda K, Ishii T, Urashima T (2011) Changes in the bovine whey proteome during the early lactation period. Anim Sci J 82: 698-706.

38. Pritchett LC, Gay CC, Besser TE, Hancock DD (1991) Management and production factors influencing immunoglobulin G1 concentration in colostrum from Holstein cows. J Dairy Sci 74: 2336-2341.

39. Godson DL, Acres SD, Haines DM (2003) Failure of passive transfer and effective colostrum management in calves. Large Animal Veterinary Rounds 3: Montreal, canada.

40. Moore M, Tyler JW, Chigerwe M, Dawes ME, Middleton JR, et al. (2005) Effect of delayed colostrum collection on colostral IgG concentration in dairy cows. J Am Vet Med Assoc 226: 1375-1377.

41. Morrill KM, Conrad E, Lago A, Campbell J, Quigley J, et al. (2012) Nationwide evaluation of quality and composition of colostrum on dairy farms in the United States. J Dairy Sci 95: 3997-4005.

42. Conneely M, Berry DP, Sayers R, Murphy JP, Lorenz I, et al. (2013) Factors associated with the concentration of immunoglobulin G in the colostrum of dairy cows. Animal 7: 1824-1832.

43. Bernabucci U, Basiricò L, Morera P (2013) Impact of hot environment on colostrum and milk composition. Cell Mol Biol (Noisy-le-grand) 59: 67-83.

44. Jaster EH (2005) Evaluation of quality, quantity, and timing of colostrum feeding on immunoglobulin G1 absorption in Jersey calves. J Dairy Sci 88: 296-302.

45. Gomes V, Madureira KM, Soriano S, Melville AM, Della LP, et al. (2011) Factors affecting immunoglobulin concentration in colostrum of healthy Holstein cows immediately after delivery. Journal of Pesq Vet Bras 31: 53-56.

46. Hansen HS (2007) Calf Management. In HS Hansen (Eds), Calf Management Proceedings Steinkjer, Norway 7-31.

47. Godden S (2008) Colostrum management for dairy calves. Vet Clin North Am Food Anim Pract 24: 19-39.

48. Haenlein GFW (2009) Bioactive Components: In YW Park (Eds), (1stedn), Iowa, USA: Wiley-Blackwell.

49. Indyk HE, Filonzi EL (2005) Determination of lactoferrin in bovine milk, colostrum and infant formulas by optical biosensor analysis. International Dairy Journal 15: 429-438.

50. Claeys WL, Verraes C, Cardoen S, Block JD, Huyghebaert A, et al. (2014) Consumption of raw or heated milk from different species: An evaluation of the nutritional and potential health benefits. Journal of Food Control 42: 188-201.

51. LaClair CE, Etzel MR (2010) Ingredients and pH are key to clear beverages that contain whey protein. J Food Sci 75: C21-27.

52. Chandrapala J, Duke MC, Gray SR, Zisu B, Weeks M, et al. (2015) Properties of acid whey as a function of pH and temperature. J Dairy Sci 98: 4352-4363.

53. Patel HA, Anema SG, Holroyd SE (2007) Methods to determine denaturation and aggregation of proteins in low- , medium- and high-heat skim milk powders. Journal of Lait 87: 251-268.

54. Dissanayake M, Ramchandran L, Donkor ON, Vasiljevic T (2013) Denaturation of whey proteins as a function of heat, pH and protein concentration. International Dairy Journal 31: 93-99.

55. Wijayanti HB, Bansal N, Deeth HC (2014) Stability of Whey Proteins during Thermal Processing: A Review. Journal of Food Science and Food Safety 13: 1235-1251.

56. Bernal V, Jelen P (1985) Thermal Stability of Whey Proteins-A Calorimetric Study. Journal of Dairy Science 68: 2847-2852.

57. Dissanayake M, Vasiljevic T (2009) Functional properties of whey proteins affected by heat treatment and hydrodynamic high-pressure shearing. J Dairy Sci 92: 1387-1397.

58. Ryan KN, Zhong Q, Foegeding EA (2013) Use of whey protein soluble aggregates for thermal stability-a hypothesis paper. J Food Sci 78: R1105-1115.

59. Havea P, Singh H, Creamer LK (2000) Formation of new protein structures in heated mixtures of BSA and alpha-lactalbumin. J Agric Food Chem 48: 1548-1556.

60. Levieux D, Levieux A, El-Hatmi H, Rigaudière JP (2006) Immunochemical quantification of heat denaturation of camel (Camelus dromedarius) whey proteins. J Dairy Res 73: 1-9.

61. Li-Chan E, Kummer A, Losso JN, Kitts DD, Nakai S, et al. (1995) Stability of bovine immunoglobulins to thermal treatment and processing. Journal of Food Research International 28: 9-16.

62. Raikos V (2010) Effect of heat treatment on milk protein functionality at emulsion interfaces. A review. Journal of Food Hydrocolloids 24: 259-265.

63. Walstra P, Geurts T, Noomen A, Jellema A, Boekel MAJSV, et al. (2005) Dairy Technology-Principles of Milk properties and Processing. In Walstra P, Geurts T, Noomen A, Jellema A, Boekel MAJSV, (eds), (2ndedn), New York, USA.

64. Saito H, Miyakawa H, Tamura Y, Shimamura S, Tomita M (1991) Potent bactericidal activity of bovine lactoferrin hydrolysate produced by heat treatment at acidic pH. J Dairy Sci 74: 3724-3730.

65. Ueno HM, Kato K, Ueda N, Matsui H, Nakajima H, et al. (2012) Native, but not thermally denatured lactoferrin solubilizes iron in the presence of bicarbonate ions. J Dairy Sci and Technol 92: 25-35.

66. Sreedhara A, Flengsrud R, Prakash V, Krowarsch D, Langsrud T, et al. (2010) A comparison of effects of pH on the thermal stability and conformation of caprine and bovine lactoferrin. Journal of International Dairy 20: 487-494.

PERMISSIONS

LIST OF CONTRIBUTORS

Sikha Bhaduri
CUNY School of Public Health, NY 10035, USA

Sasi Kumar R
Department of Agri-Business Management and Food Technology, North-Eastern Hill University, NEHU Tura Campus, Chandmari, Meghalaya, India

Mahdieh Iranmanesh and Hamid Ezzatpanah
Department of Food Science and Technology, Science and Research Islamic Azad University, Tehran, Iran

Naheed Mojgani
Biotechnology Department, Razi Vaccine and Serum Research Institute, Karaj, Iran

Torshizi MAK
Department of Poultry Science, Faculty of Agriculture, Tarbiat Modares University, Tehran, Iran

Ojo DO and Enujiugha VN
Department of Food Science and Technology, Federal University of Technology, Akure, Nigeria

Danuwat P, Rimruthai P, Phattanawan C and Peerarat D
Institute of Product Quality and Standardization, Maejo University, Nonghan, Sansai, Chiang Mai, Thailand

Yuki Kayanuma, Reiko Ueda, Yoshiro Ishimaru and Tomiko Asakura
Graduate School of Agricultural and Life Sciences, The University of Tokyo, Tokyo, Japan

Michiko Minami
Faculty of Education, Tokyo Gakugei University, Koganei, Tokyo, Japan

Arata Abe and Kazumi Kimura
Department of Neurological Sciences, Graduate School of Medicine, Nippon Medical School, Sendagi, Bunkyo-ku, Tokyo, Japan

Junko Funaki
International College of Arts and Sciences, Fukuoka Women's University, Higashi-ku, Fukuoka, Japan

Graduate School of Health and Environmental of Sciences, Fukuoka Women's University, Higashi-ku, Fukuoka, Japan

Sravan Kumar K
Department of Food Engineering and Technology, Sant Longowal Institute of Engineering and Technology, Longowal, India

Oduro I and Ellis WO
Food Science and Technology Department, College of Science, Kwame Nkrumah University of Science and Technology, Kumasi, Ghana

Carey EE
International Potato Centre (CIP), Kumasi, Ghana

Owusu-Mensah E
Food Science and Technology Department, College of Science, Kwame Nkrumah University of Science and Technology, Kumasi, Ghana
International Potato Centre (CIP), Kumasi, Ghana

Padalino L, Conte A, Lecce L, Likyova D and Del Nobile MA
University of Foggia, Services Center of Applied Research - Via Napoli 25 Foggia, Italy

Sicari V, Pellicanò TM and Poiana M
Mediterranean University of Reggio Calabria, Agricultural Department, Reggio Calabria, Italy

Ashaye OA and Olanipekun OT
Institute of Agricultural Research and Training P.M.B 5029, Moor-Plantation Ibadan, Nigeria

Ojo SO
Federal College of Agriculture, Ibadan, Nigeria

Algadi MZ
Arab Center for Nutrition, Muharraq, Bahrain

Yousif NE
Food Science and Technology, Khartoum University, Sudan

Darwish AZ
Dairy Science Department, Faculty of Agriculture, Assiut University, Egypt

Bayomy H and Rozan M
Food Science and Technology Department, Faculty of Agriculture, Damanhour University, Behira, Egypt

Al Surmi NY and El Dengawy RAH
Food Industries Department, Faculty of Agriculture, Damietta University, Damietta, Egypt

Khalifa AH
Food Science and Technology Department, Faculty of Agriculture, Assiut University, Assiut, Egypt

Nur Sofuwani ZA and Siti Aslina H
Department of Chemical and Environmental Engineering, Faculty of Engineering, Universiti Putra Malaysia, Selangor, Malaysia

Siti Mazlina MK
Department of Food and Process Engineering, Faculty of Engineering, Universiti Putra Malaysia, Selangor, Malaysia

Fadeyibi A
Department of Agricultural and Biological Engineering, Kwara State University, Malete, Ilorin, Nigeria

Osunde ZD, Agidi G and Idah PA
Department of Agricultural and Bioresources Engineering, Federal University of Technology, Minna, Nigeria

Egwim EC
Department of Biochemistry, Federal University of Technology, Minna, Nigeria

Sana Mabrouk, Yosra Braham, Houcine Barhoumi and Abderrazak Maaref
Laboratory of Interfaces and Advanced Materials (LIMA), University of Monastir, Tunisia

Qun Huang, Lei Chen, Hong-bo Song, Feng-ping An, Hui Teng and Mei-yu Xu
College of Food Science, Fujian Agriculture and Forestry University, Fuzhou, China

Ezeibekwe IO, Umeoka N and Izuka CM
Department of Plant Science and Biotechnology, Faculty of Science, Imo State University Owerri, Imo State, Nigeria

Ben Said Ines, Mezghani Imed and Chaieb Mohamed
University of Sfax, Department of Biology, Faculty of Sciences of Sfax, 3000, Sfax, Tunisia

Adele Muscolo
Department of Agriculture, Mediterranea University, Feo di Vito, 89124 Reggio Calabria, Italy

Sangeeta and Bahadur Singh Hathan
Department of Food Engineering and Technology, Sant Longowal Institute of Engineering and Technology, (SLIET), Sangrur, Punjab, India

Karine H Rebouças, Laidson P Gomes, Eduardo M Del Aguila, Vania M Flosi Paschoalin, Thais M Uekane and Claudia M Rezende
Universidade Federal do Rio de Janeiro, Instituto de Química, Avenida Athos da Silveira Ramos, Cidade Universitária - Rio de Janeiro, Brazil

Analy MO Leite
Universidade Federal do Rio de Janeiro Campus Macaé. Rua Aloísio da Silva Gomes, Macaé-RJ, Brazil

Maria Ines B Tavares
Universidade Federal do Rio de Janeiro, Instituto de Macromoléculas Professora Eloisa Mano, Brazil

Eveline L Almeida
Universidade Federal do Rio de Janeiro, Escola de Química, Brazil

Odunmbaku LA
Food Technology Department, Moshood Abiola Polytechnic, Abeokuta, Nigeria

Babajide JM and Shittu TA
Food Science and Technology Department, Federal University of Agriculture, Abeokuta, Nigeria

Eromosele CO
Chemistry Department, Federal University of Agriculture, Abeokuta, Nigeria

Jayasinghe PS and Pahalawattaarachchi V
National Aquatic Resource Research and Development Agency, Crow Island, Colombo, Srilanka

Ranaweera KKDS
Faculty of Food science, University of Sri Jayewardenepura, Nugegoda, Srilanka

Sobowale SS
Department of Food Technology, Moshood Abiola Polytechnic, Abeokuta, Ogun State, Nigeria

Awonorin SO and Shittu TA
Department of Food Science and Technology, Federal University of Agriculture, Abeokuta, Ogun State, Nigeria

Oke MO and Adebo OA
Department of Food Science and Engineering, Ladoke Akintola University of Technology, Ogbomoso, Oyo State, Nigeria
Department of Biotechnology and Food Technology, University of Johannesburg, Doornfontein 2028, South Africa

Nassar KS and Shamsia SM
Department of Food and Dairy Science and Technology, Damanhour University, Egypt

Attia IA
Department of Dairy Science and Technology, Alexandria University, Egypt

Sefa Salo
Department of Animal Science, Wachemo University, Hosanna, Ethiopia

Mengistu Urge and Getachew Animut
Department of Animal Sciences, Haramaya University, Dire Dawa, Ethiopia

Rafiq SM and Ghosh BC
ICAR (Indian Council of Agricultural Research), National Dairy Research Institute (SRS), Bengaluru, India

Singh H and Griffiths MW
Canadian Research Institute for Food Safety, Department of Food Science, University of Guelph, 43 Mc Gilvray Street, Guelph, ON N1G 2W1, Canada

Borges A
INOVA, Institute for Technological Innovation of the Azores, Road São Gonçalo, S/N 9504-540, Ponta Delgada, Portugal

Kongo JM and Ponte DJB
INOVA, Institute for Technological Innovation of the Azores, Road São Gonçalo, S/N 9504-540, Ponta Delgada, Portugal

University of the Azores, Department of Technological Sciences and Development, Ponta Delgada, Acores, Portugal

Gupta P
Department of Food Science and Nutrition, Lovely Professional University, Punjab, India

Bhat A
Division of Post-harvest Technology, Sher-e-Kashmir University of Agricultural Sciences and Technology of Jammu, Jammu, India

Emire SA
Food, Beverage and Pharmaceutical Industry Development Institute, Ministry of Industry, Addis, Ababa, Ethiopia

Buta MB
Food Process Engineering Department, Addis Ababa Science and Technology University, Ethiopia

Syed IR, Sukhcharn S and Saxena DC
Department of Food Engineering and Technology, Sant Longowal Institute of Engineering and Technology, Longowal, Punjab, India

Aroyeun SO and Jayeola CO
Cocoa Research Institute of Nigeria, Ibadan, Nigeria

Woldemariam HW
Department of Food Process Engineering, College of Biological and Chemical Engineering, Addis Ababa Science and Technology University, Addis Ababa, Ethiopia

Asres AM
Department of Food Technology and Food Process Engineering, Faculty of Chemical and Food Engineering, Bahir Dar Institute of Technology, Bahir Dar, Ethiopia

Dana E and Khodabandehlo H
Department of Chemical Engineering, Shahrood Branch, Islamic Azad University, Shahrood, Iran

Ardestani SS
Department of Chemical Engineering, Food industry, Islamic Azad university of Science and Research unit, Tehran, Iran

Geetha V, Bhavana KP, Chetana R, Gopala Krishna AG and Suresh Kumar G
Department of Traditional Foods and Sensory Science, CSIR-Central Food Technological Research Institute, Mysore, India

Tabeen Jan, Yadav KC and Sujit Borude
Department of Food Process Engineering, Sam Higginbottom Institute of Science and Technology, Allahabad, UP, India

Berber M and Alvarez VB
Department of Food Science, The Ohio State University, Columbus, OH 43210, USA

González-Quijano GK
Departamento de Graduados en Alimentos, Escuela Nacional de Ciencias Biológicas, Instituto Politécnico Nacional, Distrito Federal, México

Habtamu LD and Birhanu K
Wolaita Sodo University, PO Box 128, Wolaita Sodo, Ethiopia

Ashenafi M
Addis Abba University, College of Veterinary Medicine and Agriculture, Ethiopia

Taddese K
Ethiopian Development Research Institute, Ethiopia

Getaw T
International Food Policy Research Institute, Ethiopia

Wu Xiaoyun, Yan Ping and Ding xuezhi
Lanzhou Institute of Husbandry and Pharmaceutical Sciences, Chinese Academy of Agricultural Sciences, Lanzhou P.R. China

Xiong Lin
Laboratory of Quality and Safety Risk Assessment for Livestock Product (Lanzhou), Ministry of Agriculture, Lanzhou, China

Shimo Peter Shimo
Lanzhou Institute of Husbandry and Pharmaceutical Sciences, Chinese Academy of Agricultural Sciences, Lanzhou P.R. China
Government Chemist Laboratory Agency, Dar es Salaam, Tanzania

Index

www.ingramcontent.com/pod-product-compliance
Lightning Source LLC
Chambersburg PA
CBHW080413190526
45161CB00003B/225